Observational Astrophysics

The Graduate Series in Astronomy

Series Editors: **R J Tayler**, University of Sussex
R E White, University of Arizona

The Graduate Series in Astronomy includes books on all aspects of theoretical and experimental astronomy and astrophysics. The books are written at a level suitable for senior undergraduate and graduate students, and will also be useful to practising astronomers who wish to refresh their knowledge of a particular field of research.

Other books in the series

The front cover shows NGC 6720, the Ring nebula.
Photograph from the Lick Observatory.

The Graduate Series in Astronomy

Observational Astrophysics

Edited by R E White

Steward Observatory, University of Arizona, USA

Institute of Physics Publishing
Bristol and Philadelphia

British Library Cataloguing-in-Publication Data
A catalogue record for this book is available from the British Library

ISBN 0 7503 0201 1

Library of Congress Cataloging-in-Publication Data are available

Citation
When citing an item published in this book, authors are requested to cite the original journal of publication. The requisite information is given in the 'catchline' at the top of the title page of each article. If required, such a citation may be supplemented by the expansion 'Reprinted in:' followed by the book reference.

Series Editors: **R J Tayler**, University of Sussex
                  R E White, University of Arizona

Published by IOP Publishing Ltd, a company wholly owned by
The Institute of Physics, London.

IOP Publishing Ltd
Techno House, Redcliffe Way, Bristol BS1 6NX, UK
335 East 45th Street, New York, NY 10017-3483, USA

US Editorial Office: IOP Publishing Inc., The Public Ledger Building,
Suite 1035, Independence Square, Philadelphia, PA 19106

Printed in Great Britain by Galliard (Printers) Ltd, Great Yarmouth

Contents

Preface

In this volume, recent results of a number of currently active fields of observational research are reviewed by major players in the pertinent areas; in order of contents, they are: M Peimbert (*Planetary nebulae*); R J Cohen (*Compact maser sources*); A C Fabian and X Barcons (*Intergalactic matter*); D E Osterbrock (*Active galactic nuclei*); S J Warren and P C Hewett (*The detection of high-redshift quasars*); and D Burstein (*Large-scale motions in the universe*). Of these articles, only a single one, *Planetary nebulae*, was an active and prolific research area when I was a post-prelim graduate student 30 years ago! There wasn't any concept of the remaining five topics: they didn't exist. Neither was there any hint that there was anything so peculiar as galactic maser sources, or—heaven forbid!—that the spaces between the galaxies were anything more than a vacuum. There was some talk about the so-called 'Seyfert galaxies' being peculiar, but the Palomar observers were hot on the trail of those objects and all would soon be well. In fact, it's a tribute to the planetary nebulae researchers that they have been able to find ever-more-interesting astrophysics in the features of their *objets d'interêt*.

The explosion of astrophysics in the past quarter of a century has been due, principally, to the burgeoning international space program, as well as to the ever-larger and more sensitive ground-based facilities found worldwide which have accompanied the new thrusts into space. The space program has opened all of the windows of the electromagnetic spectrum to our gaze: the UV and X-ray satellites have revealed to us a cosmos of heretofore unimagined energetic activity; the technological advances in the regimes of the infrared, sub-millimetre, and millimetre wave radio astronomy have revealed an ever-increasing complexity of atomic and molecular gas clouds, as well as the late-breaking news of the isotropy of the expanding frontier of the Universe's 'Initial Event'. The interactions of all of the intra- and inter-galactic solid/crystalline 'stuff' with the high-energy radiation fields in which it is bathed has produced the research which is being reviewed in this volume. It can be fairly said that the nature of these reviews is truly of 'research in progress', because there will be, inevitably, more recent information which contradicts some of the remarks of our reviewers as soon as this volume sees the light of day. The real value of the articles contained herein is, therefore, in setting the stage for further work in all of the reviewed areas. Of course! That's what review articles are supposed to do! However, these articles are not a summing-up of a few fields which have been pretty much well-defined, picked over, and laid to rest: our authors are working with their colleagues at the forefront of activity in their reviewed areas. Another set of reviews of these topics will be necessary within, probably, five years. I expect to find all of the fields discussed here to have served as springboards into even more exotic, and as yet unimagined, areas. Professor Geoffrey Burbidge once commented: 'Astronomers don't find answers: they discover bigger questions!'

Voici some curtain-raisers.

R E White
Steward Observatory
University of Arizona
January 1992

Rep. Prog. Phys. **53** (1990) 1559–1619. Printed in the UK

Planetary nebulae

Manuel Peimbert

Instituto de Astronomía, Universidad Nacional Autónoma de México, Apartado Postal 70-264, México D F, 04510 Mexico

Abstract

Planetary nebulae (PN) are shells of low-density ionized gas that have been expelled by their central stars. Intermediate-mass stars, while leaving the red giant stage to become white dwarfs, undergo mass loss, producing PN.

The main emphasis of the review is on the properties of the shell and their relation to other branches of astrophysics. Recent studies on the distance scale to galactic PN, the extragalactic distance scale, and the PN birth rate are described. The relevance of the PN chemical composition to the study of stellar evolution and the chemical evolution of galaxies is discussed.

This review was received in its present form in August 1990.

0034-4885/90/121559 + 61$14.00 © 1990 IOP Publishing Ltd   1

Contents

1. Introduction

Planetary nebulae (PN) are shells of low-density ionized gas that have been ejected by their central stars. The shells are expelled at low velocity and are expanding away from the central stars with typical velocities of about 25 km s^{-1}; they are expected to dissipate in about 25 000 years. In this time interval the central star also modifies appreciably its luminosity and effective temperature. Typical densities in the shell range from 10^5 cm^{-3}, for young and compact PN, to about 1 cm^{-3}, for very extended objects or faint outer halos of objects with higher-density inner shells. The electron temperatures, T_e, for the ionized region of the shell, range from 5000 to 20 000 K. The central star surface temperatures range from 25 000 to 300 000 K. The ionized shell is easily observable in the visible spectrum since it produces bright recombination lines of H and He and also bright forbidden lines of O, N, Ne, S and Ar. An excellent review with emphasis on the physical processes taking place in the low-density ionized gas was presented by Seaton (1960).

The study of PN includes research into the properties of the ejected envelope, or nebular shell, and research into the central star, or nucleus. Very often the term PN is used to denote only the shell, but sometimes it is also used to denote the shell and the nucleus. The name planetary nebula dates from the end of the eighteenth century and is based on the disk-like greenish appearance of some bright PN seen through small telescopes, similar to the appearance of Uranus and Neptune. The name is kept for historical reasons but there is no connection between PN and planets. Figures 1 and 2 show two classic galactic PN.

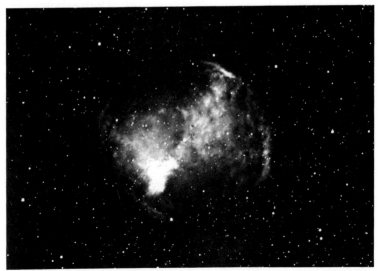

Figure 1. NGC 6853, the Dumbbell nebula, a nearby galactic PN, 330 seconds of arc in diameter. Original photograph was taken with the 3 m Shane reflector and a plate-filter combination emphasizing [N II] $\lambda\lambda$6548, 6583 and Hα. (Lick Observatory photograph.)

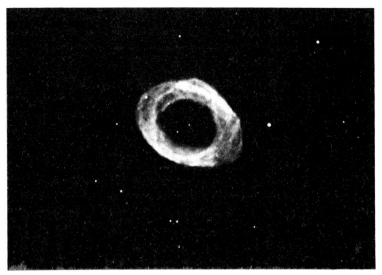

Figure 2. NGC 6720, the Ring nebula, a nearby galactic PN, 75 seconds of arc in diameter. Original photograph taken with the same equipment as figure 1. (Lick Observatory photograph.)

Intermediate mass stars (IMS), those with masses in the range $0.8 \leqslant M_i/M_\odot \leqslant 8$, where M_i is their initial mass and $M_\odot$ is the solar mass, are considered to be the progenitors of PN. IMS spend most of their life in the main sequence transforming hydrogen into helium in their nuclei; after leaving the main sequence they become red giants and develop a degenerate carbon-oxygen core. It is thought that a considerable fraction of the IMS while leaving the red giant phase and before becoming white dwarfs eject their outer envelope, forming a PN. The central star evolves toward higher surface temperatures during the shell ejection and the emergent flux of wavelength less than 912 Å becomes intense enough to ionize most or all of the ejected gas. Figure 3 is a graph of stellar luminosity as a function of stellar temperature (an HR diagram) where the direction of the evolution is represented by arrows.

Stars with $M_i \gtrsim 8\ M_\odot$ ignite carbon in a non-degenerate core; their final evolutionary stages are greatly different from those of the IMS, probably producing supernovae. Stars with $M_i \lesssim 0.8\ M_\odot$ have not had time to evolve off the main sequence during the lifetime of the Galaxy (e.g. Iben and Rood 1970), moreover they may lose their hydrogen-rich envelope before the central star becomes hot enough to ionize the ejected envelope (Renzini and Voli 1981).

From the study of the gaseous envelope of a PN it is possible to determine its chemical composition. There are two types of elements: those affected by the stellar evolution of IMS, such as H, He, C and N and those not affected by it, such as Ne, S and Ar. Data on the abundances of the elements of the first group provide constraints for stellar evolution models, while data on the abundances of the elements of the second group indicate the relative abundances of those elements at the time the progenitors of the PN were formed. The determination of the chemical composition of PN is paramount for the study of the chemical evolution of the Galaxy.

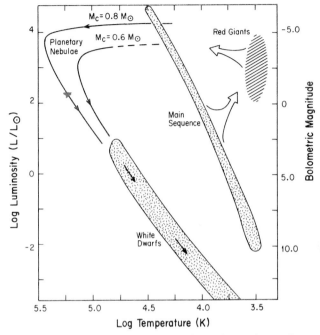

Figure 3. HR diagram showing the location of: central stars of PN, main sequence stars, white dwarf stars and red giants. The evolutionary tracks for central stars of PN for $M_c = 0.6 \, M_\odot$ and $M_c = 0.8 \, M_\odot$ by Paczynski (1971) are also shown schematically.

There are several recent books dedicated to the study of the physical conditions within ionized gaseous nebulae (Osterbrock 1989, Pottasch 1984, Aller 1984). The topic has been of considerable interest and every five years since 1967 there has been a major symposium dedicated to all aspects of PN; the last three, those of 1977, 1982 and 1987, contain very useful review articles that are still relevant to current research (Terzian 1978, Flower 1983, Torres-Peimbert 1989). A recent review on PN has been presented by Khromov (1989).

As of 1967 there were 1067 galactic objects in the PN catalogue compiled by Perek and Kohoutek (1967). This catalogue contains most of the PN observations available up to 1966. At present the sample of galactic PN has grown to about 1600. The number of known extragalactic PN reaches many hundreds and includes objects as far away as 15 Mpc from us. With present techniques it is possible to discover many thousands of them.

This review concentrates on the properties of the PN shell and their relevance to other astrophysical problems. In sections 2 and 3 we describe the ionization structure of the shell and the methods used to derive the electron temperature and the electron density. In section 4 we describe the determination of the chemical abundances. In section 5 we present a classification of PN based on their chemical composition and compare the abundances of PN with those predicted by stellar evolution theory. In section 6 we review the galactic distance determinations to PN and how they are used to determine masses, birth rates and the total number of galactic PN. In section 7 we discuss the information on abundances and on the extragalactic distance scale (the determination of the Hubble constant) that can be derived from extragalactic PN. Also in section 7 we discuss the PN birth rate in extragalactic systems and their relation with the galactic distance scale. In section 8 we discuss the relevance of PN research to the problem of galactic chemical evolution, in particular to the primordial or pregalactic He/H abundance ratio. The concluding remarks are presented in section 9.

Very important advances have been achieved during the last ten years in other areas of PN research, that will not be reviewed here, including: (a) properties and evolution of the central stars (e.g. Kaler 1985, Méndez *et al* 1988, Schönberner 1989), (b) stellar mass loss and its effects on the late stages of stellar evolution (e.g. Kwok 1987, Renzini 1989, Perinotto 1989), (c) protoplanetary nebulae (e.g. Preite-Martinez 1987 and other articles in the same book), (d) dust content of the shell (e.g. Roche 1989), (e) molecules in the shell (e.g. Zuckerman and Gatley 1988, Rodríguez 1989), and (f) multiple shells (e.g. Chu 1989).

2. Ionization structure

The ionization structure of a PN depends on the physical processes occurring within the nebula. We will discuss equilibrium structures that do not include gas motions; this approach simplifies our task considerably. The processes that determine the ionization structure and the kinetic temperature have been widely studied and are in general well understood.

2.1. *Photoionization and recombination of a hydrogen nebula: Strömgren spheres*

Typical PN envelopes are composed of H, He and trace elements such as C, N, O and Ne. The envelope is ionized by a hot central star that produces a significant photon flux above the Lyman limit, $h\nu_0 = 13.6$ eV. The envelope radiates through recombination and collisional excitation processes. We will simplify our discussion by assuming a pure hydrogen nebula of uniform density with spherical symmetry.

It is necessary to consider the transfer of UV radiation through the nebula. This radiation is the ultimate source of energy for all the physical processes at each point in the nebula. The stellar radiation flux at frequency ν impinging upon a unit area at a distance r from the exciting star is:

$$4\pi J_\nu^s(r) = \pi F_\nu(r_*)\left(\frac{r_*}{r}\right)^2 \exp(-\tau_\nu(r)) \tag{2.1}$$

where $\pi F_\nu(r_*)$, in erg cm^{-2}s^{-1}Hz^{-1}, is the stellar flux at the surface of the star of radius r_*, $\pi F_\nu(r_*) = L_\nu/4\pi r_*^2$, L_ν is the luminosity of the star at frequency ν, in erg s^{-1} Hz^{-1}, and τ_ν is the optical depth due to absorption by hydrogen atoms and is given by

$$d\tau_\nu = N(H^0)a_\nu(H^0)dr \tag{2.2}$$

where $N(H^0)$ is the number of neutral H atoms per unit volume at the distance r and $a_\nu(H^0)$, in cm^2, is the absorption cross section for the H^0 ground state. The dependence of a_ν on frequency may be represented as

$$a_\nu(H^0) = \begin{cases} 0 & \nu < \nu_0 \\ 6.3 \times 10^{-18}(\nu_0/\nu)^3 & \nu \geq \nu_0 \end{cases} \tag{2.3}$$

2.1.1. Ionization equilibrium. Equilibrium of ionization prevails if, in each volume element, the photoionization rate is equal to the recombination rate; that is if the capture rate of H^0 ionizing photons is balanced by the recombination rate of protons with free electrons. This can be expressed as

$$N(H^0) \int_{\nu_0}^{\infty} \frac{4\pi J_\nu}{h\nu} a_\nu \, d\nu = N(H^+) N_e \alpha_A(T_e) \tag{2.4}$$

where $\alpha_A(T_e)$, the recombination coefficient to all levels defined as $\alpha_A = \Sigma_{i=1}^{\infty} \alpha_i$, is related to a_ν through the Milne relation and is equal to 4×10^{-13} cm^3 s^{-1} for $T_e = $ 10 000 K. The radiation intensity, J_ν, consists of the stellar component, J_ν^s, and the 'diffuse' component, J_ν^d, i.e.

$$J_\nu = J_\nu^s + J_\nu^d \tag{2.5}$$

where J_ν^d is produced by the nebula itself upon recombination of electrons to the H ground state and in general is assumed to be isotropic. The diffuse radiation consists predominantly of photons with frequencies just above ν_0 and, to a good approximation, it can be assumed to be absorbed locally on the spot. Under this approximation, the photoionization rate by diffuse radiation is equal to the rate of recombination to the ground state at every point, consequently the ionization equilibrium condition (equation (2.4)) can be written as

$$N(H^0) \int_{\nu_0}^{\infty} \frac{4\pi J_\nu^d}{h\nu} a_\nu \, d\nu = N(H^+) N_e \alpha_1 \tag{2.6}$$

and

$$N(H^0) \int_{\nu_0}^{\infty} \frac{4\pi J_\nu^s}{h\nu} a_\nu \, d\nu = N(H^+) N_e \alpha_B \tag{2.7}$$

where $\alpha_B = \Sigma_{i=2}^{\infty} \alpha_i$ (equal to 2.6×10^{-13} cm^3 s^{-1} for $T_e = 10\,000$ K) is the recombination coefficient to all levels but the first.

The ionization structure of the nebula is obtained by solving the transfer equation (equation (2.1)) coupled with the optical depth and the ionization equilibrium equations (equations (2.2) and (2.7)). In a pure-H nebula the following relations apply:

$$N(H) = N(H^0) + N(H^+) \qquad N_e = N(H^+). \tag{2.8}$$

This problem was originally solved by Strömgren (1939), who found that H is almost completely ionized within a sphere of radius R_s and that the degree of ionization decreases rapidly at the outer edge of the ionization zone ($r = R_s$), achieving neutrality in a distance $\Delta s \sim [a_\nu N(H^0)]^{-1}$. These configurations have been named Strömgren spheres.

The Strömgren or critical radius, R_s, can be derived by assuming that within it, the rate of recombinations to all levels but the first is equal to the number of ionizing photons produced by the central star, $Q(H^0)$, i.e.

$$Q(H^0) = \int_{\nu_0}^{\infty} \frac{L_\nu}{h\nu} \, d\nu = \frac{4\pi}{3} R_s^3 N_e^2 \alpha_B. \tag{2.9}$$

If the number of ionizing photons is larger than the number of recombinations, then a fraction of the ionizing photons escape the PN and the object is considered to be optically thin; if this is not the case all the ionizing photons are trapped inside the nebula and the PN is optically thick. Under conditions of inhomogeneous density a PN could be optically thick in some directions and not in others. Optically thin and optically thick nebulae are also called density bounded and ionization bounded, respectively.

2.2. Helium

After H the next most abundant element in PN is He, its relative abundance is ~10% by number and a better approximation to the ionization structure of an actual nebula is provided by taking it into account. The ionization potential of He^0 is 24.6 eV, or 1.8 $h\nu_0$, and that of He^+ is 54.4 eV, or 4 $h\nu_0$. The photons of energy $13.6 < h\nu < 24.6$ eV can only ionize H^0, while those of energy $24.6 < h\nu < 54.4$ eV can ionize both H^0 and He^0. Different types of ionizing structures are present depending on the spectrum of the ionizing radiation and on the He abundance.

In the presence of He, the on-the-spot approximation can also be adopted, and ionization equilibrium equations can be written for H^0, He^0 and He^+. The additional complication is that some photons produced by He recombination processes are energetic enough to ionize H and have to be taken into account. The ionization cross sections reach their maximum value at the threshold energies, and also show a very rapid decrease with increasing energy. Table 1 shows the interrelation between the H and He recombination fields for typical conditions within a PN. Moreover, He^0 and He^+ have to be taken into account for the calculation of τ_ν and N_e.

Table 1. Relative absorption† of the diffuse radiation field by H^0, He^0 and He^+.

Process	H⁰	He⁰	He⁺
		Absorber	
H I Lyman continuum	1	—	—
He I Lyman continuum	0.13	0.87	—
He II Lyman continuum	0.03	0.56	0.41
He II Balmer continuum	1	—	—
He I Lyman lines	1	—	—
He II Lyman lines	0.08	0.92	—
He I Two-photon emission	1	—	—
He II Two-photon emission $(1-1.8)h\nu_0$	1	—	—
He II Two-photon emission $(1.8-3)h\nu_0$	0.11	0.89	—

$$\text{† Given in } \frac{\int J_\nu{}^d N(X^{+i}) a_\nu(X^{+i}) \mathrm{d}\nu}{\int J_\nu{}^d [N(H^0) a_\nu(H^0) + N(He^0) a_\nu(He^0) + N(He^+) a_\nu(He^+)] \mathrm{d}\nu}.$$

2.3. Heavy elements

In PN the abundance of heavy elements is very small, and thus they do not affect significantly the ionization structure of the radiation field. However, through collisional excitation, the heavy elements emit forbidden and semiforbidden lines that radiate energy away very efficiently.

The ionization equilibrium for consecutive stages of ionization, i and $i+1$, of any element X may be written as:

$$N(X^{+i}) \int_{\nu_i}^{\infty} \frac{4\pi J_\nu}{h\nu} a_\nu(X^{+i}) \, d\nu = N(X^{+i+1}) N_e \alpha_G(X^{+i}, T_e) \tag{2.10}$$

where $N(X^{+i})$ and $N(X^{+i+1})$ are the number densities of the two successive stages of ionization, $a_\nu(X^{+i})$ is the photoionization cross section for the ground level of X, with threshold ν_i, and α_G is the recombination coefficient of the ground level of X^{+i+1} to all levels of X^{+i}. Additional terms should be added to equations of this type to take into account charge exchange reactions and dielectronic recombinations.

The ionization equilibrium equations, together with the condition that the ionic abundances add up to the total element abundance.

$$N(X^0) + N(X^+) + N(X^{+2}) + \ldots = N(X) \tag{2.11}$$

completely determine the ionization equilibrium at each point.

2.4. Heating and cooling mechanisms

In the static case the temperature is fixed by the equilibrium between heating by photoionization and cooling by recombination, collisional excitation of low lying energy levels which upon de-excitation produce photons that leave the nebula, and free–free radiation or bremsstrahlung. When hydrogen absorbs an ultraviolet photon of energy $h\nu > h\nu_0$ it becomes ionized, the electron released has an initial kinetic energy given by $\frac{1}{2}mv^2 = h(\nu - \nu_0)$. The electron interacts with other electrons and ions, rapidly redistributing its kinetic energy and becoming 'thermalized'. Upon recombination the inverse process occurs and an electron with an energy equal to $\frac{1}{2}mv^2$ is captured by an ion. The balance of energy gains and losses defines the kinetic energy of the gas. Therefore to construct a model that solves for the temperature distribution of a nebula it is necessary to assess these processes at each point.

The energy input by photoionization of hydrogen (in erg cm^{-3} s^{-1}) is

$$G(H) = N(H^0) \int \frac{4\pi J_\nu}{h\nu} h(\nu - \nu_0) a_\nu(H^0) \, d\nu. \tag{2.12}$$

While the kinetic energy lost by the electron gas in recombinations (in erg cm^{-3} s^{-1}) is

$$L_R(H) = N_e N_p k T_e \beta_A(H^0, T_e) \tag{2.13}$$

where

$$\beta_A(H^0, T_e) = \sum_{n=1}^{\infty} \beta_n(H^0, T_e) \tag{2.14}$$

and

$$\beta_n = \frac{1}{kT_e} \int v\sigma_n(H^0, T_e) \tfrac{1}{2} mv^2 f(v) \, dv. \tag{2.15}$$

In addition, free–free emission is more effective in cooling the nebula than recombination. The rate of cooling is in this case

$$L_{FF}(Z) = (1.42 \times 10^{-27}) Z^2 T_e^{1/2} g_{ff} N_e N_{ion} \tag{2.16}$$

where Z is the charge of the cooling ions and g_{ff} is the mean Gaunt factor for free-free emission, for nebular conditions g_{ff} is close to 1.3 (Osterbrock 1989). The collisional excitation of low lying levels of common ions, with subsequent radiation, is the most important of all cooling processes. These ions make a significant contribution, in spite of their low abundances, because they have energy levels with excitation potential $\sim kT_e$. These processes are more laborious to calculate since for each element the excitation to several levels has to be computed for different stages of ionization. We can represent the energy lost by radiation following collisional excitation as

$$L_C = \sum_{i,k,a} N_a(X^{+k})_i \sum_{j<i} A_{ij} h\nu_{ij} \tag{2.17}$$

where i is the excited level, k is the stage of ionization, a is the atomic mass and A_{ij} is the Einstein transition probability coefficient.

At each point in the nebula the kinetic temperature is determined by the equilibrium between the heating and cooling mechanisms discussed above, i.e.

$$G = L_R + L_{FF} + L_C. \tag{2.18}$$

Given the ultraviolet spectrum of the ionizing star, as well as the density distribution and the chemical composition of the surrounding gas, it is possible to compute model ionization structures for gaseous nebulae.

3. Temperatures, densities and filling factors

The determination of physical conditions in PN such as temperature, density, ionization degree and abundance of the elements, is based on the observed line intensities and requires a large quantity of accurate atomic data. This need has created a crucial interplay between astronomical observations and atomic physics.

The observed spectra have to be corrected for interstellar reddening caused by dust grains. This is usually carried out by adopting the normal reddening law (e.g. Whitford 1958, Seaton 1979) and by comparing expected intensity ratios with observed ones, either from Balmer line ratios as computed by Brocklehurst (1971) and Hummer and Storey (1987), or from lines that originate in the same upper level and whose ratios are given by the ratios of the Einstein A values, or from radio continuum to $H\beta$ ratios. The intrinsic line intensities, $I(\lambda)$, are related to the observed line intensities, $F(\lambda)$ by

$$\log \frac{I(\lambda)}{I(H\beta)} = \log \frac{F(\lambda)}{F(H\beta)} + f(\lambda)C(H\beta) \tag{3.1}$$

where $f(\lambda)$ is the reddening function, which is practically the same for all galactic PN in the visual region, and $C(H\beta)$ is the logarithmic reddening correction at $H\beta$, which varies from object to object. Table 2 shows the average reddening law assembled by Seaton (1979).

Most of the observed emission lines are produced by recombination or by electron impact excitation. The parameters that govern their intensities are: (a) the recombination coefficients, (b) the transition probabilities, (c) the collisional excitation cross sections and (d) the ionization cross sections. Sets of H^0 and He^+ recombination coefficients for different densities and temperatures have been computed by Brocklehurst (1971) and Hummer and Storey (1987), similarly a set of recombination coefficients for He^0 has been computed by Brocklehurst (1972) for the conditions

Table 2. Mean interstellar extinction curve from Seaton (1979).

λ (Å)	$1/\lambda(\mu^{-1})$	$f(\lambda)$	λ (Å)	$1/\lambda(\mu^{-1})$	$f(\lambda)$
10 000	1.0	−0.63	2326	4.3	+1.36
8 333	1.2	−0.50	2222	4.5	+1.61
7 143	1.4	−0.39	2174	4.6	+1.71
6 250	1.6	−0.28	2128	4.7	+1.65
5 556	1.8	−0.15	2041	4.9	+1.49
5 000	2.0	−0.03	1923	5.2	+1.25
4 545	2.2	+0.08	1818	5.5	+1.14
4 167	2.4	+0.16	1667	6.0	+1.13
3 846	2.6	+0.23	1538	6.5	+1.19
3 571	2.8	+0.30	1429	7.0	+1.28
3 333	3.0	+0.38	1250	8.0	+1.61
2 857	3.5	+0.61	1111	9.0	+2.12
2 500	4.0	+0.97	1000	10.0	+2.78

present in PN. A compilation of transition probabilities, collisional excitation cross sections and ionization cross sections, together with a review of the different computational approaches to determine these parameters is given by Mendoza (1983). Many of these atomic parameters cannot be determined in terrestrial laboratories due to the very low densities present in PN shells. In fact, PN are excellent low-density laboratories to test theoretical computations of atomic parameters.

3.1. Temperatures and densities derived from forbidden lines

At low densities the relative population of the excited levels of many ions departs from the Boltzmann distribution and is given by the equilibrium between collisional excitation and de-excitation, and spontaneous emission.

The electron de-excitation rate coefficient, q_{ij}, is given by

$$q_{ij}(T_e) = \frac{8.63 \times 10^{-6} \Upsilon_{ij}(T_e)}{\omega_i T_e^{1/2}} \ (\text{cm}^3 \ \text{sec}^{-1}) \tag{3.2}$$

where ω_i is the statistical weight of the ith level, and $\Upsilon_{ij}(T_e)$ is the effective collision strength, given by

$$\Upsilon_{ij}(T_e) = \int_0^\infty \Omega(i,j) \exp(-\varepsilon_i / kT_e) \, d(\varepsilon_i / kT_e) \tag{3.3}$$

where $\Omega(i,j)$ is the collision strength and $\varepsilon_i = \frac{1}{2}mv_i^2$ is the initial kinetic energy of the colliding electron. The excitation rate coefficient is given by

$$q_{ji} = (\omega_i / \omega_j) q_{ij} \exp(-\Delta E_{ij} / kT_e) \tag{3.4}$$

where $i > j$.

3.1.1. Forbidden line densities. Ions whose excited levels are close together in energy produce line intensity ratios that are independent of temperature. These line ratios are used to determine the electron density; for example the observed forbidden line ratios of [O II] (configuration $2p^3$) and those of [S II], [Cl III], [Ar IV] and [K V] (configuration $3p^3$) are given by

$$R(N_e) = I(^4S^0_{3/2} - {}^2D^0_{5/2})/I(^4S^0_{3/2} - {}^2D^0_{3/2}) \tag{3.5}$$

and are very sensitive to the electron density.

The level populations are determined by electron collisions and spontaneous emission of radiation. At low densities R is equal to the ratio of the collision strengths $\Omega(^4S, {}^2D_{5/2})/\Omega(^4S, {}^2D_{3/2})$ and it is equal to $\frac{3}{2}$, the ratio of the statistical weights in the upper states. For large N_e, R is equal to $\frac{3}{2}$ times the ratio of the transition probabilities, $A(^4S, {}^2D_{5/2})/A(^4S, {}^2D_{3/2})$.

$R(N_e)$ provides a useful comparison between observations and theory of atomic structure; there is excellent agreement for O^+ and S^+ (e.g. Kaler *et al* 1976, Dopita *et al* 1976, Zeippen 1982, Mendoza and Zeippen 1982a, b, Stanghellini and Kaler 1989). From this comparison it was found that higher-order relativistic corrections to the magnetic dipole operator, which are usually negligible, had to be taken into account (the experimental techniques for measuring lifetimes of allowed transitions are in general not suited for forbidden transitions and consequently there have been few experimental results).

3.1.2. Forbidden line temperatures. Electron temperatures are derived from the ratio of two emission lines from the same ion whose excited levels are of very different energies. The classical ratio to derive the electron temperature in PN is given by the [O III] lines $\lambda 5007$ $^3P_2-{}^1D_2$, $\lambda 4959$ $^3P_1-{}^1D_2$ and $\lambda 4363$ $^1D_2-{}^1S_0$ (Menzel *et al* 1941, Seaton 1954). These lines are collisionally excited from the ^3P level and are easily observed in most PN. The excitation energies are $[E(^1D) - E(^3P)] = 2.49$ eV and $[E(^1S) - E(^3P)] = 5.33$ eV. Since typical electron temperatures in PN have associated energies ~ 1 eV, the line ratio is very sensitive to the electron temperature. A good approximation to the ratio is given by

$$\frac{I(4959) + I(5007)}{I(4363)} = \frac{7.73 \exp[(3.29 \times 10^4)/T_e]}{1 + 4.5 \times 10^{-4}(N_e/T_e^{1/2})}. \tag{3.6}$$

The accuracy of the atomic parameters used is within 10%, and the accuracy of the best observed line intensity ratios for the brightest PN is also within 10%. Therefore for these objects typical uncertainties of several hundred degrees in the derived electron temperatures are found.

A similar line ratio used to determine electron temperatures is given by the [N II] lines $\lambda 6583$ $^3P_2-{}^1D_2$, $\lambda 6548$ $^3P_1-{}^1D_2$, and $\lambda 5755$ $^1D_2-{}^1S_0$. This ratio is important for PN with regions of a low degree of ionization.

3.2. Density inhomogeneities

Almost all PN show structure on two-dimensional images. In addition to clumpiness, which gives them an inhomogeneous appearance, they show large-scale density gradients, filaments, shells, bipolar geometry, *etc* (e.g. Balick 1987, Balick *et al* 1987). The role that these spatial density variations play in the determination of basic parameters, like the envelope mass and the ionization structure, is paramount to the study and formation of PN.

A global density, $N_e(\text{RMS})$, can be obtained from the PN hydrogen line fluxes. The observed flux in $H(\beta)$ from a homogeneous sphere after correction for absorption is given by

$$I(H\beta) = \tfrac{1}{3}(R^3/d^2)N_e(\text{RMS})N_p(\text{RMS})\alpha(H\beta)h\nu(H\beta) \tag{3.7}$$

where R is the radius of the object, d is the distance to the observer, $\alpha(H\beta)$ is the $H\beta$ effective recombination coefficient and N_p, the proton density. To a very good approximation, the proton density is given by

$$N_p(\text{RMS}) = N_e(\text{RMS})\left(1 + \frac{N(\text{He}^+)}{N(\text{H}^+)} + 2\frac{N(\text{He}^{+2})}{N(\text{H}^+)}\right)^{-1}. \tag{3.8}$$

Similar equations can be written for other Balmer lines or for the free–free radio continuum. To solve this equation it is necessary to determine the He^+/H^+ and the $\text{He}^{+2}/\text{H}^+$ abundance ratios (see subsection 4.1). For density homogeneous PN the $N_e(\text{RMS})$ density corresponds to the real electron density.

The densities derived from forbidden line ratios and from global emissions are not identical. In general $N_e(\text{FL})$ is larger than $N_e(\text{RMS})$, indicating the presence of density inhomogeneities. A first approximation to the study of this problem is to assume that a given PN of total volume V_T has two types of regions: high-density ones, which occupy a fraction $\varepsilon = V/V_T$ of the total volume, with an average density given by $N_e(\text{FL})$, and low-density regions with negligible densities.

The filling factor, ε, is given by

$$\varepsilon = N_e^2(\text{RMS})/N_e^2(\text{FL}). \tag{3.9}$$

The concept of filling factor was introduced by Strömgren (1948) (see also Seaton and Osterbrock 1957, Osterbrock and Flather 1959, Mallik and Peimbert 1988 and references therein).

Since for a given nebula the $N_e(\text{FL})$ value is independent of distance while the $N_e(\text{RMS})$ is not, a good distance estimate is needed to derive ε (see (3.7) and (3.9)).

A useful observational quantity is the emission measure (EM) defined as

$$\text{EM} = \int N_e(\text{RMS})N_i(\text{RMS})\,dl \tag{3.10}$$

where N_i is the number of ions cm^{-3} and l is given in pc; the EM is proportional to the surface brightness of an object. The brightest PN have EM $\sim 10^7\,\text{cm}^{-6}$ pc and the faintest PN that can be detected in the red Palomar Survey Plates have EM in the range 20–50 (Peimbert *et al* 1975).

3.3. Temperature inhomogeneities

The variations in density, degree of ionization, and chemical composition in PN are likely to produce spatial temperature variations. In addition to these variations, PN are losing mass at supersonic velocities, producing shock waves within the envelopes; these shock waves also produce spatial temperature variations.

There are at least four different methods for deriving temperatures of PN: (a) intensity ratios of auroral to nebular forbidden lines, (b) intensity ratios of a recombination line i times ionized to a collisionally excited line of the same element $i+1$ times ionized like $\lambda 4267$ of C II to $\lambda 1909$ of C III, (c) intensity ratios of bound–free continuum to recombination lines of hydrogen, and (d) intensity ratios of free–free continuum to recombination lines of hydrogen. The first three are used in the UV and optical region of the spectrum, and the fourth one in the radio region.

It can be shown that in the presence of spatial temperature variations over the observed volume, the methods based on collisionally excited lines yield higher temperatures than those based on recombination lines, free-bound or free-free continua. The average temperature weighted by the square of the density over the volume considered is given by

$$T_0(N_i, N_e) = \frac{\int T(r) N_i(r) N_e(r) \, d\Omega \, dl}{\int N_i(r) N_e(r) \, d\Omega \, dl} \tag{3.11}$$

where Ω is the observed solid angle, l is the distance along the line of sight and r is the position vector; for the case of forbidden lines N_i is replaced by $N(X^{+p})$. When the temperature is derived from a parameter that depends on the αth power of the local temperature we have

$$\langle T^\alpha \rangle = \frac{\int T^\alpha(r) N_e(r) N_i \, d\Omega \, dl}{\int N_e(r) N_i(r) \, d\Omega \, dl}. \tag{3.12}$$

For the case of small fluctuations $\langle T^\alpha \rangle$ can be expanded in a Taylor series about the mean temperature, yielding a temperature

$$T_\alpha = \langle T^\alpha \rangle^{1/\alpha} \approx T_0 \left(1 + \frac{\alpha - 1}{2} t^2 \right) \qquad \alpha \neq 0, \text{ for } t^2 \ll 1. \tag{3.13}$$

In this relation terms of order higher than 2 have been neglected. Here t is the root mean square temperature fluctuation and is given by

$$t^2 = \frac{\int T^2 N_e N_i \, d\Omega \, dl - T_0^2 \int N_e N_i \, d\Omega \, dl}{T_0^2 \int N_e N_i \, d\Omega \, dl}. \tag{3.14}$$

For a value of the temperature derived from two processes that depend on different powers of the temperature, it can be shown that

$$T_{\alpha/\beta} = \left(\frac{\langle T^\alpha \rangle}{\langle T^\beta \rangle} \right)^{1/(\alpha - \beta)} \approx T_0 \left(1 + \frac{\alpha + \beta - 1}{2} t^2 \right) \qquad \alpha \neq \beta, \text{ for } t^2 \ll 1. \tag{3.15}$$

Similarly for the case of a temperature derived from the ratio of two forbidden lines, it can be shown that

$$T_{(\lambda_{nm}/\lambda_{n'm'})} \approx T_0 \left[1 + \left(\frac{\Delta E + \Delta E^*}{kT_0} - 3 \right) \frac{t^2}{2} \right] \qquad \Delta E \neq \Delta E^*, \text{ for } t^2 \ll 1 \tag{3.16}$$

where ΔE and ΔE^* represent the differences in energy between the upper levels of each of the transitions and the ground level.

For $t^2 \ll 1$ and the temperature derived from the Balmer continuum and Hβ, it is found that

$$T_{(Bac/H\beta)} = T_0(1 - 1.70t^2) \tag{3.17}$$

and for the temperature derived from radio lines and continuum:

$$T_{(\Delta_\nu T_L/T_c)} = T_0(1 - 1.42t^2) \tag{3.18}$$

while for $\lambda 4363$ and $\lambda 5007$ of O^{++} from equation (3.16) we obtain

$$T_{(4363/5007)} = T_0\left[1 + \frac{1}{2}\left(\frac{90800}{T_0} - 3\right)t^2\right]. \qquad (3.19)$$

Consequently for the temperatures calculated from optical lines of PN, the forbidden-line method emphasizes high-temperature regions while the other two methods favour the low-temperature regions.

From recent theoretical models of constant density, homogeneous chemical composition and photon ionization it is found that for PN $0.005 \leqslant t^2 \leqslant 0.03$. Obviously each nebula has a different value of t^2.

The derivation of the correct electron temperature is very important for the determination of the chemical abundances. If constant temperature is assumed, $t^2 = 0.00$, instead of $t^2 \simeq 0.02$, an underestimate of 0.3 to 0.2 dex is obtained for the relative abundances based on forbidden-to-permitted line ratios ($10^a \doteq a$ dex). The errors in dismissing spatial temperature fluctuations are negligible when both lines are formed by recombination.

In type-I PN (see subsection 5.1) it is likely that the relative importance of shock waves is higher than in the other objects, and consequently that the electron temperature derived from the 4363/5007 ratio could be considerably higher than the average temperature.

In the presence of chemical inhomogeneities the temperature differences can be very large, a factor of two or more, and sophisticated models of ionization structure are necessary to properly determine the temperature distribution (Torres-Peimbert *et al* 1990).

4. Chemical compositions

One of the most important aspects of the study of PN is the determination of their chemical abundances. The chemical composition of a PN is the result of the galactic chemical evolution prior to the formation of the progenitor star and of the modification of the initial abundances by nuclear reactions in its interior. Therefore the study of the chemical composition of PN can be used as a test of models of stellar evolution and of models of galactic chemical evolution. In this article the term chemical composition is used to describe the relative abundances of all the elements, and not the relative abundances of the molecules.

4.1. Ionic abundances

Hydrogen constitutes $\sim 90\%$, helium $\sim 10\%$ and all the other elements together $\sim 0.1\%$ of the atoms in the Universe.

The ionic abundances of a PN are always given relative to H^+ not only because H is the most abundant element but because H^+ essentially defines the ionized region. The H^+ abundances are directly derived from the Balmer line intensities that are very bright and easy to observe. The effective recombination coefficients for the Balmer line intensities have been computed by Brocklehurst (1971) and Hummer and Storey (1987) for a wide range of N_e and T_e values; these coefficients are almost independent of N_e and are roughly proportional to T_e^{-1}, therefore precise values of the electron temperature are not needed to obtain accurate H emissivities.

4.1.1. Helium. It is possible to derive the $N(\text{He}^{+2})/N(\text{H}^{+})$ ratios from equations of the type

$$\frac{N(\text{He}^{+2})}{N(\text{H}^{+})} = \frac{\alpha_{4,2}(\text{H}^{0})}{\alpha_{4,3}(\text{He}^{+})} \frac{4686}{4861} \frac{I(4686)}{I(\text{H}\beta)} \tag{4.1}$$

where the $4 \to 3$ transition corresponds to the strongest He II line in the optical region at $\lambda 4686$ Å. Equation (4.1) can be written as

$$\frac{N(\text{He}^{+2})}{N(\text{H}^{+})} = 2.28 \times 10^{-2} T_{e}^{0.14} \frac{I(4686)}{I(\text{H}\beta)} \tag{4.2}$$

where the effective recombination coefficients have been derived from the computations by Hummer and Storey (1987) interpolated in the 10 000–20 000 K electron temperature range for $N_{e} = 10\,000$ cm^{-3}. Since both H and He recombination lines are roughly proportional to T_{e}^{-1}, accurate $N(\text{He}^{+2})/N(\text{H}^{+})$ ratios can be derived even if the electron temperature is not well known.

The $N(\text{He}^{+})/N(\text{H}^{+})$ ratios can be derived from equations of the type

$$\frac{N(\text{He}^{+})}{N(\text{H}^{+})} = \frac{\alpha_{4,2}(\text{H}^{0})}{\alpha_{nm}(\text{He}^{0})} \frac{\lambda_{nm}}{4861} \frac{I(\lambda_{nm})_{\text{rec}}}{I(\text{H}\beta)} \tag{4.3}$$

where the effective recombination coefficients for helium, $\alpha_{nm}(\text{He}^{0})$, have been computed by Brocklehurst (1972). The He I line intensities are affected by radiative transfer effects and collisions from the 2^{3}S metastable level and the observed line intensities are related to the radiative recombination line intensities by

$$I(\lambda) = I(\lambda)_{\text{rec}}[1 + \gamma_{\text{ce}}(\lambda) + \gamma_{\text{rt}}(\lambda)] \tag{4.4}$$

where γ_{ce} is the collisional excitation parameter and γ_{rt} is the radiative transfer parameter. Fortunately each He I line is affected differently by the effects of radiative transfer and collisional excitation, therefore by measuring accurately different He I lines in the same nebula it is possible to evaluate each effect and to derive accurate $N(\text{He}^{+})/N(\text{H}^{+})$ ratios.

The collisional excitation parameters depend strongly on the T_{e} and N_{e} values and are given by

$$\gamma_{\text{ce}}(\lambda_{nm}, T_{e}, N_{e}) = \gamma_{\text{ce}}(\lambda_{nm}, T_{e}, \infty)/[1 + N_{c}/N_{e}(T_{e})] \tag{4.5}$$

where N_{c} is the critical density for radiative depopulation of the 2^{3}S He^{0} level and corresponds to 3230 cm^{-3} for $10\,000 \leqslant T_{e} \leqslant 20\,000$. The latest estimates of the $\gamma_{\text{ce}}(\lambda_{nm})$ values for the different helium lines are those by Clegg (1987) and Peimbert and Torres-Peimbert (1987a, b) based on the 19-state *ab initio* computation for collisions to He^{0} states with $n \leqslant 4$ by Berrington and Kingston (1987). The possibility of depopulation of the 2^{3}S He^{0} level by photoionization and by other processes, that could reduce the γ_{ce} values, has been studied by Peimbert and Torres-Peimbert (1987b) and by Clegg and Harrington (1989). The collisional excitation effects typically reduce the $N(\text{He}^{+})/N(\text{H}^{+})$ ratios by factors ranging from 1.02 to 1.20; the most affected ratios are those of PN with high N_{e} and T_{e} values.

The $2^{3}\text{S}-n^{3}\text{P}$ series with $n > 2$ is weakened by self-absorption and $\lambda\lambda 7065\ 2^{3}\text{P}-3^{3}\text{S}$, $5876\ 2^{3}\text{P}-3^{3}\text{D}$ and $4472\ 2^{3}\text{P}-4^{3}\text{D}$ are strengthened by resonance fluorescence. The self-absorption effect for $\lambda 7065$ is considerable while for $\lambda\lambda 5876$ and 4472 it is substantially smaller. From observations of the $I(7065)/I(4472)$ intensity ratios and the computations by Robbins (1968) and Cox and Daltabuit (1971) it is possible to estimate the optical thickness in the various triplet lines. For typical PN the self-absorption effect increases the intensity of $\lambda 7065$ by factors in the 1.1–3 range while the increase on $\lambda\lambda 5876$ and 4472 is smaller than a factor of 1.02. A partial energy level diagram showing the He I triplet system is presented in figure 4.

Similarly, self-absorption from the 1^1S ground state has a considerable effect on the emergent line intensities; for example $\lambda 5016$ 2^1P–3^1P is strengthened by a large factor. From observations of the $I(5016)/I(4472)$ intensity ratios (e.g. Aller and Walker 1970) and the computations by Robbins and Bernat (1973) it is possible to estimate the optical thickness in the various singlet lines, from this estimate it follows that the effect on $\lambda\lambda 6678$ 2^1P–3^1D and 4922 2^1P–4^1D is smaller than a factor of 1.01 and can be neglected.

4.1.2. Ionic abundances from other recombination lines. Faint emission lines of C, N, O and other elements have been detected in bright PN. Some are excited by resonance fluorescence, while others seem to be due to pure recombination processes.

Lines produced by fluorescence cannot be used to determine accurate abundances because radiative transfer effects dominate their intensities. A classic example is provided by the O III lines in the 2800–3800 Å range, the strongest lines being $3p^3S_1$–$3d^3P^0_2$ $\lambda 3133$ and $3p^3P_2$–$3d^3P^0_2$ $\lambda 3444$; these lines are produced by a coincidence between the He II Lyα and the [O III] $2p^2$ $^3P_2$–$3d^3P^0_2$ $\lambda 303.80$ transitions. This coincidence converts He II Lyα into O III Bowen resonance-fluorescence photons (Bowen 1928, Saraph and Seaton 1980, Osterbrock 1989).

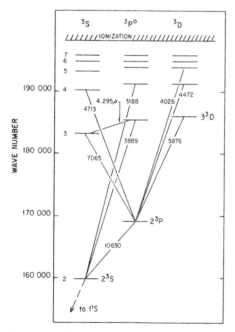

Figure 4. Partial Grotrian diagram of He I, showing the strongest optical lines of the triplet system. The line intensities of the triplet system are affected by collisional excitation and self-absorption from the metastable 2^3S level.

A typical line produced by recombination is $3d^2D-4f^2F^0 \lambda 4267$ of C^+, the C^{+2} abundance derived from this line is given by

$$\frac{N(C^{+2})}{N(H^+)} = \frac{\alpha_{4,2}(H^0)}{\alpha_{4267}(C^+)} \frac{4267}{4861} \frac{I(4267)}{I(H\beta)}$$

$$= 1.67 \times 10^{-2} T_e^{0.20} \frac{I(4267)}{I(H\beta)} \tag{4.6}$$

where $\alpha_{4267}(C^+)$ was obtained from Seaton (1978) and was interpolated for $10\,000 \leqslant T_e \leqslant 20\,000$ K.

The C^{++} abundance derived from $\lambda 4267$ is in general higher than that derived from the [C III] $3s^2\,^1S_0-3s\,3p^3\,P_2 \lambda 1907 + $[C III] $3s^2\,^1S_0-3s\,3p^3\,P_1 \lambda 1909$ collisionally excited lines; this difference is of an order of magnitude in extreme cases. General discussions of this problem have been given in the literature (e.g. Torres-Peimbert *et al* 1980, Barker 1982, French 1983, Kaler 1986, Clegg 1989, Peimbert 1989). The discrepancy could be due to spatial variations in the electron temperature produced by inhomogeneities in the C/H abundance ratio (Peimbert 1983); the regions with higher C/H abundance ratios would be at lower temperatures and would contribute preferentially to $\lambda 4267$, while the regions with lower C/H abundance ratios would be at higher temperatures and would contribute preferentially to $\lambda\lambda 1907 + 1909$. Torres-Peimbert *et al* (1990) have computed ionization structure models for NGC 4361 that are consistent with this suggestion.

The abundances of C^{+3} and C^{+4} can also be determined from recombination lines present in the visual region of the spectrum (e.g. French 1983); the effective recombination coefficients have been presented by Seaton (1978). For the line C III $4f^3F-5g^3G$, $\lambda 4070$ the contribution due to dielectronic recombination (Nussbaumer and Storey 1984) has to be taken into account.

Recombination lines of N^+ have also been used to determine N^{+2}/H^+ ratios (Wilkes *et al* 1981, Escalante 1988). Some of the N^+ lines might be affected by fluorescence effects. Escalante (1988) has found that for a group of PN the N^+ lines that originate from F levels are due only to recombination, while those that originate from P levels are affected by fluorescence in some objects. The accuracy of the N^{+2}/H^+ abundance ratios is about a factor of two due to the faintness of the N^+ lines. Non-hydrogenic effective recombination coefficients for the N^+ lines have been computed by Escalante and Victor (1990).

4.1.3. Ionic abundances from collisionally excited lines. The collisionally excited line flux $I(X^{+p}, \lambda_{nm})$ received from a nebula in ergs cm^{-2} s^{-1} is given by

$$I(X^{+p}, \lambda_{nm}) = \int N_n(X^{+p}) A_{nm} h\nu(\lambda_{nm})\, d\Omega\, dl \tag{4.7}$$

where λ_{nm} is the wavelength associated with the emission, $N_n(X^{+p})$ is the number of atoms in the excited level n of the state of ionization p in cm^{-3}, A_{nm} is the Einstein transition probability coefficient in s^{-1}, Ω is the observed solid angle and l is the depth of the nebula along the line of sight in cm. If collisional de-excitations are negligible, equation (4.7) can be written as

$$I(X^{+p}, \lambda_{nm}) = C(\lambda_{nm}) \int N(X^{+p}, r), N_e(r) T_e(r)^{-1/2} \exp[-\Delta E/kT_e(r)]\, d\Omega\, dl \tag{4.8}$$

where $N(X^{+P})$ is the total number of atoms in the pth state of ionization, $\mathbf{r}$ is the position in the nebula, ΔE is the energy difference between the ground and excited levels and C is a constant that can be determined from the atomic parameters.

Analogously, the Hβ flux of a nebula can be expressed as

$$I(H\beta) = \int N_e(r)N(H^+, r)\alpha_{4,2}(H^0)h\nu(H\beta)\, d\Omega\, dl \tag{4.9}$$

where $\alpha_{4,2}(H^0) = 1.32 \times 10^{-10} T_e^{-0.91}$ for $10\,000 \leqslant T_e \leqslant 20\,000$ K (Hummer and Storey 1987). Consequently equation (4.9) can be written as

$$I(H\beta) = D \int N_e(r)N(H^+, r)T_e(r)^{-0.91}\, d\Omega\, dl \tag{4.10}$$

where D is a constant.

For homogeneous nebulae with uniform temperature the relative abundances can be obtained from equations (4.8) and (4.10), and are given by

$$\frac{N(X^{+P})}{N(H^+)} = \frac{D}{C(\lambda_{nm})}\frac{\exp(\Delta E/kT_e)}{T_e^{0.41}}\frac{I(X^{+P}, \lambda_{nm})}{I(H\beta)}. \tag{4.11}$$

Accurate values of the electron temperature are needed to obtain accurate abundances from collisionally excited lines in the ultraviolet and visual regions of the spectrum because for them $\Delta E \geqslant 2$ eV and typical temperatures in PN correspond to $kT_e \sim 1$ eV. Alternatively, abundances derived from infrared lines are almost temperature independent since $\Delta E < 1$ eV.

In Tables 3, 4 and 5, we list some of the most important collisionally excited lines used to determined chemical abundances; notice that all the visual and infrared lines are forbidden while in the ultraviolet there are forbidden, semi-forbidden and permitted lines. A review of the relevance of infrared forbidden fine-structure emission lines for the study of PN has been given by Dinerstein (1983). Typical spectra of PN in the UV, optical and infrared regions are shown in figures 5 to 8. Feibelman et al (1988) have produced a very useful spectral atlas of PN based on the archives of the International Ultraviolet Explorer satellite.

Table 3. Ultraviolet collisionally excited lines used to determine chemical abundances.

λ (Å)	Ion	Transition
1239	N V	$^2S_{1/2}-^2P_{3/2}$
1243	N V	$^2S_{1/2}-^2P_{1/2}$
1483	[N IV]	$^1S_0-^3P_2$
1486	N IV]	$^1S_0-^3P_1$
1548	C IV	$^2S_{1/2}-^2P_{3/2}$
1550	C IV	$^2S_{1/2}-^2P_{1/2}$
1661	O III]	$^3P_1-^5S_2$
1666	O III]	$^3P_2-^5S_2$
1907	[C III]	$^1S_0-^3P_2$
1909	C III]	$^1S_0-^3P_1$
2422	[Ne IV]	$^4S_{3/2}-^2D_{3/2}$
2426	[Ne IV]	$^4S_{3/2}-^2D_{5/2}$
2854	[Ar IV]	$^4S_{3/2}-^2P_{3/2}$
2868	[Ar IV]	$^4S_{3/2}-^2P_{1/2}$

Table 4. Optical collisionally excited lines used to determine chemical abundances.

λ (Å)	Ion	Transition
3426	[Ne V]	$^3P_2-^1D_2$
3726	[O II]	$^4S_{3/2}-^2D_{3/2}$
3729	[O II]	$^4S_{3/2}-^2D_{5/2}$
3869	[Ne III]	$^3P_2-^1D_2$
4711	[Ar IV]	$^4S_{3/2}-^2D_{5/2}$
4724	[Ne IV]	$^2D_{3/2}-^2P_{3/2}$
4726	[Ne IV]	$^2D_{3/2}-^2P_{1/2}$
4740	[Ar IV]	$^4S_{3/2}-^2D_{3/2}$
5007	[O III]	$^3P_2-^1D_2$
6583	[N II]	$^3P_2-^1D_2$
6716	[S II]	$^4S_{3/2}-^2D_{5/2}$
6731	[S II]	$^4S_{3/2}-^2D_{3/2}$
7006	[Ar V]	$^3P_2-^1D_2$
7136	[Ar III]	$^3P_2-^1D_2$
9531	[S III]	$^3P_2-^1D_2$

Table 5. Infrared collisionally excited lines used to determine chemical abundances.

λ (μm)	Ion	Transition
6.98	[Ar II]	$^2P_{3/2}-^2P_{1/2}$
8.99	[Ar III]	$^3P_2-^3P_1$
10.52	[S IV]	$^2P_{1/2}-^2P_{3/2}$
12.81	[Ne II]	$^2P_{3/2}-^2P_{1/2}$
18.71	[S III]	$^2P_1-^3P_2$
24.28	[Ne V]	$^3P_0-^3P_1$
25.87	[O IV]	$^2P_{1/2}-^2P_{3/2}$
51.81	[O III]	$^3P_1-^3P_2$
57.34	[N III]	$^2P_{1/2}-^2P_{3/2}$
88.36	[O III]	$^3P_0-^3P_1$

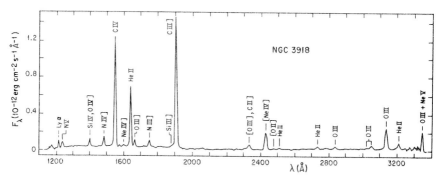

Figure 5. Ultraviolet spectrum of NGC 3918 obtained with the International Ultraviolet Explorer satellite.

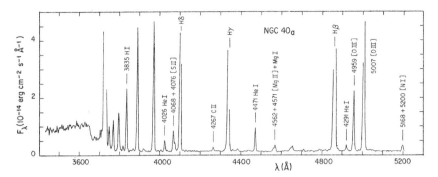

Figure 6. Blue spectrum of NGC 40 taken with the IIDS at Kitt Peak National Observatory (Clegg *et al* 1983).

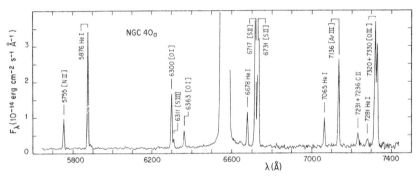

Figure 7. Red spectrum of NGC 40 taken with the same equipment as figure 6 (Clegg *et al* 1983).

4.2. Total abundances

The total abundance of a given object relative to H is given by

$$\frac{N(X)}{N(H)} = \frac{N(X^0) + N(X^+) + N(X^{+2}) + \ldots}{N(H^+)}. \tag{4.12}$$

Ideally all the ions present should be observable. Unfortunately very often not all ions have observable transitions and corrections for the unobservable stages of ionization have to be applied.

4.2.1. Ionization correction factors. To estimate the abundance of unobserved ions it is possible to rely on the abundances of observed ions with similar ionization potentials.

PN can be divided into (a) those with a low degree of ionization, i.e. those without a He^{+2} region and (b) those with a high degree of ionization, i.e. those with a He^{+2} region. In what follows we will mention the most used ionization correction factors for PN with high and low degrees of ionization.

For PN with a low degree of ionization the oxygen abundance is given by

$$\frac{N(O)}{N(H)} = \frac{N(O^+ + O^{+2})}{N(H^+)} \tag{4.13}$$

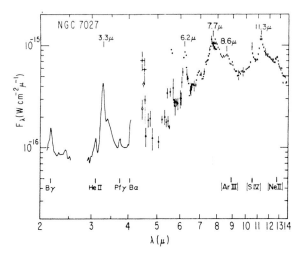

Figure 8. Infrared spectrum of NGC 7027; the data in the 4–8μ region were taken with the UCSD spectrograph on the Kuiper Airborne Observatory and the other data from ground based observatories (Rusell *et al* 1977).

the ionic O^+ and O^{+2} abundances are directly determined from strong lines. The oxygen ratios are used to determine the total abundances of N and Ne when only N^+ and Ne^{+2} abundances are available as follows:

$$\frac{N(N)}{N(H)} = \frac{N(O^+ + O^{+2})}{N(O^+)} \frac{N(N^+)}{N(H^+)} = i_{cf}(N) \frac{N(N^+)}{N(H^+)} \tag{4.14}$$

$$\frac{N(Ne)}{N(H)} = \frac{N(O^+ + O^{+2})}{N(O^{+2})} \frac{N(Ne^{+2})}{N(H^+)} = i_{cf}(Ne) \frac{N(Ne^{+2})}{N(H^+)}. \tag{4.15}$$

Equation (4.15) breaks down for the external regions and at the edges of low-density PN like NGC 7293 and NGC 6720, where there are appreciable quantities of H^0 inside the H^+ zone and the following charge transfer reactions occur

$$O^{+2} + H^0 \rightleftharpoons O^+ + H^+ \tag{4.16}$$

$$Ne^{+2} + H^0 \rightleftharpoons Ne^+ + H^+. \tag{4.17}$$

The large value of the rate coefficient $\beta(O^{+2})$ relative to $\beta(Ne^{+2})$ produces an increase of $N(Ne^{+2})/N(O^{+2})$ toward the edge of low-density PN (Péquignot *et al* 1978, Hawley and Miller 1978a, Butler *et al* 1979, Péquignot 1980). Equation (4.15) also breaks down for objects with a very low degree of ionization, those with a He^0 zone inside the H^+ region (e.g. Flower 1969, Torres-Peimbert and Peimbert 1977).

For PN with a high degree of ionization the helium abundance is given by

$$\frac{N(He)}{N(H)} = \frac{N(He^+ + He^{+2})}{N(H^+)}; \tag{4.18}$$

From these values it is possible to correct for the higher ionization stages of oxygen, which are not usually available, as follows:

$$\frac{N(O)}{N(H)} = \frac{N(He^+ + He^{+2})}{N(He^+)} \frac{N(O^+ + O^{+2})}{N(H^+)}$$

$$= i_{cf}(O) \frac{N(O^+ + O^{+2})}{N(H^+)}. \tag{4.19}$$

Equation (4.18) breaks down for objects with a very low degree of ionization which have a considerable fraction of He^0 inside the H^+ region. To estimate the He^0 amount in a given PN we need observations along different lines of sight, or a sophisticated ionization structure model.

Ionization correction factors have been proposed for other elements, nevertheless many of them are density dependent and have been of limited use.

4.2.2. Ionization structure models. Photoionization models have been used to estimate the ionization correction factors for unobserved ions. To construct a model we require: (a) atomic data to determine the ionization equilibrium at each point, (b) the integrated electron temperature, (c) line emissivities, (d) the gaseous density distribution, (e) the stellar ionizing flux distribution and (f) the dust distribution and its absorbing properties. The data for ionization cross sections and recombination probabilities can be obtained from Chapman and Henry (1971, 1972), Reilman and Manson (1979), Gould (1978) and the compilation by Osterbrock (1989). The data for charge transfer reactions and dielectronic recombinations can be obtained from Butler and Dalgarno (1980), Butler *et al* (1980) and Nussbaumer and Storey (1983, 1984). The data for transition probabilities and collision strengths have been summarized by Mendoza (1983).

There are several sets of stellar atmosphere fluxes available: the LTE models by Hummer and Mihalas (1970) and the NLTE models by Husfeld *et al* (1984) and by Clegg and Middlemass (1987). Unfortunately the observed properties of the central stars and the degree of ionization of the nebulae predicted by these models often do not match the observations. For example the stellar temperatures and gravities derived by Méndez *et al* (1981, 1988) for NGC 1535 and NGC 4361 were used together with ionization structure models by Adam and Koppen (1985) and Torres-Peimbert *et al* (1990) to try to fit the observed line intensities. These authors were not able to fit the observed nebular ionization structures, to obtain agreement they found that the stellar flux beyond 54.4 eV had to be significantly higher than those of the model atmosphere fluxes available.

On the other hand the model atmosphere fluxes by Clegg and Middlemass (1987) for stars with very different gravities and effective temperatures, but similar ratios of H^0 to He^+ ionizing photons, produce very similar ionization structures and consequently similar i_{cf} values, which implies that if the strong lines of H^0, He^0, He^+, O^+ and O^{+2} are fitted by the photoionization models the i_{cf} for the other elements are almost independent of the stellar atmosphere adopted (e.g. Torres-Peimbert *et al* 1990).

4.2.3. Accuracy of the abundance determinations. The accuracy of the ionic abundances is determined by the accuracy of the line intensities and of the atomic parameters, the relevant line intensities and atomic parameters are not only those that correspond to the ion in question but also those that are needed to determine the distributions of the electron temperature and of the electron density. The errors vary from ion to ion and from object to object. For the best observed PN the accuracy of the ionic concentrations relative to H^+ is better than 0.04 dex for He^+ and He^{+2}, 0.1 dex for N^+, O^+, O^{+2},

Ne^{+2}, S^+ and Ar^{+2} and 0.2 dex for C^+, C^{+2}, Ne^{+3}, Ne^{+4}, S^{+2}, Ar^{+3} and Ar^{+4}; similarly the accuracy of the total abundances relative to H is better than 0.04 dex for He, 0.1 dex for O, 0.15 dex for N and Ne and 0.2 dex for C, Ar and S.

5. Classification of planetary nebulae

5.1. Types of PN

From the range in kinematical properties, galactic distribution, chemical composition and mass of the shell it follows that the progenitors of PN range from population I to extreme population II objects. That is, the masses of the progenitor stars range from a few solar masses to less than one solar mass. In this review we will follow Peimbert's (1978) classification, which divides PN into four types: type I He-N rich; type II (disk or intermediate population); type III (high velocity) and type IV (halo population). From a variety of arguments related to stellar dynamics and stellar evolution it seems that this scheme is not only a chemical composition classification but that it corresponds to progenitor stars of different initial or main sequence masses M_i, with the following approximate values: type I $(2.4-8)M_\odot$, type II $(1.2-2.4)M_\odot$, type III $(1-1.2)M_\odot$ and type IV $(0.8-1.0)M_\odot$. Faúndez-Abans and Maciel (1987a) have divided type II PN into types IIa and IIb and Maciel (1989) has suggested that PN in the bulge of the Galaxy be called type V. Other classifications have been proposed by Greig (1971, 1972), Kaler (1983b), Heap and Augensen (1987), Balick (1987) and Amnuel *et al* (1989).

As mentioned before, the M_i values are in the 0.8-8 $M_\odot$ range; alternatively, after the envelope ejection most PN central stars have M_c values in the 0.54-0.8 $M_\odot$ range. The difference between M_i and M_c for each object varies from about 0.3 to a few solar masses, part of the difference is due to mass loss during the red giant stage and the rest is due to the mass of the PN shell. The shell masses range from ~ 0.02 $M_\odot$ for extreme type IV PN to ~ 1 $M_\odot$ for the most extreme type I PN. In optically thin PN all the shell mass is ionized while in optically thick PN a considerable fraction of the shell mass is in the form of H^0, He^0 and H_2.

Classifications based on chemical abundances depend on the quality of the abundance determinations, particularly for borderline objects between types I and II and between types IIa and IIb.

5.1.1. Type I PN. PN of type I have been defined as those objects with $N(He)/N(H) \geqslant$ 0.125 or $\log N(N)/N(O) \geqslant -0.3$ (Peimbert 1978, Peimbert and Torres-Peimbert 1983), average abundances have been presented in table 6. Most He-N rich PN are very filamentary and show a bipolar structure (Greig 1971, 1972, Peimbert and Torres-Peimbert 1983, Kaler 1983b); this shape has been given several names: B, bipolar, binebulous, or hourglass. Moreover, in general their spectra present very strong forbidden lines ranging from [O I], [N I], and [S II] up to [Ne V].

Based on morphological arguments, Greig (1971, 1972) divided PN into four main classes: A (annular), B (binebulous), C (centric, surface brightness increases toward the centre) and E (egg shaped). There is a strong overlap between type I PN and class B PN. Different kinematical and galactic distribution studies have shown that class B (Greig 1972, Cudworth 1974) and type I objects (Acker 1980, 1983, Dutra and Maciel 1990) as a group correspond to more massive stellar progenitors than the other types of PN.

Table 6. Chemical abundances of galactic PN, the Sun and the Orion nebula, given in $12 + \log N(X)/N(H)$.

Object	He	C	N	O	Ne	Ar	Reference
Type I PN	11.13	8.4–9.1	8.55	8.65	8.00	—	1
Type II PN	11.04	8.3–9.1	8.15	8.70	8.10	6.60	2, 3
Type III PN	11.04	—	8.00	8.40	7.80	—	4, 5
Sun	—	8.67	7.99	8.92	8.03	6.69	6, 7, 8
Orion nebula	11.01	8.57	7.68	8.65	7.80	6.65	9, 10, 11
Population I	11.01	8.62	7.84	8.75	7.98	6.65	12

1. Peimbert and Torres-Peimbert (1987a)
2. Torres-Peimbert and Peimbert (1977)
3. Aller and Czyzak (1983)
4. Barker (1978)
5. Maciel (1990)
6. Lambert (1978)
7. Lambert and Luck (1978)
8. Meyer (1985)
9. Peimbert and Torres-Peimbert (1977)
10. Torres-Peimbert *et al* (1980)
11. Peimbert (1982)
12. The present paper

There are a few type I PN for which estimates of the mass of the progenitor in the main sequence, M_i, are reported in the literature. From the assumption that NGC 2818 belongs to the galactic cluster of the same name, Peimbert and Serrano (1980) and Dufour (1984a) estimate M_i to be $2.2 \pm 0.3\ M_\odot$. NGC 2346 has a binary nucleus with a current orbital separation of $38\ R_\odot$; from the assumption that a common envelope event has been responsible for NGC 2346, Iben and Tutukov (1989) estimate that M_i was between 2 and 3 $M_\odot$ (see also Calvet and Peimbert 1983). Based on the increase in brightness (0.06 mag year^{-1}), the mass in the molecular shell ($\geqslant 2\ M_\odot$), and the mass of the central star ($M_c \sim 0.8\ M_\odot$), Kwok and Bignell (1984) estimate that $M_i \geqslant 3\ M_\odot$ for GL 618.

Several authors have found that, in general, central stars of type I PN lie on higher-mass tracks in the HR diagram than the central stars of other PN (Kaler 1983a, Pottasch 1983, 1989, Gathier and Pottasch 1985, Kaler and Jacoby 1989, 1990, Kaler *et al* 1990, Stasinska and Tylenda 1990); this result implies higher M_i values for type I PN than for other objects (see figure 9).

Sabbadin (1986a) finds that B nebulae spend most of their life as optically thick, whereas C nebulae spend most of their life as optically thin; this result supports the idea that B nebulae have higher-mass progenitors than C nebulae. Amnuel *et al* (1989) based on morphological, kinematic and other characteristics have classified PN in three groups: (a) massive, (b) intermediate mass and (c) low mass. There is a very strong overlap between type I PN and those classified as massive PN by Amnuel *et al*.

5.1.2. Type II PN. PN of type II are defined by the following properties: $N(\mathrm{He})/N(\mathrm{H}) < 0.125$, $\log N(\mathrm{N})/N(\mathrm{O}) < -0.3$, $\log N(\mathrm{O})/N(\mathrm{H}) > -3.9$, $|\Delta v_{pr}| < 60\ \mathrm{km\ s}^{-1}$, and $|z| < 0.8$ kpc, where $|v_{pr}|$ is the peculiar radial velocity relative to the galactic rotation curve and z is the distance away from the galactic plane. About 80% of the well studied PN of the solar vicinity belong to this group. Chemical abundances for large groups of

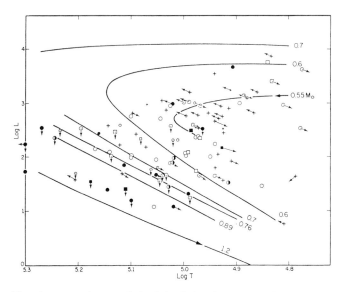

Figure 9. PN central stars on the log L–log T plane from Kaler *et al* (1990). *Filled symbols*: N/O $\geqslant$ 0.8, or He/H $>$ 0.15; *half filled*: 0.8 $>$ N/O $\geqslant$ 0.5; *open*: N/O $<$ 0.5; *crosses*: no abundance data. *Solid curves*: evolutionary tracks from Wood and Faulkner (1986) for $\phi = 0.5$, mass loss core A, except for $M_c = 1.2\ M_\odot$, which is from Paczynski (1971).

type II PN are present in the literature (e.g. Torres-Peimbert and Peimbert 1977, Kaler 1979, Aller and Czyzak 1983, Aller and Keyes 1987), average abundances are presented in table 6. PN of the solar neighbourhood have an average height above the galactic plane of 150 pc, which corresponds to progenitors of $M_i = 1.5\ M_\odot$ (Osterbrock 1973); these values are typical of intermediate population I stars.

Faúndez-Abans and Maciel (1987a) have divided type II PN in two subtypes: type IIa are those with log $N(N)/N(H) \geqslant -4.0$ and type IIb are those with log $N(N)/N(H) < -4.0$. Dutra and Maciel (1990) find that the kinematic behaviour and the average z value of type IIa PN are intermediate between those of type I and type IIb PN.

5.1.3. Type III PN. Type III PN are defined by the following properties: $|\Delta v_{pr}| > 60$ km s^{-1} and log $N(O)/N(H) > -3.9$. Kaler (1970) has defined as Population II PN those objects with $|\Delta v_{pr}| > 60$ km s^{-1} or $|z| > 0.8$ kpc. Barker (1978) studied 22 PN that fit Kaler's definition for population II, most of them are type III PN and their average abundances are presented in table 6.

5.1.4. Type IV PN. Type IV PN or halo PN are defined by having log $N(O)/N(H) \leqslant -3.9$, very often they show $|\Delta v_{pr}| > 60$ km s^{-1} and $|z| > 0.8$ kpc. There are eight known objects including a borderline case. Their abundances are presented in table 7; typical errors are 0.03 dex for He, 0.1 dex for O and less than 0.2 dex for the other elements. The prototype is K648, a PN in the globular cluster M15, with $M_i \sim 0.8\ M_\odot$ and $M_{shell} \sim 0.02\ M_\odot$ (O'Dell *et al* 1964, Peimbert 1973). For NGC 4361: $M_c \sim 0.55\ M_\odot$ and $M_{shell}(FL) \sim 0.07\ M_\odot$ (Méndez *et al* 1988, Torres-Peimbert *et al* 1990).

Table 7. Chemical abundances of type IV PN, given in $[X/H] = \log (X/H) - \log (X/H)_{popl}$.

Object	He	C	N	O	Ne	Ar	Reference
K648	−0.01	+0.08 (rec)	−1.34	−1.05	−1.28	−2.35	1, 2, 3
BB-1	−0.03	+0.47 (rec)	+0.50	−0.85	+0.02	−2.05	1, 4
H4-1	−0.02	+0.69 (rec)	−0.09	−0.35	−1.28	−1.95	1, 2
NGC 4361	+0.01	+0.66 (rec)	−0.48	−0.92	−0.41	−0.74	5
NGC 2242	−0.01	−0.23 (coll)	−0.12	−0.72	−0.20	−0.76	5
DDDM-1	−0.01	<−1.52 (coll)	−0.44	−0.65	−0.68	−0.85	6, 7
PRMG-1	−0.05	—	—	−0.65	−0.68	−0.85	8
PRTM-1	+0.03	<−1.02 (coll)	<+0.16	−0.35	−0.08	−0.45	9

1. Barker (1980)
2. Torres-Peimbert and Peimbert (1979)
3. Adams *et al* (1984)
4. Torres-Peimbert *et al* (1981)
5. Torres-Peimbert *et al* (1990)
6. Clegg *et al* (1987)
7. Barker and Cudworth (1984)
8. Peña *et al* (1989)
9. Peña *et al* (1990)

5.1.5. Extremely inhomogeneous PN. A30 and A78 show inner regions that are extremely H underabundant, while the outer regions are normal (Jacoby 1979, Hazard *et al* 1980, Jacoby and Ford 1983, Manchado *et al* 1988). The abundances of the most contaminated inner regions of A30 and A78 are presented in table 8. The standard deviations for the abundances of A30 amount to ~0.14 dex; the C abundances were derived from the C^+ $\lambda 4267$ line.

5.1.6. Galactic bulge PN. Galactic bulge PN are important for the study of general properties of the bulge: dynamics, star formation history and chemical evolution. Moreover, they are also important for comparing the luminosity of their central stars with stellar evolution predictions since the distance to the bulge is well known. Dust obscuration makes the study of bulge PN in the visual region very difficult. Several recent studies have been made to detect new PN in the galactic bulge (e.g. Kinman *et al* 1988, Pottasch *et al* 1988, Ratag *et al* 1990).

Webster (1988) has studied 65 PN toward, and probably in, the galactic bulge. She obtained He/H, O/H and N/O abundance ratios for 49 of them. Between 10 and 20 per cent of the sample are type I PN, providing evidence for a young component of

Table 8. Chemical abundances of A30 and A78, given in $12 + \log N(X)/N(H)$.

Object	He	C	N	O	Ne	Reference
Abell 30, 3	12.99	11.89	9.14	9.32	8.83	1
Abell 30	12.95	11.96	9.93	9.63	—	2
Abell 78	12.78	<11.73	9.19	9.89	9.52	1
Abell 78, 4	12.43	—	11.09	11.09	10.41	3

1. Jacoby and Ford (1983)
2. Peimbert (1983)
3. Manchado *et al* (1988)

the population of the galactic bulge. Most PN in the sample have O/H ratios similar to those of PN in the solar neighbourhood. Based on these observations Webster proposes that the star formation rate was high early in the history of the bulge and that afterwards it has become considerably smaller.

Webster (1988) also finds that M2-29 might be a type IV PN since its $N(\mathrm{O})/N(\mathrm{H})$ ratio is 2.8×10^{-5}. She also identifies in her sample a few super-oxygen-rich candidates that deserve further study.

5.2. Comparison of the observed abundances with those predicted by stellar evolution models

The evolution of the surface abundances of He, C, N and O for intermediate-mass stars during their asymptotic giant branch phase has been predicted by stellar evolution models (e.g. Iben 1975, Iben and Truran 1978, Becker and Iben 1979, 1980, Renzini and Voli 1981, Iben and Renzini 1983, Renzini 1984, Wood and Faulkner 1986). The abundances predicted by the models have been compared with observations by many authors (e.g. Kaler *et al* 1978, Peimbert 1981, 1984, 1985, Aller 1983, Kaler 1983b, Peimbert and Torres-Peimbert 1983, Torres-Peimbert 1984, Clegg 1985, Kaler and Jacoby 1989, 1990, Henry 1989, Kaler *et al* 1990).

In what follows we will briefly describe the evolutionary models by Renzini and Voli (1981) and the information derived from observational comparisons. The models consider two processes: (a) convective dredge ups, when the envelope convection extends inward mixing to the surface nuclearly processed material and (b) nuclear burning at the base of the convective envelope. Three dredge-up phases prior to the PN ejection have been considered. The first dredge-up corresponds to the inward penetration of the convective envelope when stars reach the red giant branch for the first time, during the H-burning shell phase, enhancing N at the expense of C. The second dredge-up occurs in stars with $M_i \geqslant 3\ M_\odot$, when following the ignition of the He-burning shell the convective envelope penetrates into the helium core, enhancing the N and He and reducing the O and C surface abundances. The third dredge-up occurs during the asymptotic giant branch (AGB) evolution and consists of one or several individual mixing episodes following each He-shell flash enhancing the He and C surface abundances. The evolution is greatly affected by two mass loss processes: the stellar wind during the AGB and the PN ejection.

The stellar structure computations by Renzini and Voli (1981) have been made considering two parameters: $\alpha = l/H$, the ratio of the mixing length to the pressure scale height, and η which multiplied by the Reimers' rate (1975) gives the mass loss rate during the AGB phase; the computations were made for $\eta = \frac{1}{3}$ and $\frac{2}{3}$ and $\alpha = 0$, 1, 1.5 and 2, with most of them for $\eta = \frac{1}{3}$ and $\alpha = 0$ and 1.5. It has been estimated semi-empirically that η is in the $\frac{1}{3}$–3 range and that for stars with $M_i < 2\ M_\odot$ it is in the $\frac{1}{3}$–$\frac{1}{2}$ range (e.g. Renzini 1984, Iben 1984, and references therein). The values of α and η are not well known and they may vary with stellar evolution stage, initial stellar mass and chemical composition.

5.2.1. Type I PN. The observed range of He/H values in type I PN is well accounted for by the theory and implies that these objects have progenitors with $M_i \geqslant 3\ M_\odot$. This result is in agreement with the other criteria that indicate that type I PN have the most massive progenitors. For PN with good abundance determinations the He/H and C/O observed values are in fair agreement with predictions for models with M_i in the 3-5 $M_\odot$ range with $0 < \alpha < 2$ and $\eta = \frac{1}{3}$; the exceptions are NGC 6302 (Aller *et al* 1981) and N97 (Barlow *et al* 1983), which can be matched with models for $M_i \sim 8\ M_\odot$ with $\alpha \sim 2$, and $\eta = \frac{1}{3}$, which makes them the two well observed PN with the highest M_i values known. Gómez *et al* (1989) estimate an envelope mass of 0.75 $M_\odot$ for NGC 6302, again supporting the idea of a massive progenitor. In figure 10 a radio map of NGC 6302 is presented.

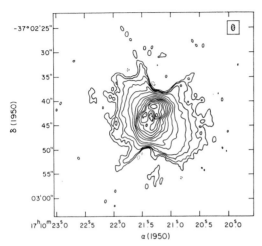

Figure 10. Continuum map at 6 cm of NGC 6302. The synthesized beam was 0.8×1.3 (arc sec) FWHM. The contour levels are: -0.003, 0.003, 0.005, 0.01, 0.02, 0.05, 0.1, 0.2, 0.3, 0.4, 0.5, 0.6, 0.7, 0.8 and 0.9 of the peak flux of 60 mJy per beam (Gómez *et al* 1989).

There is a well established positive correlation between the He/H and N/O ratios. This correlation implies that: (a) if $\eta = \frac{1}{3}$, α diminishes from 2 to 1 as M_i increases from ~ 3 to $\sim 6\ M_\odot$ or (b) if $\alpha \sim 2$, η increases with M_i. From galactic chemical evolution considerations (Peimbert 1985) the second possibility is more likely.

There is an O/H versus N/O anticorrelation that seems to be real (Peimbert and Torres-Peimbert 1983, 1987a, Henry 1989, 1990, Kaler *et al* 1990) and which cannot be explained by the available models. The O depletion reaches factors of 2-3, while the most favourable theoretical models produce depletions of a factor of 1.4 (Renzini and Voli 1981, Renzini 1984). The O depletion increases with α and the highest depletion is obtained for the model with $\alpha = 2$, $\eta = \frac{1}{3}$ and $Z = 0.02$. It is possible to explain the O depletion with a larger value of α. It is also possible that part of the correlation could be due to an underestimate of the O/H abundance caused by shock wave enhancements of $\lambda 4363$ of [O III] but no models have been made to check this suggestion.

Models predict many more bright AGB stars ($M_{bol} \lesssim -6.0$) than observed in the Magellanic Clouds (e.g. Iben and Truran 1978, Renzini and Voli 1981, Iben and Renzini 1983, Renzini 1989, and references therein). It has been suggested that this discrepancy might be solved by a radiation pressure mechanism responsible for a rapid hydrodynamical ejection of the stellar envelope (Wood and Faulkner 1986, Renzini 1989). Renzini argues that this mechanism would limit the PN enrichment due to the third dredge-up phase, apparently in contradiction with the $C/O > 1$ values present in most PN of type I (Peimbert and Torres-Peimbert 1983, 1987a).

5.2.2. Type II and type III PN. The observed range of He/H values in type II PN is well accounted for by theory and implies that those objects have progenitors with $M_i < 3 M_\odot$.

From the average C/H values it is possible to predict the average M_i value, which turns out to be somewhat higher than the average value derived from their galactic distribution. This result indicates that the third dredge-up phase is more efficient for low-mass stars than predicted by Renzini and Voli (1981); further theoretical developments indicate that this is the case (e.g. Iben and Renzini 1983), therefore the C/H and He/H predictions by Renzini and Voli (1981) for objects with $M_i < 3 M_\odot$ should be used with caution (Renzini 1983).

There are well established positive correlations between the He/H and N/O ratios and the O/H and N/O ratios. These correlations are in agreement with a secondary production of N and with different initial He, C and N abundances produced by galactic chemical evolution and galactic abundance gradients.

5.2.3. Type IV PN. Of the seven halo PN with C/O abundance determinations, five of them show C/O > 1 (see table 7). These values are difficult to explain because AGB carbon stars are not found in globular clusters, nor are they predicted by current AGB models with $M_c \leqslant 0.6 M_\odot$ (e.g. Renzini and Voli 1981, Iben and Renzini 1984). To explain C/O > 1 values, Renzini (1989) has proposed that these objects eject their envelope during one thermal pulse. Thermal pulses, also known as helium shell flashes, have often been regarded as possible triggers for the envelope ejection. The mass that could be ejected during one pulse peak for an object with $M_c = 0.54 M_\odot$ is about 0.1 $M_\odot$, a value considerably larger than the envelope mass of K648 or NGC 4361. This implies that in type IV PN the envelope ejection may be possible during just a single thermal pulse. Moreover, the inhomogeneous model for NGC 4361 with C/O higher in the inner region (Torres-Peimbert *et al* 1990) and the strong C IV emission lines shown by the central star of NGC 4361 (Méndez 1989) agree with this suggestion.

5.2.4. Extremely inhomogeneous PN. A fraction of post AGB stars is expected to experience a final thermal pulse while descending along the white dwarf cooling track (Schönberner 1979, Iben and Renzini 1983, Iben *et al* 1983). It is thought that A30 and A78 might belong to this group.

6. Distances, masses and birth rates

Two of the main parameters searched for in the study of astronomical objects are their distances and their masses. The distances of PN are needed in order to: (a) compare the central stars of PN with theoretical evolutionary tracks in the HR diagram, (b) calculate the mass of the ionized shell, (c) estimate the number of PN in the galaxy and (d) determine the distances to other galaxies. Similarly the masses of PN are needed to: (a) constrain the mass loss models of asymptotic giant branch evolution, (b) study the chemical enrichment of the interstellar medium, (c) study the shell evolution, and (d) determine the scale for statistical distances.

6.1. Distances derived from individual properties

In the seventies only a handful of PN in the solar vicinity had direct distance determinations; at present the number has grown to about 50. For most objects in the solar vicinity we only have distances based on statistical properties. A review of recent distance determinations has been given by Lutz (1989). In what follows we will mention the most important methods, not based on statistical arguments, that have been used to determine distances.

6.1.1. Binary stars and clusters. There are four binary PN nuclei with distance determinations based on the spectral type of the secondary star: NGC 246, NGC 1514, NGC 2346 and NGC 3132 (e.g. Minkowski 1965, Kohoutek 1967, Méndez 1975, Maciel and Pottasch 1980, Mallik and Peimbert 1988, and references therein). The distance to K648 has been determined under the assumption that it belongs to the globular cluster M15 (Peimbert 1973). The distance to NGC 2818 has been determined under the assumption that it belongs to the open cluster of the same name (Dufour 1984a, Pedreros 1989).

6.1.2. 21 cm absorption and emission. Gathier *et al* (1986a) have derived distances for 12 PN based on 21 cm observations: (a) in absorption in the direction of the PN, (b) in absorption of background sources within 1 degree of the PN and (c) in emission in directions close to the PN. They combined these observations with information on the galactic rotation curve and on nearby H II regions to determine the distances, the estimated accuracies are in the 0.12–0.25 dex range.

6.1.3. Visual extinction. It is possible to construct extinction versus distance diagrams by measuring the visual extinction of many stars with known distance in directions near a given PN. From a well behaved relation between extinction and distance it is possible to estimate the distance to the PN by measuring its extinction. The method was pioneered by Lutz (1973), and has been used by several authors (Kaler and Lutz 1985, Maciel 1985, Maciel *et al* 1986). Probably the most precise set of determinations based on this method was made by Gathier *et al* (1986b), who determined distances to 12 PN with estimated accuracies in the 0.04–0.18 dex range.

6.1.4. Expansion and radial velocity. By comparing radial velocities of expansion, determined spectroscopically, with measured angular expansions it is possible to estimate distances. The angular expansion is determined by comparing the position of knots, filaments, edges, and other features on two different epochs (e.g. Liller *et al* 1966, Masson 1986). This method is expected to provide reliable results for optically thin objects. For optically thick objects it provides only a lower limit to the distance because the size of the ionized region increases not only due to gas expansion but also due to the advance of the ionization front into the external neutral shell and the inner neutral pockets. The distances determined by this method are usually smaller than those determined by other methods (e.g. Maciel and Pottasch 1980, Méndez *et al* 1988), probably indicating that most of the objects to which this method has been applied are optically thick.

6.1.5. Central stars. It is possible to derive the distance to a PN from the intrinsic luminosity of the central star, since it depends on the surface gravity, g, and the effective temperature, T_{eff}. The distance to the central star in pc is given by (Méndez *et al* 1988)

$$d^2 = 3.82 \times 10^{-11} \frac{M_c F_*}{g} 10^{0.4 m_0(V)} \tag{6.1}$$

where M_c is the mass of the star in solar units, F_* is the monochromatic model atmosphere flux at $\lambda 5480$ Å, in erg cm^{-2} s^{-1} cm^{-1}, g is the surface gravity in cm s^{-2} and $m_0(V)$ is the apparent visual magnitude corrected for reddening. F_* depends strongly on T_{eff}. From high-dispersion spectra of the central stars and model atmospheres it is possible to derive g and T_{eff}. Combined errors of 0.04 dex in T_{eff} and 0.2 dex in g produce an error of 0.08 dex in the distance. Based on this method, Méndez et al (1988) have derived distances for 22 PN.

6.2. Masses

The ionized mass of a homogeneous density nebula made of H and He is given by

$$M(\text{RMS}) = (1+4y)m_H \int \frac{N_e(\text{RMS})}{(1+x)} \, dV \tag{6.2}$$

where $y = N(\text{He})/N(\text{H})$, $x = N(\text{He}^+)/N(\text{H}^+) + 2N(\text{He}^{+2})/N(\text{H}^+)$. The factor $1/(1+x)$ denotes the contribution of helium to the electron density per hydrogen atom (see equation (3.8)), and m_H is the mass of the H atom. For a spherical PN of radius R, equation (6.2) can be written as

$$M(\text{RMS}) = \frac{(1+4y)}{(1+\bar{x})} \, m_H \frac{4}{3}\pi R^3 N_e(\text{RMS}) \tag{6.3}$$

where $\bar{x}$ is the average value within the ionized volume. The linear size is related to the observed angular size in arc sec, ϕ, by

$$R(\text{pc}) = \frac{\phi(\text{arc sec})d(\text{pc})}{206265}. \tag{6.4}$$

Compilations of angular sizes can be found in the catalogue by Perek and Kohoutek (1967) or in the list by Cahn and Kaler (1971).

It is well known that PN are not homogeneous in density; in general they present complex density structures with cavities, shells, knots, filaments, etc. For non-homogeneous nebulae, $M(\text{RMS})$ corresponds to an upper limit to the real mass. To derive a more realistic mass it can be assumed that a fraction ε of the volume is filled with dense material and that the remainder of the volume is empty. The density of the dense material is expected to be the one derived from the forbidden line ratios, $N_e(\text{FL})$. Therefore the total mass in the nebula is given by

$$M(\text{FL}) = \frac{(1+4y)}{(1+\bar{x})} \, m_H V_T \varepsilon N_e(\text{FL}) \propto V N_e(\text{FL}). \tag{6.5}$$

Similarly, from equations (3.7)-(3.9) we have

$$I(\text{H}\beta) \propto V_T \varepsilon N_e^2(\text{FL})/d^2 \propto V N_e^2(\text{FL})/d^2. \tag{6.6}$$

From equations (6.5) and (6.6) it follows that

$$M(\text{FL}) \propto I(\text{H}\beta)d^2/N_e(\text{FL}) \tag{6.7}$$

therefore, $M(\text{FL})$ can be derived independently of the filling factor and the assumed geometry (Barlow 1987). Alternatively $M(\text{RMS})$ depends on the geometry and the filling factor, from equations (6.3), (6.5) and (3.9) it follows that

$$M(\text{FL}) = \varepsilon^{1/2} M(\text{RMS}). \tag{6.8}$$

Only a small fraction of the PN have reliable $N_e(\text{FL})$ and $N_e(\text{RMS})$ determinations; consequently, ε values have been derived for only a few PN and most authors have *adopted an average* ε value, usually in the 0.6–0.8 range, for large samples of PN (e.g. O'Dell 1962, Cahn and Kaler 1971, Maciel and Pottasch 1980, Daub 1982, Gathier *et al* 1986a, b, Méndez *et al* 1988). In all the cases where $M(\text{FL})$ values are given, it is possible to derive the corresponding $M(\text{RMS})$ values by means of equation (6.8).

A list of 35 PN with distances derived by direct methods and with individual ε and $M(\text{FL})$ determinations has been presented by Mallik and Peimbert (1988); the $M(\text{FL})$ values vary from ~0.005 $M_\odot$ to 0.3 $M_\odot$ and the ε values vary from 0.01 to 1, with almost half of the sample having ε values smaller than 0.16. These ε values are considerably smaller than the average value adopted in most papers.

For optically thick PN the $M(\text{RMS})$ and $M(\text{FL})$ values are lower limits to the total PN mass because the neutral and molecular components have not been considered. The presence of a neutral and a molecular component can be established by studying: (a) the variation of the degree of ionization with r, (b) the presence of the 21 cm hyperfine transition of H I in absorption and (c) the presence of molecules (e.g. Rodríguez 1989 and references therein).

The envelope masses derived for type IV PN are smaller than those derived for other types of PN, in agreement with the idea that the progenitors of type IV PN had lower masses in the main sequence (e.g. Peimbert 1973, Adams *et al* 1984, Mallik and Peimbert 1988, Torres-Peimbert *et al* 1990).

6.3. Relation between mass and radius

From observations of large samples of PN, several authors have obtained relations between masses and radii of the type

$$M(\text{RMS}) \propto R^\gamma \qquad (6.9)$$

(see table 9). Moreover, Kwok (1985), using a method based on evolution models of central stars of PN by Schönberner (1983), has obtained that $M(\text{RMS}) \propto R^{5/2}$. Mallik and Peimbert (1988), based on observations of 35 PN, have obtained that

$$M(\text{FL}) \propto R^{1.02 \pm 0.15} \qquad (6.10)$$

$$\varepsilon \propto R^{-0.91 \pm 0.18} \qquad (6.11)$$

(see figure 11). Other authors, basing their estimates on large samples of PN, have also obtained ε values which decrease with increasing R (Torres-Peimbert and Peimbert

Table 9. Exponent γ for the mass–radius relation given by $M \propto R^\gamma$.

R range (pc)	ε range	γ (FL)	γ (RMS)	Reference
0.01–0.33	1.0	—	1.43 ± 0.09	Pottasch (1980)
0.007–0.066	1.0	—	1.47 ± 0.12	Gathier *et al* (1983)
0.004–0.96	1.0	—	2.46	Phillips and Pottasch (1984)
0.032–0.27	1.0	—	1.88	Phillips and Pottasch (1984)
0.004–0.12	0.01–1.0	1.21 ± 0.10	—	Daub (1982)
0.01–0.56	0.01–1.0	1.02 ± 0.15	—	Mallik and Peimbert (1988)

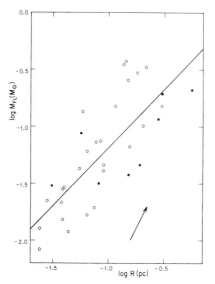

Figure 11. Plot of mass as a function of radius for PN with scale-independent distances (Mallik and Peimbert 1988). *Filled circles*: type I PN; *open circles*: the rest. The arrow indicates the change in mass if an object is 1.4 times farther away than assumed.

1977), Daub 1982, Amnuel *et al* 1984).

From equation (6.9) and the following relations:

$$M(\text{RMS}) \propto N(\text{RMS}) R^3 \qquad (6.12)$$

$$E(\text{H}\beta) \propto N^2(\text{RMS}) R^3 \propto Q(\text{H}^0) \qquad (6.13)$$

where $E(\text{H}\beta)$ is the total emitted flux in Hβ and $Q(\text{H}^0)$ is the number of hydrogen ionizing photons per unit time (see equation (2.9)), it follows that

$$Q(\text{H}^0) \propto \frac{M^2(\text{RMS})}{R^3} \propto R^{2\gamma-3} \qquad (6.14)$$

which implies that $Q(\text{H}^0)$ is independent of R for $\gamma = 1.5$.

Notice that the derived value of $Q(\text{H}^0)$ is also independent of R if $\varepsilon^{-1/2} M(\text{FL}) \propto R^{3/2}$ (equations (6.8), (6.9), (6.12) and (6.13)).

6.4. Distances derived from statistical properties

Since only a small fraction of PN in the solar vicinity have distance determinations based on individual characteristics, the so-called direct distance determinations, it has been the aim of many investigators to find a good distance scale that can be applied to all PN. This distance scale is based on the assumption that a general property has the same value for a set of PN or that a general relationship holds for a set of PN.

6.4.1. Assumption of constant mass. The Shklovsky (1956) method (see also Minkowski and Aller 1954) applies to optically thin PN, those that are completely ionized, and is based on the assumption that the ejected mass is the same for all objects. This method has been used extensively to determine distances to PN (e.g. O'Dell 1962, 1963, 1968, Seaton 1966, 1968, Webster 1969, Cahn and Kaler 1971, Osterbrock 1973, Cudworth 1974, Milne and Aller 1975, Acker 1978, Khromov 1979, Schneider and Terzian 1983). From equations (3.7), (6.4) and (6.5) it follows that the distance is given by

$$d = \left[\frac{3}{16\pi^2} \frac{\alpha(H\beta)h\nu(H\beta)}{m_H^2} \frac{(1+\bar{x})}{(1+4y)^2} \right]^{1/5} M(FL)^{2/5} \varepsilon^{-1/5} \phi^{-3/5} I(H\beta)^{-1/5}$$

$$= KM(FL)^{2/5} \varepsilon^{-1/5} \phi^{-3/5} I(H\beta)^{-1/5} \tag{6.15}$$

where K depends weakly on T_e and the He/H abundance ratio. Equation (6.15) can also be written as

$$d = KM(RMS)^{2/5} \phi^{-3/5} I(H\beta)^{-1/5}. \tag{6.16}$$

Two calibrations of equation (6.15) have been widely used: Seaton's (1968) based on Webster's (1969) observations of PN in the Magellanic Clouds and employed in the extensive study by Cahn and Kaler (1971) (hereafter to be referred to as the S distance scale) and Cudworth's (1974) based on statistical parallaxes (hereafter to be referred to as the C distance scale). The S distance scale is a factor of 1.47 smaller than the C distance scale. In many applications of this method it has been assumed that objects with $R \geq 0.08$ pc are optically thin and those with $R \leq 0.08$ pc are optically thick.

From the C distance scale it follows that

$$d(pc) = 108 \phi^{-3/5} I(H\beta)^{-1/5}. \tag{6.17}$$

Under the assumptions that $\langle T_e \rangle = 10\,000$ K, $\langle y \rangle = 0.11$ and $\langle \bar{x} \rangle = 0.13$, equation (6.16) implies that $\langle M(RMS) \rangle = 0.42\ M_\odot$. Similarly, from the S distance scale it follows that $\langle M(RMS) \rangle = 0.16\ M_\odot$.

There are two problems with the Shklovsky method that render the distances derived from equation (6.16) very uncertain: (a) the assumption that the ejected mass is the same for all objects is not valid since values of $M(RMS)$ for individual shells of PN vary from a few hundredths of a solar mass for type IV PN to almost one solar mass for the most massive PN, (b) the method applies to optically thin PN while recent results have shown that most PN in the solar vicinity are optically thick with R values as high as 0.6 pc (Pottasch 1980, Phillips and Pottasch 1984, Gathier 1987, Mallik and Peimbert 1988, Pottasch 1989, Amnuel et al 1989). For a given PN, equation (6.16) could yield errors in the distance determination as high as a factor of 4.

Molecules have been detected in many PN that were classified as optically thin based on their large R values; the presence of molecules indicates that these PN are optically thick in at least some directions. For example, Zuckerman and Gatley (1988) have detected H_2 molecules in NGC 6720 and NGC 6853 (see figures 1 and 2), these PN have R values of 0.23 and 0.21 pc, respectively and had been classified as optically thin (e.g. Cudworth 1974).

6.4.2. Assumption of a constant ionizing flux. For optically thick objects it can be assumed that the number of H ionizing photons per unit time, $Q(H^0)$, is the same for all objects (Zanstra 1931, Vorontsov-Velyaminov 1934, Minkowski 1965). Therefore from equations (2.9) and (3.7) it follows that

$$d = \left[\frac{Q(H^0)}{4\pi I(H\beta)} \frac{\alpha(H\beta)}{\alpha_B} h\nu(H\beta) \right]^{1/2}. \tag{6.18}$$

Cudworth (1974) calibrated equation (6.18) based on statistical parallaxes and for objects with $R \leqslant 0.07$ pc found that

$$d(\mathrm{pc}) = 0.0178 \, I(\mathrm{H}\beta)^{-1/2}. \tag{6.19}$$

This calibration is in very good agreement with the earlier calibration by Minkowski (1965). Acker (1978) has also used this method extensively.

Milne (1982) used this method for a sample of PN observed in radio continuum, the observed flux is due to free-free emission which is proportional to the hydrogen recombination flux; he calibrated the radio continuum analogue of equation (6.18) with the distances by Cudworth (1974) and Acker (1978). The distances found by Acker are closer to the S distance scale than to the C distance scale.

Based on a sample of PN with individual distance determinations, Mallik and Peimbert (1988) obtained, for objects in the $0.02 \leqslant R \leqslant 0.3$ pc range, the following equation:

$$d(\mathrm{pc}) = 0.0171 \, I(\mathrm{H}\beta)^{-1/2} \tag{6.20}$$

in very good agreement with the determination by Cudworth (1974).

Figure 12 presents a plot of $E(\mathrm{H}\beta)$ as a function of R for the distance-independent sample of Mallik and Peimbert (1988); filled circles denote type I PN, open circles the rest. Also this figure presents: (a) the distance scale proposed by Mallik and Peimbert for objects in the $0.02 \leqslant R(\mathrm{pc}) \leqslant 0.3$ range, (b) the optically thick distance scale by Cudworth (1974) for $R \leqslant 0.07$ pc and (c) the optically thin distance scale by Cudworth (1974) for $R > 0.07$ pc. The two largest objects in figures 11 and 12 are NGC 2818 and NGC 7293; the presence of molecules in these PN (Storey 1984) indicates that they are not optically thin in all directions and have not been so in the past. The optically thin distance scale by Cudworth (1974) might yield the proper distance for some large nebulae, not because they are optically thin but because their central star is fading, as it is in the case of NGC 7293 (Méndez *et al* 1988). For any given object the error produced by applying equation (6.20) could be as large as a factor of three, but in

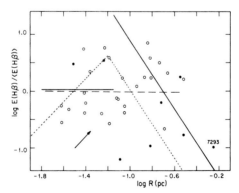

Figure 12. Plot of $\log E(\mathrm{H}\beta)$ against $\log R$ for the scale-independent sample (Mallik and Peimbert 1988). *Solid horizontal line* and *solid diagonal line*: optically thick and optically thin distance scales, respectively, by Cudworth (1974); *broken horizontal line*: distance scale by Mallik and Peimbert (1988); *dotted diagonal lines*: predicted behaviour by Kwok (1985, 1987). The arrow indicates the displacement of any given point if its distance is increased by a factor of 1.4.

general it will be smaller than a factor of two, which amounts to a difference of 0.6 in the log $E(H\beta)$ value of figure 12.

The spread in masses and $E(H\beta)$ values for a given R in figures 11 and 12 is probably real and could be due to an additional parameter, such as the mass of the central star (Mallik and Peimbert 1988, Maciel 1989, Pottasch 1989).

6.4.3. Assumption of a relationship between mass and radius. Maciel and Pottasch (1980) and Pottasch (1980) developed a method for determining distances to optically thick PN based on an empirical ionized mass to nebular radius relation. Maciel (1981a, b, 1984), Daub (1982), Phillips and Pottasch (1984) and Amnuel *et al* (1984) have also used this method to derive distances to a large number of PN. Each group obtains different distance scales due to differences in their assumptions; the most important differences are: (a) the proportionality constant in equation (6.9), (b) the value of the exponent γ in equation (6.9) and (c) the range of R values for which the method has been used.

The relations between mass and radius found by Pottasch (1980), Gathier *et al* (1983), and Mallik and Peimbert (1988) indicate that $Q(H^0)$ does not vary systematically with R, but variations of an order of magnitude in $Q(H^0)$ for a given R value are present, see figure 12, where $E(H\beta)$ is proportional to $Q(H^0)$. The relation $M(\text{RMS}) \propto R^{5/2}$ derived by Kwok (1985, 1987) should be analysed further.

From equations (6.13) and (6.14) it is seen that if $\gamma = 2.5$ then $E(H\beta) \propto R^2$; i.e. the *EM* is constant for all objects (equation (3.10)). This implies that the surface brightness is independent of R, and consequently that it is not possible to derive distances based only on the surface brightness or the total observed flux.

In figure 12 we show the behaviour predicted by Kwok (1985, 1987) for $\gamma = 2.5$, which corresponds to $EM = 7.4 \times 10^6$ cm^{-6} pc for $R \leqslant 0.065$ pc. For $R > 0.065$ pc Kwok assumes that the PN are optically thin, with $M(\text{RMS}) = 0.16\ M_\odot$, which corresponds to $\gamma = 0.0$. From this figure it is clear that γ could be equal to 2.5 for $R \leqslant 0.065$ pc. Alternatively, for optically thick objects with $R > 0.065$ the $\gamma = 2.5$ solution provides a very poor fit to the observations.

Distance scales based on relations for surface brightness, $N_e(\text{FL})$ and $N_e(\text{RMS})$ as functions of R have been proposed in the literature (Amnuel *et al* 1984, 1989, Sabbadin *et al* 1984b, Phillips and Pottasch 1984, Mallik and Peimbert 1988). Since surface brightness, $N_e(\text{FL})$ and $N_e(\text{RMS})$ are related to the mass, these scales are equivalent to those derived from mass radius relations.

6.4.4. Scale factor. To compare different distance scales it is possible to introduce a relative scale factor, k, with the normalization $k = 1$ for Seaton's distance scale. In Table 10 we present the relative sizes for some of the most frequently used distance scales. For optically thin distance scales $k \propto M(\text{RMS})^{2/5}$. For optically thick distance scales k increases with R and an average value for $0.1 \leqslant R(\text{pc}) \leqslant 0.3$ is presented. If objects smaller than 0.1 pc are considered, the spread in k values is even larger (see Gathier 1987). In this comparison the $M(\text{RMS})$ values have been computed under the assumptions that $y = 0.11$, $\bar{x} = 0.13$ and $T_e = 10\,000$ K.

Weidemann (1977) proposed a scale factor of 1.3 (see table 10) to superimpose the observed nuclei of PN over the 0.6 $M_\odot$ track on the HR diagram and to lower the PN birth rate to a value compatible with the white dwarf birth rate. To reach agreement

Table 10. Comparison of distance scales for PN with $0.10 \leq R(\text{pc}) \leq 0.30$.

$k = d/d_{\text{Seaton}}$	$M(\text{RMS})(M_\odot)$	Scale
1.55	—	Mallik and Peimbert (1988)
1.47	0.42	Cudworth (1974)
1.40	0.37	Schneider and Terzian (1983)
1.30	0.31	Weidemann (1977)
1.16	—	Maciel and Pottasch (1980)
1.00	0.16	Seaton (1968)
1.00	0.16	Cahn and Kaler (1971)
1.00	0.16	Milne and Aller (1975)
1.00	0.16	Acker (1978)
1.00	—	Gathier (1987)
0.95	0.14	Daub (1982)

with the lower white dwarf birth rates derived recently (see subsection 6.5) a similar argument to that of Weidemann would increase k by an even larger factor.

In spite of the many objects and their corresponding distances in common between the samples by Gathier (1987) and by Mallik and Peimbert (1988) there is a substantial difference between their scale factors (see table 10). The difference is due to two systematic effects: Mallik and Peimbert do not include in their sample the expansion distances determined by Liller *et al* (1966), which are smaller than those derived by other methods (see subsubsection 6.1.4) and include the distances determined by Méndez *et al* (1988), which are generally larger than those derived by other methods.

6.5. Birth rates

The determination of the local and galactic birth rates of PN are important in order to: (a) find out which is the fraction of the stars in the $0.8 \leq M_i(M_\odot) \leq 8$ range that undergo the PN phase (e.g. Tinsley 1978), and (b) to study the chemical enrichment of the interstellar medium.

6.5.1. Local birth rate. The local PN birth rate is given by

$$\dot{\rho} = \rho/\Delta t = \rho \langle v \rangle / (R_f - R_i) \tag{6.21}$$

where ρ is the density, in the solar vicinity, of PN with $R_i < R < R_f$, Δt is the time needed for R to increase from R_i to R_f, and $\langle v \rangle$ is the average velocity of expansion from R_i to R_f. Usually ρ is given in pc^{-3} and Δt in years.

In the optically thick phase $\langle v \rangle$ denotes the average velocity of the ionization front relative to the central star, v_{ion}; while in the optically thin phase $\langle v \rangle$ denotes the average velocity of expansion of matter, v_{exp}, given by the Doppler effect. The determinations of $\dot{\rho}$ have been carried out under the assumption that there is an interval in size $R_f - R_i$ where all PN are optically thin.

There are several sources of error associated with the use of equation (6.21) (see Phillips (1989) for a review). The main uncertainty is due to the adopted distance scale: since $\dot{\rho}$ is proportional to d^{-4} the $\dot{\rho}$ estimates vary, to a first approximation, like k^{-4} (see table 10). If we divide the distance scales into long, medium and short ($k \geq 1.4$, $1.3 \geq k \geq 1$, $k < 1$) the $\dot{\rho}$ values, in units of $10^{-12} \text{ pc}^{-3} \text{ y}^{-1}$, are in the ranges 0.6–1, 2–3 and 4–6, respectively (Phillips 1989).

There are two factors that increase the birth rate estimates. While most investigators have used $\langle v_{exp} \rangle = 20 \, \text{km s}^{-1}$, a careful study by Phillips (1989), primarily based on observations by Sabbadin *et al* (1983, 1984a, 1985, 1986) and Sabbadin (1984a, b, 1986b) yields $\langle v_{exp} \rangle = 26 \, \text{km s}^{-1}$ for objects with $0.1 \leq R(\text{pc}) \leq 0.6$ and $25 \, \text{km s}^{-1}$ for objects with $R \leq 0.6$ pc. As mentioned before (see subsubsection 6.4.1), most PN in the solar vicinity are optically thick, therefore in addition to the expansion of the shell, the ionized size increases further by the expansion of the ionized front into the neutral matter, consequently $\langle v_{ion} \rangle$ should be used instead of $\langle v_{exp} \rangle$ in equation (6.19). The measurement of angular expansions of PN with known distances could provide the direct determinations of v_{ion} which are needed to improve the birth rate estimates.

The $\dot{\rho}(\text{PN})$ values have often been compared with the white dwarf birth rate $\dot{\rho}(\text{WD})$. Recent results by Downes (1986) and Fleming *et al* (1986) yield $\dot{\rho}(\text{WD})$, in units of $10^{-12} \, \text{pc}^{-3} \, \text{y}^{-1}$, equal to 0.72 ± 0.25 and 0.62 ± 0.13, respectively; in very good agreement with the $\dot{\rho}(\text{PN})$ values derived from long distance scales (Alloin *et al* 1976, Phillips 1989).

It has been argued that $\dot{\rho}(\text{WD})$ should be increased to take into account those WD in binary systems. On the other hand it has been argued that $\dot{\rho}(\text{PN})$ does not include low-mass WD progenitors because the ejected nebula dissipates before the central star becomes hot enough to ionize it. The magnitude of these two effects is similar. Considering all the uncertainties involved, it can be concluded that $\dot{\rho}(\text{PN})$ and $\dot{\rho}(\text{WD})$ agree within a factor of 2 if a long distance scale is adopted.

6.5.2. Galactic birth rate. The galactic PN birth rate, $\dot{N}$, has been estimated by scaling the surface density birth rate of the solar vicinity to the overall galaxy, $\dot{N}$ is proportional to k^{-3}. Values of 0.5–1.5 PN per year have been derived for $\dot{N}$. As expected, the smaller values correspond to the longer k values (e.g. Phillips 1989 and references therein).

6.5.3. Total number of PN. The total number of PN in the Galaxy with $R \leq 0.6$ pc, N_T, has been derived by multiplying the galactic birth rate by the mean lifetime of a PN. N_T is proportional to k^{-2} and the values that have been determined are in the 10 000–30 000 range (Phillips 1989 and references therein).

7. Extragalactic planetary nebulae

Extragalactic PN allow us to study such properties of PN as: (a) the mass range of the progenitor stars based on their location and velocity distribution; (b) [O III] and Hα luminosity functions, and the masses of their envelopes, based on the common distance of the PN sample in a given galaxy; and (c) characteristics of stellar evolution models by comparing the chemical composition of the PN with that of the general interstellar medium. Moreover, extragalactic PN permit us to study properties of the host galaxy such as: (a) stellar death rate, (b) stellar mass return rate to the interstellar medium, (c) galactic chemical evolution, (d) mass distribution of the host galaxy, and (e) distance to the galaxy. Some results relevant to these problems will be discussed below.

Extragalactic PN are extremely faint and abundance determinations have been possible only for PN in Local Group galaxies, while for PN outside the Local Group only absolute fluxes in the [O III] 5007 line have been determined. These observations have been obtained with the largest telescopes and the best detectors available. Extragalactic PN will be prime targets for the new generation of large telescopes now under construction.

7.1. Magellanic Clouds

The Magellanic Cloud extragalactic planetary nebulae have been the most extensively studied PN, due to their proximity. Recent reviews on MC PN are those by Jacoby (1983), Peimbert (1984) and Barlow (1989).

7.1.1. [O III] luminosity function and number of PN. Jacoby (1980) was able to detect PN in the Magellanic Clouds (MC) spanning a range of 250 in their [O III] luminosity; he established an [O III] luminosity function and from it estimated that the total number of PN in the Small Magellanic Cloud (SMC) and in the Large Magellanic Cloud (LMC) with a limiting envelope size of $R \leq 0.64$ pc, are 285 ± 78 and 996 ± 253, respectively. The limiting envelope size comes from an expansion velocity of 25 km s^{-1} and a lifetime of 25 000 years.

Based on the MC PN luminosity function, Jacoby (1980) estimated that, for Local Group galaxies, the visual luminosity specific number is $6.1 \pm 2.2 \times 10^{-7}$ PN/$L_\odot$ and the mass specific number is $2.1 \pm 1.5 \times 10^{-7}$ PN/$M_\odot$, where the uncertainties are 1σ values derived from the averaging procedure.

7.1.2. Sizes and masses. Speckle interferometry has been used by Barlow *et al* (1986) and Wood *et al* (1986) to determine the diameter of bright compact PN in the MC. Wood *et al* (1987) used a direct imaging technique to determine the angular diameter of fainter and larger PN.

Wood *et al* (1986, 1987) determined $M(\text{RMS})$ values from the measured diameters and the observed $I(\text{H}\beta)$ fluxes of 24 PN in the MC. From a plot of mass against radius they concluded that the PN become optically thin for $R > 0.12$ pc. For those objects with $R > 0.12$ pc they found that $\langle M(\text{RMS}) \rangle = 0.32\ M_\odot$, and $\langle M(\text{FL}) \rangle = 0.27\ M_\odot$ where they assumed that $\varepsilon = 0.7$ (see equation (6.8.)).

Barlow (1987) determined $M(\text{FL})$ values for 32 PN in the MC and was the first to show that the $M(\text{FL})$ values derived from $I(\text{H}\beta)$ and $N_e(\text{FL})$ are independent of geometry and filling factor (equation (6.7)). For the 10 most dense objects, assumed to be optically thick, Barlow found that the average distance is $\langle d(\text{pc}) \rangle = 0.0332\ I(\text{H}\beta)^{-1/2}$, which is about a factor of 2 larger than those derived from the calibrations by Cudworth (1974) for PN with $R \leq 0.07$ pc and by Mallik and Peimbert (1988) for PN in the $0.02 \leq R(\text{pc}) \leq 0.3$ range (equations (6.19) and (6.20)). Nevertheless, there is a selection effect in favour of PN with large Hβ fluxes in the sample used by Barlow, therefore even if the difference in distance scales is significant it does not imply that the galactic distance scale should be revised.

From a plot of $N_e(\text{FL})$ against $M(\text{FL})$ Barlow (1987, 1989) concluded that the 12 PN with the lowest $N_e(\text{FL})$ values are optically thin and derived for them a mean mass $\langle M(\text{FL}) \rangle = 0.27 \pm 0.06\ M_\odot$. Moreover, by adopting $\varepsilon = 0.65$, and equation (6.8), Barlow derived, for this sample, $\langle M(\text{RMS}) \rangle = 0.33\ M_\odot$. This value corresponds to a distance scale factor, k, of 1.34. Nevertheless this k value is uncertain and should not be applied to galactic distances for two reasons: (a) ε might be considerably smaller than 0.65 (Mallik and Peimbert 1988), which would increase k, and (b) there is a selection effect in favour of PN with large Hβ fluxes; correcting for this effect would decrease k.

7.1.3. Chemical composition. There have been many abundance determinations for PN in the MC (e.g. Peimbert 1984 and references therein), some of the more recent ones are those by Aller *et al* (1987), Monk *et al* (1988), Peña and Ruiz (1988) and Henry *et al* (1989).

Table 11 presents the mean abundances of He, N, O and Ne for the PN samples by Monk *et al* (1988) and the mean abundances of C for the PN samples by Aller *et al* (1987), the mean values exclude type I nebulae, which are discussed below, and P25 of the sample by Aller *et al* which is C poor. Also table 11 presents the mean abundances for H II regions derived by Dufour (1984c). The standard deviations of the PN mean abundances are typically of 0.05 dex for He and of 0.2 dex for C, N, O and Ne, while for the H II regions they are about half as large. Presumably the H II region abundances are similar to the initial abundances of most PN progenitors, particularly those formed recently, and by comparing them it should be possible to estimate the enrichment produced by the evolution of the progenitor star.

The main results derived from samples that exclude type I PN are: (a) the He/H ratio is about the same in PN and in H II regions of the same galaxy, which implies that these PN do not contain significant amounts of freshly made helium (Monk *et al* 1988, Henry *et al* 1989); (b) the N/H ratios in the SMC and LMC samples are 0.94 dex and 0.84 dex higher than those of H II regions in the same galaxies, which implies that PN progenitors are producing substantial amounts of N; (c) by comparing the PN N abundances with the H II region C abundances, it follows that in PN most of the initial C has been converted into N, alternatively the N/H excess in PN of type II in the solar vicinity (see table 6) amounts to only about one third of the C abundance of population I objects, these results are in agreement with the computations by Renzini and Voli (1981), where it is expected that the smaller the initial metallicity, the higher the fraction of original C that is transformed into N (Monk *et al* 1988); (d) the C/H ratios in the SMC and LMC samples are 1.49 dex and 0.66 dex higher than those of H II regions in the same galaxies, which implies that C produced by the triple α reaction has been brought up to the surface by the third dredge-up phase (Renzini and Voli 1981, Aller and Czyzak 1983); (e) the O/H, Ne/H and Ar/H ratios in PN are similar to those of H II regions, indicating that IMS are not important sources of O, Ne and Ar.

Table 12 lists abundances for a group of type I PN. The C abundances were derived from UV collisionally excited lines. The He/H abundance ratio of N67 has not been corrected for the effect of collisional excitation from the 2^3S He^0 state, the correction would reduce the ratio by about 0.1 dex.

Monk *et al* (1988) have classified MC PN as type I using only the criterion that N/O $\geqslant$ −0.3 dex. They did not use the He/H abundance criterion because the initial He/H ratios are much lower than those found in the Galaxy, so a significantly larger He enhancement would have been necessary before a MC PN could satisfy the He/H

Table 11. Mean abundances for PN and H II regions in the Magellanic Clouds, given in $12 + \log N(X)/N(H)$. The mean values for the PN samples exclude type I nebulae.

Sample	He	C	N	O	Ne	Reference
SMC PN	10.92	8.65	7.44	8.26	7.36	1, 2
SMC H II	10.90	7.16	6.46	8.02	7.22	3
LMC PN	10.94	8.56	7.81	8.49	7.64	1, 2
LMC H II	10.93	7.90	6.97	8.43	7.64	3

1. Monk *et al* (1988)
2. Aller *et al* (1987)
3. Dufour (1984c)

Table 12. Abundances for type I PN in the Magellanic Clouds given in $12 + \log N(X)/N(H)$.

Object	He	C	N	O	Ne	Reference
N67 (SMC)	11.17:	6.27	7.47:	7.15	6.54	1
N66 (LMC)	11.07	7.45	8.17	8.26	7.60	2
N97 (LMC)	11.09	7.47	8.44	8.09	7.48	1, 2, 3, 4
N102 (LMC)	11.13	7.49	8.39	8.12	7.39	1, 2, 4

1. Aller *et al* (1987)
2. Peña and Ruiz (1988)
3. Barlow *et al* (1983)
4. Monk *et al* (1988)

abundance criterion for galactic PN. Nevertheless the LMC type I PN are extremely He–N rich since they show an average He/H enrichment of 0.17 dex and an average N/H enrichment of 1.36 dex, relative to the values found in H II regions of the LMC. Their very high He/H and N/H ratios combined with their very low C/O ratios are in agreement with models by Renzini and Voli (1981) for $\alpha \sim 2$ and $M_i \sim 8\,M_\odot$, where α is the ratio of the convective mixing length to the pressure scale height. The PN in table 12 have similar C/O and N/O ratios to those of NGC 6302 (Aller *et al* 1981, Barral *et al* 1982), which is probably the most extreme type I PN known in the Galaxy.

The O/H mean value of the LMC type I PN is smaller than that of the LMC H II regions. Furthermore, the (N+O)/H mean value of the LMC type I PN is similar to that of the LMC H II regions, probably indicating that the ON cycle has contributed to the N enrichment (Aller *et al* 1987, Monk *et al* 1988, Peña and Ruiz 1988, Henry *et al* 1989).

Monk *et al* (1988) find that 14% of the PN in the SMC and 20% in the LMC are of type I; these values are similar to those found in the solar neighbourhood. Maran *et al* (1982), on the evidence of the large C abundances of PN compared to those of H II regions (table 11), concluded that most of the C enrichment in the MC is due to PN.

Henry *et al* (1989) determined the average net yield of N due to PN for the LMC and the SMC and compared it with the yield needed to explain the current interstellar (H II region) level of N, assuming a simple, closed system model for chemical evolution of galaxies. They conclude that while PN have contributed significantly to the N enrichment in the LMC, they have made only a small contribution to it in the SMC.

7.2. Chemical abundances in other Local Group galaxies

The chemical composition of PN in galaxies of the Local Group, excluding the Magellanic Clouds and the Galaxy, has been reviewed by Ford (1983).

Table 13 presents abundances for PN of the Local Group of galaxies. The He/H abundance ratios have not been corrected for the effect of collisional excitation from the 2^3S He^0 state, this correction is in general smaller than 0.04 dex, but for the NGC 185 and NGC 6822 PN it can be as high as 0.10 dex.

The O/H ratio in the Fornax PN is relatively high, considering that the galaxy is metal poor, it probably implies that there are two stellar populations: an old metal-poor population characterized by Fornax's four globular clusters that predominates, and an intermediate-age population characterized by the PN listed in table 13 and the carbon stars present (Frogel *et al* 1982, Ford 1983).

Table 13. Chemical abundances of extragalactic PN and H II regions, given in $12 + \log N(X)/N(H)$.

Object	He	N	O	Ne	Reference
PN (Fornax)	11.08	7.41	8.51	—	1
PN (NGC 185)	11.32:	>8.14	7.88	6.81:	2, 3
PN (NGC 6822)	11.27:	8.79	8.12	6.91	4
H II (NGC 6822)	10.92	6.52	8.24	7.58	5
PN (M31-290)	11.20:	7.81	8.55	7.79	6
PN (M31-363)	11.21:	7.88	8.67	8.11	6
H II (M31-BA 685)	10.93	7.39	8.51	7.66	6
PN (M31-372)	—	<7.71	8.06	7.49	6

1. Danziger *et al* (1978)
2. Jenner and Ford (1978)
3. Ford (1983)
4. Dufour and Talent (1980)
5. Lequeux *et al* (1979)
6. Jacoby and Ford (1986)

The NGC 185 PN in table 13 is of type I, probably with an initial stellar mass $M_i \geq 3\ M_\odot$. This indicates that its progenitor is relatively young and that it might be associated with the OB stars present in this galaxy.

The NGC 6822 PN in table 13 is also of type I. Its N/H enrichment relative to the NGC 6822 H II regions is very high, of 2.27 dex which probably constitutes a record. Its O/H ratio is slightly smaller than those of the NGC 6822 H II regions, in agreement with other type I PN in other galaxies. Its Ne/O ratio is smaller than that of the NGC 6822 H II regions and of the other type I PN known. This latter result is difficult to explain and could be due to errors in the abundance determinations; if real it would imply that the Ne/O ratio is not homogeneous in the ISM or that its progenitor star produced O. Both possibilities are unlikely and the object should be reobserved.

In the spiral galaxy M31, Jacoby and Ford (1986) have determined the chemical composition of three PN: M31-290, M31-363 and M31-372. These objects are at projected distances from the nucleus of 3.5, 18 and 33 kpc. From their radial velocities it is found that only the second one might belong to the disk while the other two belong to the halo. The O/H ratios of the PN M31-363 and of the H II region BA 685 are similar, which implies that disk PN are good tracers of the ISM O/H ratios. The O/H values of the two halo PN, M31-290 and M31-372, are considerably higher than the O/H ratios derived for galactic halo PN and imply that the chemical evolution of M31 was different from that of the Galaxy. M31-290 seems to correspond to a metal-rich halo population and M31-372 to a relatively metal-poor one; this division has also been found from the study of globular clusters in M31. Highly accurate observations of PN in M31 are needed to study further its chemical evolution: more data on disk PN would permit us to evaluate abundance gradients, and more data on halo PN would permit us to study its early chemical evolution.

7.3. Extragalactic distance scale

Jacoby and collaborators (Jacoby 1989, Jacoby *et al* 1989, 1990, Ciardullo 1989a, b, Ford *et al* 1989) have presented the physical rationale for using the planetary nebula [O III] luminosity function (PNLF) to derive highly accurate extragalactic distances.

The PNLF was established by observing galaxies with two filters: an [O III] filter corrected for the peculiar velocity of the galaxy and an off-band [O III] filter. Both images were blinked to detect stellar objects emitting in the [O III] 5007 Å line. Objects that were extended or that appeared in the off-band image were rejected as PN candidates.

Distances to the galaxies were obtained by adjusting the shape of a given PNLF to that of M31. The M31 PNLF has a sharp cutoff in the bright end of the observed luminosity function and a plateau about one magnitude below the sharp cutoff (the PNLF of other galaxies show a very similar behaviour). They assumed that the size of the galaxy affects only the number of PN, but not the shape of the PNLF.

Jacoby (1989) simulated [O III] luminosity functions for Gaussian M_c distributions with different central values, $\langle M_c \rangle$, and widths, b. He obtained an excellent match to the observations of the M31 PNLF for a model based on helium-burning central stars with $\langle M_c \rangle = 0.61 \ M_\odot$ and b $= 0.02 \ M_\odot$.

For the Sb spiral galaxy M81 Jacoby *et al* (1989) derived a PNLF distance of 3.50 ± 0.40 Mpc, which is in excellent agreement with the distance derived from I-band observations of Cepheids, indicating that the accuracy of the PNLF method is as good as the Cepheids' one. Ciardullo *et al* (1989a) have applied the PNLF method to three galaxies of the Leo I Group with different Hubble types: NGC 3379 an E0 elliptical, NGC 3377 and E6 elliptical, and NGC 3384 S0 spiral; the derived distances are 9.8, 10.3 and 10.1 Mpc with a formal 1σ error of about 10%

Jacoby *et al* (1990) applied the PNLF method to six galaxies of the Virgo Cluster, with elliptical or S0 Hubble types (M49, M60, M84, M85, M86, M87), to derive their distances. They found a very small dispersion in the distances and an average distance of 14.7 ± 1.0 Mpc. By merging the six PNLF into a single one, a distance of 14.9 ± 0.3 Mpc is derived, the 1σ uncertainty is less than 2%. This result illustrates the power of the method when large samples of bright PN are available.

Table 14 lists the distances to the Virgo Cluster of galaxies derived by different methods and different observers.

Table 14. Distances to the Virgo cluster of galaxies.

Method	Distance (Mpc)	Reference
Mean of six methods	11.9 ± 0.6	de Vaucouleurs (1985)
Luminosity fluctuations	13.9 ± 1.2	Tonry *et al* (1989)
$L - \sigma - \Sigma$ relation	14.4 ± 1.6	Pierce (1989)
IR Tully–Fisher relation	14.6 ± 0.8	Aaronson *et al* (1986)
PN luminosity function	14.7 ± 1.0	Jacoby *et al* (1990)
H II region luminosities	15.1 ± 1.0	Melnick *et al* (1988)
Novae	18.2 ± 3.0	Capaccioli *et al* (1990)
Globular cluster luminosity function	21.9 ± 2.2	Harris (1988)
Mean of six methods	21.9 ± 0.9	Sandage and Tammann (1990)

To determine the Hubble constant, H_0, from the distance to the Virgo cluster it is necessary to determine the infall velocity of the Local Group of galaxies toward Virgo. A recent determination of this velocity by Sandage and Tammann (1990) of $168 \pm 50 \ \mathrm{km \ s^{-1}}$, combined with the observed redshift of Virgo of $976 \pm 45 \ \mathrm{km \ s^{-1}}$ yields $H_0 = 52 \pm 2 \ \mathrm{km \ s^{-1} \ Mpc^{-1}}$.

On the other extreme, from the infall velocity of the Local Group toward Virgo by Sandage and Tamman (1990) and the distance by de Vaucouleurs (1985), a value of $H_0 = 96 \pm 5$ km s^{-1} Mpc^{-1} is derived. The distance to Virgo derived by Jacoby *et al* (1990) and the infall velocity by Sandage and Tammann implies that $H_0 = 77 \pm 6$ km s^{-1} Mpc^{-1}, safely in the middle between de Vaucouleurs' and Sandage and Tammann's determinations. Notice that there are other estimates in the literature of the infall velocity of the Local Group toward Virgo; adopting that of Kraan-Korteweg (1985), the distance to Virgo given by Jacoby *et al* would imply that $H_0 = 81 \pm 6$ km s^{-1} Mpc^{-1}, while adopting that of Aaronson *et al* (1986), the distance given by Jacoby *et al* would imply that $H_0 = 94 \pm 6$ km s^{-1} Mpc^{-1}. It is beyond the scope of this review to discuss the error estimates of the different Hubble constant determinations.

7.4. Birth rate and total number of PN

Table 15 presents the PN birth rate per solar luminosity, $\dot{\xi}$, and the total number of PN, N_T, for 16 galaxies.

For M31 the $\dot{\xi}$ and N_T values were obtained from: a solar bolometric absolute magnitude, $M(\odot)_{bol} = +4.75$ (Allen 1973), a visual absolute magnitude for the galaxy,

Table 15. PN birth rates per unit luminosity, $\dot{\xi}$, and total numbers of PN, N_T, for the Galaxy and other galaxies. $(B-V)_0$ is the intrinsic colour index and M_{bol} is the absolute bolometric magnitude.

Object	$(B-V)_0$	M_{bol}	$\dot{\xi}$ (10^{-12} yr^{-1} $L_\odot^{-1}$)	N_T (10^3)	References
M49	0.98	−23.29	2.7 ± 0.6	11.1 ± 2.5	1
M87	0.97	−23.08	3.4 ± 0.6	11.5 ± 2.0	1
M60	1.00	−22.85	2.6 ± 0.8	7.1 ± 2.2	1
M86	0.96	−22.78	5.4 ± 0.8	13.9 ± 2.1	1
M84	0.95	−22.60	6.8 ± 1.3	14.8 ± 2.8	1
M85	0.88	−22.31	8.0 ± 1.2	13.3 ± 2.0	1
M31	0.80	−21.96	6.6 ± 1.2	8.0 ± 1.5	2, 3, 4
M81	0.84	−21.88	8.4 ± 1.8	9.4 ± 2.0	5, 6
NGC 3379	0.89	−21.57	8.5 ± 1.7	7.2 ± 1.4	7
NGC 3384	0.84	−21.25	15.0 ± 3.0	9.4 ± 1.9	7
The Galaxy	0.53:	−21.2:	12.0 ± 3.0	7.2 ± 1.8	4, 8
The Galaxy	—	—	—	9.1 ± 3.3	9
NGC 3377	0.79	−20.69	15.0 ± 3.8	5.6 ± 1.4	7
LMC	0.43	−19.0:	12.6 ± 2.7	1.0 ± 0.25	4, 9, 10
SMC	0.44	−17.2:	19.2 ± 5.3	0.29 ± 0.08	4, 9, 10
M32	0.85	−17.14	10.1 ± 2.5	0.14 ± 0.04	3, 4
NGC 205	0.75	−16.99	14.5 ± 4.2	0.18 ± 0.05	3, 4

1. Jacoby *et al* (1990)
2. Freeman (1970)
3. Ciardullo *et al* (1989b)
4. Peimbert (1990)
5. Brandt *et al* (1972)
6. Jacoby *et al* (1989)
7. Ciardullo *et al* (1989a)
8. de Vaucouleurs and Pence (1978)
9. Jacoby (1980)
10. Hindman (1967)

$M_V = -21.2$ (Freeman 1970), a bolometric correction, BC $= -0.8$ mag and an estimated number of 970 PN for a region of M31 with $M_{bol} = -19.68$ (Ciardullo *et al* 1989b).

In figures 13 and 14 we plot $\dot{\xi}$ against M_{bol} and $\dot{\xi}$ against B-V using values for the sample in table 15, excluding the Galaxy (Peimbert 1990). It is found that there is a strong correlation between M_{bol} and $\dot{\xi}$, in the sense that the brighter the galaxy, the smaller the $\dot{\xi}$ value. From the same sample it is also found that there is a strong correlation between B-V and $\dot{\xi}$ in the sense that the redder the galaxy, the smaller the $\dot{\xi}$ value.

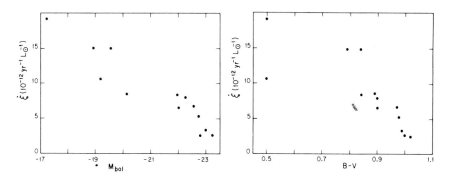

Figure 13. Plot of PN birth rate per solar luminosity, $\dot{\xi}$, against bolometric magnitudes, M_{bol}, for a group of fifteen galaxies.

Figure 14. Plot of $\dot{\xi}$ against the colour index, B-V, for the sample of objects in figure 13.

The $\dot{\xi}$ value for the Galaxy in table 15 was estimated from the $(\dot{\xi}, M_{bol})$ relation presented in figure 13 (a higher and less precise $\dot{\xi}$ value would have been obtained from the relationship between B-V and $\dot{\xi}$, see figure 14). The N_T value was obtained from the M_V value by de Vaucouleurs and Pence (1978) and the $\dot{\xi}$ value in table 15.

It is possible to compare $\dot{\xi}$ with the stellar death rate per solar bolometric luminosity, $\dot{s}$. Renzini and Buzzoni (1986) computed $\dot{s}$ values as a function of time from models with a single burst of star formation and three widely different initial mass functions. They found that the $\dot{s}$ values are extremely insensitive to the population age or to the initial mass function. For a stellar population with an age of 1×10^{10} years the $\dot{s}$ values derived by Renzini and Buzzoni are $\sim 20 \times 10^{-12}$ yr^{-1} $L_\odot^{-1}$. These values apply to elliptical galaxies dominated by a very old stellar population.

For spiral and irregular galaxies, models with continuous star formation have to be considered to determine the $\dot{s}$ values. To compare spiral and irregular galaxies with the models by Renzini and Buzzoni (1986), an average age for the stellar content has to be estimated. In these galaxies the stellar death rate and the total luminosity are dominated by the younger generations of stars, and an average age, weighted by the higher luminosity of the younger generations, of about 10^9 years should be used. Moreover, $1-2 \times 10^9$ years is the time that a star with $M_i \sim 1.5 \, M_\odot$ spends in the main sequence and corresponds to the $\langle M_i \rangle$ derived from the height of PN above the galactic plane. An age of $\sim 10^9$ years would reduce $\dot{s}$ to $\sim 15 \times 10^{-12}$ yr^{-1} $L_\odot^{-1}$ (Renzini and Buzzoni 1986).

For the LMC, the SMC, M81, M31 and the Galaxy the $\dot\xi$ values are very similar to the $\dot s$ values derived from the models by Renzini and Buzzoni (1986), considering all the uncertainties that enter into both types of determinations. In particular, for the Galaxy $\dot\xi \sim (\frac{2}{3})\dot s$; this result implies that the majority of the intermediate-mass stars undergo the PN phase.

Moreover, for the Galaxy the N_T value derived from the $\dot\xi$ value is concordant with the N_T values derived from the PN surface density of the solar neighbourhood and long distance scales like those of Cudworth (1974) and Mallik and Peimbert (1988) (see section 6).

For all the galaxies in table 15 a value of $15 \lesssim \dot s\,(10^{-12}\,\mathrm{yr}^{-1}\,L_\odot^{-1}) \lesssim 20$ can be expected and, if anything, slightly higher $\dot s$ values for the more luminous galaxies. Therefore the decrease of $\dot\xi$ with B-V and with bolometric luminosity is not due to a decrease in $\dot s$ and should be explored further.

Most of the PN luminosity functions used for the determinations of $\dot\xi$ in table 15 are complete only at the high-luminosity end; the completeness limit of the observations extends only ~ 2.5, 1.5, 1.1 and 0.8 mag below the bright-end cutoff for M31, M81, the Leo I Group, and the Virgo Cluster, respectively. The shape of the entire PNLF is obtained by scaling the observed upper end with the PNLF for the Magellanic Clouds derived by Jacoby (1980). Since the PNLF spans about eight magnitudes (Ciardullo *et al* 1989b), the $\dot\xi$ values represent the upper end of the PNLF. Therefore, strictly speaking, it can only be said that the number of bright PN decreases with increasing luminosity of the galaxy.

There are at least three possible causes for the decrease in the number of bright PN with the increase of B-V and luminosity: (a) an increase of the heavy-element abundances, (b) a decrease of $\langle M_c \rangle$ due to an age effect, and (c) a decrease of the fraction of intermediate-mass stars that produce luminous PN. In what follows we will analyse these possibilities.

There is a well known positive correlation between the total mass of the galaxy and the heavy-element abundances that includes irregular, spiral and elliptical galaxies (e.g. Lequeux *et al* 1979, Mould 1984, Garnett and Shields 1987). From the relatively close relationship between mass and luminosity, a positive correlation between luminosity and heavy-element abundances is also expected. Even if a higher heavy-element abundance produces a decrease of the [O III] luminosity, the expected effect is very small (Jacoby 1989) and cannot explain the correlation present in figures 13 and 14.

It is also possible that the $\langle M_c \rangle$ decreases with the luminosity of the galaxy and consequently the luminosities of all PN. This possibility seems unlikely because the three galaxies of the Leo I Group, at practically the same distance, have different $\dot\xi$ values and the six galaxies of the Virgo Cluster, also at practically the same distance, also have different $\dot\xi$ values.

The third possibility is that there are two stellar populations in each galaxy: a relatively young one, with an average age of ~ 1-2×10^9 years, and an old one, with an average age of $\sim 10^{10}$ years. The observed bright end of the PNLF would be due to the younger population with $\langle M_c \rangle \sim 0.61\,M_\odot$, while the older population would seldom produce PN or would produce PN with $\langle M_c \rangle \lesssim 0.57\,M_\odot$, which are considerably fainter (e.g. Jacoby 1989). The brighter the galaxy, the more important the old population relative to the young population. The older population is expected to produce fainter and fewer PN per star due to the following reasons: (a) lower luminosity of the central star, (b) smaller mass of the shell and (c) longer stellar evolutionary times that might prevent the star from becoming hot enough to ionize the nebula before it has dissipated.

8. Chemical evolution of galaxies

To fit the observed abundances in the interstellar medium (ISM) of the Galaxy and of other galaxies into a consistent picture it is necessary to construct models of galactic chemical evolution. Such models predict the time variation of the chemical composition of the ISM depending on assumptions for: (a) the mass distribution of the recently formed stars, the so-called initial mass function, (b) the time dependence of the stellar birth rate, (c) the chemical composition of the material ejected by the stars during their evolution, and (d) the large-scale mass flows, like infall from the halo, outflow to the intergalactic medium or radial flows within the galaxy. The observed abundances are used as tests for the input assumptions, since these physical parameters are not generally known.

Massive stars, those with $M_i \gtrsim 8\ M_\odot$, produce supernovae of types Ib and II, while some binary stars with $M_i \lesssim 8\ M_\odot$ produce SN of type Ia. The ISM enrichment of elements heavier than N, like O, Ne, S, Ar and Fe is mainly due to SN. Intermediate-mass stars, those with masses in the $0.8 \lesssim M_i/M_\odot \lesssim 8$ range, undergo the PN phase and enrich the ISM with He, N and C (e.g. Kaler 1979, Peimbert and Serrano 1980, Torres-Peimbert *et al* 1980, Aller and Czyzak 1983, Clegg 1989, Barlow 1989, and references therein). Stars with $M_i \lesssim 0.8\ M_\odot$ do not enrich the ISM with heavy elements because they are still transforming H into He in their nuclei without producing any ejecta. It is important to know which is the relative contribution of PN and SN to the enrichment of any given element.

PN can be used to study the chemical composition of the ISM at the time they were formed, particularly the abundances of elements heavier than Ne, like S and Ar that are not affected by the evolution of the PN progenitors. PN can also be used to determine the pregalactic He abundance and their He, C and N contribution to the enrichment of the ISM.

8.1. Evolution of the halo of the Galaxy

Type IV PN are located in the halo and probably were born in it; from the study of those elements not affected by stellar evolution it is possible to estimate the chemical composition of the halo at the time the progenitor stars were formed.

From studies of K648, H4-1 and BB-1, it has been found that Ar, Fe and S are underabundant by about two orders of magnitude, while O and Ne are underabundant by about one order of magnitude relative to the solar vicinity (see table 7). Two possibilities have been discussed in the literature to explain the different underabundances: (a) that the enrichment of O and Ne in the ISM has proceeded faster than that of Ar, Fe and S; and (b) that the O and Ne excesses relative to Ar, Fe and S were produced by the progenitors of the PN themselves (Peimbert 1973, 1981, Hawley and Miller 1978b, Torres-Peimbert and Peimbert 1979, Barker 1980, 1983, Clegg *et al* 1987, Clegg 1989, Torres-Peimbert *et al* 1990).

Clegg (1989) suggested that possibly most of the Fe, Ar and S atoms are made by supernovae of type Ia while most of the O atoms are made by supernovae of types Ib and II; the delay in the Fe, Ar and S enrichment of the ISM relative to that of O is due to the smaller mass of the progenitors of SN of type Ia relative to those of types Ib and II. Similarly, Torres-Peimbert *et al* (1990) have suggested that most of the Ne enrichment of the ISM is due to SN of types Ib and II.

Figure 15 presents a plot of the [O/Ar] against [Ar/H] abundance ratios for the type IV PN (see table 7) as well as those for the Orion nebula, the Sun, and type II PN (see table 6). From this figure it follows that the eight halo PN show significant deficiencies relative to Orion, the Sun, and type II PN. Moreover, based on their chemical composition, the eight PN can be divided into two groups: one composed by K648, BB-1 and H4-1 and the other one composed by the other five.

Based on their distances to the plane of the galaxy, z, and on their velocities relative to the local standard of rest, v_{lsr} (see table 16), it seems that DDDM-1 and PRMG-1

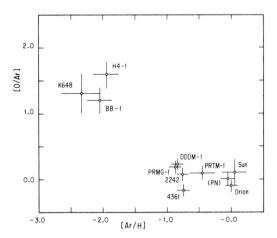

Figure 15. Plot of [O/Ar] against [Ar/H] abundances for Halo PN, where $[A/B] = \log (A/B) - \log (A/B)_{popI}$. We have also plotted the abundances for the Sun, the Orion nebula and an average for type II PN.

Table 16. Velocities relative to the local standard of rest and distances from the plane of the Galaxy for type IV PN.

Object	v_{lrs} (km s^{-1})	z (kpc)	Reference
K648	-128 ± 17	11.8	1, 2
BB-1	-192 ± 10	10.5	1, 3
H4-1	-133 ± 12	6.0	1, 3
NGC 4361	$+8 \pm 2$	0.83	1, 3
NGC 2242	$+15 \pm 21$	1.09	3, 4, 5
DDDM-1	-285 ± 20	14.9	6, 7
PRMG-1	—	5.1	8
PRTM-1	—	3.0	9

1. Schneider *et al* (1983)
2. O'Dell *et al* (1964)
3. Torres-Peimbert *et al* (1990)
4. Maehara *et al* (1986)
5. Huchra (1984)
6. Barker and Cudworth (1984)
7. Clegg *et al* (1987)
8. Peña *et al* (1989)
9. Peña *et al* (1990)

belong to the halo and NGC 2242 and NGC 4361 to the thick disk (see the discussion by Freeman (1987) on the different components of the galaxy). If DDDM-1 and PRMG-1 were formed in the halo it follows that the chemical enrichment of the ism was not significant between the formation of the metal-rich halo stars and the formation of the metal-poor stars that now belong to the thick disk.

8.2. Abundance gradients across the disk of the Galaxy

The presence of O/H abundance gradients across the disks of spiral galaxies based on the study of H II regions is now well established (e.g. Pagel and Edmunds 1981, Shields 1990 and references therein). Studies of galactic H II regions show a similar O/H gradient to those of other spiral galaxies (Peimbert *et al* 1978, Hawley 1978, Shaver *et al* 1983). Gradients in He/H, N/O and S/O in the Galaxy and other spiral galaxies have been reported by some authors but not by others.

PN present several advantages over H II regions for the study of chemical abundance gradients across the disk of the Galaxy because: (a) a considerably larger sample of objects is available; (b) PN show a larger scatter in the direction perpendicular to the plane, it is therefore possible to observe them in the optical domain over a larger range of galactocentric distances; (c) in general small PN have higher surface brightness than galactic H II regions observable in the optical domain. On the other hand they also present several disadvantages: (a) the initial He, C, N and O could have been modified by the evolution of their parent stars, (b) their orbits may be non-circular, (c) the stellar progenitors have a large age spread.

Abundance gradients derived from PN have been reported in the literature by many authors (D'Odorico *et al* 1976, Aller 1976, Torres-Peimbert and Peimbert 1977, Barker 1978, Peimbert and Serrano 1980, Faúndez-Abans and Maciel 1986, 1987a, b). In table 17 we present abundance gradients derived by Faúndez-Abans and Maciel (1986) for PN of type II, see also figure 16. The PN results are similar to those derived for H II

Table 17. Solar neighbourhood abundance gradients given in $\Delta \log(X/H)/\Delta R$, with R in kiloparsecs, (from Peimbert *et al* 1978, Shaver *et al* 1983, Faúndez-Abans and Maciel 1986).

Ratio	H II regions (kpc^{-1})	Planetary nebulae (kpc^{-1})
He/H	-0.02 ± 0.01	-0.019 ± 0.003
O/H	-0.07 ± 0.015	-0.072 ± 0.012

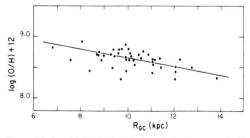

Figure 16. Log $(O/H) + 12$ plotted against galactocentric distance, R_{GC}, for type II PN (Faúndez-Abans and Maciel 1986).

regions and imply that most of the PN in these samples show orbits of relatively low eccentricity and that their ages are small compared with the age of the disk of the Galaxy.

Maciel and Faúndez-Abans (1985) have derived a well defined electron temperature gradient from type II galactic PN given by $\sim 600 \pm 120$ K kpc^{-1}; this value is about 1.5 times larger than that derived from galactic H II regions (Lichten *et al* 1979, Garay and Rodríguez 1983, Shaver *et al* 1983). The difference could be real since there are many different factors affecting each gradient. The main factor producing the T_e gradient in the galactic disk is the radial decrease of the abundance of the heavy elements that are responsible for the cooling of the PN shells.

There are two effects that could affect the derived values of the PN abundance and electron temperature gradients: (a) for a larger PN distance scale the derived gradients are smaller because they are proportional to k^{-1}, (b) the presence of objects in the sample with elliptical orbits tends to smooth out the gradients.

Several models of galactic chemical evolution have been proposed to explain the abundance gradients present in the ISM (e.g. Pagel 1989, Matteucci and Francois 1989, Matteucci *et al* 1989). These models explore several hypotheses: (a) variations in the initial mass function, (b) variations in the gas fraction, (c) variations in the ratio of star formation rate to inflow rate and (d) gas flows across the disk of the Galaxy. There are several possible combinations of hypotheses that can explain the gradients and other observational constraints are needed to produce a unique model.

8.3. The pregalactic helium abundance and the $\Delta Y/\Delta Z$ ratio

The fractional abundances of hydrogen, helium and all the other elements, normalized to unit mass, are denoted by X, Y and Z, respectively. The pregalactic, or primordial helium abundance, Y_p, is an important parameter for models of galactic chemical evolution and its value has several cosmological implications (e.g. Boesgaard and Steigman 1985, Pagel and Simonson 1989). To determine Y_p from the observed Y value of a given PN it is necessary to estimate two effects: the helium enrichment of matter previous to the formation of the progenitor star due to galactic evolution, ΔY_{GE}, and the helium enrichment due to the stellar evolution of its progenitor, ΔY_{SE}, i.e.

$$Y = Y_p + \Delta Y_{GE} + \Delta Y_{SE} = Y_i + \Delta Y_{SE} \qquad (8.1)$$

where Y_i is the initial helium abundance of the star, and ΔY_{SE} is given by

$$\Delta Y_{SE} = \Delta Y^1 + \Delta Y^2 + \Delta Y^3 \qquad (8.2)$$

where the terms on the right-hand side are due to the three dredge-up episodes (see subsection 5.2).

Type I PN are not good candidates to determine Y_P for the following reasons: (a) their ΔY_{SE} values are higher than those of other PN, probably due to their higher M_i values and the effects of the second dredge-up, (b) they are expected to have higher ΔY_{GE} values than those of other PN due to their younger ages.

In what follows we will discuss the Y_p and $\Delta Y_{GE}/\Delta Z_{GE}$ determinations presented in table 18. The first determinations of Y_p and $\Delta Y_{GE}/\Delta Z_{GE}$ based on PN were made by D'Odorico *et al* (1976) under the following considerations: (a) they obtained the ΔY^1 values from Torres-Peimbert and Peimbert (1971), (b) they assumed that oxygen constitutes 45% by mass of the heavy elements, Z, at the time the PN were formed and that the O/H ratio is not affected by the evolution of the PN progenitor, (c) they adopted a value of 0.055 for the mean square temperature fluctuation, t^2, to determine

Table 18. Pregalactic helium abundance by mass, Y_p, and the relative helium to heavy elements enrichment ratio produced by galactic evolution, $\Delta Y_{GE}/\Delta Z_{GE}$.

Sample	Y_p	$\Delta Y_{GE}/\Delta Z_{GE}$	ΔY^1	ΔY^3	Reference
7 type II	0.227	2.95	0.006	0	D'Odorico *et al* (1976)
6 type II	0.227 ± 0.012	3.0	0.016	0.013	Peimbert (1983)
16 type II + III	0.220 ± 0.010	3.6 ± 0.8	0.016	0.013	Peimbert (1983)
3 type IV	0.218 ± 0.03	—	0.02	0.037	Peimbert (1983)
4 type IV	0.23 ± 0.003	—	0.015	0.026	Clegg (1989)
H II + type IIb	0.227 ± 0.003	6.1 ± 0.6	0	0	Maciel and Leite (1990)
H II + type IIb	0.230 ± 0.003	5.2 ± 0.5	0.010	0	Maciel and Leite (1990)
H II	0.232 ± 0.004	4.0 ± 1	—	—	Pagel (1987a)
H II	0.229 ± 0.004	7.0 ± 2	—	—	Pagel and Simonson (1989)
H II	0.230 ± 0.006	3.8 ± 0.7	—	—	Torres-Peimbert *et al* (1989)

the chemical abundances. At present values of $t^2 \leqslant 0.03$ are favoured for PN (Torres-Peimbert *et al* 1980, Harrington *et al* 1982).

Peimbert (1983) determined Y_p from a sample of six type II PN of intermediate and high degree of ionization (i.e. without neutral helium in the H^+ zone). The procedure followed was: (a) ΔY^1 and ΔY^3 were taken from stellar evolution models by Renzini and Voli (1981), the ΔY^1 value for $M_i = 1.45\ M_\odot$ was adopted and ΔY^3 was computed from the difference between the observed C/O ratio and the interstellar C/O ratio, (b) it was assumed that O constitutes 45% by mass of Z_i, and that the O/H ratio is not affected by the evolution of the PN progenitor, (c) that $\Delta Y_{GE} = 3\Delta Z_{GE} = 3\Delta Z_i$ and (d) a value of $t^2 = 0.00$ was adopted.

Peimbert (1983) derived the Y_p and $\Delta Y_{GE}/\Delta Z_{GE}$ values in table 18 from the sample of 16 PN of types II and III of Peimbert and Serrano (1980) based on the stellar evolution models by Renzini and Voli (1981), the ΔY^1 value for $M_i = 1.45\ M_\odot$, the average ΔY^3 determined from the six best observed type II PN and $t^2 = 0.00$. For $t^2 = 0.02$ the ratio derived is $\Delta Y_{GE}/\Delta Z_{GE} = 3.0 \pm 0.7$ and the primordial helium is $Y_p = 0.223$, while for $t^2 = 0.035$ the derived quantities are $\Delta Y_{GE}/\Delta Z_{GE} = 2.2 \pm 0.5$ and $Y_p = 0.225$.

For type IV PN, ΔY_{GE} can be neglected and therefore $Y_p = Y_i$ (see equation (8.1)). Peimbert (1983) and Clegg (1989) determined Y_p from three and four type IV PN, respectively. The ΔY^1 and ΔY^3 values were also obtained from the computations by Renzini and Voli (1981) and from the observed C abundances. The main source of error is the large correction due to ΔY^3, moreover the computations by Renzini and Voli were made for stars with higher heavy-element abundances, consequently a new set of stellar evolution models is needed to improve the accuracy of these determinations.

Maciel (1988) and Maciel and Leite (1990) combined the abundances for type IIb PN with those for galactic H II regions by Shaver *et al* (1983) and for extragalactic H II regions by Pagel (1987a) to derive Y_p. For their sample of type II PN they found that $Z = 3.88\ Z(16)$, where $Z(16)$ is the O abundance by mass, while from H II regions it is found that $Z = 2.22\ Z(16)$, the difference is mainly due to freshly made C by the PN progenitor. To compare the $\Delta Y_{GE}/\Delta Z_{GE}$ ratio derived from PN with that derived from H II regions it is necessary to compare with $Z(16)$ which is not expected to be affected by the evolution of the PN progenitor. Therefore we have defined $\Delta Z_{GE} = 2.22\ Z(16) = Z_i$, in agreement with the other entries for ΔZ_{GE} in table 18. Consequently the $\Delta Y_{GE}/\Delta Z_{GE}$ ratio is different from that defined by Maciel (1988) and Maciel and Leite (1990), but it can be derived from their data since they also present a relation for Y against $Z(16)$. The Y_p values by Maciel and Maciel and Leite depend mainly on the O-poor H II regions, while the $\Delta Y_{GE}/\Delta Z_{GE}$ values depend mainly on the PN. By increasing the $\Delta Y^1 + \Delta Y^3$ value adopted by Maciel and Leite to ~ 0.03, a value of $\Delta Y_{GE}/\Delta Z_{GE} \sim 3.4$ is obtained, in excellent agreement with the H II region value (e.g. Peimbert 1986, Torres-Peimbert *et al* 1989), and with the value derived by Peimbert (1983) from PN.

To compare the PN results with those for H II regions, we present in table 18 three recent Y_p and $\Delta Y_{GE}/\Delta Z_{GE}$ determinations. Pagel and Simonson (1989) argue that their $\Delta Y_{GE}/\Delta Z_{GE}$ value is probably too large, being biased by a few nitrogen-rich objects. It is no surprise that the Y_p values by Maciel and Leite (1990) are very similar to that derived by Pagel (1987a) since, as mentioned before, the Y_p values by Maciel and Leite depend mainly on the data by Pagel for O-poor H II regions. The Y_p values by Peimbert (1983) and by Clegg (1989), derived exclusively from PN, are not as accurate as those derived from H II regions, but are in agreement with them.

The $Y_p = 0.23 \pm 0.01$ value is in agreement with standard big-bang nucleosynthesis with three families of light neutrinos ($m_\nu < 1$ MeV) and implies that the ratio of the nucleon density to the critical density required to close the universe, $\Omega_N = \rho_N/\rho_C$, is given by $0.007 \leqslant \Omega_N h(50) \leqslant 0.05$, where $h(50)$ is the present value of the Hubble parameter in units of 50 km s^{-1} Mpc^{-1} (e.g. Boesgaard and Steigman 1985, Pagel and Simonson 1989, Denegri *et al* 1990, Steigman 1990, Olive *et al* 1990). Therefore if most of the matter in the universe is baryonic, then the standard big-bang model coupled with the Y_p value implies an open universe.

8.4. Interstellar medium enrichment

8.4.1. The $\Delta Y_{GE}/\Delta Z_{GE}$ enrichment ratio. From models of galactic chemical evolution, Serrano and Peimbert (1981), Chiosi and Matteucci (1982), Schild and Maeder (1985) and Mallik and Mallik (1985) have obtained for $\Delta Y_{GE}/\Delta Z_{GE}$ values of 3.1, 2, 1 and 0.6, respectively. The differences are due to the adopted initial mass functions and stellar evolution models.

Schild and Maeder (1985) and Mallik and Mallik (1985) have considered the possibility that stars with initial masses above a critical mass, $M(\text{BH})$, produce black holes without enriching the ISM with heavy elements. Schild and Maeder have suggested that $M(\text{BH}) \sim 50\ M_\odot$, for which $\Delta Y_{GE}/\Delta Z_{GE} \sim 1.4$. To be able to explain the observed $\Delta Y_{GE}/\Delta Z_{GE}$ values, Mallik and Mallik conclude that stellar evolution models still required considerable improvements or that $M(\text{BH})$ is very small, of the order of $20\ M_\odot$; for the latter $M(\text{BH})$ value they obtain $\Delta Y_{GE}/\Delta Z_{GE} \sim 2.1$.

An important contribution to the enrichment of the ISM by He-rich objects, like type I PN and supernovae of the Crab nebula type (e.g. Péquignot and Dennefeld 1983, Davidson and Fesen 1985, Henry 1986, MacAlpine *et al* 1989) would help to explain the high $\Delta Y_{GE}/\Delta Z_{GE}$ values presented in table 18.

8.4.2. Nitrogen and helium. PN of type I show very high He/H and N/H overabundances relative to those of the ISM, probably due to the three dredge-up phases (see subsection 5.2); while PN of type II show moderate N/H overabundances, of about a factor of three to four, and very small He/H overabundances, probably due to the first and third dredge-up phases.

Peimbert (1987) has estimated that the He enrichment of the ISM due to PN of type I is five to six times larger than that produced by PN of types II and III. This result is based on the following assumptions: (a) an ISM helium abundance by mass of 0.28, (b) a production rate of type I PN of 20% relative to the total PN production rate, and (c) an average envelope mass for type I PN three times higher than that for PN of types II and III. Similarly, Peimbert has estimated that the N enrichment produced by PN of type I is about three times larger than that produced by PN of types II and III.

The N enrichment of the ISM is due to intermediate-mass stars and massive stars. Moreover, it could have had a primary or a secondary origin, where primary elements are those that are directly synthesized in the star from H and He, and secondary elements are those synthesized from heavy elements that were already present in the star when it was formed. Different models of galactic chemical evolution have been made to study the N enrichment of the ISM, nevertheless there is still an open discussion on the relative importance of different stellar mass ranges for N production and on the fraction of N that is of secondary origin (e.g. Edmunds and Pagel 1978, Peimbert and Serrano 1980, Serrano and Peimbert 1983, White and Audouze 1983, Matteucci 1986, Diaz and Tosi 1986, Forieri 1986, Pagel 1986, 1987b, Peimbert 1987, Tosi 1988a, b).

8.4.3. Carbon. Tinsley (1978) considered that the observed solar C/O ratio and estimates of nucleosynthesis in massive stars indicate that PN are a major source of C enrichment of the ISM. Further work showed that most of the C in the ISM comes from intermediate mass stars (e.g. Dufour 1984b, 1985, Mallik and Mallik 1985, Sarmiento and Peimbert 1985, Matteucci 1986).

Sarmiento and Peimbert (1985) found that intermediate mass stars produce from 60 to 80% of the C present in the ISM and that massive stars produce the rest. Their result is based on three independent determinations: (a) the variation of C/H in the ISM of galaxies with different O/H ratio; (b) the comparison of the solar C/O ratio with the C/O ratio predicted by Arnett and Thielemann (1984) for a single star with an $8 M_\odot$ He-core, corresponding to a 20–25 $M_\odot$ main sequence star; and (c) the comparison of the solar-vicinity ISM C/H ratio with the predicted C/H ratio from the models by Renzini and Voli (1981).

SN of type Ia are not important sources of C production since carbon deflagration models produce very high Fe/C ratios. Even if these objects were responsible for all of the Fe in the solar vicinity their contribution to the C abundances would be negligible (Nomoto 1984, Woosley *et al* 1984, Mallik and Mallik 1985). The contribution of novae to the C abundance of the ISM is also negligible (Sarmiento and Peimbert 1985).

9. Concluding remarks

PN correspond to one of the main stages of stellar evolution. The PN birth rate in the Galaxy and other galaxies indicates that most IMS go through the PN stage. Therefore a complete stellar evolution theory should contemplate the transition from red giants to white dwarfs and the relation between Miras, OH/IR stars, protoplanetary nebulae, PN and white dwarfs.

Stellar evolution computations for a wide variety of masses and chemical composition exploring the parameters α and η, the ratio of the mixing length to the pressure scale height and the mass loss rate during the asymptotic giant branch phase, are needed to produce a better fit with the observations.

From future observations it will be possible to place the central stars of Magellanic Cloud PN in the luminosity–temperature diagram, providing us with a very powerful tool to test stellar evolution predictions.

The ejection mechanism of the shell should be better understood. It is hoped that the study of the shell evolution can lead us to an improved knowledge of the ejection mechanism. Advances on the interaction of the stellar wind with the previous ejected shell are being made, as well as on the study of multiple shells and their relation to stellar evolution. Observations of molecules with higher angular resolution and of atomic hydrogen in the shells will also help us to advance in the study of the shell's evolution.

The relation between binarity, morphology of the nebulae and chemical composition should be studied further.

The determination of the chemical composition of PN with known M_i values and with known kinematical properties will permit us to evaluate the chemical composition at the time PN were formed as well as the chemical enrichment of the Galaxy produced by the PN themselves. This type of study is beginning to be carried out for other galaxies and appears very promising.

Inhomogeneous models with a varying C/H ratio as a function of distance to the central star, coupled with stellar atmospheric abundances will provide strong constraints for stellar evolution models.

PN have been used as test particles to study dynamical aspects of external galaxies and the Galaxy. These studies should be extended further.

The data on extragalactic PN have already yielded important results for the study of the properties of PN and for the general properties of the Universe. It is expected that the study of extragalactic PN will flourish in the next few years.

Acknowledgements

Over the years I have profited enormously by discussions on PN with: L H Aller, G Haro, R Minkowsky, C R O'Dell, D E Osterbrock, M J Seaton and S Torres-Peimbert. I am greatly indebted to J Fierro, J B Kaler, W J Maciel, M Peña, S R Pottasch, A Sarmiento, Y Terzian and S Torres-Peimbert for critical readings of drafts of this article and for constructive suggestions.

Addendum

In the last year there have been many investigations devoted to PN, in particular a large number of them have produced significant advances on our knowledge of the chemical composition of PN (Barlow 1991; Clegg 1991; Dufour 1991; Köppen *et al* 1991; Maciel 1991; Perinotto 1991). In what follows I will discuss some of the most interesting recent results related to the chemical composition of halo, Galactic bulge and Magellanic Clouds PN.

Halo planetary nebulae

A recent review on the abundances of ten halo PN, type IV PN, has been presented by Peimbert (1991). It is found that about half of them show excesses in the O/Ar and

Ne/Ar ratios relative to the solar neighbourhood values; these excesses could have been produced by their progenitor stars or could be the result of galactic chemical evolution.

The computations of Thielemann *et al* (1991) of nucleosynthesis in supernovae show that the O/Ar and Ne/Ar yields are higher, by more than an order of magnitude, in the 25 $M_\odot$ model than in the 13 $M_\odot$ model. Since 25 $M_\odot$ stars live less than 13 $M_\odot$ stars it is conceivable that, early in the evolution of the halo, stars with $M \sim 25\ M_\odot$ contributed more to the enrichment of the interstellar medium than stars with $M \sim 13\ M_\odot$. Moreover it is also possible that due to poor mixing in the ISM, some stars could have been made of O/Ar and Ne/Ar rich material. On the other hand the Ne/O range predicted by the models of Thielemann *et al* (1991) amounts to about a factor of two, while the observed range in halo PN comprises about two orders of magnitude; consequently part of the observed Ne/O range seems to be due to nuclear reactions in the PN progenitors and not to galactic chemical evolution.

Galactic bulge planetary nebulae

The Galactic bulge PN have one main advantage with respect to other samples of Galactic PN: the uncertainties in their distances are considerably smaller and consequently better estimates of the progenitor masses and shell masses can be made. Several groups have carried out independent studies of Galactic bulge PN (Dopita *et al* 1990; Ratag *et al* 1991a, b, c; Stasinska *et al* 1991a, b; Tylenda *et al* 1991).

Terndrup (1988) made a study of the stars in the Galactic nuclear bulge and from their main sequence turn-off determined that their age is in the 11–14 Gyr range which corresponds to $0.9 \leq M_i\ /M_\odot \leq 1.2$ (see also Ratag *et al* 1991c). These masses are similar to those of PN of types III and IV of the solar neighbourhood and are considerably smaller than the M_i masses of PN of types I and II.

Ratag *et al* (1991c) have estimated the absolute luminosities of the central stars for a sample of 103 PN in the direction of the bulge assuming that they are located in the centre of the Galaxy. By comparing these luminosities with theoretical evolutionary tracks they have determined their masses and find that more than 90% have $M_c < 0.60\ M_\odot$. These masses are in general smaller than those of the Méndez *et al* (1988) sample for PN of the solar neighbourhood. Tylenda *et al* (1991) find $<M_c> = (0.593 \pm 0.036)\ M_\odot$ for their bulge sample and $<M_c> = (0.615 \pm 0.036)\ M_\odot$ for their solar neighbourhood sample. These results indicate that bulge PN originate from less massive central stars than disk PN.

From the chemical abundances determination of their bulge sample Ratag *et al* (1991a, b) find that the O/H, Ne/H, S/H and Ar/H ratios are similar to those of the solar neighbourhood sample studied by Aller and Keyes (1987). Furthermore these ratios are smaller than predicted from extrapolations of the solar neighbourhood abundance gradients to the Galactic centre, the differences amount to ~ 0.4 dex. This result also indicates that the PN in the bulge represent a different population to that of the disk PN.

Some bulge PN show N/O excesses with respect to type II PN of the solar neighbourhood (Ratag *et al* 1991a, b). The second dredge-up phase produces N/O excesses in models with $M_i \geq 2.4\ M_\odot$; these excesses have been identified with type I PN. But not only type I PN show N/O excesses, there are at least two type IV PN with N/O ≥ 0.5: BB-1 and M2-29 (see table 7 and Peña *et al* 1991). Therefore it is also possible for PN with $M_i \leq 1.0\ M_\odot$ to have N/O values ≥ 0.5, probably due to additional CN cycling with N of primary origin or ON cycling with N of secondary origin. Due to the smaller M_i values of the bulge PN relative to the disk PN it follows that the N/O excesses in some of the bulge PN could be due to their similarity with type IV PN and not to type I PN. It should be mentioned that some PN in the direction of the bulge are foreground objects like M2-29 (Peña *et al* 1991).

It is also possible that the N/O excesses could be due to differences in the chemical evolution of the interstellar medium. Evidence for a different chemical evolution of the bulge and the disk is provided by their very different effective yields (*e.g.* Peimbert 1986a and references therein).

Magellanic Clouds planetary nebulae

The Magellanic Clouds also provide PN samples at known distances. Reviews on this subject have been presented by Barlow (1991) and Clegg (1991).

Dopita and Meatheringham (1991) have placed 44 central stars of PN of both Clouds in the HR diagram. Most of the central stars in their sample have M_c masses between 0.55 and 0.7 $M_\odot$. They find that type I PN have higher M_c values than the other PN reaching up to 1.2 $M_\odot$. They also find that type I PN are deficient in O; they interpret this result as evidence for ON processing in these objects.

References

Aaronson M, Bothun G, Mould J, Huchra J, Schommer R A and Cornell M E 1986 *Astrophys. J.* **302** 536–63
Acker A 1978 *Astron. Astrophys. Suppl. Ser.* **33** 367–81
—— 1980 *Astron. Astrophys.* **89** 33–40
—— 1983 *IAU Symp. 103 Planetary Nebulae* ed D R Flower (Dordrecht: Reidel) p 241
Adam J and Koppen J 1985 *Astron. Astrophys.* **142** 461–75
Adams S, Seaton M J, Howarth I D, Aurriere M and Walsh J R 1984 *Mon. Not. R. Astron. Soc.* **207** 471–89
Allen C W 1973 *Astrophysical Quantities* (London: Athlone)
Aller L H 1976 *Publ. Astron. Soc. Pacific* **88** 574–84
—— 1983 *IAU Symp. 103 Planetary Nebulae* ed D R Flower (Dordrecht: Reidel) pp 1–13
—— 1984 *Physics of Thermal Gaseous Nebulae* (Dordrecht: Reidel)
Aller L H and Czyzak S J 1983 *Astrophys. J. Suppl. Ser.* **51** 211–48
Aller L H and Keyes C D 1987 *Astrophys. J. Suppl. Ser.* **65** 405–28
Aller L H, Keyes C D, Maran S P, Gull T R, Michalitsianos A G and Stecher T P 1987 *Astrophys. J.* **320** 159–77
Aller L H, Ross J E, O'Mara B J and Keyes C D 1981 *Mon. Not. R. Astron. Soc.* **197** 95–106
Aller L H and Walker M F 1970 *Astrophys. J.* **161** 917–45
Alloin D, Cruz-González C and Peimbert M 1976 *Astrophys. J.* **205** 74–81
Amnuel P R, Guseimov O H, Novruzova H I and Rustamov Yu S 1984 *Astrophys. Sp. Sci.* **107** 19–50
Amnuel P R, Guseimov O H and Rustamov Yu S 1989 *Astrophys. Sp. Sci.* **14** 21–88
Arnett W D and Thielemann F K 1984 *Stellar Nucleosynthesis* ed C Chiosi and A Renzini (Dordrecht: Reidel) pp 145–50
Balick B 1987 *Astron. J.* **94** 671–8
Balick B, Reston H L and Icke V 1987 *Astrophys. J.* **94** 1641–52
Barker T 1978 *Astrophys. J.* **220** 193–209
—— 1980 *Astrophys. J.* **237** 482–5
—— 1982 *Astrophys. J.* **253** 167–73
—— 1983 *Astrophys. J.* **270** 641–4
Barker T and Cudworth K M 1984 *Astrophys. J.* **278** 610–4

Barlow M J 1987 *Mon. Not. R. Astron. Soc.* **227** 161–83
—— 1989 *IAU Symp. 131 Planetary Nebulae* ed S Torres-Peimbert (Dordrecht: Kluwer) pp 319–34
—— 1991 *IAU Symp. 148 The Magellanic Clouds* ed R Haynes and D Milne (Dordrecht: Kluwer) pp 291–8
Barlow M J, Adams S, Seaton M J, Willis A J and Walker A R 1983 *IAU Symp. 103 Planetary Nebulae* ed D R Flower (Dordrecht: Reidel) p 538
Barlow M J, Morgan B L, Standley C and Vine H 1986 *Mon. Not. R. Astron. Soc.* **223** 151–72
Barral J F, Cantó J, Meaburn J and Walsh J R 1982 *Mon. Not. R. Astron. Soc.* **199** 817–32
Becker S A and Iben I, Jr 1979 *Astrophys. J.* **232** 831–53
—— 1980 *Astrophys. J.* **237** 111–29
Berrington K and Kingston A E 1987 *J. Phys. B: At. Mol. Phys.* **29** 6631
Boesgaard A M and Steigman G 1985 *Ann. Rev. Astron. Astrophys.* **23** 319–78
Bowen I S 1928 *Astrophys. J.* **67** 1–15
Brandt J C, Kalinowski J K and Roosen R G 1972 *Astrophys. J. Suppl.* **24** 421–48
Brocklehurst M 1971 *Mon. Not. R. Astron. Soc.* **153** 471–90
—— 1972 *Mon. Not. R. Astron. Soc.* **157** 211–27
Butler S E, Bender C F and Dalgarno A 1979 *Astrophys. J.* **230** L59–61
Butler S E and Dalgarno A 1980 *Astrophys. J.* **241** 838–43
Butler S E, Heil T G and Dalgarno A 1980 *Astrophys. J.* **241** 442–7
Calvet N and Peimbert M 1983 *Rev. Mexicana Astron. Astrof.* **5** 319–28
Cahn J H and Kaler J B 1971 *Astrophys. J. Suppl. Ser.* **22** 319–68
Capaccioli M, Cappellaro E, Della Valle M, D'Onofrio M, Rosino L and Tuatto M 1990 *Astrophys. J.* **350** 100–18
Chapman R D and Henry R J W 1971 *Astrophys. J.* **168** 169–171
—— 1972 *Astrophys. J.* **173** 243–5
Chiosi C and Matteucci F 1982 *Astron. Astrophys.* **105** 140–8
Chu Y H 1989 *IAU Symp. 131 Planetary Nebulae* ed S Torres-Peimbert (Dordrecht: Kluwer) pp 105–15
Ciardullo R, Jacoby G H and Ford H C 1989a *Astrophys. J.* **344** 715–25
Ciardullo R, Jacoby G H, Ford H C and Neill J D 1989b *Astrophys. J.* **339** 53–69
Clegg R E S 1985 *Production and Distribution of C, N, O Elements* ed I J Danziger, F Matteucci and K Kjar (Garching: European Southern Observatory) pp 261–75
—— 1987 *Mon. Not. R. Astron. Soc.* **229** 31–39
—— 1989 *IAU Symp. 131 Planetary Nebulae* ed S Torres-Peimbert (Dordrecht: Kluwer) pp 139–56
—— 1991 *IAU Symp. 145 Evolution of Stars: The Photospheric Abundance Connection* ed G Michaud and A Tutukov (Dordrecht: Kluwer) pp 387–97
Clegg R E S and Harrington J P 1989 *Mon. Not. R. Astron. Soc.* **239** 869–83
Clegg R E S and Middlemass D 1987 *Mon. Not. R. Astron. Soc.* **228** 759–78
Clegg R E S, Peimbert M and Torres-Peimbert S 1987 *Mon. Not. R. Astron. Soc.* **224** 761–79
Clegg R E S, Seaton M J, Peimbert M and Torres-Peimbert S 1983 *Mon. Not. R. Astron. Soc.* **205** 417–34
Cox D P and Daltabuit E 1971 *Astrophys. J.* **167** 257–9
Cudworth K M 1974 *Astron. J.* **79** 1384–95
Danziger I J, Dopita M A, Hawarden T G and Webster B L 1978 *Astrophys. J.* **220** 458–66
Daub C T 1982 *Astrophys. J.* **260** 612–24
Davidson K and Fesen R A 1985 *Ann. Rev. Astron. Astrophys.* **23** 119–46
Denegri D, Sadoulet B and Spiro M 1990 *Rev. Mod. Phys.* **62** 1–42
de Vaucouleurs G 1985 *ESO Workshop on The Virgo Cluster of Galaxies* ed O-G Richter and B Binggeli (Garching: European Southern Observatory) p 413
de Vaucouleurs G and Pence W D 1978 *Astron. J.* **83** 1163–73
Diaz A I and Tosi M 1986 *Astron. Astrophys.* **158** 60–6
Dinerstein H L 1983 *IAU Symp. 103 Planetary Nebulae* ed D R Flower (Dordrecht: Reidel) pp 79–88
D'Odorico S, Peimbert M and Sabbadin F 1976 *Astron. Astrophys.* **47** 341–4
Dopita M A, Henry J P, Tuohy I R, Webster B L, Roberts E H, Byun Y-I, Cowie L L and Songalia A 1990 *Astrophys. J.* **365** 640–7
Dopita M A, Mason D J and Robb W D 1976 *Astrophys. J.* **207** 102–9
Dopita M A and Meatheringham S J 1991 *Astrophys. J.* **377** 480–91
Downes R A 1986 *Astrophys. J. Suppl. Ser.* **61** 569–84
Dufour R J 1984a *Astrophys. J.* **287** 341–52
—— 1984b *Bull. Astron. Soc. Pacific* **16** 888–9
—— 1984c *IAU Symp. 108 Structure and Evolution of the Magellanic Clouds* ed S van den Bergh and K S de Boer (Reidel: Dordrecht) pp 353–61

—— 1985 *Future of Ultraviolet Astronomy based on Six Years of IUE Research* ed J M Mead, R D Chapman and Y Kondo (NASA CP) pp 107–10

—— 1991 *Pub. A.S.P.* **103** 857–60

Dufour R J and Talent D L 1980 *Astrophys. J.* **235** 22–9

Dutra C M and Maciel W J 1990 *Rev. Mexicana Astron. Astrof.* **21** 264–7

Edmunds M G and Pagel B E J 1978 *Mon. Not. R. Astron. Soc.* **185** 77p–80p

Escalante V 1988 *Thesis* Harvard University

Escalante V and Victor G A V 1990 *Astrophys. J. Suppl. Ser.* **73** 513–53

Faúndez-Abans M and Maciel W J 1986 *Astron. Astrophys.* **158** 228–32

—— 1987a *Astron. Astrophys.* **183** 324–6

—— 1987b *Astrophys. Space Sci.* **129** 353–60

Feibelman W A, Oliversen N A, Nichols-Bohlin J and Garhart M P 1988 *International Ultraviolet Explorer, Spectral Atlas of Planetary Nebulae, Central Stars, and Related Objects* (*NASA Reference Publication* **1203**) (Greenbelt, MD: NASA)

Fleming T A, Liebert J and Green R F 1986 *Astrophys. J.* **308** 176–89

Flower D R 1969 *Mon. Not. R. Astron. Soc.* **146** 243–63

—— 1983 *IAU Symp. 103 Planetary Nebulae* ed D R Flower (Dordrecht: Reidel)

Ford H C 1983 *IAU Symp. 103 Planetary Nebulae* ed D R Flower (Dordrecht: Reidel) pp 443–60

Ford H C, Ciardullo R, Jacoby G H and Hui X 1989 *IAU Symp. 131 Planetary Nebulae* ed S Torres-Peimbert (Dordrecht: Kluwer) pp 335–50

Forieri C 1986 *Spectral Evolution of Galaxies* ed C Chiosi and A Renzini (Dordrecht: Reidel) pp 473–5

Freeman K C 1970 *Astrophys. J.* **160** 811–30

—— 1987 *Ann. Rev. Astron. Astrophys.* **25** 603–32

French H B 1983 *Astrophys. J.* **273** 214–8

Frogel J A, Blanco V M, McCarthy M F and Cohen J G 1982 *Astrophys. J.* **252** 133–46

Garay G and Rodríguez L F 1983 *Astrophys. J.* **266** 263–70

Garnett D R and Shields G A 1987 *Astrophys. J.* **317** 82–101

Gathier R 1987 *Astron. Astrophys. Suppl. Ser.* **71** 245–53

Gathier R and Pottasch S R 1985 *Production and Distribution of C, N, O Elements* ed I J Danziger, F Matteucci and K Kjar (Garching: European Southern Observatory) pp 307–12

Gathier R, Pottasch S R and Goss W M 1986a *Astron. Astrophys.* **157** 191–203

Gathier R, Pottasch S R, Goss W M and van Gorkom J H 1983 *Astron. Astrophys.* **128** 325–34

Gathier R, Pottasch S R and Pel J W 1986b *Astron. Astrophys.* **157** 171–90

Gómez Y, Moran J M, Rodríguez L F and Garay G 1989 *Astrophys. J.* **345** 862–70

Gould R J 1978 *Astrophys. J.* **219** 250–61

Greig W E 1971 *Astron. Astrophys.* **10** 161–74

—— 1972 *Astron. Astrophys.* **18** 70–8

Harrington J P, Seaton M J, Adams S and Lutz J H 1982 *Mon. Not. R. Astron. Soc.* **199** 517–64

Harris W E 1988 *A.S.P. Conf. Series no 4, The Extragalactic Distance Scale* ed S van den Bergh and C J Prichet (Provo: Brigham Young University Press) pp 231–254

Hawley S A 1978 *Astrophys. J.* **224** 417–36

Hawley S A and Miller J S 1978a *Publ. Astron. Soc. Pacific* **90** 39–44

—— 1987b *Astrophys. J.* **220** 609–13

Hazard C, Terlevich R, Morton D C, Sargent W L W and Ferland G 1980 *Nature* **285** 463–4

Heap S R and Augensen H J 1987 *Astrophys. J.* **313** 268–83

Henry R B C 1986 *Publ. Astron. Soc. Pacific* **98** 1044–8

—— 1989 *Mon. Not. R. Astron. Soc.* **241** 453–68

—— 1990 *Astrophys. J.* **356** 229–40

Henry R B C, Liebert J and Boroson T A 1989 *Astrophys. J.* **339** 872–88

Hindman J V 1967 *Australian J. Phys.* **20** 147–71

Huchra J 1984 *Preprint 2067 Center for Astrophysics*

Hummer D G and Mihalas D 1970 *Mon. Not. R. Astron. Soc.* **147** 339–54

Hummer D G and Storey P J 1987 *Mon. Not. R. Astron. Soc.* **224** 801–20

Husfeld D, Kudritzki R P, Simon K P and Clegg R E S 1984 *Astron. Astrophys.* **134** 139–46

Iben I, Jr 1975 *Astrophys. J.* **196** 525–47

—— 1984 *Astrophys. J.* **277** 333–54

Iben I, Jr, Kaler J B, Truran J W and Renzini A 1983 *Astrophys. J.* **264** 605–12

Iben I, Jr and Renzini A 1983 *Ann. Rev. Astron. Astrophys.* **21** 271–342

—— 1984 *Phys. Rep.* **105** 329

Iben I, Jr and Rood R T 1970 *Astrophys. J.* **159** 605–17
Iben I, Jr and Truran J W 1978 *Astrophys. J.* **220** 980–95
Iben I, Jr and Tutukov A V 1989 *IAU Symp. 131* ed S Torres-Peimbert (Dordrecht: Kluwer) pp 505–22
Jacoby G H 1979 *Publ. Astron. Soc. Pacific* **91** 754–60
—— 1980 *Astrophys. J. Suppl.* **42** 1–18
—— 1983 *IAU Symp. 103 Planetary Nebulae* ed D R Flower (Dordrecht: Reidel) pp 427–42
—— 1989 *Astrophys. J.* **339** 39–52
Jacoby G H, Ciardullo R and Ford H 1990 *Astrophys. J.* **356** 332–49
Jacoby G H, Ciardullo R, Ford H C and Booth J 1989 *Astrophys. J.* **344** 704–14
Jacoby G H and Ford H C 1983 *Astrophys. J.* **266** 298–308
—— 1986 *Astrophys. J.* **304** 490–500
Jenner D C and Ford H C 1978 *IAU Symp. 76 Planetary Nebulae* ed Y Terzian (Dordrecht: Reidel) p 246
Kaler J B 1970 *Astrophys. J.* **160** 887–913
—— 1979 *Astrophys. J.* **228** 163–78
—— 1983a *IAU Symp. 103 Planetary Nebulae* ed D R Flower (Dordrecht: Reidel) pp 245–57
—— 1983b *Astrophys. J.* **271** 188–220
—— 1985 *Ann. Rev. Astron. Astrophys.* **23** 89–117
—— 1986 *Astrophys. J.* **308** 337–46
Kaler J B, Aller L H, Czyzak S J and Epps H W 1976 *Astrophys. J. Suppl. Ser.* **31** 163–86
Kaler J B, Iben I, Jr and Becker S A 1978 *Astrophys. J.* **224** L63–6
Kaler J B and Jacoby G H 1989 *Astrophys. J.* **345** 871–80
—— 1990 *Astrophys. J.* **362** 491–502
Kaler J B and Lutz J H 1985 *Publ. Astron. Soc. Pacific* **97** 700–6
Kaler J B, Shaw R A and Kwitter K B 1990 *Astrophys. J.* **359** 392–418
Khromov G 1979 *Astrofisika* **15** 269–84
—— 1989 *Space Sci. Rev.* **51** 339–423
Kinman T D, Feast M W and Lasker B M 1988 *Astron. J.* **95** 804–20
Kohoutek L 1967 *Bull. Astron. Inst. Czech.* **18** 103–13
Köppen J, Acker A and Stenholm B 1991 *Astron. Astrophys.* **248** 197–208
Kraan-Korteweg R C 1985 *ESO Workshop on The Virgo Cluster of Galaxies* ed O-G Richter and B
    Binggeli (Garching: European Southern Observatory) p 397
Kwok S 1985 *Astrophys. J.* **290** 568–77
—— 1987 *Phys. Rep.* **156** 111–46
Kwok S and Bignell R C 1984 *Astrophys. J.* **276** 544–50
Lambert D L 1978 *Mon. Not. R. Astron. Soc.* **182** 249–72
Lambert D L and Luck R E 1978 *Mon. Not. R. Astron. Soc.* **183** 79–100
Lequeux J, Peimbert M, Rayo J F, Serrano A and Torres-Peimbert S 1979 *Astron. Astrophys.* **80** 155–66
Lichten S M, Rodríguez L F and Chaisson E J 1979 *Astrophys. J.* **229** 524–32
Liller M H, Welther B L and Liller W 1966 *Astrophys. J.* **144** 280–90
Lutz J H 1973 *Astrophys. J.* **181** 135–45
—— 1989 *IAU Symp. 131 Planetary Nebulae* ed S Torres-Peimbert (Dordrecht: Kluwer) pp 65–72
MacAlpine G M, McGaugh S S, Mazzarella J M and Uomoto A 1989 *Astrophys. J.* **342** 364–73
Maciel W J 1981a *Astron. Astrophys.* **98** 406–7
—— 1981b *Astron. Astrophys. Suppl. Ser.* **44** 123–5
—— 1984 *Astron. Astrophys. Suppl. Ser.* **55** 253–8
—— 1985 *Rev. Mexicana Astron. Astrof.* **10** 199–202
—— 1988 *Astron. Astrophys.* **200** 178–84
—— 1989 *IAU Symp. 131 Planetary Nebulae* ed S Torres-Peimbert (Dordrecht: Kluwer) pp 73–82
—— 1990 private communication
—— 1991 *Elements and the Cosmos* ed R Terlevich, M G Edmunds, B E J Pagel and R F Carswell
    (Cambridge: Cambridge University Press) in press
Maciel W J and Faúndez-Abans M 1985 *Astron. Astrophys.* **149** 365–71
Maciel W J, Faúndez-Abans M and de Olivera M 1986 *Rev. Mexicana Astron. Astrof.* **12** 233–9
Maciel W J and Leite C C M 1990 *Rev. Mexicana Astron. Astrof.* **21** 197–200
Maciel W J and Pottasch S R 1980 *Astron. Astrophys.* **88** 1–7
Maehara H, Okamura S, Noguchi T, He X-T and Liu J 1986 *Second Japan–China Workshop on Stellar
    Activities and Observational Techniques* (Kyoto: University of Tokyo) p 71
Mallik D C V and Mallik S V 1985 *J. Astrophys. Astron.* **6** 113–30
Mallik D C V and Peimbert M 1988 *Rev. Mexicana Astron. Astrof.* **16** 111–21

Manchado A, Pottasch S R and Mampaso A 1988 *Astron. Astrophys.* **191** 128–36
Maran S P, Aller L H, Gull T R and Stecher T P 1982 *Astrophys. J.* **253** L43–7
Masson C R 1986 *Astrophys. J.* **302** L27–30
Matteucci F 1986 *Publ. Astron. Soc. Pacific* **98** 973–8
Matteucci F, Franco J, Francois P and Treyer M A 1989 *Rev. Mexicana Astron. Astrof.* **18** 145–52
Matteucci F and Francois P 1989 *Mon. Not. R. Astron. Soc.* **239** 885–904
Melnick J, Terlevich R and Moles M 1988 *Mon. Not. R. Astron. Soc.* **235** 297–313
Méndez R H 1975 *Astrophys. J.* **199** 411–7
—— 1989 private communication
Méndez R H, Kudritzki R P, Gruschinke J and Simon K P 1981 *Astron. Astrophys.* **101** 323–31
Méndez R H, Kudritzki R P, Herrero A, Husfeld D and Groth H G G 1988 *Astron. Astrophys.* **190** 113–36
Mendoza C 1983 *IAU Symp. 103 Planetary Nebulae* ed D R Flower (Dordrecht: Reidel) pp 143–72
Mendoza C and Zeippen C J 1982a *Mon. Not. R. Astron. Soc.* **198** 127–39
—— 1982b *Mon. Not. R. Astron. Soc.* **199** 1025–32
Meyer J-P 1985 *Astrophys. J. Suppl.* **57** 151–71
Menzel D H, Aller L H and Hebb M H 1941 *Astrophys. J.* **93** 230–5
Milne D K 1982 *Mon. Not. R. Astron. Soc.* **200** 51p–54p
Milne D K and Aller L H 1975 *Astron. Astrophys.* **38** 183–96
Minkowski R 1965 *Galactic Structure* ed A Blaauw and M Schmidt (Chicago: University of Chicago Press) pp 321–43
Minkowski R and Aller L H 1954 *Astrophys. J.* **120** 261–4
Monk D J, Barlow M J and Clegg R E S 1988 *Mon. Not. R. Astron. Soc.* **234** 583–624
Mould J R 1984 *Publ. Astron. Soc. Pacific* **96** 773–8
Nomoto K 1984 *Stellar Nucleosynthesis* ed C Chiosi and A Renzini (Dordrecht: Reidel) pp 205–37
Nussbaumer H and Storey P J 1983 *Astron. Astrophys.* **126** 75–9
—— 1984 *Astron. Astrophys. Suppl.* **56** 293–312
O'Dell C R 1962 *Astrophys. J.* **135** 371–84
—— 1963 *Astrophys. J.* **138** 67–78
—— 1968 *IAU Symp. 34 Planetary Nebulae* ed D E Osterbrock and C R O'Dell (Dordrecht: Reidel) pp 361–75
O'Dell C R, Peimbert M and Kinman T D 1964 *Astrophys. J.* **140** 119–29
Olive K A, Schramm D N, Steigman G and Walker T P 1990 *Phys. Lett. B* in press
Osterbrock D E 1973 *Mém. Soc. R. Sci. Liege* Ser 6 **5** 391–402
—— 1989 *Astrophysics of Gaseous Nebulae and Active Galactic Nuclei* (California: University Science Books)
Osterbrock D E and Flather E 1959 *Astrophys. J.* **129** 26–43
Paczynski B 1971 *Acta Astron.* **21** 417–35
Pagel B E J 1986 *Advances in Nuclear Astrophysics* ed E Vangioni-Flan, J Adouze, M Casse, J P Chieze and J Tran Thanh Van (France: Frontieres) p 53
—— 1987a *A Unified View of the Macro- and the Micro-Cosmos* ed A de Rujula, D V Nanopoulos and P A Shaver (Singapore: World Scientific) pp 399–423
—— 1987b *The Galaxy* ed G Gilmore and B Carswell (Dordrecht: Reidel) p 341
—— 1989 *Rev. Mexicana Astron. Astrof.* **18** 161–72
Pagel B E J and Edmunds M G 1981 *Ann. Rev. Astron. Astrophys.* **19** 77–113
Pagel B E J and Simonson E A 1989 *Rev. Mexicana Astron. Astrof.* **18** 153–9
Pedreros M 1989 *Astron. J.* **98** 2146–55
Peimbert M 1973 *Mém. Soc. R. Sci. Liege* Ser 6 **5** 307–16
—— 1978 *IAU Symp. 76 Planetary Nebulae* ed Y Terzian (Dordrecht: Reidel) pp 215–23
—— 1981 *Physical Processes in Red Giants* ed I Iben, Jr and A Renzini (Dordrecht: Reidel) pp 409–20
—— 1982 *Ann. N.Y. Acad. Sci.* **295** 24–42
—— 1983 *Primordial Helium* ed P A Shaver, D Kunth and K Kjar (Garching: European Southern Observatory) pp 267–79
—— 1984 *IAU Symp. 108 Structure and Evolution of the Magellanic Clouds* ed S van den Bergh and K S de Boer (Dordrecht: Reidel) pp 363–74
—— 1985 *Rev. Mexicana Astron. Astrof.* **10** 125–34
—— 1986 *Publ. Astron. Soc. Pacific* **98** 1057–60
—— 1986a *Star Forming Dwarf Galaxies and Related Objects* ed D Kunth, T X Thuan and J T T Van (Editions Frontieres) pp 403–15

—— 1987 *Planetary Nebulae and Protoplanetary Nebulae: from IRAS to ISO* ed A Preite-Martinez (Dordrecht: Reidel) pp 91–100

—— 1989 *IAU Symp. 131 Planetary Nebulae* ed S Torres-Peimbert (Dordrecht: Kluwer) pp 577–87

—— 1990 *Rev. Mexicana Astron. Astrof.* **237** 454–60

—— 1991 *Elements and the Cosmos* ed R Terlevich, M G Edmunds, B E J Pagel and R F Carswell (Cambridge: Cambridge University Press) in press

Peimbert M, Rayo J F and Torres-Peimbert S 1975 *Rev. Mexicana Astron. Astrof.* **1** 289–97

Peimbert M and Serrano A 1980 *Rev. Mexicana Astron. Astrof.* **5** 9–18

Peimbert M and Torres-Peimbert S 1977 *Mon. Not. R. Astron. Soc.* **179** 217–34

—— 1983 *IAU Symp. 103 Planetary Nebulae* ed D R Flower (Dordrecht: Reidel) pp 233–42

—— 1987a *Rev. Mexicana Astron. Astrof.* **14** 540–58

—— 1987b *Rev. Mexicana Astron. Astrof.* **15** 117–23

Peimbert M, Torres-Peimbert S and Rayo J F 1978 *Astrophys. J.* **220** 516–24

Peña M and Ruiz M T 1988 *Rev. Mexicana Astron. Astrof.* **16** 55–62

Peña M, Ruiz M T, Maza J and González L E 1989 *Rev. Mexicana Astron. Astrof.* **17** 25–30

Peña M, Ruiz M T, Torres-Peimbert S and Maza J 1990 *Astron. Astrophys.* **257** 654–60

Peña M, Torres-Peimbert S and Ruiz M T 1991 *Pub. A.S.P.* **103** 865–73

Péquignot D 1980 *Astron. Astrophys.* **81** 356–8

Péquignot D, Aldrovandi S M V and Stasinska G 1978 *Astron. Astrophys.* **63** 313–24

Péquignot D and Dennefeld M 1983 *Astron. Astrophys.* **120** 249–62

Perek L and Kohoutek L 1967 *Catalog of Galactic Planetary Nebulae* (Prague: Czechoslovakian Academy of Science)

Perinotto M 1989 *IAU Symp. 131 Planetary Nebulae* ed S Torres-Peimbert (Dordrecht: Kluwer) pp 293–300

—— 1991 *Astrophys. J. Suppl. Ser.* **76** 687–713

Phillips J P 1989 *IAU Symp. 131 Planetary Nebulae* ed S Torres-Peimbert (Dordrecht: Kluwer) pp 425–34

Phillips J P and Pottasch S R 1984 *Astron. Astrophys.* **130** 91–6

Pierce M J 1989 *Astrophys. J. Lett.* **344** L57–60

Pottasch S R 1980 *Astron. Astrophys.* **89** 336–41

—— 1983 *IAU Symp. 103 Planetary Nebulae* ed D R Flower (Dordrecht: Reidel) pp 391–409

—— 1984 *Planetary Nebulae* (Dordrecht: Reidel)

—— 1989 *IAU Symp. 131 Planetary Nebulae* ed S Torres-Peimbert (Dordrecht: Kluwer) pp 481–92

Pottasch S R, Bignell C, Olling R and Zijlstra A A 1988 *Astron. Astrophys.* **205** 248–56

Preite-Martinez A 1987 *Planetary Nebulae and Protoplanetary Nebulae: from IRAS to ISO* ed A Preite-Martinez (Dordrecht: Reidel)

Ratag M A, Pottasch S R, Dennefeld M and Menzies J W 1991a *Astron. Astrophys. Suppl. Ser.* in press

—— 1991b *Astron. Astrophys.* in press

Ratag M A, Pottasch S R and Waters L B F M 1991c *Astron. Astrophys.* submitted

Ratag M A, Pottasch S R, Zijlstra A A and Menzies J 1990 *Astron. Astrophys.* **28** 525–60

Reilman R F and Manson S T 1979 *Astrophys. J. Suppl. Ser.* **40** 815–80

Reimers D 1975 *Mém. Soc. R. Sci. Liege* Ser 6 **8** 369–82

Renzini A 1983 *IAU Symp. 103 Planetary Nebulae* ed D R Flower (Dordrecht: Reidel) pp 267–80

—— 1984 *Stellar Nucleosynthesis* ed C Chiosi and A Renzini (Dordrecht: Reidel) pp 99–114

—— 1989 *IAU Symp. 76 Planetary Nebulae* ed Y Terzian (Dordrecht: Reidel)

Renzini A and Buzzoni A 1986 *Spectral Evolution of Galaxies* ed C Chiosi and A Renzini (Dordrecht: Reidel) pp 195–235

Renzini A and Voli M 1981 *Astron. Astrophys.* **94** 175–93

Robbins R R 1968 *Astrophys. J.* **151** 511–29

Robbins R R and Bernat A P 1973 *Mém. Soc. R. Sci. Liege* Ser 6 **5** 263–74

Roche P F 1989 *IAU Symp. 131 Planetary Nebulae* ed S Torres-Peimbert (Dordrecht: Kluwer) pp 117–27

Rodríguez L F 1989 *IAU Symp. 131 Planetary Nebulae* ed S Torres-Peimbert (Dordrecht: Kluwer) pp 129–37

Russell R W, Soifer B T and Willner S P 1977 *Astrophys. J.* **217** L149–53

Sabbadin F 1984a *Mon. Not. R. Astron. Soc.* **209** 889–94

—— 1984b *Mon. Not. R. Astron. Soc.* **210** 341–58

—— 1986a *Astron. Astrophys.* **160** 31–8

—— 1986b *Astron. Astrophys. Suppl. Ser.* **64** 579–89

Sabbadin F and Bianchini A 1983 *Astron. Astrophys. Suppl. Ser.* **52** 395–8

Sabbadin F, Bianchini A and Hamzaoglu E 1984a *Astron. Astrophys.* **136** 200–5

Sabbadin F, Gratton R G, Bianchini A and Ortalani S 1984b *Astron. Astrophys.* **136** 181–92
Sabbadin F, Ortolani S and Bianchini A 1985 *Mon. Not. R. Astron. Soc.* **213** 563–73
Sabbadin F, Strafella F and Bianchini A 1986 *Astron. Astrophys. Suppl. Ser.* **65** 259–65
Sandage A and Tammann G A 1990 *Astrophys. J.* in press
Saraph H E and Seaton M J 1980 *Mon. Not. R. Astron. Soc.* **193** 617–29
Sarmiento A and Peimbert M 1985 *Rev. Mexicana Astron. Astrof.* **11** 73–81
Schild H and Maeder A 1985 *Astron. Astrophys.* **143** L7–10
Schneider S E and Terzian Y 1983 *Astrophys. J.* **274** L61–4
Schneider S E, Terzian Y, Purgathofer A and Perinotto M 1983 *Astrophys. J. Suppl.* **52** 399–423
Schönberner D 1979 *Astron. Astrophys.* **79** 108–14
—— 1983 *Astrophys. J.* **272** 708–14
—— 1989 *IAU Symp. 131 Planetary Nebulae* ed S Torres-Peimbert (Dordrecht: Kluwer) pp 463–71
Seaton M J 1954 *Mon. Not. R. Astron. Soc.* **114** 154–71
—— 1960 *Rep. Prog. Phys.* **23** 313–54
—— 1966 *Mon. Not. R. Astron. Soc.* **132** 113–35
—— 1968 *Astrophys. Lett.* **2** 55–8
—— 1978 *IAU Symp. 76 Planetary Nebulae* ed Y Terzian (Dordrecht: Reidel) pp 131–7
—— 1979 *Mon. Not. R. Astron. Soc.* **187** 73p–76p
Seaton M J and Osterbrock D E 1957 *Astrophys. J.* **125** 66–83
Serrano A and Peimbert M 1981 *Rev. Mexicana Astron. Astrof.* **5** 109–24
—— 1983 *Rev. Mexicana Astron. Astrof.* **8** 117–29
Shaver P A, McGee R X, Newton L M, Danks A C and Pottasch S R 1983 *Mon. Not. R. Astron. Soc.* **204** 53–112
Shields G A 1990 *Ann. Rev. Astron. Astrophys.* **28** 525–60
Shklovsky I R 1956 *Soviet Astron.* **33** 222–35
Stanghellini L and Kaler J B 1989 *Astrophys. J.* **343** 811–27
Stasinska G, Fresneau A, da Silva Gameiro G F and Acker A 1991a *Astron. Astrophys.* in press
Stasinska G and Tylenda R 1990 *Astron. Astrophys.* in press
Stasinska G, Tylenda R, Acker A and Stenholm B 1991b *Astron. Astrophys.* in press
Steigman G 1990 *Fundamental Symmetries in Nuclei and Particles* ed P Vogel (Singapore: World Scientific) in press
Storey J W V 1984 *Mon. Not. R. Astron. Soc.* **206** 521–7
Strömgren B 1939 *Astrophys. J.* **89** 526–47
—— 1948 *Astrophys. J.* **188** 242–75
Terndrup D M 1988 *Astron. J.* **96** 884–908
Terzian Y 1978 *IAU Symp. 76 Planetary Nebulae* ed Y Terzian (Dordrecht: Reidel)
Thielemann F-K, Nomoto K and Hashimoto M 1991 *Supernovae* ed J Audouze, S Bludman, R Mochkovitch and J Zinn-Justin (Elsevier Science Publishers) in press
Tinsley B M 1978 *IAU Symp. 76 Planetary Nebulae* ed Y Terzian (Dordrecht: Reidel)
Tonry J L, Ajhar E A and Luppino G A 1989 *Astrophys. J. Lett.* **346** L57–60
Torres-Peimbert S 1984 *Stellar Nucleosynthesis* ed C Chiosi and A Renzini (Dordrecht: Reidel) pp 3–14
—— 1989 *IAU Symp. 131 Planetary Nebulae* ed S Torres-Peimbert (Dordrecht: Kluwer)
Torres-Peimbert S and Peimbert M 1971 *Bol. Obs. Tonantzintla y Tacubaya* **6** 101–11
—— 1977 *Rev. Mexicana Astron. Astrof.* **2** 181–207
—— 1979 *Rev. Mexicana Astron. Astrof.* **4** 341–50
Torres-Peimbert S, Peimbert M and Daltabuit E 1980 *Astrophys. J.* **238** 133–9
Torres-Peimbert S, Rayo J F and Peimbert M 1981 *Rev. Mexicana Astron. Astrof.* **6** 315–9
Torres-Peimbert S, Peimbert M and Fierro J 1989 *Astrophys. J.* **345** 186–95
Torres-Peimbert S, Peimbert M and Peña M 1990 *Astron. Astrophys.* **233** 540–52
Tosi M 1988a *Astron. Astrophys.* **197** 33–46
—— 1988b *Astron. Astrophys.* **197** 47–51
Tylenda R, Stasinska G, Acker A and Stenholm B 1991 *Astron. Astrophys.* **246** 221–30
Vorontsov-Velyaminov B A 1934 *Z. Astrophys.* **8** 195–207
Webster B L 1969 *Mon. Not. R. Astron. Soc.* **143** 79–95
—— 1988 *Mon. Not. R. Astron. Soc.* **230** 377–401
Weidemann V 1977 *Astron. Astrophys.* **61** L27–30
White S D M and Audouze J 1983 *Mon. Not. R. Astron. Soc.* **203** 603–18
Whitford A E 1958 *Astron. J.* **63** 201–7
Wilkes B J, Ferland G J, Hanes D and Truran J W 1981 *Mon. Not. R. Astron. Soc.* **197** 1–6

Wood P R, Bessell M S and Dopita M A 1986 *Astrophys. J.* **311** 632–6

Wood P R, Meatheringham S J, Dopita M A and Morgan D H 1987 *Astrophys. J.* **320** 178–81

Wood P R and Faulkner D J 1986 *Astrophys. J.* **307** 659–74

Woosley S E, Axelrod T S and Weaver T A 1984 *Stellar Nucleosynthesis* ed C Chiosi and A Renzini (Dordrecht: Reidel) pp 263–93

Zanstra H 1931 *Zs. f. Astrophys.* **2** 329

Zeippen C J 1982 *Mon. Not. R. Astron. Soc.* **198** 111–25

Zuckerman B and Gatley I 1988 *Astrophys. J.* **324** 501–15

Rep. Prog. Phys. **52** (1989) 881-943. Printed in the UK

Compact maser sources

R J Cohen

University of Manchester, Nuffield Radio Astronomy Laboratories, Jodrell Bank, Macclesfield, Cheshire SK11 9DL, UK

Abstract

The past decade has seen tremendous growth in the study of cosmic maser sources. Radio interferometers have fully resolved the structure of the sources on all angular scales, and have enabled large numbers of sources to be studied in detail for the first time. We now have a far clearer picture of the disposition of hydroxyl, water and silicon monoxide masers in star-forming regions and in circumstellar envelopes than was possible ten years ago. There have also been major new developments, including the discovery of extremely powerful megamasers in the nuclei of distant galaxies, the discovery of strong 12 GHz methanol masers in star-forming regions and the discovery of widespread 89 GHz hydrogen cyanide masers in circumstellar envelopes. Large numbers of new maser sources have been found using the all-sky infrared survey by the IRAS satellite to provide candidates. Maser sources of different types can be identified by their characteristic infrared emission. IRAS follow-up surveys have also revealed new types of circumstellar maser associated with stars in unusual stages of their evolution.

On the theoretical front there have been major advances in our understanding of saturation and competitive gain. New mechanisms for producing polarisation in saturated masers have been found, and new pump mechanisms have been proposed for several maser lines, including the first satisfactory pump for strong water masers. This review summarises these and other important developments, concentrating mainly on the new observations and the new types of astrophysical investigation which these have made possible. The use of masers for astronomical distance measurements is also discussed.

This review was received in December 1988.

65

Contents

1. Introduction

Cosmic masers are the most compact molecular line sources studied by radio astronomers. They provide unique information on dense dusty regions of space ranging from comets to star-forming clouds, circumstellar envelopes and active galactic nuclei. Their powerful beams are also useful probes of the interstellar medium. The molecules that can be studied with a radio telescope are only trace constituents of the interstellar gas, which is mainly hydrogen and helium. Abundances of only one part in 10^8 are typical even for simple molecules such as formaldehyde (Mann and Williams 1980). This means that large regions are usually needed to produce a detectable signal. In masers, however, the molecular lines are strongly amplified by stimulated emission. This enables us to study in great detail compact regions which could not be detected by normal means. In star-forming regions, for example, the normal thermal emission from molecular lines at millimetre wavelengths cannot be studied on angular scales less than about one second of arc, because of sensitivity limits. However, the maser emission from excited regions close to the young star can be mapped with milliarcsecond precision. (One milliarcsecond is approximately the angle subtended by a man on the Moon.) This has opened up far-reaching prospects for distance measurements and for detailed astrophysical investigations. Masers are becoming an important astrophysical tool.

The golden age began in 1980-1. That epoch saw the first aperture synthesis maps of hydroxyl masers in star-forming regions (Reid *et al* 1980) and in circumstellar shells (Booth *et al* 1981), and the measurement of proper motions of water masers in star-forming regions (Genzel *et al* 1981a). Wide-ranging reviews by Reid and Moran (1981) and Elitzur (1982) summarise work done up to and including that period. The present review charts the progress that has been made since then, concentrating on the strong maser lines of hydroxyl, water, silicon monoxide and methanol. The review specifically excludes weak masers such as low-frequency recombination lines (Dupree and Goldberg 1969), molecular lines of CH (Rydbeck *et al* 1973), H_2CO (Forster *et al* 1980, Whiteoak and Gardner 1983), NH_3 (Madden *et al* 1986) and other species, and also the weak hydroxyl masers in comets (Despois *et al* 1981). All the masers discussed here have high intensities and high photon rates: indeed this is part of their challenge. The instruments and techniques of radio interferometry that have transformed their study are not discussed here. Details can be found in Thompson *et al* (1986) and in references given below.

1.1. Historical background

The discovery of cosmic masers was so unexpected that radio astronomers thought that they had found a new interstellar species, 'mysterium' (Weaver *et al* 1965). The identification as hydroxyl (OH) masers was soon cleared up, but an element of mystery remained. The powerful line emission came from previously hidden regions: clouds of gas and dust surrounding young stars. Astronomers could only guess at the physical conditions there. A series of pioneering experiments established the compact sizes of the maser sources and the non-thermal nature of the emission (Davies *et al* 1967,

Moran *et al* 1968). However, it was almost impossible to establish the physical processes at work, and evaluate the many competing theories (Litvak 1972).

The discovery of water vapour (H_2O) masers by Cheung *et al* (1969) added new problems. These are more powerful and more compact than hydroxyl (OH) masers (Burke *et al* 1972). The brightest Galactic sources have an output power in the 22 GHz line equal to the entire bolometric luminosity of the Sun. They also cover a wide range of velocity, many hundreds of kilometres per second in some cases. Although it was not obvious at the time, this was early evidence for the phenomenon of energetic outflow from young stars. The star-forming region Orion-KL was found to contain other strong masers of methanol (CH_3OH) and silicon monoxide (SiO) (Barrett *et al* 1971, Snyder and Buhl 1974). Both these discoveries were fortuitous, it should be added. The nature of strong masers in star-forming regions is still not firmly established today.

In contrast, the study of circumstellar masers, which are associated with highly evolved stars, gives the impression of a well planned campaign. A new class of infrared stars had been found which radiate primarily at infrared wavebands. Wilson and Barrett (1968) searched them for OH emission and discovered that many radiate strongly in the OH 1612 MHz line. These maser sources soon acquired the name OH-IR sources. The OH line is twin-peaked, which led to the suggestion that the masers arise in an expanding shell of gas and dust surrounding the central star. The fact that the stars pulsate was crucial in diagnosing the problem. Theory and observation rapidly converged on a model in which the OH masers are excited by infrared radiation (Elitzur *et al* 1976). The detailed predictions of the model were beautifully confirmed in several experiments reviewed by Elitzur (1982). The OH-IR sources are now proving to be a valuable tool in Galactic research (Habing 1987). Other widespread circumstellar masers of H_2O (Knowles *et al* 1969, Schwartz and Barrett 1970) and SiO (Buhl *et al* 1974, Kaifu *et al* 1975) were discovered in an equally methodical way but have turned out to be more difficult to interpret.

Some of the earlier excitement has returned recently with the discovery of mega-masers. These have an enormous output of $\sim 10^{29}$ W which is generated in gas clouds in the nuclei of certain 'active' galaxies (Baan *et al* 1982, Baan 1985). The great distances of these galaxies and the fact that their nuclei are dust-enshrouded present the same problem encountered in star-forming regions. The megamasers are unique probes of very unusual regions, but they have yet to be understood properly.

1.2. Classification

Masers have several properties that distinguish them from other molecular line sources studied by radio astronomers. The most fundamental is their high specific intensity. A direct measurement of this requires interferometry, which is not always possible. Secondary indicators must sometimes be used therefore, including non-equilibrium line ratios, narrow linewidths, polarisation and variability. OH masers in star-forming regions have all these attributes, but other masers do not generally have them all. The physical processes that are responsible for these unusual properties are discussed in § 2. Physically, the maser regions are different in having higher densities (typically 10^6-10^{10} cm^{-3}) than the bulk of molecular gas in the interstellar medium, and secondly they are close to some powerful energy source which drives their radio emission. For the purposes of this review it seems most natural to classify the masers according to the central source. Masers in circumstellar envelopes are treated first, in § 3. These are sometimes termed stellar masers because the maser molecules are formed from material recently lost by the evolving star. More than one Earth mass of material is thrown off per year in some cases. Section 4 describes masers in star-forming regions. These trace molecular gas close to young stars. They are sometimes termed interstellar masers since they are concentrations within normal interstellar clouds. Masers in external galaxies are treated separately in § 5. A few are extragalactic counterparts of the types of maser we know in our own Galaxy, but most are qualitatively and quantitatively different. They occur under extreme conditions not found in our Galaxy, and they are orders of magnitude more powerful.

Mention should be made of a classification scheme for OH masers developed by Turner (1970, 1979), which still occurs in the literature. This is based on regularities in the OH ground-state line ratios, line shapes and polarisation of galactic sources. Sources in which the OH mainlines (1665 and 1667 MHz) dominate are type I, and sources in which the OH satellite lines (1612 or 1720 MHz) dominate are type II in this scheme. Given the wide range of maser lines now known, such a classification seems unduly restrictive.

A more powerful way to recognise masers in different physical environments is by their infrared emission. The all-sky survey by the Infrared Astronomical Satellite IRAS has made it possible to select candidate stellar, interstellar or megamaser sources purely on the basis on their infrared 'colours', that is the flux ratios in different infrared wavebands. Many hundreds of new maser sources have been detected in this way, with detection rates of 20–60%. The techniques are widely used, but have yet to be formalised into a classification scheme for maser sources.

1.3. *Astronomical units and abbreviations*

For the benefit of non-astronomers some of the units and abbreviations employed in this review are summarised below. Further explanation may be found in Allen (1973).

1 astronomical unit (AU) = 1.496×10^{11} m
(mean Earth–Sun distance)

1 parsec (pc) = 3.086×10^{16} m
(distance at which 1 AU subtends 1 arcsec)

1 solar mass ($M_\odot$) = 1.989×10^{30} kg

1 solar luminosity ($L_\odot$) = 3.827×10^{26} W

1 Jansky (Jy) = 10^{-26} W m^{-2} Hz^{-1}
(unit of spectral flux density S_ν)

The specific intensity of radiation, I_ν W m^{-2} Hz^{-1} sr^{-1} is sometimes expressed in Jy per beam area (the worker's unit). It can also be specified in terms of brightness temperature

$$T_b = \frac{c^2}{2k\nu^2} I_\nu \quad \text{K},$$

the temperature at which a black body would produce the observed intensity I_ν (in the Rayleigh–Jeans approximation).

By convention radio astronomers measure (Doppler) radial velocities in terms of the frequency shift of a spectral line

$$V_{rad} = \frac{\nu_0 - \nu_{obs}}{\nu_0} c$$

where ν_0 is the rest frequency. Positive radial velocities are directed away from the observer. Radial velocities are usually corrected for Earth motion, and referred to either the barycentre of the solar system (heliocentric) or the local standard of rest (LSR), which is the rest frame defined by nearby stars.

The following abbreviations for instruments recur throughout this article:

IRAS: Infrared Astronomical Satellite (Neugebauer et al 1984)

MERLIN: multi-element radio linked interferometer network (Davies et al 1980, Thomasson 1986)

VLA: Very Large Array (Thompson et al 1980)

VLBI: very long baseline interferometry (Thompson et al 1986).

2. Maser physics

Masers are the result of microwave amplification by stimulated emission of radiation. They arise naturally in space when interstellar molecules are forced to interact with systems at several different temperatures (e.g. radiation fields, hydrogen molecules or particle streams) and are unable to achieve a Maxwell–Boltzmann distribution of energies. In some cases the level populations become inverted. Stimulated emission, which is the inverse of absorption, can then provide gain factors of up to $\sim 10^{10}$ in a single pass through the excited region of molecular gas. The powerful beams of molecular line radiation that emerge from compact maser sources are an astonishing spectacle of nature. The following sections introduce some of the main physical processes which are at work.

2.1. Radiative transfer

The interaction between radio waves and neutral gas in the interstellar medium is governed by the equation of radiative transfer

$$\frac{dI_\nu}{ds} = -\kappa_\nu I_\nu + \varepsilon_\nu \tag{2.1}$$

where I_ν is the specific intensity of the radiation, ε_ν is the emissivity and κ_ν is the absorption coefficient of the gas. Consider for simplicity a two-level system. In this case the emissivity is

$$\varepsilon_\nu = n_2 A_{21} \frac{h\nu_{21}}{4\pi} f(\nu) \tag{2.2}$$

where n_2 is the number density in the upper level, A_{21} is the Einstein coefficient giving the probability of spontaneous emission, ν_{21} is the mean frequency of the emitted photons and $f(\nu)$ is the normalised line profile. The absorption coefficient is

$$\kappa_\nu = (n_1 B_{12} - n_2 B_{21}) \frac{h\nu_{21}}{c} f(\nu) \tag{2.3}$$

where B_{12} is the Einstein coefficient for absorption and B_{21} is the coefficient for stimulated emission. The Einstein coefficients are related by the equations

$$A_{21} = B_{21} \frac{8h\nu^3}{\pi c^3} \qquad (2.4)$$

and

$$B_{12} = \frac{g_2}{g_1} B_{21} \qquad (2.5)$$

where g_i are the statistical weights of the upper and lower states. Note that the ratio of stimulated emission to spontaneous emission increases as λ^3, and so is $\sim 10^{18}$ times larger for radio waves than for visible light.

The effect of stimulated emission is to reduce the absorption coefficient κ_ν (equation (2.3)). At radio wavelengths the effect is enormous. To see why, we need to consider the level populations. These are usually specified by an *excitation temperature* T_{ex}, the temperature at which the populations would be in the Boltzmann ratio

$$\frac{n_2}{n_1} = \frac{g_2}{g_1} \exp\left(-h\nu_{21}/kT_{ex}\right). \qquad (2.6)$$

At radio frequencies $h\nu_{21}/k$ is of order 0.1 K, so for gas in local thermodynamic equilibrium the factor $h\nu_{21}/kT_{ex}$ is very small and stimulated emission almost cancels absorption. The net absorption coefficient is thus the difference between two much larger terms which are almost equal. As a result the radiative transfer through the gas is very sensitive to small changes in the level populations from their LTE values. Remarkable effects can be produced by quite small population shifts. In particular if the populations become inverted ($n_2/n_1 > g_2/g_1$) then the absorption coefficient becomes negative. The gas then adds to the radiation field, rather than subtracting from it, the radiation becomes more and more intense as it passes through the gas, and a maser is formed. Population inversion or pumping can arise naturally in interstellar space because of the low gas densities and long collision times. Detailed pumping processes are described later.

The solution to the equation of transfer is readily found to be

$$I_\nu = I_\nu(0) \exp(-\tau_\nu) + \int_0^{\tau_\nu} \frac{j_\nu}{\kappa_\nu} \exp[-(\tau_\nu - \tau_\nu')]d\tau_\nu' \qquad (2.7)$$

where $I_\nu(0)$ is the intensity of radiation entering the gas (from $s = 0$) and τ_ν is the optical depth defined by $d\tau_\nu = \kappa_\nu \, ds$. Normally the optical depth would be a positive number, and the radiation reaching the observer would be the background radiation $I_\nu(0)$ attenuated by the factor $\exp(-\tau_\nu)$, plus the emission (and absorption) from the cloud itself. For a maser, however, the absorption coefficient is negative, the optical depth is negative and so is the excitation temperature. In this case radiation is amplified exponentially. For many celestial masers τ is of order -25, leading to gain factors of order 10^{10}. All this gain occurs in a single pass through the maser cloud. Exponential growth along the line of sight continues so long as the pumping processes are able to maintain the population inversion against the growing losses by stimulated emission. Such a maser is said to be *unsaturated*.

Exponential gain in an unsaturated maser leads to several striking effects which are observed in cosmic masers: line narrowing, beaming and rapid variability. If the normalised line profile function $f(\nu)$ is constant throughout the maser and gaussian in form then

$$\tau_\nu = \tau_{\nu_{21}} \exp[-(\nu - \nu_{21})^2/2\sigma^2]$$

where σ is the dispersion. Exponential gain amplifies the centre of the line much more than the wings, leading to an emergent line profile which is still gaussian but reduced in width by the factor $(1 - \tau_{\nu_{21}})^{1/2}$ (Litvak *et al* 1966, Goldreich and Kwan 1974a). This factor is about five for a typical strong cosmic maser. In a similar way small path differences through a maser cloud become greatly distorted by exponential gain. Thus maser spot sizes are generally smaller than the clouds producing them. For example, radiation passing exactly through the centre of a spherical maser cloud is amplified much more than radiation following a path further out in the cloud, so the emergent radiation is beamed. The apparent angular size of the spherical cloud is reduced by the factor $(1 - \tau_{\nu_{21}})^{1/2}$ in this case (Goldreich and Keeley 1972).

Finally let us consider time variations. The gain of an unsaturated maser depends exponentially on the population inversion and the path length, through equations (2.3) and (2.8). Any variations in either parameter are amplified exponentially, and the resulting changes in maser output are much more rapid than the underlying physical changes that they reflect. Thus the exponential growth in unsaturated masers leads naturally to the narrow lines, the compact spot sizes and the rapid variability that are observed in Galactic sources.

2.2. Saturation

There are limits to exponential growth, even in space. The radiation from cosmic masers usually grows to be so intense that stimulated emission begins to reduce the population inversion and hence the gain (equation (2.3)). The maser is then said to saturate, although, strictly speaking, only specific regions of the maser cloud are saturated.

The phenomenon of saturation has been treated analytically by Goldreich and Keeley (1972) for two simple cases: uniform spherical masers and cylindrical masers (long thin tubes). Their calculation is purely classical: each molecule is assumed to radiate at a precise frequency independently of the presence of other molecules and the strength of the radiation field. Then the level populations of a two-state maser are governed by the rate equations

$$\frac{dn_2}{dt} = -n_2 A_{21} - (n_2 - n_1)B_{21}J + P_2(n - n_1 - n_2) - \Gamma_2 n_2 \qquad (2.8)$$

and

$$\frac{dn_1}{dt} = n_2 A_{21} + (n_2 - n_1)B_{21}J + P_1(n - n_1 - n_2) - \Gamma_1 n_1 \qquad (2.9)$$

where n is the total density of the maser species, P_i is the pump rate into level i from all levels other than 1 and 2 (including radiative, chemical and collisional processes), Γ_i is the loss rate from level i and

$$J = \frac{1}{4\pi} \int\int I_\nu \, d\nu \, d\Omega \qquad (2.10)$$

is the number of maser photons crossing unit volume per second (from any direction). It has been assumed, without loss of generality, that the statistical weights of the maser states are equal. For a steady state

$$(n_2 - n_1)(2B_{21}J + \Gamma_2) + 2n_2A_{21} = (P_2 - P_1)(n - n_1 - n_2) + n_1(\Gamma_1 - \Gamma_2). \qquad (2.11)$$

The term n_2A_{21} is negligible in comparison with the others. Thus the population inversion depends on the differences between the pump and loss rates, shown on the right-hand side of this equation, and on the losses by stimulated emission and other processes (the term $2B_{21}J + \Gamma_2$). The maser saturates wherever

$$J \sim \frac{\Gamma_2}{2B_{21}} = \frac{\Gamma_2}{A_{21}} \frac{4h\nu^3}{\pi c^3}.$$

Excessive losses Γ_2, for example by high collision rates, can also destroy the population inversion and quench the maser.

The solutions of the rate equations are then substituted into the equation of radiative transfer, which is effectively one-dimensional in the cases considered by Goldreich and Keeley (1972). Only the amplification of spontaneous emission is considered in detail. What renders the problem non-trivial is that the local population inversion, and hence the local gain, is determined by the integral J of radiation travelling in all directions. For example, the solution for the thin tube sketched in figure 1 contains an unsaturated core where growth is exponential and saturated ends where the growth is approximately linear. Most of the maser gain occurs across the unsaturated core region. The initial growth of spontaneous emission travelling into the tube is hindered

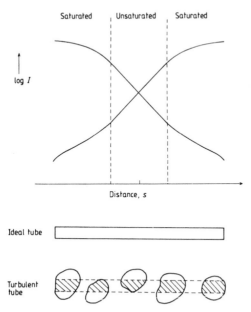

Figure 1. Variation of intensity with distance along a tube maser. Astrophysical masers are likely to be incomplete tubes, as shown schematically in the lower part of the diagram, because of turbulence in the interstellar gas.

by the much stronger beam leaving the tube in the opposite direction: *both* beams are affected by saturation. For large values of the tube length l the unsaturated core approaches a minimum size (independent of l), and the output intensity becomes proportional to l^3. Nearly all inversions lead to maser photons in this case, and the output is simply proportional to the volume of the cylinder.

The solution for the partially saturated sphere is qualitatively similar to this. There is an unsaturated core surrounded by a saturated region in which the intensity grows linearly. The apparent size of the spherical maser is approximately that which the unsaturated core would have if viewed directly (i.e. the core diameter reduced by the factor $(1 - \Delta\tau_{core})^{1/2}$). As the radius r of the sphere is increased, the core radius decreases steadily and the maser output increases as r^3. One qualitative difference from the case of the cylinder is that the core of the spherical maser becomes saturated for very large radius. The fully saturated sphere has a saturated core in which the radiative intensity is constant, surrounded by a saturated region in which the intensity grows linearly (Goldreich and Keeley 1972).

Idealised though these models are, they give insight into the kinds of saturation behaviour to be expected from cosmic masers. The cylindrical model may not be too far removed from the real case of masers in star-forming regions. In these turbulent regions an intense maser beam can build up only if the maser radiation encounters sufficient excited molecules with the same frequency (i.e. the same velocity component along the direction of propagation). The resulting maser takes the form of chance alignments of velocity-coherent cells, which can be thought of as an incomplete tube.

For masers in circumstellar envelopes, where turbulence is much lower, new saturation phenomena arise because of the three-dimensional nature of the maser region. Each volume element is crossed by maser radiation travelling in all directions, and each beam is competing for the limited number of inversions available. For these masers too the gas density and pump rate may vary systematically with radial distance, introducing further complications. The effects of radial gradients in spherically symmetric masers have been considered by Litvak (1973) and Bettweiser and Misselbeck (1977). The latter authors show that the apparent radius of a saturated spherical shell maser is smaller than the true size of the gas density distribution, a result we will return to later.

The problem of competition between different beams in a three-dimensional saturated maser has been approached in simple stages by Alcock and Ross (1985a, b, 1986a, b). In their 'four-stream' model radiation is allowed to propagate in only two orthogonal directions, parallel to the x and y axes of a rectangular maser cloud. This gives four competing beams travelling N, S, E and W. The problem is treated numerically. Figure 2 shows the results for the case of a partially saturated maser with a 'background' source: the rectangular maser cloud illuminated on its western edge by a beam of intensity I_0. The easterly beam is increased in intensity at the expense of beams travelling in other directions. What is interesting is the relative smallness of the increase. Saturation is very effective at turning off the weaker beams, but does not increase the dominant beam very much. On the basis of these calculations one would not expect to observe great differences between the intensities of maser beams which did or did not intersect a background source, unless the background was comparable in intensity with the maser itself. Strong galactic masers have beams which are $\sim 10^5$ or more times intense than any background source (star or HII region), so the maser amplification of the background sources is likely to be a relatively small effect. On the other hand the OH megamasers arise in galactic nuclei which have synchrotron

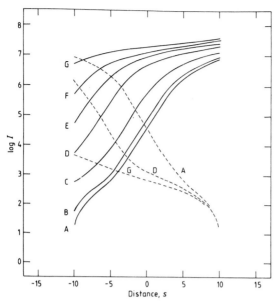

Figure 2. Variation of intensity along the centre line of a square maser which is illuminated along one edge ($s = -10$) by a background source I_0 (Alcock and Ross 1985a). As the background is increased from zero to 5×10^6 the output at the far end of the maser ($s = 10$) increases by less than an order of magnitude, whereas the output of the backwards travelling beam (at $s = -10$) is reduced by over three orders of magnitude. The background intensities are as follows: (A) 0, (B) 50, (C) 500, (D) 5000, (E) 5×10^4, (F) 5×10^5 and (G) 5×10^6.

radio continuum emission of comparable intensity to the masers, so the effects of background amplification might well be important there. Alcock and Ross (1986b) also consider the problem of inhomogeneously pumped masers, and find that in their four-stream model the strongest maser beam is emitted from the cloud edge that has the highest pump rate. The application of this and other results of their modelling to circumstellar maser shells is discussed further in § 3.8.

The effect of saturation on the maser linewidth has been considered only for very simplified cases in the literature. For a tube maser in which the spontaneous emission and the pumping have the same gaussian profile saturation halts the line narrowing discussed in § 2.1 and causes the line to rebroaden to its full (thermal) value (Goldreich and Kwan 1974a and references therein). This is the model widely used by observers. If, however, the maser levels are connected by other transitions whose rates exceed the stimulated emission rate across the maser transition, then rebroadening does not occur: instead the maser line continues to narrow through saturation (Goldreich and Kwan 1974a).

All the treatments so far mentioned neglect the homogeneous (natural) linewidth of the maser transition. The justification for this is that cosmic masers are very large and the inhomogeneous linewidth greatly exceeds the homogeneous linewidth. Hence the total output is incoherent, as has been verified experimentally (Evans *et al* 1972). However, Field and Richardson (1984) have shown that the local coherence effects due to the homogeneous linewidth cannot be ignored in saturated regions. As the stimulated emission rate increases so the lifetime in the upper state decreases, and hence the homogeneous linewidth increases. More photons (covering a greater frequency range) can then cause stimulated emission, so saturation will set in at lower intensities than in the classical case (equation (2.14)). Field and Richardson (1984) find that the classical treatment overestimates the saturation intensity by a factor of about five in a typical case.

2.3. Pumping

The theory of maser pumping is discussed extensively in the literature (e.g. Elitzur 1982), and so only the basic principles will be outlined here. The main point to grasp is that relatively small departures from a Maxwell–Boltzmann distribution are all that is required to produce strong maser action. These small population shifts can come about because of inequalities in the radiative transition rates or collision cross sections connecting different states, or even through chemical processes. Calculating the partition function to the necessary accuracy is sometimes more of a job for an accountant than for an astrophysicist!

Consider for example the excitation of OH. The rotational energy levels are split by lambda doubling and hyperfine interaction to give the energy levels shown in figure 3. The splitting has been greatly exaggerated in the diagram for reasons of clarity. Purely radiative processes can lead to population inversion if the collision rate is very low compared with the rotational transition rates and if some of the rotational transitions are optically thick. This is true in the cool circumstellar envelopes of OH-IR sources. The OH 1612 MHz pumping scheme proposed by Elitzur *et al* (1976) is

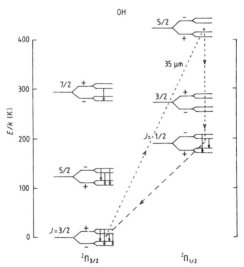

Figure 3. Part of the rotational energy level diagram of OH. The splittings due to lambda doubling and hyperfine interaction are not shown to scale. The astronomically observed maser transitions are indicated by arrows. The broken lines show the pumping scheme for 1612 MHz masers in OH-IR sources (Elitzur *et al* 1976). The broken transition $^2\Pi_{1/2}\, J = \frac{1}{2} \to$ $^2\Pi_{3/2}\, J = \frac{3}{2}$ must be optically thick to produce 1612 MHz inversion.

illustrated schematically in figure 3. Warm dust surrounding the source radiates strongly near 35 μm wavelength and so excites the OH from the ground state $^2\Pi_{3/2} J = \frac{3}{2}$ to the excited state $^2\Pi_{1/2} J = \frac{5}{2}$. The cascade to the ground state proceeds preferentially down the $^2\Pi_{1/2}$ ladder via optically thin transitions, but if the final transition $^2\Pi_{1/2} J = \frac{1}{2} \rightarrow$ $^2\Pi_{3/2} J = \frac{3}{2}$ is optically thick then the $F = 1$ components of the ground state are repopulated faster than the $F = 2$ components. This leads to inversion of the ground-state populations across the 1612 MHz transition. The scheme requires at least four 35 μm pump photons to produce one 1612 MHz maser photon. The ratio of 4:1 is achieved for saturated masers. The model of Elitzur *et al* treats the interaction between maser radiation and level populations in the Sobolev approximation. It is assumed that I_ν depends only on the local value of the source function and on the probability that an emitted photon will escape without further interaction. The theory is thus inherently unable to address the questions of saturation due to source geometry (§§ 2.2 and 3.8).

A purely radiative pump of this kind can be regarded as converting pump photons of frequency ν_p to maser photons of frequency ν with an efficiency η which depends on details of the pump scheme. Clearly for radiative pumping to be a viable model the number of available pump photons should exceed the number of maser photons. Another general point is that a purely radiative pump boosts the brightness temperature of the maser radiation by a factor of at most $\eta(\nu_p/\nu)^2$. For the OH 1612 MHz masers this is a factor of $\sim 10^7$, and so brightness temperatures of up to 10^9 K can be expected from maser regions at a temperature of ~ 100 K. Beaming can increase the brightness temperature by a further factor $4\pi/\Delta\Omega$, where $\Delta\Omega$ is the beam solid angle.

Another process that can boost the brightness temperature of radiatively pumped masers is the non-local effect of line overlap (Litvak 1969, Lucas 1980). The hyperfine splittings of the OH energy levels correspond to only modest Doppler velocities of a few km s^{-1} at the infrared rotational transition frequencies. Systematic motions or turbulence within a maser region can bring these hyperfine components into Doppler alignment and allow one part of the cloud to pump another. Indications are that line overlap can be very effective in the expanding envelopes of OH-IR sources (Bujarrabal *et al* 1980b).

Purely radiative pumping is only possible at relatively low densities. At higher densities, population inversion can come about through collisional effects, such as the unequal cross sections of the upper and lower lambda doublet states, or through some more general interaction between radiative and collisional processes. A full treatment of the radiative and collisional excitation of OH has been attempted recently by Field and Gray (1988). They treat the saturation problem in the semi-classical approximation and account explicitly for the effects of maser radiation in all the rotational levels. Effects of saturation in one transition can strongly influence the growth of other maser lines, a phenomenon termed 'competitive gain' by Field (1985). These calculations yield explicitly the intensities of excited OH masers as well as ground-state OH masers. In principle the maser line ratios could one day be used to diagnose physical conditions in the maser regions very accurately, including not only the temperature, gas density and OH abundance but also the infrared radiation field.

A good example of the interaction between radiative and collisional processes is provided by the pumping of water molecules in circumstellar envelopes (Cooke and Elitzur 1985). Water is an asymmetric top molecule, with a complicated rotational spectrum as shown in figure 4. The 22 GHz maser transition is between the 6_{16} and 5_{23} excited states. Hence water masers are expected only in warm regions where these states are appreciably populated. Because of the selection rules for radiative decay,

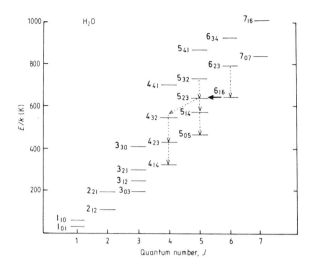

Figure 4. Part of the rotational energy level diagram for ortho H_2O. The 22 GHz maser transition (6_{16}–5_{23}) is indicated by the bold arrow. The broken lines show some possible radiative decay routes. The 6_{16} state has only one possible decay route whereas 5_{23} has two. Other maser transitions are also predicted (e.g. Cooke and Elitzur 1985).

levels like 6_{16} at the bottom of a J ladder have only one route of decay, to the $J-1$ ladder, whereas levels above, such as 5_{23}, can decay either down the J ladder or sideways to the $J-1$ ladder. Hence the 'backbone' levels like 6_{16} tend to be overpopulated relative to adjacent levels under a fairly wide range of conditions, up to a maximum density ($\sim 10^{11}$ cm^{-3}) at which collisions become frequent enough to thermalise the level populations. Masers are predicted not only for the 6_{16}–5_{23} line at 22 GHz, but also for the similar transition 4_{14}–3_{21}, and others (Cooke and Elitzur 1985).

Finally we return to OH for an example of chemical pumping. Andresen *et al* (1984) have demonstrated in the laboratory that the photodissociation of cold water vapour leads to selective population of the lambda doublet states of OH. This is because the dissociation process is planar, and the transition state is symmetric, corresponding to the unpaired pi-lobe of the OH being perpendicular to the plane of rotation of the OH. Provided the gas density is not too high, radiative decay to the ground state thus tends to overpopulate the upper half of the $^2\Pi_{3/2}$ lambda doublet and anti-invert the $^2\Pi_{1/2}$ lambda doublet. Andresen (1986) discussed the possibility that this mechanism contributes to the excitation of OH masers in star-forming regions. He shows that a young OB star provides more than enough UV photons to dissociate water at the necessary rate, but he has yet to find a renewable source of H_2O molecules that would maintain OH masers over their observed lifetimes.

2.4. Polarisation

The primary mechanism that produces the polarisation of cosmic maser emission is the Zeeman effect. Theoretical treatment of the problem is very complicated but fortunately the results have a simple physical interpretation. The results depend on the size of the Zeeman splitting in relation to the maser linewidth, the stimulated emission rate and the rate of cross-relaxation via non-maser (infrared) transitions (Goldreich *et al* 1973a, b). If the Zeeman splitting exceeds the linewidth then large circular polarisations are generated. This case applies to the OH 18 cm masers in a typical field of a few mG. Because the Zeeman splitting dominates it is actually rare to see a complete Zeeman pattern in these lines: any gas providing gain in one hand of circular polarisation cannot provide gain in the other hand of polarisation because of the large Zeeman shift. The occurrence of strong circular polarisation thus sets a lower limit on the magnetic field strength, even when no Zeeman pattern can be identified (Deguchi and Watson 1986a). If the Zeeman splitting is similar to the linewidth then large circular polarisation still occurs, but Zeeman patterns are more likely since the conditions for right- or left-hand gain are no longer completely exclusive. The 5 cm lines of excited OH seem to be ideal in this respect (§ 4.3). If the Zeeman splitting is much smaller than the linewidth, as for the 6 cm lines of excited OH, then no polarisation is expected for an unsaturated maser. The theory of Goldreich *et al* thus accounts well for the observed circular polarisation properties of the OH maser lines. Linear polarisation is also produced by the Zeeman effect, but it can be masked by Faraday rotation within the source (Goldreich *et al* 1973a). Cross-relaxation via infrared transitions can also suppress the growth of linear polarisation unless the stimulated emission rate exceeds the relaxation rate (Goldreich *et al* 1973b). Theory and observation are again in agreement for the OH masers.

The high degree of linear polarisation seen in some SiO and H_2O masers has proved more difficult to understand. For both these molecules any likely Zeeman splitting is far less than the linewidth. Goldreich *et al* showed that linear polarisation of up to 30% can be produced provided that the masers are saturated and provided that the Zeeman splitting exceeds the stimulated emission rate. More recent calculations show that even higher degrees of linear polarisation up to 60% can be produced for heavily saturated masers, but the fields needed are considerable. For example, field strengths of ~1 G would be needed to produce this level of polarisation in SiO and H_2O masers (Deguchi *et al* 1986, Deguchi and Watson 1986b, Western and Watson 1984). In order to explain the highest polarisations that are observed in SiO masers some additional factor must also be invoked, such as the effects of source geometry or anisotropic pumping (Western and Watson 1983). The unfortunate corollary is that the linear polarisation cannot be a reliable guide to the magnetic field in such cases.

3. Masers in circumstellar envelopes

Circumstellar masers are among the best understood cosmic masers because their physical environment is understood, in outline at least. Towards the end of their lives stars develop pulsational instabilities which can lead to extensive mass loss. The stars then build up dense envelopes. As the material cools, dust particles and molecules form. In oxygen-rich envelopes the high density of gas and molecules together with the intense radiation field provide ideal conditions for maser excitation. Powerful SiO, H_2O and OH masers occur in various zones around the star, depending on their excitation requirements. The SiO masers are rotational transitions of vibrationally excited SiO, with excitation temperatures of ~2000 K and above. These states are only populated close to the star. The H_2O maser line at 22 GHz is a transition between rotationally excited states (6_{16}–5_{23}), each of which lies some ~650 K above the ground state, so H_2O masers are found at greater radii than SiO masers. Finally the OH masers at 1.6 GHz are due to hyperfine splitting of the lambda doublet in the rotational ground state. These have the lowest excitation of all and are found at the greatest radii. This excitation sequence has long been recognised (e.g. Olnon 1977, Elitzur 1981) and has now been confirmed by observations of the star VX Sagittarii (Chapman and Cohen 1986).

Before discussing the characteristics of the different masers it will be helpful to review what is known about the central stars and their circumstellar envelopes.

3.1. OH-ɪʀ stars and their envelopes

More than a thousand stellar maser sources are known. Their properties and their distribution in the Galaxy are described in detail by Herman and Habing (1985a) and Habing (1987). The stars are long-period variables with periods ranging from 200 to about 2000 days. Some are identified optically as Mira-type variables or supergiants, usually of spectral class M. Not all stars of these types have masers, however. Empirically it is found that the detectability of OH and H_2O masers is related to the shape of the visual light curve (Bowers and Kerr 1977, Vardya 1987). Other long-period variable stars are completely obscured at optical wavelengths, either by their envelopes of dust or by foreground dust clouds within the Galaxy. These are the OH-ɪʀ stars proper, which are identified by their characteristic OH and infrared emission.

All the long-period variable stars are of great interest from the point of view of stellar evolution because they are in transition from the red giant phase (asymptotic giant branch) to the planetary nebula phase (Iben and Renzini 1983, Herman and Habing 1985a). The stars are also important sources of material for the interstellar medium, providing dust grains (Rowan-Robinson 1986 and references therein) and oxygen-rich material from the stellar core, which is dredged up by convection before the onset of mass loss (Iben and Renzini 1983).

From the point of view of their maser emission the stars are all rather similar. They have typical mass $\sim 1\ M_\odot$, luminosity $\sim 10^4\ L_\odot$ and surface temperature ~ 2800 K†. The important variable is the mass loss rate, $\dot M$, which ranges from $\sim 10^{-8}$ to $\sim 10^{-4}\ M_\odot\ \mathrm{yr}^{-1}$. Techniques for estimating $\dot M$ are discussed by Knapp (1986). It is found empirically that $\dot M$ varies with stellar period P. This gives a natural progression from Mira-type variables (low P and $\dot M$) through to supergiants and OH-ɪʀ stars, which have the highest $\dot M$ and P (Engels *et al* 1983). Along this sequence the infrared emission varies systematically, with heavier mass loss leading to larger far-infrared fluxes and cooler colour temperatures. Figure 5 shows results obtained by Olnon *et al* (1984) using the ɪʀᴀs satellite. This striking sequence is now being used widely and successfully to find new masers associated with point sources in the ɪʀᴀs database (Engels *et al* 1984, Lewis *et al* 1985, Sivagnanam and Le Squeren 1986). The ratios of the different maser lines also vary systematically along this sequence. As mass loss rate increases (left to right in figure 5) so too does the ratio of OH to H_2O maser luminosities (Bowers and Hagen 1984) and the ratio of OH 1612 to OH 1667 MHz emission (Kirrane 1987). Thus H_2O masers are found at the bottom left of this diagram and OH 1612 MHz masers at the top right.

The processes by which mass loss is initiated near the star are not fully understood. Observations by Hinkle *et al* (1984) reveal cyclical motions of cool gas in the outer

---

† However, there are a few spectacular counter-examples which are much hotter, including the F8 supergiant IRC 10 420 (Giguere *et al* 1976) and the A2 supergiant associated with the reflection nebula Roberts 22 (Allen *et al* 1980).

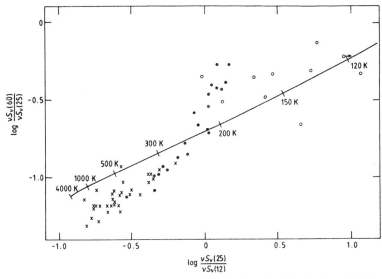

Figure 5. IRAS colour–colour diagram showing the distinctive far-infrared colours of OH-IR sources. The open circles indicate OH-IR sources with small amplitudes of variation and the full circles indicate OH-IR sources with larger amplitudes. The crosses indicate optically visible Mira variables and near-infrared sources with OH maser emission. The curve shows the IRAS colours of black bodies at various temperatures (Olnon *et al* 1984).

layers of the star, with velocities of ~ 20 km s^{-1}, which are sufficient to take the material several stellar radii above the photosphere. Once the gas temperature falls to ~ 1000 K then dust grains condense out. From this point on, the physical processes are more straightforward. The dust grains are driven away from the star by radiation pressure and drag the gas with them. Goldreich and Scoville (1976) have modelled the variation of density, temperature and chemical composition of the material with radius, for an envelope optically thin to the driving radiation. The molecular abundance is high in two regions, one near the star where densities are high, and an outer photodissociation zone where the background UV radiation breaks up the molecules. The OH 1612 MHz masers are believed to lie in the photodissociation zone (§ 3.3) and the H_2O masers are believed to lie in the inner high-density region (§ 3.5). SiO masers probably lie close to the transition region where dust forms (§ 3.6).

In the following sections the different maser lines are discussed in order of increasing excitation, which is also the order of their discovery. First, however, we introduce a simple geometrical model for the maser emission from a thin shell, a model that has guided much thinking on the subject.

3.2. Thin-shell model

Many properties of circumstellar masers can be understood in terms of the simple thin-shell model which is described here (e.g. Olnon 1977, Reid *et al* 1977 and references therein). We assume that the maser emission occurs in a uniformly expanding thin shell centred on the star, with radius r and expansion velocity V_E, and consider an element of the shell. Let θ be the angle between the expansion velocity vector and the line of sight to the observer. Then the radial velocity of this element will be $V = V_* - V_E \cos\theta$, where V_* is the radial velocity of the star, and the projected separation of the element from the stellar position will be $a = r \sin\theta$, as shown in figure 6. Hence for any element of the shell the radial velocity V and projected separation a satisfy

$$\left(\frac{a}{r}\right)^2 + \left(\frac{V_* - V}{V_E}\right)^2 = 1. \tag{3.1}$$

The radial velocity ranges from $V_* - V_E$ on the near side of the shell to $V_* + V_E$ on the far side, and each intermediate radial velocity corresponds to a particular slice through the shell orthogonal to the line of sight, as indicated in figure 6. In fact, equal velocity intervals correspond to equal areas on the spherical surface of the shell.

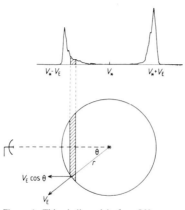

Figure 6. Thin-shell model of an OH-IR maser source. For a uniformly expanding thin shell each radial velocity in the OH spectrum corresponds to a slice through the maser shell perpendicular to the line of sight. The radial velocity V and radius a of such a slice are related by equation (3.1).

The twin-peaked OH 1612 MHz spectrum characteristic of the OH-IR sources comes about because velocity coherence is required through the shell for maser amplification to occur. Suppose that the circumstellar gas has an expansion velocity

$$V_E = f(r) \tag{3.2}$$

which depends only on the distance r from the centre of the star. Then for a shell of finite thickness the radial velocity of a given element of shell varies with distance x along the line of sight, and

$$\frac{\partial V}{\partial x} = -f'(r)\cos^2\theta - \frac{f(r)}{r}\sin^2\theta \tag{3.3}$$

where θ is defined as before. If we neglect the effects of saturation then the maser gain depends only on path length, and the maximum maser gain occurs at angles

θ for which $|\partial V/\partial x|$ is a minimum. This can happen in four ways, depending on the value of the logarithmic velocity gradient:

$$\varepsilon = \frac{f'(r)r}{f(r)} = \frac{d\ln V_E}{d\ln r}. \tag{3.4}$$

(a) If $\varepsilon > 1$ then $|\partial V/\partial x|$ has its minimum value at $\theta = \pm \pi/2$. The brightest masers will be seen in a ring around the central star, and the integrated maser emission from the shell will peak at the stellar velocity V_*.

(b) If $\varepsilon = 1$ then $|\partial V/\partial x|$ is independent of θ, and we expect to see a uniform disc of masers and a flat-topped profile.

(c) If $0 \le \varepsilon < 1$ then $|\partial V/\partial x|$ has its minimum value at $\theta = 0$ and π. In this case we see the familiar twin-peaked profile, with peaks at $V_* - V_E$ and $V_* + V_E$ from the very front and the very back of the maser shell. This is the case applicable to the OH 1612 MHz masers, which lie in the outer layers of the circumstellar envelope where the gas has almost reached its terminal velocity ($f'(r) \to 0$).

(d) If $\varepsilon < 0$ then $|\partial V/\partial r|$ has its minimum value when

$$\cos^2 \theta = \frac{1}{1 - \varepsilon}. \tag{3.5}$$

The brightest masers occur in a ring, as in case (a), but now there are two velocity peaks at each point on the ring and the integrated emission profile is twin-peaked. This case was proposed by Reid et al (1980) for the source W3OH.

The four cases are discussed by Chapman and Cohen (1985). Nearly all OH-IR sources have line profiles that can be classified according to this scheme. According to the thin-shell model the shapes of the spectra and the angular distributions of the maser emission should be closely interrelated. The evidence for this is discussed in the following sections. A critique of the model is given in § 3.8.

3.3. OH 1612 MHz masers

The OH 1612 MHz maser emission from OH-IR sources provides dramatic confirmation of the thin-shell model. It was recognised early that the 1612 MHz spectra have the twin-peaked shape characteristic of an expanding shell. Typically the two peaks might be separated by 30 km s^{-1}, corresponding to an expansion velocity of 15 km s^{-1}. Individual sources exhibit a range of profile shapes, as illustrated in figure 7; however, there is no statistical difference between the shapes and intensities of the red-shifted and blue-shifted peaks (Johansson et al 1977). Observations at high velocity resolution reveal beautifully detailed fine structure in some 1612 MHz spectra, with components as narrow as 0.1 km s^{-1} full width to half power (Fix 1987, Cohen et al 1987).

3.3.1. Shell structure. Radio interferometer maps of OH-IR sources confirm that the 1612 MHz maser emission comes from an expanding shell in nearly every case (Booth et al 1981, Bowers et al 1983, Herman et al 1985, Diamond et al 1985b, and others). Typical shell radii are 10^{16}–10^{17} cm. Representative maps of the source OH127.8 at several radial velocities are shown in figure 8. The emission at the extreme velocities comes from barely resolved caps at the front and back of the shell, whereas emission at intermediate velocities shows different slices through the shell. The lumpy and incomplete structure of this maser shell is typical. Complete shells are almost never seen. This is not surprising in view of the special requirements on the production and

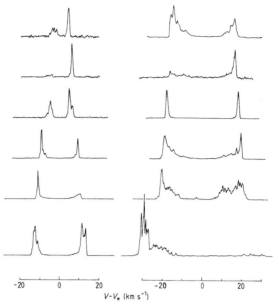

$V-V_*$ (km s^{-1})

Figure 7. Montage of 1612 MHz spectra of OH-IR sources, illustrating the range in expansion velocities and the wide range in profile shapes.

excitation of OH, and the effects of turbulence, which can destroy the velocity coherence necessary for maser amplification. Turbulence is estimated to be ~1 km s^{-1} in the maser region (Diamond *et al* 1985b).

Although the intensity of the 1612 MHz emission is rather irregular the kinematics of the shell are very regular. The angular displacement of maser 'hot spots' from the stellar position varies with velocity, following closely the predictions of the thin-shell model (equation (3.1)). The shell radius can be estimated even in cases where the emission at the stellar velocity, which would show the full shell size, is too weak to be mapped. However, the shell radius is systematically underestimated when the shell is of comparable size to the instrumental beam, as discussed by Chapman (1985) and Herman and Habing (1987).

The OH 1612 MHz maser shells are thought to represent a real increase in OH abundance at the particular radius where H_2O molecules are photodissociated by the external UV radiation (Goldreich and Scoville 1976). If the mass loss is steady and spherically symmetric then the photodissociation radius r_{OH} should increase with the mass loss rate (Huggins and Glassgold 1982). Figure 9 shows that this is indeed the case. It is also predicted that r_{OH} should decrease when the UV intensity increases, but this is a weaker effect and has yet to be observed. As a corollary to this, OH-IR stars lying well out of the Galactic plane may be expected to have non-spherical OH shells on account of the non-uniform UV radiation field.

Two of the OH-IR sources mapped so far show radical departures from a simple shell geometry. One is the proto-planetary nebula Vy2-2, which has a single OH 1612 MHz emission peak, and where it appears that only a narrow filament remains of the original OH envelope (Seaquist and Davis 1983). The other is the source

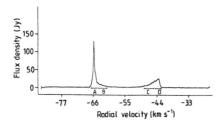

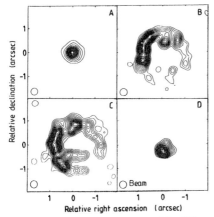

Figure 8. OH 1612 MHz spectrum of the OH-IR source OH127.8, and MERLIN maps showing the angular distribution of the maser emission over four velocity ranges. Adapted from Booth *et al* (1981).

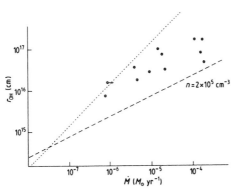

Figure 9. The OH maser shell radius of an OH-IR source is correlated with the stellar mass loss rate $\dot{M}$. The broken line shows a line of constant density 2×10^5 cm^{-3} assuming isotropic mass loss at 15 km s^{-1}. The dotted line shows the condition (3.6) for 1612 MHz pumping. OH 1612 MHz masers are predicted to lie above the broken line but below the dotted line in this diagram. Adapted from Bowers (1985).

OH19.2 − 1.0 which has a multipeaked 1612 MHz spectrum and a regular but non-spherical distribution of masers. Chapman (1988) interprets this source in terms of a biconical geometry of the kind seen in 'dumb-bell nebulae'. In view of the many asymmetries seen in planetary nebulae it is perhaps surprising that most OH-IR sources, which are supposedly the precursors of planetary nebulae, show such simple and apparently regular shell structure.

3.3.2. Pumping. The derivation of the infrared pumping scheme for 1612 MHz masers around OH-IR sources, and its subsequent confirmation by observations, is one of the success stories of maser theory (Elitzur *et al* 1976, Elitzur 1982). The scheme, which has already been described in § 2.3, involves radiative excitation of OH rotational states by IR photons at 35 and 53 μm, followed by a radiative cascade to the ground state. For the 1612 MHz inversion to occur the final 119 μm transition to the ground state must be optically thick. This is a condition on the OH column density. For a steady spherically symmetric mass loss it translates into the condition

$$\frac{r}{10^{16} \text{ cm}} < 2.1 \left(\frac{M}{10^{-6} \, M_\odot \, \text{yr}^{-1}} \right) \left(\frac{10 \text{ km s}^{-1}}{V} \right) \tag{3.6}$$

(Elitzur 1982, his equation (4.7)). On the other hand the maser will be quenched if the density n_{H_2} exceeds $\sim 4 \times 10^5$ cm^{-3}. Together these two requirements for maser excitation specify a triangular region in the r_{OH}–$\dot{M}$ plane within which 1612 MHz maser shells should lie, for steady spherically symmetric mass loss. The two criteria are plotted in figure 9 for a typical expansion velocity of 15 km s^{-1}. There is surprisingly good agreement with the measured values of r_{OH} and $\dot{M}$. In particular these two excitation requirements show why it is that 1612 MHz masers are found only in the stars with greatest mass loss. The mass loss rate must exceed $\sim 3 \times 10^{-7} \, M_\odot \, \text{yr}^{-1}$, otherwise the OH column density necessary to invert the 1612 MHz transition will be achieved only at densities that are so high that the maser is quenched. It is interesting that these pumping considerations alone can account for the observed increase of r_{OH} with $\dot{M}$.

The pumping scheme predicts a specific ratio of four infrared photons to one maser photon when the maser saturates. This has been confirmed by Werner *et al* (1980), who found the ratio to be constant from source to source, and for each individual source as it varied through the stellar cycle. Subsequent studies of a larger sample of objects using the IRAS database reveal more scatter in ratio. The pumping efficiency varies from a few per cent for Mira variables to the canonical 25% for the more luminous OH-IR sources (Dickinson 1987). A few sources whose spectra rise steeply into the far infrared appear to have pumping efficiencies above the theoretical limit, but this may be explained by optical depth effects (photon trapping) at $\sim 35 \, \mu$m.

3.3.3. Variability. The 1612 MHz spectra of individual OH-IR sources are generally stable over many years, but their intensity varies systematically following the far-infrared emission (Harvey *et al* 1974). The amplitude varies by typically a factor of two. The linear relation between the radio and infrared emission is strong evidence for infrared pumping and saturated masers. The masers vary in phase with the infrared pump, but because of the light travel time across the maser shell there is a measurable phase lag of some weeks between the blue- and red-shifted peaks, which come from the near and far sides of the shell respectively. Accurate monitoring of the 1612 MHz emission thus yields the shell diameter (Jewell *et al* 1980, Herman and Habing 1985b). By combining phase-lag measurements with maps of the maser shell the distance to the source can be found by simple geometry (§ 6.1). For a uniformly expanding thin shell the phase lag should vary linearly with velocity across the spectrum. In principle this enables a check to be made on the spherical symmetry of the shell, but in practice this is hampered by the weakness of the emission midway between the peaks.

A small minority of sources do not follow this regular pattern of cyclic variability. Some sources show no appreciable variation at all. It has been suggested that they are in an advanced stage of great mass loss immediately preceding the formation of planetary nebulae (Habing 1987). At the other extreme, two sources have undergone rapid and violent variations. The Mira variable U Orionis underwent a spectacular maser outburst in 1974 involving the lines at 1667, 1665 and 1612 MHz (Cimerman 1979, Jewell *et al* 1981, and others). In the case of the F8 supergiant IRC 10 420, the 1612 MHz spectrum has undergone a more gradual but equally dramatic change in shape and intensity (Benson *et al* 1979).

3.3.4. VLBI hotspots. Most of the 1612 MHz emission from OH-IR sources is resolved out in VLBI experiments, but a fraction (typically 10%) of the flux is detected in the form of compact hotspots. These have angular sizes of ~ 20 milliarcsec and brightness temperatures of $\sim 10^9 - 10^{10}$ K (Reid *et al* 1977, Benson and Mutel 1979, Bowers *et al* 1980, Norris *et al* 1984). The brightness temperatures are lower than those of OH masers in star-forming regions, but they are still bright enough to be of interest in terms of the maser pump. The radiative pump alone can only achieve mean brightness temperatures of up to 10^9 K, as discussed in § 2.3. The extra intensity must be due to beaming, with implied beaming factors of about ten. It is sometimes suggested that the hotspots represent amplified 'images' of the central star, a result that is potentially of great astronomical interest (e.g. Norris *et al* 1984). However, the comprehensive (if crude) VLBI data of Bowers *et al* (1980) show no clear preference for blue-shifted rather than red-shifted hotspots. This is an area where good modern data are needed on a wide sample of sources.

3.3.5. Polarisation. The early studies by Wilson *et al* (1970) and Wilson and Barrett (1972) suggested that 1612 MHz maser emission from OH-IR sources is essentially unpolarised. Recently, however, there has been a growing awareness that many 1612 MHz spectra of OH-IR sources contain polarised components. The flaring sources U Orionis and IRC 10 420 were found to contain highly polarised 1612 MHz masers (Reid *et al* 1977, Fix 1979, Benson *et al* 1979). VLBI measurements show that the 1612 MHz masers of IRC 10 420 could be up to 100% circularly polarised (Mutel *et al* 1979). More recently the 1612 MHz spectra of supergiants have been found to contain many narrow circularly polarised components which are apparent only when high velocity resolution (~ 0.1 km s^{-1}) is employed (Cohen *et al* 1987). Further unpublished observations suggest that narrow circularly polarised components are also present in the 1612 MHz spectra of many well known OH-IR sources (Cohen, in preparation). The significance of these results is that they indicate a substantial magnetic field in the maser regions. Field strengths of ~ 1 mG are needed to account for the circular polarisation in terms of Zeeman splitting, following the argument of Deguchi and Watson (1986a). Measurements combining high velocity resolution with high angular resolution may enable more precise estimates of the field strengths to be made.

3.4. OH mainline masers

Up until now OH mainline masers have been less well studied than OH 1612 MHz masers, for the simple reason that they are weaker. In the catalogue by Engels (1979), for example, mainline masers are four times weaker than 1612 MHz masers, on average. This is to be expected from consideration of the infrared pumping mechanisms, which predict efficiencies of 1–10% for OH mainlines, compared with 25% for the 1612 MHz line (§§ 3.3 and 3.4.2). Searches of nearby Mira-type stars indicate that ~30% have significant OH mainline emission (Bowers and Kerr 1977, Bowers 1985 and references therein). There is a tendency for the 1667 MHz emission to occur more often than the 1665 MHz emission (Wilson and Barrett 1972, Slootmaker *et al* 1985). When both mainlines occur together their profiles are usually, but not always, similar in shape (Olnon *et al* 1980).

3.4.1. Variability. OH mainline emission varies systematically throughout the stellar cycle, but the variations are larger and less regular than those of the OH 1612 MHz masers (Harvey *et al* 1974, Fillit *et al* 1977). For OH Miras there is generally a phase lag of 20–30 days between the OH variations and the optical variations (Fillit *et al* 1977, Le Squeren and Sivagnanam 1985). The amplitudes of the OH mainline variations do not always repeat from cycle to cycle. Individual peaks in the spectra can change substantially on this timescale both in intensity and in polarisation (Booth 1969, Robinson *et al* 1970, Sullivan and Kerstholt 1976, Nguyen-Q-Rieu *et al* 1979, Morris and Bowers 1980, and references therein). Extreme cases of variability have been observed in the sources U Orionis and R Leonis, where the whole mainline profile has changed on a timescale of one year (Pataki and Kolena 1974, Cimerman 1979, Fix 1979, Jewell *et al* 1981, Le Squeren and Sivagnanam 1985, and references therein). There were no comparable infrared or optical counterparts to these events. It is not yet known how frequently such events occur or what their significance is in terms of the evolution of the central star.

Monitoring of OH mainline emission has so far yielded only one phase-lag measurement of shell size: Bowers and Morris (1984) have measured a size $\geq 9 \times 10^{16}$ cm for the masers associated with the peculiar bipolar nebula OH 0739-14.

3.4.2. Pumping models. The agreement between the phases of OH mainline and infrared emission suggests that the masers are radiatively pumped. Several mechanisms have been explored. Elitzur (1978) pointed out that since the spacing of the OH lambda doublets increases with frequency, inversion or anti-inversion of the ground-state lambda doublet might be produced by suitable radiation fields. If the circumstellar dust is hot enough and optically thin then the radiation field will selectively excite the upper half of the ground-state doublet and the cascade to the ground state will lead to inversion. Calculations by Bujarrabal *et al* (1980a) show that the cascade reinforces the inversion because of asymmetries in the matrix of rotational transition probabilities. Pumping efficiencies of ~1% are predicted, with a firm upper limit of 3%. This has been confirmed by Dickinson (1987) using IRAS data on a large sample of sources. The inversion can be further strengthened by infrared line overlap in the expanding circumstellar envelope (Lucas 1980, Bujarrabal *et al* 1980b). In this case the pumping efficiency can reach 10%, and brightness temperatures of 10^{12} K can be achieved, given reasonable assumptions about beaming. According to these pumping models mainline masers should occur in warmer and less dense regions of the circumstellar envelope than 1612 MHz masers.

3.4.3. Angular structure. VLBI measurements have yielded angular sizes for OH mainline masers associated with only two sources, U Orionis and VY Canis Majoris. The corresponding brightness temperatures are 10^{10} K and 10^{12} K respectively (Fix *et al* 1980, Benson and Mutel 1982). Three other sources studied by Reid *et al* (1977) were not detected. Interferometer baselines of order 100 km are more suitable for studying the overall structures of the mainline regions. The sources that have been mapped so far using MERLIN and the VLA divide naturally into two classes. Sources with a classical twin-peaked mainline profile show resolved shell structure (thin-shell case (*c*)). Examples include the sources IRC 10 420 and OH127.8 (Bowers 1984, Diamond *et al* 1985a). The mainline masers are of comparable extent to the 1612 MHz masers but appear to occupy complementary zones of the circumstellar envelope. The infrared pumping schemes imply that the mass loss is far from symmetrical in these cases. Sources in the second group have more variable and irregular line profiles, frequently with circular polarisation. The distribution of masers is more irregular and more compact, as in VX Sagittarii for example (thin-shell case (*b*)). One source, U Orionis, displays an incomplete ring of mainline masers (thin-shell case (*a*)) (Chapman and Cohen 1985, 1986). There is still much work to be done to improve the statistics on mainline masers.

3.4.4. Polarisation. Mira-type variables and supergiants frequently show 50–100% polarisation of their OH mainline emission. Circular polarisation is most common, whereas linear polarisation occurs in only ~10% of sources (Olnon *et al* 1980). Zeeman groups have been identified for several sources, in some cases on the basis of accurate spectroscopy (e.g. Claussen and Fix 1982) and in other cases by interferometric measurements which showed the close coincidence in the sky of pairs of features of opposite circular polarisations (e.g. Benson and Mutel 1982). The magnetic fields implied by these measurements are typically ~3 mG. A field of this strength has an energy density comparable with the kinetic energy density of the outflowing gas, and so may be expected to influence the envelope dynamics, perhaps leading to some of the asymmetries observed in the envelope structures (Chapman and Cohen 1986). If the magnetic field strengths are assumed to be typical of a widespread field of solar type (with a $1/r^2$ dependence), extrapolation back to the stellar surface yields estimates of 1–100 G for the stellar magnetic field (Reid *et al* 1979, Chapman and Cohen 1986). Although these estimates are highly uncertain at present, it may be possible in future to use several OH lines to measure the magnetic field at a number of different locations in the circumstellar envelope, and so investigate the field geometry.

3.5. H_2O masers

Circumstellar H_2O masers are widespread. For example, Bowers and Hagen (1984) have estimated that 75% of Mira variables in the solar neighbourhood have H_2O maser emission. Engels *et al* (1986) detected H_2O emission from 33% of a sample of OH-IR sources, with a 100% success rate for the strongest OH emitters. This suggests that the detection of H_2O masers in most OH-IR sources is also simply a matter of sensitivity. The H_2O lineshapes have been classified into two types by Engels *et al* (1986). Type A have a centrally concentrated profile covering a narrow velocity range, and type B have the characteristic twin-peaked emission profile, with a brighter blue-shifted peak on average. Usually the H_2O emission is centred on the stellar velocity and covers a smaller velocity range than the OH 1612 MHz emission (Engels *et al* 1986).

Linear polarisation has been detected in two sources at the level of 3% (Knowles and Batchelor 1978).

3.5.1. Variability. A characteristic feature of circumstellar H_2O masers is their strong time variability. The H_2O lineshapes vary on timescales of months to years, in some cases changing from a single-peaked to a twin-peaked type of spectrum (Engels *et al* 1986). There have been many monitoring programmes to investigate the relationship between the H_2O maser variations and the underlying stellar variations (Schwartz *et al* 1974, Cox and Parker 1979, Hagen 1979, Berulis *et al* 1983, Nyman *et al* 1986, Gómez Balboa and Lépine 1986, and references therein). It is generally agreed that the H_2O emission follows the stellar cycle, but it does so very erratically. The integrated H_2O emission varies by typically an order of magnitude more than the stellar luminosity, and the variations do not repeat from cycle to cycle. There are both short- and long-term changes in H_2O luminosity and profile shape, and in some cases there are clear anticorrelations between the H_2O and the optical or infrared variations (e.g. Spencer *et al* 1979). Gómez Balboa and Lépine (1986) have presented evidence for superperiods spanning several stellar cycles, and have attempted to model these in terms of shock waves propagating into the circumstellar envelope. In other cases rapid H_2O flares have been recorded that have no optical or infrared counterpart. The fastest of them can double or halve their intensity in only 2–3 weeks (Berulis *et al* 1983).

Because of the irregular variations in H_2O intensity, attempts to determine the phase of the H_2O variations relative to the optical and infrared variations have had mixed success. Schwartz *et al* (1974) found no phase-lag greater than ± 30 days, but recent studies indicate that the H_2O lags the optical emission by typically 0.1–0.2 period (e.g. Gómez Balboa and Lépine 1986). Phase lags of this order have also been measured between H_2O and infrared variations, either by following sources through several cycles (e.g. Berulis *et al* 1984) or by examining data on many sources statistically (Engels *et al* 1986). In the case of the supergiant S Persei careful monitoring showed that different spectral features had different phase lags of up to a few hundred days (Cox and Parker 1978). Taken together these results strongly suggest that radiative pumping alone cannot be responsible for the H_2O maser variations, and that collisional processes are involved, as discussed below.

A number of variability studies have revealed frequency drifts of individual features in the H_2O spectrum (Sullivan 1973, Berulis *et al* 1983). They correspond to typically 0.6 km s^{-1} over one year. Shifts of this size can be accounted for by redistribution of maser gain among the different hyperfine components of the H_2O spectrum, without the need for acceleration of the gas (Sullivan 1973). However, a dynamical explanation for the observations is not altogether ruled out either. Berulis *et al* (1983) have made the interesting suggestion that velocity shifts might be produced by a change in maser excitation. Because of the velocity gradient in the expanding envelope any change in radius of excitation will automatically be accompanied by a change in line-of-sight velocity. To distinguish between the various possible explanations it will be necessary to carry out VLBI measurements at several epochs spaced over a year and measure the changes in angular structure, if any, which accompany the frequency shifts.

3.5.2. Structure. Circumstellar H_2O masers are distributed over regions typically 10^{15} cm in extent, centred on the stellar position to a precision of 0.1 arcsec (Johnston *et al* 1985). Changes in the H_2O lineshape are sometimes accompanied by dramatic changes in the angular structure, as shown in figure 10. No source has yet been mapped repeatedly to see how these variations are related to the stellar cycle.

In maps published to date circumstellar H_2O masers appear as discrete maser spots each with a characteristic velocity and a velocity width of ~1 km s^{-1}. The maser spots are unresolved by the VLA or MERLIN at ~50 milliarcsec resolution, but well resolved by VLBI at 1 milliarcsec resolution. Interferometer baselines of ~200 km giving ~10 milliarcsec resolution are needed to reveal the detailed structure of these regions. The VLBI hotspots contain typically 20% of the flux and have brightness temperatures of ~3×10^{11} K, which are considerably lower than those found for H_2O masers in star-forming regions (Spencer *et al* 1979). The H_2O masers associated with supergiants appear to be systematically larger than those associated with Mira-type variables and to be spread over a larger region (Diamond *et al* 1987).

The kinematics of the H_2O masers are not so regular as those of the OH 1612 MHz masers. Nevertheless Chapman and Cohen (1986) and Diamond *et al* (1987) have found velocity patterns that can be explained in terms of a 'thick shell' region around the star, in which the gas is accelerated to approximately escape speed. The logarithmic velocity gradient in these H_2O regions is of order unity.

The maser scale sizes and velocity gradient can be related to the variability timescale, following Rosen *et al* (1978) and others. For an H_2O region ~10^{15} cm in extent with a velocity range of ~10 km s^{-1}, the path length contributing to a single maser feature ~1 km s^{-1} in width must be of order 10^{14} cm. The brightness temperatures (10^{11}–10^{12} K) and likely input temperature (1000 K) imply a maser gain of ~e^{20}. Hence the path length corresponding to unit optical depth (gain e) is ~5×10^{12} cm. The most rapid variations involve a change by a factor e in ~3 weeks, and so could be produced by a disturbance propagating at ~30 km s^{-1}. This speed is too high for turbulence in neutral gas—the case considered by Rosen *et al*—but it is not an unreasonable speed for a shock wave in the circumstellar envelope.

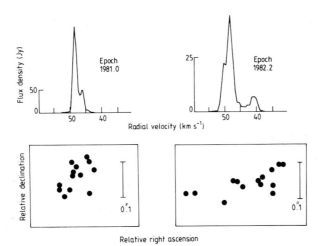

Figure 10. H_2O 22 GHz spectra of the Mira variable R Aquilae at two epochs one year apart, and schematic maps showing the locations of the individual H_2O maser spots as measured by the VLA. Compiled from Johnston *et al* (1985).

3.5.3. Pumping. Results from variability studies appear to rule out pumping by radiative processes alone, even though there are sufficient near-infrared photons available to excite the H_2O vibration-rotation bands. Most observers have concluded that collisional pumping is most likely. A detailed collisional pumping scheme has been worked out by Cooke and Elitzur (1985) using cross sections of Green (1980) for H_2O and He. Their model has two features of general interest. The first is that the H_2O 'backbone' rotational levels defined by de Jong (1973) tend to be overpopulated under a wide range of conditions involving collisional excitation and radiative decay. Thus as well as predicting a strong $6_{25}-5_{16}$ inversion the model predicts inversions of several other transitions that it may ultimately be possible to observe, albeit from space (see table 1 of Cooke and Elitzur). The second point is that the inner boundary of the maser region is set by collisional quenching, which occurs when the collision rate approaches the rate at which maser photons are lost. For spherically symmetric mass loss this leads to a quenching radius

$$r_1 \propto \dot{M}^{2/3} V_E^{-1}$$

where $\dot{M}$ is the mass loss rate and V_E the expansion velocity. The published maps of H_2O masers support this result (Cohen 1987). However, it must be stressed that the correlation is at present due almost entirely to the difference between the supergiants and the Mira variables. More data are needed to establish the result firmly.

In view of the success of the model of Cooke and Elitzur (1985) in explaining the luminosities and sizes of the H_2O masers it would be extremely valuable to extend the model to incorporate the stellar pulsations, which will affect the collision rate through temperature and density changes.

3.6. SiO masers

SiO masers are potentially the most interesting of all the circumstellar masers since they trace the regions where mass loss is established. Indeed SiO may be an important coolant in these regions (Muchmore *et al* 1987). The SiO molecule has a simple and well understood vibrational-rotational spectrum. Masers have been detected in the $V = 0, 1, 2$ and 3 vibrational states and as high as $J = 6$-5 in the rotational ladders (Scalise and Lépine 1978, Jewell *et al* 1987, and references therein). There is a degree of similarity in the polarisation and velocity structure of lines from the same rotational transition of different vibrational states (Lane 1982, Barvainis and Predmore 1985). On the other hand there is a lack of correspondence between different rotational transitions of the same vibrational state.

The SiO lines are centred on the stellar velocity and cover a rather small velocity range, typically 10 km s^{-1}. The velocity extent does not depend strongly on either the mass loss rate or the terminal velocity of expansion (Jewell *et al* 1984, Nyman and Olofsson 1986).

3.6.1. Variability. SiO shows the strongest variability of all the circumstellar masers. There is a general correlation between the integrated SiO maser flux and the optical light curve of the star (Lane 1982, Nyman and Olofsson 1986, and references therein). However, the SiO lines that have been monitored in detail show none of the regularity of the optical and infrared emission. The peak intensity varies wildly throughout the cycle, by factors of up to 100:1, and the shape of the spectrum also varies from cycle to cycle. Time-averaged profiles are centred on the stellar velocity to within ~ 1 km s^{-1} provided data are averaged over several cycles (Nyman and Olofsson 1986, Snyder *et al* 1986). Nyman and Olofsson (1986) found that the $V = 1$, $J = 2$-1 line lags behind the optical light curve at maximum but is in phase with the near-infrared emission. For two stars, χ Cygni and R Leonis, the $V = 2$, $J = 2$-1 line was found to have a more sharply peaked maximum which was delayed with respect to the $V = 1$, $J = 2$-1 line and the near-infrared continuum (Olofsson *et al* 1985). These studies and those by Lane (1982) suggest that simultaneous monitoring of several vibrational and rotational transitions will be very fruitful.

3.6.2. Angular structure. Up to now SiO masers have only been mapped using VLBI. Up to half of the flux is detected in these experiments in the form of 'hotspots', with brightness temperatures ranging up to 2×10^{10} K (Moran *et al* 1979, Lane 1982). Measured spot sizes are from 2×10^{13} cm to 4×10^{14} cm. The most complete published data are in the thesis by Lane (1982). She found the SiO maser hotspots to be distributed over regions ~ 0.1 arcsec in angular extent, some 4–6 times the stellar diameter. The results for the star VX Sagittarii are reproduced in figure 11. The maps of $J = 1$-0, $V = 1$

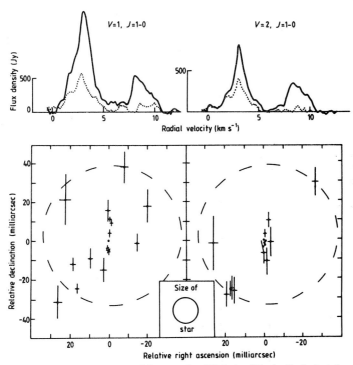

Figure 11. SiO 43 GHz spectra of the supergiant VX Sagittarii in the $V = 1$, $J = 1$-0 and $V = 2$, $J = 1$-0 maser lines, together with VLBI maps showing the relative positions of the different maser spots. The broken lines indicate an expanding shell model which was fitted to the data, and the inset shows the size of the supergiant star. Adapted from Lane (1982).

and $J = 1-0$, $V = 2$ lines for this source are in fair agreement with each other. The data are consistent with the location of the SiO masers in an expanding circumstellar shell of radius $6R^*$, which is indicated by the broken circle in each map. It would be very interesting to know where the remaining 50% of the emission originates.

3.6.3. Pumping. The basic physics behind SiO masers was explained long ago by Kwan and Scoville (1974). When the vibrational transitions of SiO become optically thick then vibrational de-excitation rates ($V \to V-1$) decrease for higher J values. Thus vibrational de-excitation can lead to overpopulation of the higher J levels and so to masers. There are many practical difficulties, however, as discussed by Langer and Watson (1984). Most of the SiO is in rotational states higher than those observed by radio astronomers. Calculations must necessarily involve many different energy levels, and collision cross sections must be known accurately. Assuming a steady symmetric wind and treating the IR radiation and the radio radiation in the Sobolev approximation, Langer and Watson were able to reproduce a number of SiO maser properties (line ratios etc) but failed by several orders of magnitude to reproduce the high photon rates observed. Rates of up to 10^{46} maser photons/s are measured. The models involved maser excitation at densities of $n_{\mathrm{H_2}} \sim 10^{10}$ cm^{-3} and temperatures of ~ 1000 K. Alcock and Ross (1986c) suggest that the smooth wind model is inappropriate. They interpret the VLBI observations of SiO masers in terms of highly beamed maser emission from dense cloudlets with $n_{\mathrm{H_2}} \sim 10^{12}$ cm^{-3}. They need a strong pump to provide inversion at the high rate implied by these densities. This still leaves the pumping problem unsolved. Although the physics is believed to be understood we are as yet unable to model the observations accurately.

3.6.4. Polarisation. SiO masers are generally linearly polarised. Polarisation of 15% is common (Troland *et al* 1979) and levels of up to 80% have been reported (Miller *et al* 1984). No circular polarisation has been detected down to $\sim 3\%$. The linear polarisation varies strongly across the line profile. In some cases individual spectral features can be followed for up to a stellar period, during which time they maintain their velocity, linewidth and polarisation (Clark *et al* 1985). This suggests that the emission regions are physically stable entities in the circumstellar envelope. They are, however, disrupted near stellar maximum. Western and Watson (1983) have shown that small anisotropies in pumping can lead to the high degrees of linear polarisation observed. Linear polarisation of 20–30% can also be produced by maser regions with planar geometry. Thus it seems that the polarisation of SiO masers can be accounted for by causes other than magnetic fields, as indeed might be expected, given the very low magnetic dipole moment of SiO.

3.7. Other masers

Apart from the widespread masers of OH, H_2O and SiO many other maser lines or suspected masers have been reported in the literature. Surprisingly there have been few reports of excited OH masers: 5 cm emission from NML Cygnii (Zuckerman *et al* 1972) and the hybrid object Vy2-2 (Jewell *et al* 1985), and 6 cm emission from AU Geminorum (Claussen and Fix 1981). OH 1720 MHz maser emission has been observed in only one highly variable source, V1057 Cyg (Winnberg *et al* 1981 and references therein). Another line seen in only a single source so far is the SiS maser reported by

Henkel *et al* (1983). Of more general interest is the newly detected strong maser at 89 GHz from vibrationally excited HCN (Guilloteau *et al* 1987). This is the $J = 1$-0 transition of HCN in the $V = (0\,2°\,0)$ vibrational state, some 2050 K above the ground state. It has been detected only in carbon-rich envelopes, and not in the oxygen-rich envelopes traced by OH, H_2O and SiO masers. A strong narrow line is seen near the stellar velocity, blue-shifted by 1–4 km s^{-1}. Some 20% of carbon stars searched to date have been detected (Lucas *et al* 1988). Rather strikingly, there are no detections among the stars with the highest mass loss rates ($\geqslant 10^{-5}\,M_\odot\,\mathrm{yr}^{-1}$). Because this line is fairly widespread it may prove to be a useful probe of carbon-rich circumstellar envelopes.

3.8. Outlook

The preceding sections may have given the impression that the problem of circumstellar masers is essentially solved, apart from the details. That is not true. There are signs that all is not right in our understanding of even the apparently simple OH 1612 MHz maser shells. Alcock and Ross (1986a) have pointed out defects and inconsistencies in the standard theory of radiative transfer in a thin spherical shell. When saturation and beaming are properly accounted for then the emission at the stellar velocity is predicted to be far weaker than is generally observed. Furthermore the emission at this velocity should appear to come from an annulus only ~1% of its radius thick, whereas MERLIN and VLA results indicate much thicker shells. Alcock and Ross suggest that a clumpy outflow can resolve the difficulty. Certainly the MERLIN maps have a very blobby appearance, but it is not clear whether this is due to irregularities in velocity (turbulence), OH density or maser excitation. A further difficulty is that the standard pump model for the 1612 MHz masers is not sophisticated enough to apply to these questions. It is not a full treatment of the radiative transfer problem. A more rigorous development of the model is urgently needed. For the other maser lines (OH mainline, H_2O and SiO) no clear observational picture has yet emerged, and we must await new instrumental developments. Aperture synthesis arrays of 100–200 km would seem to be ideal for studying these lines.

As maser maps have become more accessible to the astronomical community their astrophysical usefulness has grown, in spite of the difficulties in interpreting them! The use of masers for distance determination is discussed in § 6. First attempts have been made to use maser lines to probe the three-dimensional structure of circumstellar envelopes. Figure 12 shows data on OH, H_2O and SiO masers around the star VX Sagittarii which were used to investigate the velocity field of the expanding envelope (Chapman and Cohen 1986). A surprising result from this study is that the outflow velocity appears to increase more slowly with radius than had been assumed hitherto. One possible explanation for this is that the opacity of dust grains to the driving radiation might increase with radius out to some 50 stellar radii. Another area requiring investigation is the role of the stellar magnetic field.

The number of known stellar maser sources has trebled in recent years following the IRAS mission, and we now have a growing number of maser sources that do not fit the previously known categories. There are single-peaked OH masers which might be proto-planetary nebulae like Vy2-2 (Pottasch *et al* 1987), and an even larger number of sources with multiple peaks, like OH19.2 − 1.0 (Chapman 1988). Another group of sources have a broad plateau of OH maser emission like the hybrid object OH231.8 + 4.2 (te Lintel Hekkert *et al* 1988). Even more exciting has been the detection of H_2O and

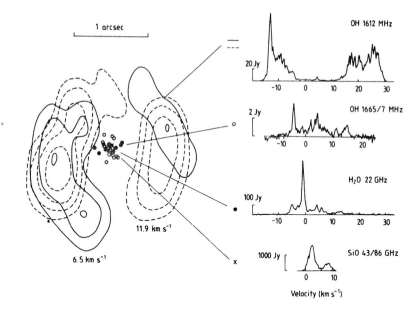

Figure 12 Schematic view of the maser emission from the circumstellar envelope of VX Sagittarii (Chapman and Cohen 1986). Spectra of the different maser lines of OH, H_2O and SiO are shown on the right, and on the left are shown the locations of the different masers in the circumstellar envelope. For the OH 1612 MHz masers only the emission near the stellar velocity is shown, to make the shell structure clear.

OH maser emission from the apparently oxygen-rich envelope of a carbon star (Little-Marenin *et al* 1988). It is likely that the maser emission will continue to play a key role in identifying such unusual phases of stellar evolution.

4. Masers in star-forming regions

Stars are born deep inside dense clouds of gas and dust from which visible light cannot escape. The problem of how they form is still unsolved. The only clues are in the radio, millimetre and infrared radiation that escapes from their surroundings. Most of this radiation is produced by thermal processes, such as thermal continuum emission from hot dust, molecular line emission from gas in LTE, and thermal bremsstrahlung from ionised gas. The non-thermal emission from interstellar masers brings complementary information from compact regions very close to the young star, regions too small to be detected by other means. The picture they present is tantalising and incomplete. We do not see regular shells of maser emission as for circumstellar masers: instead we see a collection of discrete maser spots. This may be due to turbulence, to inhomogeneities in the temperature, density, chemical composition or excitation of the gas, or most probably to some combination of these factors. Nevertheless masers are important probes of star-forming regions because of the unique information they provide, because they can be detected to great distances and because they can be studied with unparalleled precision.

In order to set maser observations in context we first summarise what has been learned about star-forming regions from recent radio, mm-wave and infrared observations. Further details can be found in reviews by Weliachew and Lucas (1988), Shu *et al* (1987) and Lada (1987).

4.1. Overview

Regions of star formation are recognised by activity on many scales. At one extreme are the giant molecular clouds (GMCs), which contain the bulk of the molecular gas in the Galaxy. GMCs are the sites where massive stars form. A typical GMC is $3 \times 10^5 \, M_\odot$ in mass and 30 pc in extent. There is a hierarchy of cloud sizes and masses ranging from GMCs through dark clouds and the condensations or clumps within GMCs ($3 \times 10^3 \, M_\odot$ and 3 pc) down to dense cloud cores which may be as small as $1 \, M_\odot$ and 0.1 pc. These cores often show signs of disc structure, such as flattening and systematic patterns in radial velocity.

Evidence for star formation within molecular clouds takes several forms. Young stars heat their surroundings and so can be identified as compact infrared sources. The more massive stars also ionise their surroundings to produce HII regions, regions of ionised gas (mainly hydrogen and helium) that expand and drive shock waves into the surrounding neutral gas (Yorke 1986). HII regions embedded deep within a molecular cloud are recognised by their radio continuum emission as well as their infrared emission (Habing and Israel 1979, Crawford and Rowan Robinson 1986). The youngest HII regions observed are the ultracompact HII regions, which have sizes of less than ~ 0.1 pc and electron densities n_e exceeding $\sim 10^4 \, \text{cm}^{-3}$. Most young stars also undergo a phase of energetic mass loss, during which a powerful stellar wind interacts violently with the surrounding gas. This produces the so-called energetic molecular outflows (Lada 1985). The outflow is often bipolar with the outflow axis orthogonal to the dense disc or core in which the young star lies. Another related phenomenon is the occurrence of Herbig–Haro (HH) objects, collisionally excited gas detected as optical nebulosity with a characteristic optical spectrum. These HH objects have motions of $\sim 200 \, \text{km s}^{-1}$.

Masers fit into this scheme as the most compact neutral regions that can be observed in the immediate vicinity of a young star or protostar. They occur at distances of only 100–10 000 AU, that is, at solar-system distances. The only other regions that can be studied on this scale at present are the ultracompact HII regions. An example of a star-forming region showing all these signs of activity is Cepheus A. Figures 13 and 14 give a schematic view of some of the different phenomena that are observed in this region.

Because of their high photon rates the masers in such regions are signposts to star formation which can be seen to great distances. Table 1 lists the distances at which Cepheus A could be detected using various tracers. Data for three other star-forming regions are also shown, since no one region could be called typical. With ground-based telescopes the H_2O masers are generally the most readily detectable tracers, followed by the OH masers. The drawback of masers as signposts to star formation is that they are only found in association with the most massive stars ($>8 \, M_\odot$ for OH masers, and $>1 \, M_\odot$ for H_2O masers), and then not always.

Two major goals of star-formation studies are to understand the sequence of events leading from molecular clouds to main sequence stars, and to identify protostars that are still accreting material and that have not yet commenced nuclear fusion. As a

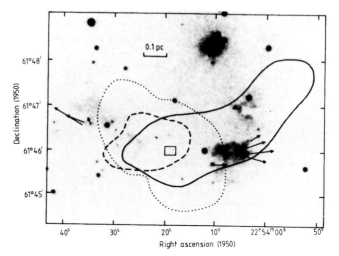

Figure 13. Activity on the arcminute scale associated with star formation in the Cepheus A region. The full and broken curves trace the red- and blue-shifted lobes of a bipolar outflow mapped in CO by Rodriguez *et al* (1980). The dotted curve traces a dense disc, mapped in NH$_3$ by Ho *et al* (1982), which is believed to constrain the outflow. The molecular line data have been superimposed on an optical image from the Palomar Sky Survey. Arrows indicate schematically the proper motions of some of the Herbig–Haro objects (Lenzen 1988). The tangential speeds range up to 250 km s^{-1}. The box indicates the core region where the masers are situated. It is shown enlarged in figure 14.

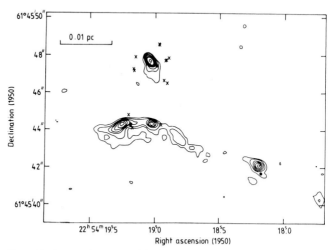

Figure 14. Arcsecond structure of the Cepheus A region, enlarged from the box in figure 13. At the centre of the outflow lie OH ($\times$) and H$_2$O ($\bullet$) masers which surround compact HII regions and young stars. Contours trace the radio continuum emission from the compact HII regions. Adapted from Brebner (1988).

Table 1. Range of detectability for different tracers of star formation.

	Orion-KL	Cepheus A	W3OH	W49
H_2O maser	~1 Mpc (flare)	~30 kpc	~100 kpc	~3 Mpc
OH maser	~10 kpc	~10 kpc	~100 kpc	~1 Mpc
IR 12 μm	~5 kpc	~3 kpc	~30 kpc	~200 kpc
IR 100 μm†	~200 kpc	~100 kpc	~300 kpc	~3 Mpc
Compact HII region	~1 kpc	~1 kpc	~100 kpc	~1 Mpc
Actual distance	0.5 kpc	0.7 kpc	2 kpc	14 kpc

† Balloon or satellite needed.

small part of this problem we need to understand where masers fit into the evolutionary sequence. Are they associated with the accretion phase as well as the outflow phase, and do these two phases overlap? What are the lifetimes of the different masers, and in what order do they occur? The uncertainties in our physical understanding of star formation of course restrict our attempts to model and interpret the associated masers. The present state of our ignorance is reviewed in the following sections. Recent reviews by Elitzur (1986) and Genzel (1986) should also be referred to.

4.2. 18 cm OH masers

The best studied OH masers in star-forming regions are the mainline masers at 1665 and 1667 MHz. They are usually the strongest OH masers in these regions, with the 1665 MHz line stronger in 90% of cases (Caswell and Haynes 1987). Satellite line emission at 1612 or 1720 MHz is generally weaker and has only been detected in ~15% of sources. Some 200 sources are known at present. Their distribution in the Galaxy is discussed by Caswell and Haynes (1987), and their association with far-infrared sources by Cohen *et al* (1988) and references therein. The OH mainline emission is usually strongly polarised. It typically covers a velocity range of ~10 km s^{-1}, although weaker high-velocity masers have now been found in several sources (Cohen and Willson 1981, Caswell and Haynes 1987, Brebner 1988). Accurate positions (better than 1 arcsec) have been measured for about half the known sources (Forster and Caswell 1987 and others).

The longest VLBI baselines are needed to resolve the structure of individual OH maser spots. Typically they have angular sizes of ~1 milliarcsec, corresponding to ~10^{14} cm, velocity widths of 0.3 km s^{-1} (FWHM) and brightness temperatures of ~10^{12} K (Reid *et al* 1980). The spots are spread over OH maser regions some 10^{16}–10^{17} cm in extent. Reid *et al* (1980) first noted a tendency for clustering of spots on a scale of ~3×10^{15} cm, and this has now been confirmed for many sources (Gaume and Mutel 1987, Baart *et al* 1986, and references therein).

4.2.1. Association with compact HII regions.
It is probably safe to assume that all interstellar OH masers are associated with compact HII regions (Turner 1982). The literature is fairly confusing on this point, however. One problem is that the effects of observational sensitivity are not always taken into account properly (see table 1). In the most recent study Gaume and Mutel (1987) found that 20% of the maser clusters in their sample did not have a detectable HII region counterpart at 15 GHz. However,

the regions concerned all lie at distances greater than 1 kpc, so it seems likely that the HII regions are simply very compact ones below the detection level of that survey. The case of W3OH is discussed further in § 4.4. HII regions can help establish suitable conditions for OH masers in at least two ways: by driving shocks into the surrounding molecular gas (Elitzur and de Jong 1978) and by photodissociation (Sternberg 1986). The chemistry in shocked regions and photodissociation zones leads to the enhanced OH abundances of $\sim 10^{-4}$ which all theories of OH masers require (Elitzur 1986).

The presence of a compact HII region is a necessary, but not a sufficient condition for OH maser emission. Only $\sim 20\%$ of all compact HII regions are known to be associated with OH masers (Habing and Israel 1979). Where compact HII regions cluster together, as is often the case, the OH masers are generally associated with the most compact regions, which are younger and denser (Ho *et al* 1983). And finally when the HII region expands to a diameter of ~ 0.1 pc the conditions for OH maser excitation disappear (Habing and Israel 1979). This sets a lifetime of $\sim 10^4$ yr on the OH maser phase.

There has been much debate over the exact location of OH masers relative to their associated HII regions. Garay *et al* (1985) and others have argued that the OH masers occur mainly in projection on the face of the HII regions, whereas Ho *et al* (1983) and others have argued that OH masers generally cluster around the edges of the HII regions. The recent study by Gaume and Mutel (1987) seems to reconcile these differences. The new data show that most OH masers lie around the edges of HII regions, but that the surface density of OH masers actually peaks across the face of the HII region. Gaume and Mutel (1987) found no correlation between the fluxes of the OH masers and their associated HII regions, or between the fluxes of individual masers and their distance from the HII region. Thus the amplification of the continuum background does not appear to affect the OH maser intensity significantly.

Baart and Cohen (1985) noted a tendency for the OH masers to lie at greater distances from more luminous young stars, following a power law $r_{OH} \propto L_*^{1/2}$. This needs to be re-investigated using the more comprehensive data now available. The implication of the result is that OH masers occur in regions of roughly constant temperature ~ 100 K.

4.2.2. Dynamics. There has been even more controversy about the dynamics of OH maser regions than about their relation to HII regions. The radial velocity field of OH maser regions is generally as chaotic as the angular distribution of spots. Nevertheless accretion, outflow and rotation have all been proposed at various times. One of the first questions to settle is whether the OH maser regions are gravitationally bound to the central star and HII region. When this test has been made the answer has generally been that they are not bound (Baart and Cohen 1985, Gaume and Mutel 1987, Brebner 1988, and references therein). The observed velocity dispersion would require a central mass of typically 100–1000 $M_\odot$ to bind the masers, whereas the central stars are usually only ~ 10–30 $M_\odot$ in mass. A qualification that needs to be made to this result is that Zeeman splitting can increase the apparent velocity dispersion (§ 4.2.5). After allowing for this, it is possible that a few maser sources might be gravitationally bound to their central star; however, the majority are not.

The question was approached in a different way by Garay *et al* (1985), who measured HII recombination line velocities of compact HII regions associated with OH masers. They found that the HII regions were generally blue-shifted with respect to the OH masers, and concluded that the OH masers were infalling. Some of the very large blue shifts would require a central mass of $\sim 1000\ M_\odot$! A weak point in their argument is the assumption that the hydrogen recombination lines give the velocity of the central star. Two independent pieces of evidence suggest that they are in fact biased towards blue shifts. Compact HII regions are often optically thick at the frequencies where the recombination lines were measured. This would bias the HII velocity towards the expanding near side of the HII region (which is blue-shifted). Direct evidence for this bias is observed in the source W3OH, where the recombination line velocity systematically approaches the OH maser velocity at high frequencies, where optical depths are less (Welch and Marr 1987). Further evidence of a bias towards blue shifts comes from comparing the velocities of the HII regions with those of the molecular cloud cores in which they are embedded. The recombination line velocities measured by Garay *et al* (1985) are systematically blue-shifted with respect to the molecular cloud cores, whereas the OH maser velocities agree with the molecular cloud velocities (Cohen 1985).

On balance it appears that most interstellar OH masers are located in regions of expansion around compact HII regions. However, it is also possible that a few maser sources might be in an accretion phase. Ultimately we must look to measurements of proper motion of the OH masers to resolve the issue in these doubtful cases.

A few OH maser sources have a regular distribution or velocity field which may indicate disc structure. One of the best examples is the source G35.2−0.7N (Brebner *et al* 1987). Here the OH masers form the innermost part of a large molecular disc of $\sim 150\ M_\odot$ which is orthogonal to a bipolar molecular outflow. The masers reproduce the elongation and position angle of the molecular disc, but on a scale 100 times smaller. This shows that dense neutral clouds near the young star can survive the onset of energetic mass loss. Brebner (1988) has studied a sample of OH maser sources associated with bipolar molecular outflows. The OH masers lie at the centre of the outflow, and there is a general tendency for the maser distribution to be orthogonal to the outflow direction. The OH masers in her sample are all younger than average, being associated with ultracompact HII regions less than 0.01 pc in diameter. It may be that further evolution destroys the evidence for disc structure.

4.2.3. Variability. Most published studies of OH variability have been based on single-telescope observations which show the total emission from all masers in the source (e.g. Sullivan and Kerstholt 1976). In interpreting such data one should be aware that several maser spots may have contributed to the emission at any given velocity. Some general conclusions can be drawn. The characteristic timescale for variation is generally between 1 and 5 yr, but timescales as short as one week have been observed in a few sources (Cohen and Brebner 1985 and references therein). There is no evidence for periodicity; however, some OH maser flares have been observed to recur at similar velocities and presumably at similar locations (e.g. Elldér 1973). Flare activity is often much more pronounced in a single OH transition and even a single polarisation. This must be telling us something very specific about the source, but we lack the tools to interpret such observations at present.

The shortest timescales for variability can be explained if the masers are unsaturated and very compact (Cohen and Brebner 1985). MERLIN observations of several sources show that the most highly variable masers are indeed confined to small localised regions often with very large velocity ranges (e.g. 20–30 km s^{-1}) (Brebner 1988). This suggests that turbulence plays an important part in determining the timescale of maser variability.

4.2.4. Pumping. Theories of the pumping of interstellar OH masers are in a state of some disarray. Until recently collisions with H_2 or electron streams were thought to be the most likely excitation mechanism (Turner 1982, Elitzur 1982). However, experiments by Andresen *et al* (1984) showed that collisions with H_2 actually lead to cooling, not inversion. The energetically lower Λ-doublet state is that in which the unpaired π-electron of the OH is directed towards the H_2. There has been confusion over an electronic phase term in the interaction potential, a point that has since been clarified by Dixon *et al* (1985). The revised conclusions are that collisions with H_2 cannot directly invert the OH mainlines, but could contribute to inversion of the 6 cm excited OH lines.

Andresen (1986) has revived interest in a chemical pump for OH. His laboratory measurements show that photodissociation of cold H_2O leads to inversion of the 18 cm Λ-doublet via radiative cascade. The difficulty with such a chemical pump is well known. Andresen himself estimates that a strong Galactic OH maser could operate for no more than an hour before exhausting the available supply of H_2O molecules. It is unlikely that H_2O can be reformed quickly enough to maintain this pump, particularly as the collisions necessary for molecule formation have the effect of destroying the population inversion.

Radiative pumps were thought to be almost eliminated (Elitzur 1982), but results from IRAS indicate that radiative excitation of OH by far-infrared (FIR) photons is much more significant than appreciated hitherto. OH masers are now known to be associated with strong sources of FIR radiation, and there is a correlation between OH and FIR flux densities. The peak OH flux density does not exceed 10% of the IRAS 60 μm flux density (Cohen *et al* 1988, Moore *et al* 1988). This shows that FIR pumping is indeed a feasible mechanism which needs further investigation.

It may be that the answer lies in some combination of collisional and FIR excitation and de-excitation. Modern computers are bringing such a comprehensive model within sight. At the same time the data against which to test new models are accumulating. Gaume and Mutel (1987) have compiled statistics on frequency of occurrence of the four 18 cm lines in maser clusters and the relative intensities of the lines. When data on the excited OH lines are added it may be possible to characterise the different types of region and identify the underlying physical conditions using such a model.

4.2.5. Polarisation. One of the characteristics of OH masers in star-forming regions is their strong circular polarisation, which often reaches 100%. Linear polarisation is less common and is typically $\sim$30% (Weaver *et al* 1968). Although the polarisation is due to Zeeman splitting it is rare to see a complete Zeeman pattern. The conditions along a line of sight that give velocity coherence in one polarisation are generally wrong for the opposite polarisation (§ 2.4). Fortunately there are a few beautiful exceptions, including the source Cepheus A, which has Zeeman doublets at 1665 and 1667 MHz (Wouterlout *et al* 1980, Cohen *et al* 1984), and Orion-KL, which has a 1665 MHz doublet and a matching 1612 MHz group in which four of the six σ-components can be recognised (Hansen 1982, Hansen and Johnston 1983). Zeeman groups are identified by their close positional coincidence and sometimes by their correlated variability. When such a group is detected it provides a direct measure of the magnetic field strength (typically 1–10 mG) and direction (towards or away from the observer). When Zeeman patterns are not seen the occurrence of circular polarisation can still be used to estimate the magnetic field strength. The field must be strong enough for Zeeman splitting to exceed the maser linewidth (Deguchi and Watson

1986a). This corresponds to a field of 1 mG or more for the OH mainlines. These field strengths are of great astrophysical interest because the magnetic energy density that they imply is comparable with the kinetic and gravitational energy densities in the maser regions (Baart *et al* 1986, Myers and Goodman 1988). Thus the magnetic field will have a significant influence on the dynamics of the region around the young star.

OH masers also give information on the field direction. A Zeeman pair gives not only the field strength but also the sense of the field, either towards or away from the observer. In sources such as W3OH or W75N with several widely separated Zeeman pairs the field is found to be surprisingly uniform across the maser region (Baart *et al* 1986, Garcia-Barreto *et al* 1988). For sources showing linear polarisation it is possible to estimate the orientation of the field to the line of sight. First steps in this direction have been taken by Garcia-Barreto *et al* (1988), who made a full polarisation study of the 1665 MHz lines from W3OH. By assuming that several elliptically polarised components are σ-components, they derive a three-dimensional model of the magnetic field. It will be extremely interesting to extend such studies to other sources, particularly in view of the early suggestion by Davies (1974) that the magnetic field in OH maser sources is parallel to the general Galactic magnetic field, and the more recent result by Cohen *et al* (1984) that bipolar molecular outflows are preferentially aligned parallel to the Galactic field.

4.3. Excited OH masers

Excited OH masers have been somewhat neglected for several reasons. They are scarce, rather weak, and except for the $^2\Pi_{1/2}\,J=\frac{1}{2}$ lines they occur in frequency bands at which few radio astronomy observatories are well equipped. This is unfortunate since these lines may well hold the key to the excitation of OH masers in star-forming regions. Recent work on the $^2\Pi_{1/2}\,J=\frac{1}{2}$ lines suggests that a revival of interest may, however, be underway.

Maser emission has been detected from several excited states of OH. The strongest and most common excited OH masers are the $^2\Pi_{3/2}\,J=\frac{5}{2}$ masers at 6 GHz, with about 30 known sources in star-forming regions (Knowles *et al* 1976, Guilloteau *et al* 1984). The mainlines at 6030 and 6035 MHz are strongest, with the latter always stronger than the former (Rickard *et al* 1975, Guilloteau *et al* 1984). Possible maser emission in the 6049 MHz satellite line has been reported, but no 6016 MHz emission sources are known (Guilloteau *et al* 1984). In general the 6 GHz lines are much weaker than the 1.6 GHz ground-state lines, but there are some exceptions, including one source, M17, which radiates more photons at 6 GHz than at 1.6 GHz (Knowles *et al* 1976). The 6 GHz lines are strongly circularly polarised, with some linear polarisation. The 6035 MHz line often shows an overall separation in frequency between pairs of left and right circularly polarised features, which is suggestive of global Zeeman splitting throughout the source. Zeeman splitting in W3OH has been elegantly confirmed in VLBI measurements by Moran *et al* (1978), who were able to identify ten Zeeman pairs. The data indicate a magnetic field of $\sim$6 mG which is surprisingly uniform across the source. Brightness temperatures of individual maser spots are $\sim 10^{10}$ K. The 6035 MHz masers coincide fairly well with some, but not all, of the ground-state maser clusters.

$^2\Pi_{1/2}\,J=\frac{1}{2}$ masers at 4.7 GHz are weaker and less common than $^2\Pi_{3/2}\,J=\frac{5}{2}$ masers, but they have the advantage that they lie close to well established radio astronomy bands. They have therefore been rather better studied. Only a dozen sources are known at present (Gardner and Martin-Pintado 1983). The 4765 MHz masers are strongest, followed by 4660 MHz masers, and the 4750 MHz line has yet to be detected in maser emission. (There are only three lines since the $F = 0 \to 0$ transition is forbidden.) The 4.7 GHz masers are essentially unpolarised (Zuckerman *et al* 1968). Four sources have been studied interferometrically. VLBI observations of W3OH give brightness temperatures of 10^8–10^9 K for individual maser spots (Baudry *et al* 1988). The spots occur in three tight clusters which coincide fairly well in position and velocity with known clusters of ground-state masers. Two of the three clusters are also close to 6035 MHz masers, suggesting that the excitations of the $^2\Pi_{1/2}$ and $^2\Pi_{3/2}$ masers are not totally dissimilar. VLA observations of the 4660 MHz and 4750 MHz lines show the more extended distribution of warm OH in which these maser clusters are embedded (Guilloteau *et al* 1985). VLA observations of 4765 MHz masers in three other sources also show a reasonable correspondnece with previously known ground-state masers (Gardner *et al* 1983, 1987, Palmer *et al* 1984). The implications of these results for OH maser pumping theories are briefly discussed by Baudry *et al* (1988). However, it must be said that of the currently available data only the VLBI observations of W3OH have the necessary angular and frequency resolution to separate out the individual masers and allow a proper comparison between different lines. More data of this quality are urgently needed.

Only four sources have been detected in the $^2\Pi_{3/2} J = \frac{7}{2}$ excited state of 13 GHz (Turner *et al* 1970, Balister *et al* 1976, Matthews *et al* 1986). In all cases only the $F = 4 \to 4$ transition was detected. The emission from W3OH shows circular polarisation and a frequency splitting between the left and right components. If this is caused by the Zeeman effect then the frequency separation implies a magnetic field of 6 mG directed away from us, which is in excellent agreement with results from the $^2\Pi_{3/2} J = \frac{3}{2}$ and $J = \frac{5}{2}$ transitions (Baudry *et al* 1981).

No masers have been conclusively detected in other excited states of OH, including $^2\Pi_{1/2} J = \frac{3}{2}$ at 8 GHz (Ball *et al* 1970a), $^2\Pi_{1/2} J = \frac{5}{2}$ at 8 GHz (Ball *et al* 1971) and $^2\Pi_{3/2} J = \frac{9}{2}$ at 24 GHz (Baudry *et al* 1981).

4.4. H_2O masers

The strong H_2O masers associated with young stars are spectacular. They have the greatest photon rates, the widest velocity ranges, the greatest variability and the most compact spot sizes of all the galactic masers. Their enormous intensities make H_2O masers the most readily detectable signposts to the formation of massive stars. About 400 sources are known at present. From a pilot survey of 2 deg^2 (Matthews *et al* 1985) it can be estimated that the Galaxy contains several thousand H_2O masers brighter than 1 Jy. To find them all in an unbiased radio survey would be a truly heroic task. Most recent searches have been based instead on infrared surveys, using the close association between H_2O masers and far-infrared sources (Jaffe *et al* 1981, Moorwood and Salinari 1983, Wouterlout and Walmsley 1986, Wilking and Claussen 1987, Wouterlout *et al* 1988, and references therein). Searches for H_2O masers near young visible stars or optical nebulosity have generally had lower detection rates (Haschick *et al* 1983, Rodriguez *et al* 1987), but they are valuable for finding H_2O masers of very low luminosity.

4.4.1. Angular structure. H_2O maser sources found in single-dish surveys are usually resolved by interferometers into well separated clusters of H_2O maser spots. These clusters, termed 'centres of activity' by Genzel *et al* (1978), have physical sizes of

10^{14}–10^{16} cm, whereas the whole H_2O maser complex may span 10^{17} cm or more (Walker *et al* 1982, Forster and Caswell 1987). The centres of activity persist for many years, although individual masers within them have shorter lifetimes. It is still not certain that each centre of activity marks the birthplace of a young star. Some authors have argued that H_2O masers can be excited (collisionally) at large distances from their associated star, in which case a maser cluster would simply indicate a dense molecular condensation in the general vicinity of a young star (Forster *et al* 1978). Ultimately it will require far-infrared observations of high sensitivity and high angular resolution to determine whether each maser centre has its own internal energy source.

It seems more certain that strong H_2O masers are associated with outflow from young stars. It is often difficult to confirm this directly in other molecular lines because of the sensitivity problems discussed in § 4.1. However, there is indirect evidence for outflow. The wide velocity ranges and compact sizes of H_2O maser clusters show that they are unbound. In fact expansional motions of H_2O masers have been directly measured by VLBI in a few sources. In some cases the proper motions approximate to outflow from a single central source (Genzel *et al* 1981a, Reid *et al* 1988). In other cases the proper motions are seemingly random, and individual centres of activity do not show any simple systematic motion relative to each other (Genzel *et al* 1981b, Schneps *et al* 1981). The transverse motions are generally comparable with the radial motions deduced from the Doppler shift, and comparison between the two enables the distance of the source to be estimated (§ 6.2).

VLBI measurements are needed to resolve individual H_2O maser spots. They are typically < 1 milliarcsec in angular extent, with projected physical sizes of 10^{13}–10^{14} cm, typical linewidths of 1 km s^{-1} and brightness temperatures of up to 10^{15} K (Johnston *et al* 1977, Matveenko *et al* 1983).

Attempts to relate the angular distribution of H_2O maser sources to the structure of the star-forming region have had rather mixed success. There have been many suggestions of disc or shell structure, based on the fact that the strongest emission tends to occur near the centre of the profile, and the fact that some sources show symmetric radial velocity structure in their profiles (Knowles and Batchelor 1976, Genzel *et al* 1978, Grinin and Grigorev 1983, and others). However, interferometer maps rarely support such a simple model. The source W49N is one of the few to show a clear kinematic pattern. In this case the red-shifted and blue-shifted masers are displaced from each other, indicating bipolar outflow (Walker *et al* 1982). A statistical analysis of the 386 masers in this source also showed evidence for the hyperfine structure of the H_2O spectrum. Masers close together on the sky have preferred velocity separations of 0.6, 2.7 and 5.5 km s^{-1} in accordance with the hyperfine structure (Walker 1984).

Interest has been growing recently in H_2O maser sources at the other end of the luminosity range. Since the H_2O luminosity correlates roughly with the infrared luminosity of the embedded star (Moorwood and Salinari 1983) these are also the H_2O masers associated with the stars of lowest mass. The H_2O spectra are simpler and more variable than those of strong sources. When several masers are seen the velocity and spatial separations again indicate that the masers are not bound to the central star (Rodriguez and Canto 1983, Rodriguez *et al* 1987, Wilking and Claussen 1987). The separations of some of these weak H_2O masers from their stars also pose pumping problems as acute as those for much stronger sources.

4.4.2. Association with OH masers and compact HII regions. H_2O masers are often the only measurable indicator of star formation in distant parts of the Galaxy. This poses severe problems of interpretation when one comes to consider their relation to OH masers, compact HII regions, FIR sources and other tracers of star formation. Consider the case of W3OH, a source which is often taken as a prototype. Figure 15 shows the location of the OH masers, which cluster mainly around the HII region, and the H_2O masers, which lie well off to one side. The large separation between the OH and H_2O masers was a puzzle for many years and led to many interesting speculations about the nature of H_2O masers. The problem of pumping H_2O masers at such apparently large distances from the central star was so acute that new theories were developed (e.g. Norman and Silk 1979). Now, however, it appears that there is actually a second young star at the position of the H_2O masers. Turner and Welch (1984) have detected a warm dense concentration of molecular gas, marked by broken contours in figure 15, and Gaume and Mutel (1987) have detected weak OH masers. Presumably an ultracompact HII region also awaits detection. However, it could be many years before

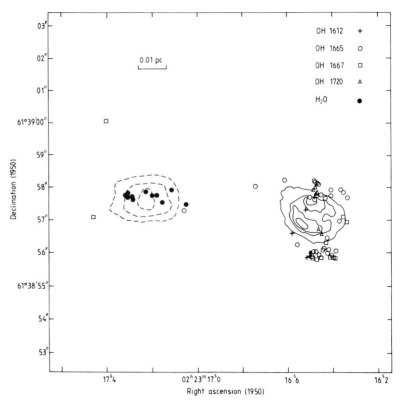

Figure 15. Schematic view of the star-forming region W3OH, showing the two active centres. Full contours show the radio continuum emission from the main HII region, which is surrounded by powerful OH masers (Gaume and Mutel 1987). Broken contours show the HCN emission from the warm molecular envelope surrounding a second much younger star which is associated with powerful H_2O masers and weak OH masers. (Turner and Welch 1984, Reid and Moran 1981).

far-infrared detectors achieve the angular resolution and sensitivity necessary to distinguish the second source from its more evolved neighbour.

Now that the second source has been recognised it is clear that the H_2O masers actually lie closer to it (in projection) than do the weak OH masers found by Gaume and Mutel (1987). This configuration, which also occurs in Cepheus A (figure 14), is reminiscent of the situation of OH and H_2O masers in circumstellar envelopes. However, one cannot answer the question of whether this is the general case until one has the means to identify all the young stars in a given region. The distribution of OH and H_2O masers alone does not uniquely define the positions of the young stars. Consider as a second example Cepheus A (figure 14). At its present distance we recognise four centres of maser activity, each associated with a compact HII region. If this source were only three times as far away the HII regions would be undetectable and the structure of the region would be open to gross misinterpretation. Against this background it appears that we have a long way to go before the relationship of H_2O masers to OH masers and compact HII regions and FIR sources is clarified. However, readers may wish to consult Genzel and Downes (1977), Forster and Caswell (1987) and others for a less pessimistic view.

4.4.3. Variability. The strong variability of H_2O masers became apparent soon after their discovery (Knowles *et al* 1969). Individual components in the spectra vary in intensity, frequency and linewidth on timescales that can be as short as days (Sullivan 1973). The amount of linear polarisation (when present) also varies. Observed lifetimes of individual H_2O masers range from weeks to years. There have been many cases of 'flare' activity, including the celebrated outburst in Orion-KL which exceeded 10^6 Jy (Abraham *et al* 1981). A summary of the most rapid variations so far recorded is given by Rowland and Cohen (1986).

Because of the large numbers of masers that are sometimes present in a source at similar velocities, blending problems can be severe. It is usually necessary to use an interferometer to study the behaviour of individual masers. Interferometric studies have established that individual maser spots can change their frequency, linewidth and intensity without any appreciable change in their angular position (Rowland and Cohen 1986). The frequency changes so far observed in this way are small, corresponding to Doppler shifts of ~ 0.5 km s^{-1}. These are small enough to be accounted for by redistribution of maser gain among the hyperfine components of the H_2O spectrum, as originally suggested by Sullivan (1973). Linewidth changes are sometimes related to intensities through a power law $\Delta V_{1/2} \propto I^{-1/2}$, which indicates saturated masers (Mattila *et al* 1985, Rowland and Cohen 1986). However, some masers vary without appreciable changes in linewidth, and the flare in Orion-KL was actually accompanied by an increase in linewidth (Abraham *et al* 1981).

Some examples have been reported of correlated variations in spatially separated masers (Gammon 1976, White 1979, Rowland and Cohen 1986). In each case several masers underwent essentially identical intensity variations, separated by a few days. The possibility of the masers amplifying a common flickering background source can be ruled out since only some masers of a cluster were affected. This leaves a common pumping 'event' as the likely cause of the correlation. The near-synchronous behaviour of the separated masers then implies that the pumping event was radiatively coupled. This observation is a crucial test of pumping theories, and therefore it is important to repeat the experiment.

4.4.4. Pumping. The problems of pumping H_2O masers in star-forming regions are well known. First there is the high luminosity of the strongest sources. The H_2O maser photon rates of events like the Orion-KL flare far exceed the continuum photon rates in any Doppler-equivalent band. Purely radiative pumps thus appear to be eliminated entirely, unless excitation by line radiation is important. The high brightness temperatures, sometimes exceeding 10^{15} K, also eliminate radiative pumps. The third consideration is the specific pump power. Strelnitskij (1984) estimates that typically 10^3 inversions per second per cubic centimetre are required for a representative sample of bright sources, not just the most luminous it should be added. He shows that most pumping models fail by several orders of magnitude to meet this requirement. For example, the scheme proposed by Cooke and Elitzur (1985) for circumstellar H_2O masers fails because the collision rates cannot be increased indefinitely without thermalising the level populations (§ 3.5.3). The same difficulty arises with the hot-dust-cold-gas scheme proposed by Goldreich and Kwan (1974b) and revived by Tarter and Welch (1986). This particular model has some attractive features, however. It relies on cloud-cloud collisions to supply infrared radiation which shines further into the cloud and excites the H_2O molecules. Collisions with cool H_2 molecules complete the cycle. The model predicts realistic lifetimes for H_2O maser features. It is well able to reproduce the most rapid timescales for variability, and it can also account for correlated variations between spatially separated masers. Nevertheless it fails by several orders of magnitude to achieve the high brightness temperatures and specific pump powers of the brightest sources.

The only type of model that can meet all the power requirements is one involving collisions between different types of particle. Strelnitskij (1984) and Kylafis and Norman (1987) consider a model in which H_2O molecules are excited by collisions with cold H_2 molecules and with hot electron streams. This type of model has the advantage that thermalisation is impossible! The higher the collision rates the greater the inversion (provided only that the ratio of collisions with ions and neutrals is maintained). Kylafis and Norman (1987) discuss the conditions that are required for maser action and argue that these conditions occur naturally in the magnetic precursors of magneto-hydrodynamic shocks. They also point out the possible application of the model to the powerful H_2O masers in external galaxies. Some details of the model may need to be confirmed when accurate collision cross sections for H_2O and H_2 molecules are available: data available at present are for H_2O and He. Nevertheless the general viability of this scheme is not in doubt. The model provides no natural explanation for correlated variability, and it remains to be seen whether this is a serious limitation or not.

4.4.5. Polarisation. H_2O masers in star-forming regions show some linear polarisation but circular polarisation has not been detected to a limit of 1% in the best cases (Bologna *et al* 1975). The true incidence of linear polarisation is difficult to establish since all large-scale studies have used single-telescope data, with the attendant problems of blending. One quarter of sources studied in this way show linear polarisation, but three quarters of the strongest sources are polarised in at least one feature (Knowles and Batchelor 1978). Typical levels of polarisation are only a few per cent, but much higher levels of polarisation have been detected in some masers, ranging up to 50% in W75S (Bologna *et al* 1975) and at least 60% in the Orion-KL flare (Abraham *et al* 1981, Matveenko *et al* 1983). The high polarisation of the Orion-KL flare is close to the maximum that can be produced by a magnetic field alone, and it implies that the masers are heavily saturated. A magnetic field of at least 0.1 G would be necessary (Deguchi and Watson 1986b).

4.5. Methanol masers

Interstellar methanol (CH_3OH) was first detected at 834 MHz by Ball *et al* (1970b) and has now been detected in well over 100 different transitions (Johansson *et al* 1984, Menten *et al* 1986b, Batrla *et al* 1987, and references therein). It is a widespread and abundant molecule with a particularly rich microwave spectrum (Lees *et al* 1973, Sutton and Herbst 1988). The complexity of the spectrum is due to the interaction between rotation and torsional motion or hindered internal rotation. In simple terms there are three potential wells for the OH relative to the CH_3. This leads to three types of methanol (E, A^+ and A^-). Both the ground and the first torsionally excited states have been observed in the interstellar medium. Part of the energy level diagram for E-type methanol in the torsional ground state is reproduced in figure 16.

The first indication of maser activity came from observations of narrow linewidths and time variability in the $\Delta J = 0$, $K = 2 \rightarrow 1$ (E) series of lines at 25 GHz from Orion-KL (Chui *et al* 1974, Barrett *et al* 1975). Interferometric observations of the $J = 6$ and $J = 7$ transitions showed the emitting regions to be less than ~ 10 arcsec in extent with brightness temperatures exceeding ~ 1000 K (Matsakis *et al* 1980). The methanol masers lie in the same region as the OH and H_2O masers but show no detailed correspondence with these or with other active sites in the Orion-KL region.

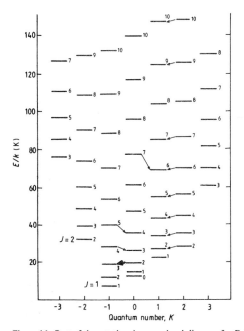

Figure 16. Part of the rotational energy level diagram for E-type methanol in the torsional ground state. The astronomically observed maser transitions are indicated by arrows. The strong 12 GHz maser (2_0-3_{-1}) is indicated by a bold arrow.

Until recently Orion-KL was the only known methanol maser source, despite extensive searches (e.g. Buxton *et al* 1977). Then several new sources and new maser lines were discovered in rapid succession. The 9_2-10_1 A^+ transition at 23 GHz was detected as a powerful maser in W3OH and two other sources by Wilson *et al* (1984). From interferometer measurements the individual masers were found to have angular sizes less than 1 arcsec and brightness temperatures exceeding 2×10^4 K, and to be located in the same cloud as the OH masers (Menten *et al* 1985). An even stronger maser, the 2_1-3_0 E transition, was found in W3OH (Wilson *et al* 1985), mm-wave masers were reported in Sgr B2 and other galactic sources (Morimoto *et al* 1985, Plambeck and Wright 1988), and a sensitive survey of the 25 GHz $\Delta J = 0$, $K = 2 \rightarrow 1$ series of lines revealed several sources of strong narrow-line emission similar to Orion-KL (Menten *et al* 1986a). Curiously these last sources, termed Class A by Batrla *et al* (1987), are sometimes located at large distances (~2 pc) from the known OH and H_2O masers and compact HII regions in their vicinity. The real breakthrough came when Batrla *et al* (1987) discovered widespread and powerful masers in the 2_0-3_{-1} E transition at 12.2 GHz.

The 2_2-3_{-1} E methanol masers are comparable in their photon rate with galactic OH and H_2O masers, and promise to be a valuable new probe of star-forming regions. More than 40 sources are known at the time of writing and all are closely associated with compact HII regions and OH masers. They have been found in 30–40% of the OH/HII regions searched to date (Batrla *et al* 1987, Norris *et al* 1987, Koo *et al* 1988, Kemball *et al* 1988b). The velocity ranges of the methanol and OH masers are generally similar. Several of the 2_0-3_1 E methanol sources have flux densities of order 1000 Jy. Small amounts of linear polarisation have been detected in some sources (Koo *et al* 1988).

Interferometer maps by Norris *et al* (1988) show that the methanol 12 GHz masers are spread over a region of similar angular extent to their OH maser counterparts. However, there is no detailed correspondence in position and velocity between the two maser species. Zeeman splitting of the OH lines will of course make such a comparison difficult. In several cases the methanol masers lie in arcs or lines, which Norris *et al* speculate might trace shock fronts. The first VLBI maps of the 12 GHz masers confirm the close association between OH and methanol masers (Menten *et al* 1988). The methanol maser spots cluster on the same scale as the OH, and absolute positions of the two types of cluster coincide. Evidently we are seeing different masers excited in adjacent regions of the same gas cloud. Individual 12 GHz maser spots have angular diameters less than 2 milliarcsec and brightness temperatures exceeding 2×10^{10} K. The strength of the methanol 12 GHz line and the complexity of the spectrum may offer possibilities for proper motion studies. The advantages of the new line over the OH 18 cm lines are its greater flux density and the greatly reduced effects of interstellar scattering at the higher frequency.

Clues as to the excitation of methanol masers come from the multi-transition study of E-type methanol masers by Menten *et al* (1986a). These authors were led to consider far-infrared excitation to explain the 9_2-9_1 E transition, which is 117 K above the ground state (figure 16). The far-infrared selection rule $\Delta K = \pm 1$ but $\Delta K \neq 0$ has the effect that radiative excitation and decay leads to a total change of $\Delta K = \pm 2$ or 0. Hence $\Delta K = 1$ transitions become extremely sensitive to small differences between the effective temperatures of the even-K and the odd-K ladders. Menten *et al* (1986a) argued that far-infrared excitation via the 50 μm torsional bands can pump the series of $K = 2 \rightarrow 1$ E masers that are observed, and that other K(even) $\rightarrow K$(odd) E masers (such as 2_0-3_{-1})

might be expected. Their prediction was confirmed within a year. In further support of infrared pumping it is notable that the IRAS 60 μm flux densities of the known methanol maser sources handsomely exceed the flux densities of the maser lines (all transitions), indicating that sufficient 50 μm photons are available for pumping (Kemball *et al* 1988b). However, given the complexity of the methanol spectrum, it may be some time before a detailed theoretical study of the far-infrared and collisional excitation is made.

4.6. SiO masers

The situation regarding SiO masers in star-forming regions is similar to that of methanol masers a few years ago. Until recently there was only one known example, Orion-KL. All other known SiO masers were circumstellar. Lines up to the $V = 2$ vibrational state and the $J = 3$-2 rotational state have been reported, including a possible $V = 0$, $J = 2$-1 maser in the isotopic species ^{29}SiO (Olofsson *et al* 1981).

The SiO maser emission from Orion-KL bears such strong resemblances to that from OH-IR sources that some authors have questioned the nature of the central infrared source, IRc2, now firmly believed to be a young star. The SiO masers coincide closely with IRc2 to within the experimental errors (Wright and Plambeck 1983) and produce a twin-peaked emission line characteristic of an expanding shell. The masers have a compact distribution some 10^{15} cm in extent, similar to that of circumstellar SiO masers (Lane 1982). VLBI measurements reveal maser hotspots with brightness temperatures of up to 5×10^9 K, which are very similar to those of circumstellar SiO masers (Lane 1982). A suspected three-year periodicity in the emission has also been reported (Nyman and Olofsson 1986). Thus on the basis of the SiO maser emission alone it would be difficult to distinguish the Orion-KL source from an OH-IR source, despite their very different evolutionary states.

One striking feature peculiar to Orion-KL is its polarisation. Linear polarisation of some 5% occurs across the profile with a beautifully regular pattern. The position angle rotates regularly across one peak and then decreases symmetrically across the other peak (Barvainis and Predmore 1985). This regular behaviour can be modelled in terms of a maser disc (Barvainis 1984). It is unlike anything seen in circumstellar SiO masers.

Ukita *et al* (1987) have recently reported two new examples of SiO masers in the star-forming regions W51N and Sgr B2. These new sources show none of the regularity of Orion-KL. Clearly we still have much to learn about the phenomenon.

4.7. Interstellar scattering

The small spot sizes of masers in star-forming regions make them potentially valuable probes of the intervening interstellar medium. Irregularities in the density of free electrons scatter radio waves and increase the apparent angular diameter of compact radio sources such as masers. The scattering effects scale roughly as λ^2 and so should be most readily detected in OH masers at 18 cm wavelength. Early VLBI measurements showed that OH maser spot sizes are indeed greater in more distant sources (Burke *et al* 1968), and more recent VLBI data confirm this (Kent and Mutel 1982, Dennison *et al* 1984, Kemball *et al* 1988a, and references therein). The sizes of the OH maser spots are consistent with predictions based on measurements of interstellar scattering in pulsars and compact extragalactic radio continuum sources. VLBI data on a larger number of OH maser sources should ultimately lead to a better model for the distribution of scattering electrons in the Galactic plane. Other strong maser lines of excited OH, methanol and water may also allow the wavelength dependence of the broadening to be examined.

Turbulence in the interstellar medium should also produce other scattering effects, including intensity variations (which may be difficult to distinguish from intrinsic maser variations), and small shifts in the apparent positions of maser spots. It is remarkable that H_2O proper motion measurements already set useful limits on the size of this latter effect, in terms of the deviations of H_2O maser spots from rectilinear motion (Gwinn *et al* 1988).

5. Extragalactic masers

The study of extragalactic masers is still in its infancy but holds great promise. One exciting prospect is to use VLBI to measure the motions of masers in nearby galaxies and so determine their distances, as discussed later (§ 6). Another challenge is to understand the recently discovered megamasers. These are the most powerful cosmic masers known and they reside in a new class of active galaxy.

5.1. OH masers in external galaxies

At the distances of even the nearest galaxies normal Galactic OH masers would be barely detectable with modern receivers. There have been several searches (e.g. Radhakrishnan 1967, Roberts 1967, Fix and Mutel 1985) but so far only one normal extragalactic maser has been found. This is the OH-IR source in the Large Magellanic Cloud (Wood *et al* 1986). The other extragalactic OH masers that have been discovered are all freaks in one way or another.

Extragalactic OH was first detected in absorption against strong radio continuum sources in the nuclei of the two galaxies M82 and NGC253 (Weliachew 1971). Superimposed on the broad absorption lines are narrow emission features which were correctly interpreted as maser emission by Whiteoak and Gardner (1973) and Nguyen-Q-Rieu *et al* (1976). These masers are two orders of magnitude more powerful than the strongest Galactic OH masers. In their OH properties they are closer to type I OH-IR sources than to OH-HII sources. They have no detectable polarisation and emit most strongly in the (red-shifted) 1667 MHz line. Whiteoak and Gardner (1973) concluded that the masers are qualitatively different from Galactic sources, and that the emission is produced by maser amplification of the nuclear radio continuum source. The two galaxies NGC253 and M82 share other unusual properties: high infrared luminosities, powerful radio continuum emission, large UV fluxes and x-ray luminosities. These are thought to indicate a 'starburst', a period of intense activity during which many massive stars were formed (Rieke *et al* 1980).

Despite the strong similarities in the integrated properties of these extragalactic masers, radio interferometer maps have revealed astonishing differences between them. The masers in M82 have been resolved into two compact objects each less than 8 pc in extent (Weliachew *et al* 1984). The implied brightness temperatures exceed $\sim$500 K. The masers coincide with two of the compact radio continuum sources in the obscured nucleus (Kronberg *et al* 1985). Both masers are unpolarised. It has been suggested that each could be composed of some hundreds of 'normal' polarised masers, like those seen in Galactic OH-HII regions (Weliachew *et al* 1984). VLBI observations combining high angular resolution with high velocity resolution will be necessary to establish if this is so. It seems unlikely at present because of the dominance of the 1667 MHz line.

The distribution of OH in NGC253 has been mapped by Turner (1985). The narrow 1667 MHz emission spike detected by Whiteoak and Gardner (1973) is unresolved and coincides with the radio nucleus. This may be similar to the M82 masers: the brightness temperature, corrected for beam dilution, is at least 1000 K. In addition to the maser spike broad emission is detected at all velocities! The emission cannot be distinguished in single-telescope measurements because of blending between emission and absorption at different positions within the single-telescope beam. Turner's observations separate the emission and absorption regions spatially, and reveal broad anomalous OH emission in all four ground-state lines. The broad emission at 1667, 1665 and 1612 MHz is distributed in a spectacular plume extending to a projected distance of 1.5 kpc from the nucleus, whereas the emission at 1720 MHz comes from a separate region on the opposite side of the nucleus, as shown in figure 17. Two isolated regions of 1612 MHz emission are also seen north of the nucleus, with no counterparts in the other lines. The apparent brightness temperatures of the extended emission regions are typically 10 K, and they bear no obvious relation to the 18 cm continuum emission. From this Turner concludes that the excitation temperature $|T_{ex}| \gg T_c$ and hence that $|\tau| \ll 1$. As yet there is no direct evidence for population inversion ($\tau < 0$), but it seems likely that these are the largest masers known to man.

Turner suggests that the source of the anomalous excitation of these OH regions is the powerful far-infrared emission from the nucleus of NGC253. Pumping via collisions, near-infrared or ultraviolet radiation can be ruled out because the rates are orders of magnitude short of what is required. This leaves far-infrared pumping as the only plausible mechanism, despite the fact that none of the far-infrared pumping schemes proposed for Galactic OH can explain the relative intensities of all the anomalously excited lines in NGC253 (Turner 1985).

5.2. OH megamasers

One of the most surprising developments in maser studies has been the discovery of OH megamasers. These are OH 18 cm masers which trace a new type of active galactic nucleus. As their name implies, megamasers are $\sim 10^6$ times more powerful than the strongest OH masers in our own Galaxy. The prototype is in Arp 220 (IC4553), a peculiar galaxy with a double radio source in its nucleus. The OH megamaser was discovered during a search for OH absorption by Baan *et al* (1982). Attempts to detect other megamasers in galaxies with radio or optical similarities to Arp 220 had only limited success (Baan 1985). The breakthrough came when it was realised that the megamaser galaxies are first and foremost infrared galaxies with enormous far-infrared luminosities of 10^{11}-$10^{12} L_\odot$. Using the IRAS database and optical redshifts it is possible to select infrared-luminous galaxies for a much better targeted radio search. This approach has been exploited very successfully at Nancay (e.g. Bottinelli *et al* 1985, 1986, 1987). A related method uses the distinctive far-infrared colours of the megamaser galaxies (Unger *et al* 1986). Radio searches based on the IRAS colours achieve a success rate of 20–30%, which is similar to that of OH-IR source surveys. For example, the 52 objects colour-selected by Staveley-Smith *et al* (1987) are known at present to contain ten megamasers, two powerful extragalactic masers and four further galaxies detected in OH absorption.

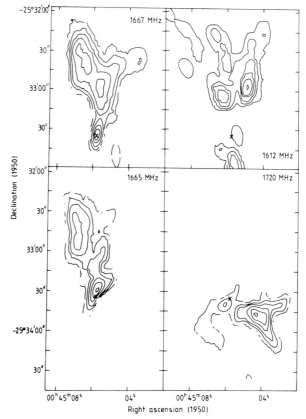

Figure 17. Integrated emission from high-velocity OH emission plumes in the nucleus of the galaxy NGC253 (Turner 1985). Emission occurs in all four ground-state lines. The cross marks the position of the galactic nucleus. At the distance of NGC253 10 arcseconds corresponds to a projected separation of 150 pc.

OH megamasers are characterised by broad OH mainlines, with typical velocity widths of ~ 300 km s^{-1} and OH luminosities of $\sim 10^3\, L_\odot$. Because of these extraordinary luminosities OH megamasers can be detected to great distances. Already two have been found which are red-shifted from 18 cm to 21 cm wavelength. Emission from the red-shifted 1667 MHz line is usually strongest, and there is no detectable polarisation. Other OH lines have been measured in a few cases: red-shifted 1612 MHz and 1720 MHz lines in emission from Arp 220 (Baan and Haschick 1987), rotationally excited OH 5 cm lines in absorption in Arp 220 (Henkel *et al* 1986a) and 6 cm lines in absorption in Arp 220 and four other galaxies (Henkel *et al* 1987). More effort may be expected in this area once the search programmes are completed. The megamaser galaxies all have non-thermal radio continuum sources in their nuclei, with similar flux density to the OH main lines. This is generally assumed to mean that the masers have low gain and amplify the nuclear continuum radiation. However, VLBI data to support this assumption are still awaited.

OH maps have been published for only two megamasers to date. In Arp 220 the OH mainline emission is from an elongated region in the nucleus some 500 pc in extent (Baan and Haschick 1984, Norris *et al* 1985). A regular velocity gradient across the region suggests that we are seeing OH maser emission from an edge-on nuclear disc of molecular clouds executing normal galactic rotation. However, given the unusual morphology of the galaxy, and the problems of identifying a central velocity, other more exotic explanations (outflow or accretion for example) cannot be excluded. A weaker feature at high velocity does not participate in this regular motion. For Arp 220 the OH masers follow the radio continuum distribution of the active nucleus. In Markarian 273 this is not so (Schmelz *et al* 1987). Here the OH emission comes from an unresolved region no more than ~100 pc in extent, which coincides with the northern component of the double radio source in the nucleus of this Seyfert galaxy. It seems likely that as with Arp 220 the broad linewidth is at least partly due to large-scale motions of molecular clouds in the galactic nucleus, but VLBI observations are needed to establish this. The OH emission from Markarian 273 contains high-velocity wings as well as a broad central peak (Staveley-Smith *et al* 1987). These wings extend to velocities at least 700 km s^{-1} away from the systemic velocity of the galaxy. Such motions cannot arise from normal galactic rotation but may be related to the Seyfert activity in the nucleus.

To date there are no measurements of the distribution of 18 cm satellite line emission or excited OH absorption in any of the megamaser galaxies. These could prove to be key diagnostics of the megamaser pump. The fact that OH megamasers occur in the most powerful infrared galaxies we know of points directly to infrared pumping. The implied pumping efficiency of ~0.1–1.0% is similar to that for OH mainline masers in OH-IR sources and in star-forming regions (Unger *et al* 1986). Indeed megamasers are similar in many ways to OH-IR mainline sources, scaled up by ~10^8 in luminosity and ~10^4 in linear extent. This is not so fanciful as might be first thought. All the extragalactic OH mainline masers have predominantly unpolarised 1667 MHz emission, and there is a progression in maser luminosities from the galactic sources through the powerful extragalactic sources M82, NGC253 and more recently NGC520 (Baan *et al* 1985), with 10^{-3}–10^{-2} $L_\odot$, to the so-called 'dwarf' megamasers with 1–10 $L_\odot$ (NGC4418, NGC3690, IRAS 1510+0724), and finally to the true megamasers with ~1000 $L_\odot$. Thus it seems natural to seek a unified explanation of all these masers in terms of a single infrared pump which operates over successively larger regions of dust and gas as the luminosity of the central source is larger (Mirabel and Sanders 1987).

Many authors consider the central non-thermal radio source in megamaser galaxies to play a fundamental role. Typically the OH line to 18 cm continuum ratio is of order unity, implying equal photon rates. Henkel *et al* (1987) have proposed a pumping scheme in which the non-thermal radio emission transfers OH across the Λ-doublets in rotationally excited states. This leads to 18 cm maser inversion and 6 cm anti-inversion. Clearly interferometer maps of the 6 cm absorption lines will be a crucial test of this model, and indeed of the OH excitation.

It is difficult to separate the problem of the OH megamasers from that of understanding the infrared-luminous galaxies with which they are associated. The galaxies have very disturbed appearances indicative of tidal interactions or mergers. The nuclei are obscured by large dense clouds of gas and dust ~1 kpc in extent, within which most of the infrared power originates (Scoville *et al* 1986, Sargent *et al* 1987). These phenomenal concentrations of gas have masses of 10^9–10^{10} $M_\odot$, which would normally be spread over an entire galaxy (Sanders *et al* 1988). It seems likely that the gas fuels the nuclear activity in some way, but the nature of the central engine is still an open question. More multiwavelength studies such as those by Mirabel *et al* (1988) and Martin *et al* (1988) will be essential steps towards understanding this new phenomenon.

Postscript. About 20 OH megamasers have been reported to date, and undoubtedly many more await discovery. It is unfortunate that search programmes are increasingly hampered by radio interference. The large red-shifts of the megamaser galaxies take them out of the frequency bands protected for radio astronomy. No account of megamasers would be complete without mention of GLONASS, the Russian global navigation satellite system, which occupies a frequency band below the OH mainlines. The spread-spectrum of the GLONASS transmissions causes disruption to measurements of megamasers over ~ 100 MHz. The satellite system is visible worldwide. When it is complete, ground-based observations of many OH megamasers will be impossible.

5.3. H_2O masers and supermasers

Several nearby galaxies are now known to contain H_2O masers similar to the strong H_2O masers in our Milky Way (Churchwell *et al* 1977, Huchtmeier *et al* 1978, 1988, Henkel *et al* 1986b, Whiteoak and Gardner 1986, and references therein). Most were found at the positions of known HII regions and are believed to trace the formation of massive stars. Some have already shown variability. With current instrumentation it is possible to detect such masers at distances up to ~ 10 Mpc (Ho *et al* 1987). For very nearby galaxies, such as M33, several masers have been detected, and their luminosity distribution can be compared with that of H_2O masers in our own Galaxy (Whiteoak and Gardner 1986).

A small number of galaxies have been found to contain H_2O masers two orders of magnitude more powerful than any in our Galaxy (dos Santos and Lépine 1979, Gardner and Whiteoak 1982, Claussen *et al* 1984, Henkel *et al* 1984, Haschick and Baan 1985). These are sometimes termed supermasers. They have broad linewidths ($\Delta V > 100$ km s^{-1}), large luminosities ($L_{H_2O} > 100$ $L_\odot$) and no detectable polarisation. So far they have only been found in the nuclei of 'active' galaxies (classified as either Seyferts or LINERS). The galaxies are luminous far-infrared sources, but there is no distinctive FIR signature as for OH megamasers (Henkel *et al* 1986b), nor have any H_2O masers been found in the OH megamaser galaxies. In the three sources that have been mapped at the VLA the H_2O masers were unresolved (physical size < 1 pc) and coincident with a compact radio continuum source in the nucleus (Claussen and Lo 1986, Haschick and Baan 1986). This is remarkable in the case of NGC1068, where most of the molecular clouds are concentrated in a large ring 3 kpc in diameter around the nucleus, but very little molecular gas lies near the nucleus itself (Myers and Scoville 1987). The compact nature of all three maser sources rules out an early suggestion that they consist of ~ 100 W49N-type masers associated with a 'starburst'. A starburst origin is also ruled out by the correlated variability which has been observed across the spectrum of NGC1068 (Claussen and Lo 1986), which implies a common maser pump for most if not all of the emission.

It is not yet clear that the powerful H_2O masers truly represent a new phenomenon. Only two orders of magnitude separate them from W49N in our own Galaxy. The H_2O maser in the Circinus galaxy is only one order of magnitude more powerful than W49N. On the other hand the occurrence of dense neutral clouds so close to an active nucleus might easily lead to physical conditions qualitatively different from those in star-forming regions. A common factor in the two environments is probably the interaction between dense clouds and a high-velocity wind. The detection of infrared lines from shocked H_2 provides direct evidence for such an interaction in NGC1068 (Hall *et al* 1981). Collisional pumping such as has been proposed by Kylafis and Norman (1986) might easily be adapted to these powerful extragalactic masers, and would require only a modest scaling up in size. It remains to be seen whether VLBI measurements can exclude this possibility.

6. Masers and the cosmic distance scale

The determination of cosmic distances is of fundamental importance to astronomy. Whereas the angular positions of stars and galaxies are known to very high precision their distances are often accurate to no more than 50%. Masers offer several possibilities for distance measurement. The advantage of these new methods is that they are independent of the assumptions and calibration difficulties of older established techniques. With some refinement the maser techniques may give distances to 5% or better by direct means.

6.1. OH-IR sources

The distance to an OH-IR source can be found by comparing the linear size of the OH maser shell, determined from phase-lag measurements, with the angular size obtained from radio interferometer maps (§ 3.3). There are published distances for 15 sources to date (Herman *et al* 1985, Diamond *et al* 1985b). The best measurements have a precision of ~5%. A further ~100 sources are being monitored for phase lags at Dwingeloo and Hartebeesthoek. These are long-term programmes requiring regular observations over many years. GLONASS transmissions at 1612 MHz present a mounting problem.

Some possible sources of systematic error need further consideration (Herman and Habing 1987). The OH-IR method assumes a spherically symmetric, uniformly expanding shell. Two simple tests will show how well these assumptions are met. Firstly the phase lag should increase linearly with velocity across the OH spectrum (§ 3.3). This is a difficult measurement to make because the OH emission is very weak near the central (stellar) velocity, and reliable data may be obtained across less than half the velocity range. Results to date are somewhat disappointing in this regard (Herman and Habing 1985b, their appendix B). It may be that more sensitive measurements with a large telescope will ultimately be needed. Secondly the roundness of the shell can be checked from the radio maps both in the plane of the sky (channel maps at constant velocity) and in three dimensions by examining the fit to a thin-shell model. Large-scale asymmetries in the OH envelope of the source OH26.5 + 0.6 were found in this way by Baud (1981). These two tests should also detect other more subtle sources of error, such as the effects of beam size on apparent shell size (§ 3.3.1) and the effects of saturation and beaming (§§ 2.2 and 3.8). In the latter case the problem is that phase lags are determined mainly using emission from the front and back of the shell, which is beamed radially, whereas angular sizes are determined from emission nearer the stellar velocity, which is directed tangentially to the shell and which may arise at a different distance from the star. It remains to be seen whether these possible sources of error can be detected and eliminated.

The OH-IR technique already has a contribution to make to the determination of the Galactic distance scale. There is a large concentration of OH-IR sources in the Galactic centre (Winnberg *et al* 1985), and the first attempts are underway to apply the phase-lag technique to a sample of these sources, and so obtain the distance to the Galactic centre. This distance can also be found in a less direct way by studying OH-IR sources in the Galactic disc (Herman *et al* 1985), and by studying OH Mira variables, some of which are already used for distance estimation through their period–luminosity relationships (Feast 1987). The detection of the first OH-IR source in the Magellanic Clouds (Wood *et al* 1986) suggests that the method may one day find even wider application.

6.2. *H_2O proper motions*

One of the highlights of VLBI has been the detection of proper motions of H_2O masers in star-forming regions. By comparing these proper motions, which are measured in milliarcsec yr^{-1}, with the radial velocities, the distance can be found using classical techniques of optical astronomy. If the proper motions are random the dispersion in radial velocities is compared with the dispersion in transverse velocities. The error therefore goes as $N^{-1/2}$, where N is the number of maser features. This is the method of *statistical parallax* (Trumpler and Weaver 1953). It has been applied successfully to the sources W51 Main and W51 North, giving distances to ~20% (Genzel *et al* 1981b, Schneps *et al* 1981). If the proper motions are not random but show systematic effects then in principle a more accurate result should be obtainable by modelling the source. Expansion motions of H_2O masers in the sources Orion-KL and Sgr B2 North have been modelled using the method of *expanding cluster parallax* (Genzel *et al* 1981a, Reid *et al* 1988), again yielding distances to ~20% accuracy. Instrumental sources of error are discussed carefully by these authors and do not appear to be a limitation. At present the accuracy is limited by other factors.

One problem is that of blending and variability. Maser clusters may not always be resolved even with VLBI arrays, and the observations may not be as frequent as one would ideally like. This could make it difficult to distinguish between proper motions and variability. Blending could then contribute non-gaussian noise to the proper motion measurements. An uncertainty that is harder to quantify is that introduced by modelling the source. The H_2O maser outflow from Orion-KL was modelled in terms of two *ad hoc* components: a low-velocity flow at constant expansion velocity and a high-velocity flow in which the expansion velocity increases with distance from the source. The model has little physical basis and takes no account of the bipolar nature of the molecular outflow. Yet it is hard to see how to do better at present! Until a physical model is found for the H_2O masers and the outflow region, it may well be this modelling problem that limits the accuracy of the distance estimates.

The VLBI distance determinations are an impressive technical achievement. However, their very difficulty means that the method cannot be widely applied at present. The main VLBI contribution to the distance scale problem is likely to come from accurate measurements of a few well chosen sources. One such source is W49, the most powerful H_2O maser in our Galaxy. By good fortune W49 lies at almost the same Galactic radius as the Sun, and furthermore it contains many hundreds of H_2O masers. VLBI data on this source are expected to yield a good value for the distance of the Galactic centre (Reid *et al* 1987). Finally it should be mentioned that the technique of measuring distances from maser proper motions could in principle be applied to other strong maser lines from OH, CH_3OH and SiO.

6.3. Extragalactic masers

The extragalactic distance scale is uncertain to within a factor of almost two, so the prospect of measuring distances to masers in other galaxies is particularly exciting. Broadly speaking, the opportunities to repeat Galactic-type experiments are limited. Extragalactic OH-IR sources will in general be too weak, and their OH shells so compact that interstellar scattering may be a serious problem. The internal proper motions of H_2O maser sources in even the nearest galaxies will be reckoned in microarcsec yr^{-1} and their detection would require an interferometer with baselines far longer than any available on Earth.

The important new possibility that arises is to measure the proper motion of H_2O maser sources due to their bulk orbital motion in an external galaxy (Moran 1984, Reid 1984). These orbital motions are typically ten times larger than the internal motions. The H_2O masers in star-forming regions in the Milky Way form a very thin disc, with rather small deviations from circular rotation ($\sim 5\%$). In an external galaxy the rotation velocities can be measured using the 21 cm line of atomic hydrogen and the inclination angle of the disc can be measured from the optical image. Measurements of the orbital proper motions of two or more H_2O maser sources in the disc of the external galaxy would thus yield the distance to the galaxy, and also the three-dimensional velocity of the galactic centre of mass. Such measurements will probably require an interferometer with baselines larger than the Earth's diameter in order to reach galaxies of interest for the cosmic distance scale problem. However, orbital proper motions of H_2O masers in the nearest galaxies could in principle be detected with current ground-based arrays.

The possible use of OH megamasers and H_2O supermasers for distance estimation should also be considered. These masers are detectable to large distances of up to ~ 1000 Mpc, so a direct measurement of their distance would be of cosmological interest. The correlated variability seen across the H_2O maser spectrum of NGC1068 (Claussen and Lo 1986) suggests that some variant of the phase-lag technique may be applicable. This possibility needs further investigation. In a similar way any variation in the far-infrared emission from the nuclei of OH megamaser galaxies would be expected to produce OH-IR phase-lag effects which might also be harnessed to the problem of distance measurement.

7. Summary

The past decade has been an exciting time in which to study cosmic masers. This may be judged from the number of new areas that have opened up, for example the use of IRAS to find large numbers of masers and new types of maser sources, the use of masers for distance measurements, the discovery of megamasers in distant galaxies, and the discovery of several strong maser lines, notably the methanol 12 GHz maser in star-forming regions. Few if any of these developments could have been predicted ten years ago. Above all it has been the decade in which we saw masers properly for the first time, through radio interferometer maps.

Nowhere is the instrumental progress more striking than in the case of circumstellar masers. We now have the means to map in detail the locations of many different types of maser in a given envelope, and so build up a comprehensive picture of the envelope structure and study the different physical processes occurring there. At present our physical understanding decreases rather sharply as we work our way inwards to the regions just above the stellar surface where mass loss is established. More observations are needed of the SiO and H_2O masers in these regions. The OH 1612 MHz masers in the outer layers of the envelope are well understood now, in general, and their use for estimating the distances of OH-IR sources looks set to make a major impact on the problem of determining the Galactic distance scale. OH-IR sources are also becoming increasingly important in Galactic structure studies, as they are one population of stars which can be readily detected throughout the Galaxy. The IRAS survey has led already to the detection of hundreds of new OH-IR sources, and the final total is expected to be some thousands. IRAS follow-up observations have also uncovered new classes of maser sources which appear to be transitional objects in short-lived evolutionary states. In this and in many other ways modern developments in the study of circumstellar masers continue to enrich and extend our knowledge of the late stages of stellar evolution.

At the other end of the evolutionary scale the use of masers to investigate the star-forming regions is still at a rudimentary level. We are far from having a basic model of a protostellar envelope, and we are also short of good pumping models. However, the physical conditions in maser regions are gradually being constrained by observations of the dense cloud cores within which the masers are embedded. Milli-metre-wave interferometers can now map these cores on an arcsecond scale, and so provide complementary information to the masers, which trace the milliarcsecond structure. The newly discovered 12 GHz maser line of methanol is an important addition to the list of maser probes. The methanol line itself may not provide an instant solution to the problems of star formation, but it adds a new piece to the sparsely filled jigsaw, and in combination with OH, H_2O and SiO masers helps to build up a more comprehensive picture of the neutral regions immediately surrounding young stars. We can look forward to the day when proper motions will be measured for all these different maser species, not just H_2O masers as at present. One important question that masers are uniquely able to address is the role of magnetic fields in star-forming regions. Present results are limited but highly suggestive of a connection between the large-scale Galactic magnetic field and the local field on the small scale traced by masers. Important new developments in this area may be expected in the near future once full polarisation mapping of maser lines becomes the established procedure.

The IRAS survey has made an enormous impact on the study of all types of cosmic maser, but above all on the study of extragalactic masers. The number of known OH megamasers trebled in little over a year once their distinctive infrared properties were recognised. Megamasers promise to be a valuable probe of active galactic nuclei on a milliarcsecond scale, and they can be detected to cosmologically interesting distances. At present we desperately need a theoretical model that can explain the well defined OH and infrared properties of megamaser galaxies. No such regularities have yet emerged in the properties of the luminous H_2O masers in active galactic nuclei.

The discovery of so many new maser lines and new settings for maser activity has emphasised that the non-equilibrium processes leading to cosmic masers affect many energy levels of a molecule. In general we need to measure all the available transitions if we are to understand fully the distribution and excitation of the gas. The revival of interest in excited OH masers is a hopeful sign in this regard. On a theoretical front we are beginning to face up to the difficulties of the multitransition problem, and deal realistically with the effects of saturation and competitive gain, at least in the Sobolev approximation. In the longer term a more sophisticated treatment of the radiative transfer in the non-local case is needed.

Finally we should not forget the instruments and techniques that have dictated progress in the last decade. The VLA and MERLIN have not only laid bare the arcsecond structure of cosmic masers, they have also made masers accessible to an ever-widening community of astronomers. The Very Long Baseline Array currently nearing completion in the USA promises to effect a similar revolution for VLBI in the coming decade. This will allow routine investigation of maser proper motions, detailed mapping of individual maser spots and other milliarcsecond investigations. Beyond that we may look to dedicated radio telescopes in space, and Earth-orbiting VLBI systems which will extend the new maser distance-measuring techniques to remote galaxies.

Addendum

Since this review was completed many hundreds of papers on cosmic masers have appeared. This addendum is an attempt to highlight some of the most significant new developments.

A.1. New masers

A surprisingly large number of astrophysical maser transitions have been discovered since 1988. Arguably the most important is the 10_{29}–9_{36} submillimetre water maser at 321 GHz (Menten *et al* 1990a). This is widespread in star-forming regions and circumstellar envelopes, and will be a prime target for the submillimetre interferometers of the future. The transition is between two rotational states which are some 1860 K above the ground state. Another widespread water maser is the 3_{13}–2_{20} line at 183 GHz (Cernicharo *et al* 1990). Further water maser transitions have been reported by Menten and Melnick (1989), who found the first vibrationally excited water maser, and by Menten *et al* (1990b). Interestingly the latter authors found the 325 GHz submillimetre water maser on the basis of theoretical modelling of water masers by Neufeld and Melnick (1991). The model includes several hundred rotational levels of the ground vibrational state, but is still based on collisional cross-sections for H_2O–He, and not H_2O–H_2 and H_2O–H.

The number of known methanol masers in star-forming regions has continued to increase steadily (Batrla and Menten 1988; Haschick *et al* 1989, 1990b). A comprehensive review of their properties has been given by Menten (1991). At the present time the most important methanol maser transitions appear to be the 7_0–6_1 A^+ line at 44 GHz, and the 5_1–6_0 A^+ line at 6.6 GHz (Menten, preprint). These two lines are also representative of the two classes of methanol maser source. Class I sources, formerly called Class A (§4.5), have simple spectra with a narrow velocity width of ~ 1 km s^{-1} centred on the systemic velocity of the region. Their positions can be offset by a parsec from the compact HII regions and bright infrared sources. Observations of the DR21 region suggest that these masers are produced by interaction between a high-velocity outflow and the dense ambient gas (Plambeck and Menten 1990). The pumping of Class I masers is now understood in outline at least. Excitation of methanol under a wide range of conditions leads to overpopulation of the backbone levels $K = -1$ for E-type methanol and $K = 0$ for A-type methanol. This can explain the numerous maser transitions observed in Class I sources (Menten 1991). The 5_1–6_0 A^+ line at 6.6 GHz was predicted to be in absorption in these sources, and this has now been observed (Menten, preprint). Class II sources, formerly called Class B, have complex spectra similar to those of OH 18 cm masers, with which they are closely associated. They also coincide with compact HII regions. Class II sources do not exhibit maser action in any of the maser lines characteristic of Class I sources, and vice versa. At present there is no accepted pumping scheme for Class II sources. More than 80 of them have already been detected by Menten (preprint) in the 6.6 GHz line, some with fluxes second only to those of the 22 GHz H_2O masers. It is clear that the 6.6 GHz maser line will have an important role to play in future.

There are several other remarkable new masers which are not so widespread. Cohen *et al* (1991) have reported the first strong maser in the $OH^2\Pi_{1/2}$ $F=1-1$ transition at 4750 GHz (§4.3). The maser source radiates 30 times as many photons in this excited line as it does in the corresponding ground-state lines. Pratap *et al* (1991) have made the first VLBI observations of interstellar NH_3 masers. The NH_3 (9,6) masers in the star-forming region W51 were found to be very compact, with angular sizes less than 0.1 milliarcsec and brightness temperatures exceeding 10^{13} K. This is totally unexpected. Equally unexpected was the discovery of time-variable maser radio recombination lines at millimetre wavelengths by Martin-Pintado *et al* (1989). The lines changed drastically in shape on a timescale of six months. The brightness temperatures are estimated to be at least 10^6 K. Finally Lucas and Cernicharo (1989) have reported a further strong HCN maser transition in circumstellar envelopes (§3.7).

A.2. Circumstellar masers

There have been three particular highlights in the observations of circumstellar masers. One was the elegant experiment by Reid and Menten (1990) to measure the position of the star W Hydrae relative to its circumstellar H_2O masers. The masers were used as a phase-reference to correct wideband radio continuum measurements for atmospheric phase irregularities, so enabling the stellar continuum to be detected and its position to be measured very precisely relative to the masers. The star was found to lie exactly at the centre of the maser distribution, thus confirming what has hitherto been only a working assumption. The data support the model by Chapman and Cohen (1986) in which H_2O masers lie in a region of strong acceleration (§3.5). A second highlight was the detection of OH maser proper motions in the circumstellar envelope of U Orionis (Chapman *et al* 1991). The bulk of the OH maser emission comes from a ring (§3.4.3) which is expanding at a few milliarcsec per year, corresponding to a transverse velocity of 8 km s^{-1}. This completely contradicts the standard model of a stationary OH photodissociation zone through which material flows steadily. The new data suggest that the mass-loss from the star has recently increased, and that the material is in filaments. There are no comparable data yet on other sources. A third highlight was the detection of circular polarization in circum-stellar SiO masers by Barvainis *et al* (1987) (§3.6.4), and the subsequent investigation by McIntosh *et al* (1989) into the circumstellar magnetic field configuration. These pioneering results suggest that the magnetic field is ordered close to the star. The field strength is ~ 30 G in the SiO maser region, and appears to fall inversely as the square of the radial distance. Further studies of this kind covering other stars and examining time variations should be very fruitful.

Other developments pointing the way forward include the first spectral line images made with the extended MERLIN array and the VLBA. The MERLIN measurements by

Shepherd *et al* (1990) were the first to resolve shell structure in the OH 1612 MHz masers around two protoplanetary nebulae, thereby establishing the observational link between OH−IR sources and planetary nebulae. The VLBA measurements by van Langevelde and Diamond were of OH−IR sources close to the galactic centre. They showed that the effects of interstellar scattering are severe, and may limit the use of these stars for measuring the galactic centre distance (§6.1).

A.3. Masers in star-forming regions

Much excitement has been generated by the large numbers of new maser transitions found in star-forming regions (§A.1). There have also been major advances in the study of the long-established OH and H_2O masers. The debate over the nature of OH masers in star-forming regions (§4.2.2) appears to be nearing resolution. Proper motion studies of two sources have independently shown that the OH masers are moving away from the central HII regions (Bloemhof *et al* 1991; Cohen and Migenes 1991). In one source the expansion is accompanied by magnetic field decay on the same timescale (Cohen *et al* 1990). A new form of rapid variability in OH masers was reported by Clegg and Cordes (1991). The fluctuations are at the level of 10% on a timescale of minutes. One interesting possibility is that the fluctuations reflect fluctuations in the level populations in the unsaturated core of the OH maser (Elitzur 1991b).

A particularly surprising result was the detection of circular polarization of H_2O 22 GHz masers (Fiebig and Güsten 1989) (§4.4.5). The magnetic fields deduced are rather high at ~ 50 mG, but they follow the increase of field strength with gas density which was known previously. However it is not yet clear how reliably the magnetic field strength can be inferred from the observed circular polarization. Nedoluha and Watson (1990, 1991) have shown that the amount of circular polarization depends in a complicated way on the stimulated emission rate and the rate of cross-relaxation among the different energy states (and hyperfine states), as well as on the Zeeman frequency. The circular polarization can even be completely reversed compared to that given by the standard treatment. The general problem of polarization of maser radiation has been addressed by Elitzur (1991a), who has extended the work of Goldreich *et al* (1973a) to arbitrary angular momenta of the maser transition states. This work may have wide application beyond the field of cosmic masers.

Finally Elitzur *et al* (1989) have produced a comprehensive model for H_2O masers in star-forming regions. In their model the masers are velocity-coherent filaments which form in dense post-shock gas. The model incorporates a full treatment of the radiative transfer in filamentary masers (Elitzur *et al* 1991). Beaming plays an important part in accounting for the high brightness temperatures of the sources, which have always been difficult to model.

A.4. Extragalactic masers

Progress in the study of extragalactic masers has been steady but not spectacular. There are now 50 known OH megamasers, including some with substantial redshifts from 18 cm to 21 cm wavelength (Kazès and Baan 1991 and references therein). The most luminous source discovered so far has an isotropic OH luminosity of almost $10^4\ L_\odot$, and could in principle be detected at much greater redshifts (Staveley-Smith *et al* 1989). This raises the possibility that we will one day observe OH gigamasers. Five OH megamasers have now been mapped at high angular resolution (Martin *et al* 1989; Chapman *et al* 1990; Baan and Haschick 1990; Montgomery 1990). In all cases the OH maser region is coincident with, and sometimes smaller than, the radio continuum nucleus. The three nearest megamasers show velocity gradients indicative of a rotating nuclear disc (§5.2). A consistent picture is steadily emerging of unsaturated low-gain amplification of the strong radio continuum nucleus, with infrared radiation providing the dominant excitation (Wilson and Henkel 1990; Burdyuzha and Vikulov 1990).

There have been important VLBI observations of extragalactic H_2O masers. The H_2O supermaser in the nucleus of the galaxy NGC3079 has been observed by Haschick *et al* (1990a) and shown to have a projected size of only 7×10^{14} m, with the individual maser spots being unresolved ($< 1.5 \times 10^{14}$ m). The small size of the region implies a very high pump rate per unit volume. It seems likely that strong beaming must be invoked to alleviate the problem. Greenhill *et al* (1990) have presented maps of H_2O masers in a spiral arm of the nearby galaxy M33. The source structure is similar to that of the W49 star-forming complex in our own Galaxy. The measurements have sufficient accuracy to provide the first epoch for a proper motion study (§6.2).

References

Abraham Z, Cohen N L, Opher R, Raffaelii J C and Zisk S H 1981 *Astron. Astrophys.* **100** L10–3
Alcock C and Ross R R 1985a *Astrophys. J.* **290** 433–44
—— 1985b *Astrophys. J.* **299** 763–8
—— 1986a *Astrophys. J.* **305** 837–51
—— 1986b *Astrophys. J.* **306** 649–54
—— 1986c *Astrophys. J.* **310** 838–41
Allen C W 1973 *Astrophysical Quantities* (London: Athlone)
Allen D A, Hyland A R and Caswell J L 1980 *Mon. Not. R. Astron. Soc.* **192** 505–19
Andresen P 1986 *Astron. Astrophys.* **154** 42–54
Andresen P, Ondrey G S, Titze B and Rothe E W 1984 *J. Chem. Phys.* **80** 2548–69
Baan W A 1985 *Nature* **315** 26–31
Baan W A and Haschick A D 1984 *Astrophys. J.* **279** 541–9
—— 1987 *Astrophys. J.* **318** 139–44
—— 1990 *Astrophys. J.* **364** 65–76
Baan W A, Haschick A D, Buckley D and Schmelz J T 1985 *Astrophys. J.* **293** 394–9
Baan W A, Wood P A D and Haschick A D 1982 *Astrophys. J.* **260** L49–52
Baart E E and Cohen R J 1985 *Mon. Not. R. Astron. Soc.* **213** 641–55
Baart E E, Cohen R J, Davies R D, Norris R P and Rowland P R 1986 *Mon. Not. R. Astron. Soc.* **219** 145–55
Balister M, Gardner F F, Knowles S H and Whiteoak J B 1976 *Proc. Astron. Soc. Aust.* **3** 59–61
Ball J A, Dickinson D F, Gottlieb C A and Radford H E 1970a *Astron. J.* **75** 762–3
Ball J A, Gottlieb C A, Lilley A E and Radford H E 1970b *Astrophys. J.* **162** L203–10
Ball J A, Gottlieb C A, Meeks M L and Radford H E 1971 *Astrophys. J.* **163** L33–4
Barrett A H, Ho P and Martin R N 1975 *Astrophys. J.* **198** L119–22
Barrett A H, Schwartz P R and Waters J W 1971 *Astrophys. J.* **168** L101–6
Barvainis R 1984 *Astrophys. J.* **279** 358–62
Barvainis R, McIntosh G and Predmore C R 1987 *Nature* **329** 613–5
Barvainis R and Predmore C R 1985 *Astrophys. J.* **288** 694–702
Batrla W, Matthews H E, Menten K M and Walmsley C M 1987 *Nature* **326** 49–51
Batrla W and Menten K M 1988 *Astrophys. J.* **329** L117–20
Baud B 1981 *Astrophys. J.* **250** L79–83
Baudry A, Diamond P J, Booth R S, Graham D and Walmsley C M 1988 *Astron. Astrophys.* **201** 105–12

Baudry A, Walmsley C M, Winnberg A and Wilson T L 1981 *Astron. Astrophys.* **102** 287–92

Benson J M and Mutel R L 1979 *Astrophys. J.* **233** 119–26

—— 1982 *Astrophys. J.* **253** 199–207

Benson J M, Mutel R L, Fix J D and Claussen M J 1979 *Astrophys. J.* **229** L87–90

Berulis I I, Gladyshev A S, Lekht E E, Paschenko M I, Rudnitskii G M, Sorochenko R L and Khozov G V 1984 *Sov. Inf. Astron. Council* **56** 92–6

Berulis I I, Lekht E E, Paschenko M I and Rudnitskii G M 1983 *Sov. Astron.* **27** 179–85

Bettweiser E and Misselbeck G 1977 *Astron. Astrophys.* **61** 567–74

Bloemhof E E, Reid M J and Moran J M 1991 *Atoms, Ions and Molecules: New Results in Spectral Line Astrophysics* ed. A D Haschick and P T P Ho (San Francisco: Astronomical Society of the Pacific) pp 393–7

Bologna J M, Johnston K J, Knowles S H, Mango S A and Sloanaker R M 1975 *Astrophys. J.* **199** 86–91

Booth R S 1969 *Nature* **224** 783–4

Booth R S, Kus A J, Norris R P and Porter N D 1981 *Nature* **290** 382–4

Bottinelli L, Fraix-Burnet D, Gouguenheim L, Kazès I, Le Squeren A M, Patey I, Rickard L J and Turner B E 1985 *Astron. Astrophys.* **151** L7–9

Bottinelli L, Gouguenheim L, Le Squeren A M and Martin J M 1986 *IAU Circular No* 4231

—— 1987 *IAU Circular No* 4357

Bowers P F 1984 *Astrophys. J.* **279** 350–7

—— 1985 *Mass Loss from Red Giants* (Dordrecht: Reidel) pp 189–209

Bowers P F and Hagen W 1984 *Astrophys. J.* **285** 637–47

Bowers P F, Johnston K J and Spencer J H 1983 *Astrophys. J.* **274** 733–54

Bowers P F and Kerr F J 1977 *Astron. Astrophys.* **57** 115–23

Bowers P F and Morris M 1984 *Astrophys. J.* **276** 646–52

Bowers P F, Reid M J, Johnston K J, Spencer J H and Moran J M 1980 *Astrophys. J.* **242** 1088–101

Brebner G C 1988 *PhD Thesis* University of Manchester

Brebner G C, Heaton B, Cohen R J and Davies S R 1987 *Mon. Not. R. Astron. Soc.* **229** 679–89

Buhl D, Snyder L E, Lovas F J and Johnson D R 1974 *Astrophys. J.* **192** L97–100

Bujarrabal V, Destombes J L, Guibert J, Marlière-Demuynck C, Nguyen-Q-Rieu and Omont A 1980a *Astron. Astrophys.* **81** 1–7

Bujarrabal V, Guibert J, Nguyen-Q-Rieu and Omont A 1980b *Astron. Astrophys.* **84** 311–6

Burdyuzha V V and Vikulov K A 1990 *Mon. Not. R. Astron. Soc.* **244** 86–92

Burke B F *et al* 1972 *Sov. Astron.* **16** 379–82

Burke B J, Moran J M, Barrett A H, Rydbeck O, Hansson B, Rogers A E E, Ball J A and Cudaback D D 1968 *Astron. J.* **73** S168–9

Buxton R B, Barrett A H, Ho P T P and Schneps M H 1977 *Astron. J.* **82** 985–8

Caswell J L and Haynes R F 1987 *Aust. J. Phys.* **40** 215–38

Cernicharo J, Thum C, Hein H, John D, Garcia P and Mattiocio F 1990 *Astron. Astrophys.* **231** L15–8

Chapman J M 1985 *PhD Thesis* University of Manchester

—— 1988 *Mon. Not. R. Astron. Soc.* **230** 415–27

Chapman J M and Cohen R J 1985 *Mon. Not. R. Astron. Soc.* **212** 375–84

—— 1986 *Mon. Not. R. Astron. Soc.* **220** 513–28

Chapman J M, Cohen R J and Saikia D J 1991 *Mon. Not. R. Astron. Soc.* **249** 227–40

Chapman J M, Staveley-Smith L, Axon D J, Unger S W, Cohen R J, Pedlar A and Davies R D 1990 *Mon. Not. R. Astron. Soc.* **244** 281–90

Cheung A C, Rank D M, Townes C H, Thornton D D and Welch W J 1969 *Nature* **221** 626–8

Chui M F, Cheung A C, Matsakis D, Townes C H and Cardiasmenos A G 1974 *Astrophys. J.* **187** L19–21

Churchwell E, Witzel A, Huchtmeier W, Pauliny-Toth I, Roland J and Sieber W 1977 *Astron. Astrophys.* **54** 969–71

Cimerman M 1979 *Astrophys. J.* **228** L79–82

Clark F O, Troland T H and Miller J S 1985 *Astrophys. J.* **289** 756–64

Claussen M J and Fix J D 1981 *Astrophys. J.* **250** L77–8

—— 1982 *Astrophys. J.* **263** 153–9

Claussen M J, Heiligman G M and Lo K Y 1984 *Nature* **310** 298–300

Claussen M J and Lo K Y 1986 *Astrophys. J.* **308** 592–9

Clegg A W and Cordes J M 1991 *Astrophys. J.* **374** 150–68

Cohen N L and Willson R F 1981 *Astron. Astrophys.* **96** 230–4

Cohen R J 1985 *Cosmical Gas Dynamics* (Utrecht: VNU Science) pp 223–35

—— 1987 *IAU Symp. No 122* (Dordrecht: Reidel) pp 229–39

Cohen R J, Baart E E and Jonas J L 1988 *Mon. Not. R. Astron. Soc.* **231** 205–27

Cohen R J and Brebner G C 1985 *Mon. Not. R. Astron. Soc.* **216** 51P–56P

Cohen R J, Brebner G C B and Potter M M 1990 *Mon. Not. R. Astron. Soc.* **246** 3P–6P

Cohen R J, Downs G, Emerson R, Grimm M, Gulkis S, Stevens G and Tarter J 1987 *Mon. Not. R. Astron. Soc.* **225** 491–8

Cohen R J, Masheder M R W and Walker R N F 1991 *Mon. Not. R. Astron. Soc.* **250** 611–6

Cohen R J and Migenes V 1991 *Bull. Am. Astron. Soc.* **23** p825

Cohen R J, Rowland P R and Blair M M 1984 *Mon. Not. R. Astron. Soc.* **210** 425–38

Cooke B and Elitzur M 1985 *Astrophys. J.* **295** 175–82

Cox G C and Parker E A 1978 *Mon. Not. R. Astron. Soc.* **183** 111–8

—— 1979 *Mon. Not. R. Astron. Soc.* **186** 197–215

Crawford J and Rowan Robinson M 1986 *Mon. Not. R. Astron. Soc.* **221** 923–9

Davies J G, Anderson B and Morison I 1980 *Nature* **288** 64–6

Davies R D 1974 *IAU Symp. No 60* (Dordrecht: Reidel) pp 275–92

Davies R D, Rowson B, Booth R S, Cooper A J, Gent H, Adgie R L and Crowther J H 1967 *Nature* **213** 1109–10

Deguchi S and Watson W D 1986a *Astrophys. J.* **300** L15–8

—— 1986b *Astrophys. J.* **302** 750–6

Deguchi S, Watson W D and Western L R 1986 *Astrophys. J.* **302** 108–19

de Jong T 1973 *Astron. Astrophys.* **23** 297–313

Dennison B, Thomas M, Booth R S, Brown R L, Broderick J J and Condon J J 1984 *Astron. Astrophys.* **135** 199–212

Despois D, Gérard E, Crovisier J and Kazès I 1981 *Astron. Astrophys.* **99** 320–40

Diamond P J, Johnston K J, Chapman J M, Lane A P, Bowers P F, Spencer J H and Booth R S 1987 *Astron. Astrophys.* **174** 95–102

Diamond P J, Norris R P and Booth R S 1985a *Mon. Not. R. Astron. Soc.* **216** 1P–5P

Diamond P J, Norris R P, Rowland P R, Booth R S and Nyman L-A 1985b *Mon. Not. R. Astron. Soc.* **212** 1–21

Dickinson D F 1987 *Astrophys. J.* **313** 408–12

Dixon R N, Field D and Zare R N 1985 *Chem. Phys. Lett.* **122** 310–4

dos Santos P M and Lépine J R D 1979 *Nature* **278** 34–5

Dupree A K and Goldberg L 1969 *Astrophys. J.* **158** L49–53

Elitzur M 1978 *Astron. Astrophys.* **62** 305–9

—— 1981 *Physical Processes in Red Giants* (Dordrecht: Reidel) pp 363–82

—— 1982 *Rev. Mod. Phys.* **54** 1225–60

—— 1986 *Masers, Molecules and Mass Outflows in Star Forming Regions* ed. A D Haschick (Massachusetts: Haystack Observatory) pp 299–319

—— 1991a *Astrophys. J.* **370** 407–18

—— 1991b *Astrophys. J.* **370** L45–8

Elitzur M and de Jong T 1978 *Astron. Astrophys.* **67** 323–32

Elitzur M, Goldreich P and Scoville N 1976 *Astrophys. J.* **205** 384–96

Elitzur M, Hollenbach D J and McKee C F 1989 *Astrophys. J.* **346** 983–90

Elitzur M, McKee C F and Hollenbach D J 1991 *Astrophys. J.* **367** 333–49

Elldér J 1973 *Research Lab. of Electronics and Onsala Space Observatory, Research Report No 116*

Engels D 1979 *Astron. Astrophys. Suppl.* **36** 337–45

Engels D, Habing H J, Olnon F M, Schmid-Burgk J and Walmsley C M 1984 *Astron. Astrophys.* **140** L9–12

Engels D, Kreysa E, Schultz G V and Sherwood W A 1983 *Astron. Astrophys.* **124** 123–38

Engels D, Schmid-Burgk J and Walmsley C M 1986 *Astron. Astrophys.* **167** 129–44

Evans N J, Hills R E, Rydbeck O E H and Kollberg E 1972 *Phys. Rev.* A **6** 1643–7

Feast M W 1987 *The Galaxy* (Dordrecht: Reidel) pp 1–25

Fiebig D and Güsten R 1989 *Astron. Astrophys.* **214** 333–8

Field D 1985 *Mon. Not. R. Astron. Soc.* **217** 1–29

Field D and Gray M D 1988 *Mon. Not. R. Astron. Soc.* **234** 353–72

Field D and Richardson I M 1984 *Mon. Not. R. Astron. Soc.* **211** 799–811

Fillit R, Proust D and Lépine J R D 1977 *Astron. Astrophys.* **58** 281–6

Fix J D 1979 *Astrophys. J.* **232** L39–42

—— 1987 *Astron. J.* **92** 433–9

Fix J D and Mutel R L 1985 *Astron. J.* **90** 736–7

Fix J D, Mutel R L, Benson J M and Claussen M L 1980 *Astrophys. J.* **241** L95−8

Forster J R and Caswell J L 1987 *IAU Symp. No 115* (Dordrecht: Reidel) pp 174−5

Forster J R, Goss W M, Wilson T L, Downes D and Dickel H R 1980 *Astron. Astrophys.* **84** L1−3

Forster J R, Welch W J, Wright M C H and Baudry A 1978 *Astrophys. J.* **221** 137−44

Gammon R H 1976 *Astron. Astrophys.* **50** 71−7

Garay G, Reid M J and Moran J M 1985 *Astrophys. J.* **289** 681−97

Garcia-Barreto J A, Burke B F, Reid M J, Moran J M, Haschick A D and Schilizzi R T 1988 *Astrophys. J.* **326** 954−66

Gardner F F and Martin-Pintado J 1983 *Astron. Astrophys.* **121** 265−70

Gardner F F and Whiteoak J B 1982 *Mon. Not. R. Astron. Soc.* **201** 13P−15P

Gardner F F, Whiteoak J B and Palmer P 1983 *Mon. Not. R. Astron. Soc.* **205** 297−302

—— 1987 *Mon. Not. R. Astron. Soc.* **225** 469−80

Gaume R A and Mutel R L 1987 *Astrophys. J. Suppl.* **65** 193−253

Genzel R 1986 *Masers, Molecules and Mass Outflows in Star Forming Regions* ed. A D Haschick (Massachusetts: Haystack Observatory) pp 233−45

Genzel R and Downes D 1977 *Astron. Astrophys. Suppl.* **30** 145−68

Genzel R, Downes D, Schneps M H, Moran J M, Kogan L R, Kostenko V I, Matveyenko L I and Rönnäng B 1981b *Astrophys. J.* **247** 1039−51

Genzel R, Reid M J, Moran J M and Downes D 1981a *Astrophys. J.* **244** 884−902

Genzel R *et al* 1978 *Astron. Astrophys.* **66** 13−29

Giguere P T, Woolf N J and Webber J C 1976 *Astrophys. J.* **207** L195−8

Goldreich P and Keeley D A 1972 *Astrophys. J.* **174** 517−25

Goldreich P, Keeley D A and Kwan J Y 1973a *Astrophys. J.* **179** 111−34

—— 1973b *Astrophys. J.* **182** 55−66

Goldreich P and Kwan J 1974a *Astrophys. J.* **190** 27−34

—— 1974b *Astrophys. J.* **191** 93−100

Goldreich P and Scoville N 1976 *Astrophys. J.* **205** 144−54

Gómez Balbao A M and Lépine J R D 1986 *Astron. Astrophys.* **159** 166−74

Green S 1980 *Astrophys. J. Suppl.* **42** 103−41

Greenhill L J, Moran J M, Reid M J, Gwinn C R and Menten K M 1990 *Astrophys. J.* **364** 513−26

Grinin V P and Grigorev S A 1983 *Sov. Astron. Lett.* **9** 244−6

Guilloteau S, Baudry A, Walmsley C M, Wilson T L and Winnberg A 1984 *Astron. Astrophys.* **131** 45−57

Guilloteau S, Baudry A and Walmsley C M 1985 *Astron. Astrophys.* **153** 179−88

Guilloteau S, Omont A and Lucas R 1987 *Astron. Astrophys.* **176** L24−6

Gwinn C R, Moran J M, Reid M J and Schneps M H 1988 *Astrophys. J.* **330** 817−27

Habing H J 1987 *The Galaxy* (Dordrecht: Reidel) pp 173−97

Habing H J and Israel F P 1979 *Ann. Rev. Astron. Astrophys.* **17** 345−85

Hagen W 1979 *Publ. Astron. Soc. Pacific* **91** 165−7

Hall D N B, Kleinmann S G, Scoville N Z and Ridgway S T 1981 *Astrophys. J.* **248** 898−905

Hansen S S 1982 *Astrophys. J.* **260** 599−603

Hansen S S and Johnston K J 1983 *Astrophys. J.* **267** 625−9

Harvey P M, Bechis K P, Wilson W J and Ball J A 1974 *Astrophys. J. Suppl.* **27** 331−57

Haschick A D and Baan W A 1985 *Nature* **314** 144−6

—— 1986 *Masers, Molecules and Mass Outflows in Star Forming Regions* ed. A D Haschick (Massachusetts: Haystack Observatory) pp 249−54

Haschick A D, Baan W A and Menten K M 1989 *Astrophys. J.* **346** 330−5

Haschick A D, Baan W A, Schnepps M H, Reid M J, Moran J M and Güsten R 1990a *Astrophys. J.* **356** 149−55

Haschick A D, Menten K M and Baan W A 1990b *Astrophys. J.* **354** 556−67

Haschick A D, Moran J M, Rodriguez L F and Ho P T P 1983 *Astrophys. J.* **265** 281−9

Henkel C, Batrla W and Guesten R 1986a *Astron. Astrophys.* **168** L13−5

Henkel C, Guesten R and Baan W A 1987 *Astron. Astrophys.* **185** 14−24

Henkel C, Guesten R, Downes D, Thum C, Wilson T L and Bierman P 1984 *Astron. Astrophys.* **141** L1−3

Henkel C, Matthews H E and Morris M 1983 *Astrophys. J.* **267** 184−90

Henkel C and Wilson T L 1990 *Astron. Astrophys.* **229** 431−40

Henkel C, Wouterlout J G A and Bally J 1986b *Astron. Astrophys.* **155** 193−9

Herman J H, Baud B, Habing H J and Winnberg A 1985 *Astron. Astrophys.* **143** 122−35

Herman J and Habing H J 1985a *Phys. Rep.* **124** 255−314

—— 1985b *Astron. Astrophys. Suppl.* **59** 523−55

—— 1987 *The Late Stages of Stellar Evolution* (Dordrecht: Reidel) pp 55–71
Hinkle K H, Scharlach W W G and Hall D N B 1984 *Astrophys. J. Suppl.* **56** 1–17
Ho P T P, Haschick A D, Vogel S N and Wright M C H 1983 *Astrophys. J.* **265** 295–303
Ho P T P, Martin R N, Henkel C and Turner J L 1987 *Astrophys. J.* **320** 663–6
Ho P T P, Moran J M and Rodriguez L F 1982 *Astrophys. J.* **262** 619–35
Huggins P J and Glassgold A E 1982 *Astron. J.* **87** 1828–35
Huchtmeier W K, Eckart A and Zensus A J 1988 *Astron. Astrophys.* **200** 26–8
Huchtmeier W K, Witzel A, Kühr H, Pauliny-Toth I I and Roland J 1978 *Astron. Astrophys.* **64** L21–4
Iben I and Renzini A 1983 *Ann. Rev. Astron. Astrophys.* **21** 271–342
Jaffe D T, Guesten R and Downes D 1981 *Astrophys. J.* **250** 621–30
Jewell P R, Batrla W, Walmsley C M and Wilson T L 1984 *Astron. Astrophys.* **130** L1–4
Jewell P R, Dickinson D F, Snyder L E and Clemens D P 1987 *Astrophys. J.* **323** 749–55
Jewell P R, Schenewerk M S and Snyder L E 1985 *Astrophys. J.* **295** 193–94
Jewll P R, Webber J C and Snyder L E 1980 *Astrophys. J.* **242** L29–31
—— 1981 *Astrophys. J.* **249** 118–23
Johansson L E B, Andersson C, Elldér J, Friberg P, Hjalmarson A, Hoglund B, Irvine W M, Olofsson H and Rydbeck G 1984 *Astron. Astrophys.* **130** 227–256
Johansson L E B, Andersson C, Goss W M and Winnberg A 1977 *Astron. Astrophys.* **54** 323–34
Johnston K J, Knowles S H, Moran J M, Burke B F, Lo K Y, Papadopolous G D, Read B R and Hardebeck E G 1977 *Astron. J.* **82** 403–7
Johnston K J, Spencer J H and Bowers P F 1985 *Astrophys. J.* **290** 660–70
Kaifu N, Buhl D and Snyder L E 1975 *Astrophys. J.* **195** 359–66
Kazès I and Baan W A 1991 *Astron. Astrophys.* **248** L15–8
Kemball A J, Diamond P J and Mantovani F 1988a *IAU Symp. No 129* (Dordrecht: Reidel) pp 255–6
Kemball A J, Gaylard M J and Nicolson G D 1988b *Astrophys. J.* **331** L37–40
Kent S R and Mutel R L 1982 *Astrophys. J.* **263** 145–52
Kirrane T M 1987 *MSc Thesis* University of Manchester
Knapp G R 1986 *Mitt. Astron. Gesell.* **67** 111–31
Knowles S H and Batchelor R A 1976 *Mon. Not. R. Astron. Soc.* **174** 69P–73P
—— 1978 *Mon. Not. R. Astron. Soc.* **184** 107–17
Knowles S H, Caswell J L and Goss W M 1976 *Mon. Not. R. Astron. Soc.* **175** 537–55
Knowles S H, Mayer C H, Cheung A C, Rank D M and Townes C H 1969 *Science* **163** 1055–7
Koo B C, Williams D R W, Heiles C and Backer D C 1988 *Astrophys. J.* **326** 931–40
Kronberg P P, Biermann P and Schwab F R 1985 *Astrophys. J.* **291** 693–707
Kwan J and Scoville N 1974 *Astrophys. J.* **194** L97–101
Kylafis N D and Norman C 1986 *Astrophys. J.* **300** L73–7
—— 1987 *Astrophys. J.* **323** 346–57
Lada C J 1985 *Ann. Rev. Astron. Astrophys.* **23** 267–317
—— 1987 *IAU Symp. No. 115* (Dordrecht: Reidel) pp 1–18
Lane A P 1982 *PhD Thesis* University of Massachusetts
Langer S H and Watson W D 1984 *Astrophys. J.* **284** 751–68
Le Squeren A M and Sivagnanam P 1985 *Astron. Astrophys.* **152** 85–8
Lees R M, Lovas F J, Kirchoff W H and Johnson D R 1973 *J. Phys. Chem. Ref. Data* **2** 205–14
Lenzen R 1988 *Astron. Astrophys.* **190** 269–74
Lewis B M, Eder J and Terzian Y 1985 *Nature* **313** 200–2
Little-Marenin I R, Benson P J and Dickinson D F 1988 *Astrophys. J.* **330** 828–34
Litvak M 1969 *Astrophys. J.* **156** 471–92
—— 1972 *Atoms and Molecules in Astrophysics* ed. T R Carson and M J Roberts (New York: Academic) pp 201–76
—— 1973 *Astrophys. J.* **182** 711–30
Litvak M M, McWhorter A L, Meeks M L and Zeiger H J 1966 *Phys. Rev. Lett.* **17** 821–6
Lucas R 1980 *Astron. Astrophys.* **84** 36–9
Lucas R and Cernicharo J 1989 *Astron. Astrophys.* **218** L20–2
Lucas R, Guilloteau S and Omont A 1988 *Astron. Astrophys.* **194** 230–6
Madden S C, Irvine W M, Matthews H E, Brown R D and Godfrey P D 1986 *Astrophys. J.* **300** L79–84
Mann A P C and Williams D A 1980 *Nature* **283** 721–5
Martin J M, Bottinelli L, Dennefeld M, Gouguenheim L, Handa T, Le Squeren A M, Nakai N and Sofue Y 1988 *Astron. Astrophys.* **195** 71–5

Martin J M, Bottinelli L, Dennefeld M, Gougenheim L and Le Squeren A M 1989 *Astron. Astrophys.* **208** 39–46

Martin-Pintado J, Thum C and Bachiller R 1989 *Astron. Astrophys.* **222** L9–11

Matsakis D N, Cheung A C, Wright M C H, Askne J I H, Townes C H and Welch W J 1980 *Astrophys. J.* **236** 481–91

Matthews H E, Baudry A, Guilloteau S and Winnberg A 1986 *Astron. Astrophys.* **163** 177–84

Matthews H E, Olnon F M, Winnberg A and Baud B 1985 *Astron. Astrophys.* **149** 227–38

Mattila K, Holsti N, Toriseva M, Anttila R and Malkamaki L 1985 *Astron. Astrophys.* **145** 192–200

Matveenko L I, Romanov A M, Kogan L R, Moiseev I G, Sorochenko R L and Timofeev V V 1983 *Sov. Astron. Lett.* **9** 240–2

McIntosh G C, Predmore C R, Moran J M, Greenhill L J, Rogers A A E and Barvainis R 1989 *Astrophys. J.* **337** 934–44

Menten K M 1991 *Atoms, Ions and Molecules: New Results in Spectral Line Astrophysics* ed. A D Haschick and P T P Ho (San Francisco: Astronomical Society of the Pacific) pp 119–36

Menten K M, Johnston K J, Wilson T L, Walmsley C M, Mauersberger R and Henkel C 1985 *Astrophys. J.* **293** L83–5

Menten K M and Melnick G J 1989 *Astrophys. J.* **341** L91–4

Menten K M, Melnick G J and Phillips T G 1990a *Astrophys. J.* **350** L41–4

Menten K M, Melnick G J, Phillips T G and Neufeld D A 1990b *Astrophys. J.* **363** L27–31

Menten K M, Reid M J, Moran J M, Wilson T L, Johnston K J and Batrla W 1988 *Astrophys. J.* **333** L83–6

Menten K M, Walmsley C M, Henkel C and Wilson T 1986a *Astron. Astrophys.* **157** 318–28

Menten K M, Walmsley C M, Henkel C, Wilson T L, Snyder L E, Hollis J M and Lovas F J 1986b *Astron. Astrophys.* **169** 271–80

Miller J S, Clark F O and Troland T H 1984 *Astrophys. J.* **287** 892–8

Mirabel I F, Kazès I and Sanders D B 1988 *Astrophys. J.* **324** L59–62

Mirabel I F and Sanders D B 1987 *Astrophys. J.* **322** 688–93

Montgomery A S 1990 *PhD Thesis* University of Manchester

Moore T J T, Cohen R J and Mountain C M 1988 *Mon. Not. R. Astron. Soc.* **231** 887–95

Moorwood A F M and Salinari P 1983 *Astron. Astrophys.* **125** 342–54

Moran J 1984 *Nature* **310** 270–1

Moran J M, Ball J A, Predmore C P, Lane A P, Huguenin G R, Reid M J and Hansen S S 1979 *Astrophys. J.* **231** L67–71

Moran J M, Burke B F, Barrett A H, Rogers A E E, Ball J A, Carter J C and Cudaback D D 1968 *Astrophys. J.* **152** L97–100

Moran J M, Reid M J, Lada C J, Yen J L, Johnston K J and Spencer J H 1978 *Astrophys. J.* **224** L67–71

Morimoto M, Ohishi M and Kanzawa T 1985 *Astrophys. J.* **288** L11–5

Morris M and Bowers P F 1980 *Astron. J.* **85** 724–37

Muchmore D O, Nuth J A and Stencel R E 1987 *Astrophys. J.* **315** L141–6

Mutel R L, Fix J D, Benson J M and Webber J C 1979 *Astrophys. J.* **228** 771–9

Myers P C and Goodman A A 1988 *Astrophys. J.* **326** L27–30

Myers S T and Scoville N Z 1987 *Astrophys. J.* **312** L39–43

Nedoluha G E and Watson W D 1990 *Astrophys. J.* **361** L53–5

—— 1991 *Astrophys. J.* **367** L63–7

Neufeld D A and Melnick G J 1991 *Astrophys. J.* **368** 215–30

Neugebauer G *et al* 1984 *Astrophys. J.* **278** L1–6

Nguyen-Q-Rieu, Laury-Micoulant C, Winnberg A and Schultz G V 1979 *Astron. Astrophys.* **75** 351–64

Nguyen-Q-Rieu, Mebold U, Winnberg A, Guibert J and Booth R 1976 *Astron. Astrophys.* **53** 467–9

Norman C and Silk J 1979 *Astrophys. J.* **228** 197–205

Norris R P, Baan W A, Haschick A D, Diamond P J and Booth R S 1985 *Mon. Not. R. Astron. Soc.* **213** 821–31

Norris R P, Booth R S, Diamond P J, Nyman L-Å, Graham D A and Matveyenko L I 1984 *Mon. Not. R. Astron. Soc.* **208** 435–46

Norris R P, Caswell J L, Gardner F F and Wellington K J 1987 *Astrophys. J.* **321** L159–62

Norris R P, McCutcheon W H, Caswell J L, Wellington K J, Reynolds J E, Peng R S and Kesteven M J 1988 *Nature* **335** 149–51

Nyman L-Å, Johansson L E B and Booth R S 1986 *Astron. Astrophys.* **160** 352–66

Nyman L-Å and Olofsson H 1986 *Astron. Astrophys.* **158** 67–82

Olnon F 1977 *PhD Thesis* University of Leiden

Olnon F M, Baud B, Habing H J, de Jong T, Harris S and Pottasch S R 1984 *Astrophys. J.* **278** L41–3

Olnon F M, Winnberg A, Matthews H E and Schultz G V 1980 *Astron. Astrophys. Suppl.* **42** 119–33

Olofsson H, Hjalmarson Å and Rydbeck O E H 1981 *Astron. Astrophys.* **100** L30–3

Olofsson H, Rydbeck O E H and Nyman L-Å 1985 *Astron. Astrophys.* **150** 169–79

Palmer P, Gardner F F and Whiteoak J B 1984 *Mon. Not. R. Astron. Soc.* **211** 41P–44P

Pataki L and Kolena J 1974 *Bull. Am. Astron. Soc.* **6** 340

Plambeck R L and Menten K M 1990 *Astrophys. J.* **364** 555–60

Plambeck R L and Wright M C H 1988 *Astrophys. J.* **330** L61–5

Pottasch S R, Bignelli C and Zijlstra A 1987 *Astron. Astrophys.* **177** L49–52

Pratap P, Menten K M, Reid M J, Moran J M and Walmsley C M 1991 *Astrophys. J.* **373** L13–5

Radhakrishnan V 1967 *Aust. J. Phys.* **20** 203–4

Reid M J 1984 *Quasat—a VLBI Observatory in Space* (Paris: European Space Agency) Publication SP-213 pp 181–4

Reid M J, Haschick A D, Burke B F, Moran J M, Johnston K J and Swenson G W 1980 *Astrophys. J.* **239** 89–111

Reid M J and Menten K M 1990 *Astrophys. J.* **360** L51–4

Reid M J and Moran J M 1981 *Ann. Rev. Astron. Astrophys.* **19** 231–76

Reid M J, Moran J M, Leach R W, Ball J A, Johnston K J, Spencer J H and Swenson G W 1979 *Astrophys. J.* **227** L89–92

Reid M J, Muhleman D O, Moran J M, Johnston K J and Schwartz P R 1977 *Astrophys. J.* **214** 60–77

Reid M J, Schneps M H, Moran J M, Gwinn C R, Genzel G, Downes D and Rönnäng B 1987 *IAU Symp. No 115* (Dordrecht: Reidel) pp 554–5

—— 1988 *Astrophys. J.* **330** 809–16

Rickard L J, Zuckerman B and Palmer P 1975 *Astrophys. J.* **200** 6–21

Rieke G H, Lebofsky M J, Thompson R I, Low F J and Tokunga A T 1980 *Astrophys. J.* **238** 24–40

Roberts M 1967 *Astrophys. J.* **148** 931–3

Robinson B J, Caswell J L and Goss W M 1970 *Astrophys. Lett.* **7** 79–84

Rodriguez L F and Canto J 1983 *Rev. Mex. Astron. Astrof.* **8** 163–73

Rodriguez L F, Haschick A D, Torrelles J M and Myers P C 1987 *Astron. Astrophys.* **186** 319–21

Rodriguez L F, Ho P T P and Moran J M 1980 *Astrophys. J.* **240** L149–52

Rosen B R, Moran J M, Reid M J, Walker R C, Burke B F, Johnston K J and Spencer J H 1978 *Astrophys. J.* **222** 132–9

Rowan-Robinson M 1986 *Mon. Not. R. Astron. Soc.* **219** 737–49

Rowland P R and Cohen R J 1986 *Mon. Not. R. Astron. Soc.* **200** 233–51

Rydbeck O E H, Elldér J and Irvine W M 1973 *Nature* **246** 466–8

Sanders D B, Scoville N Z, Sargent A I and Soifer B T 1988 *Astrophys. J.* **324** L55–8

Sargent A I, Sanders D B, Scoville N Z and Soifer B T 1987 *Astrophys. J.* **312** L35–8

Scalise E and Lépine J R D 1978 *Astron. Astrophys.* **65** L7–8

Schmelz J T, Baan W A and Haschick A D 1987 *Astrophys. J.* **321** 225–32

Schneps M H, Lane A P, Downes D, Moran J M, Genzel R and Reid M J 1981 *Astrophys. J.* **249** 124–33

Schwartz P R and Barrett A H 1970 *Astrophys. J.* **159** L123–7

Schwartz P R, Harvey P M and Barrett A H 1974 *Astrophys. J.* **187** 491–6

Scoville N Z, Sanders D B, Sargent A I, Soifer B T, Scott S L and Lo K Y 1986 *Astrophys. J.* **311** L47–50

Seaquist E R and Davis L E 1983 *Astrophys. J.* **274** 659–65

Shepherd M C, Cohen R J, Gaylard M J and West M E 1990 *Nature* **344** 522–4

Shu F H, Adams F C and Lizano S 1987 *Ann. Rev. Astron. Astrophys.* **25** 23–81

Sivagnanam P and Le Squeren A M 1986 *Astron. Astrophys.* **168** 374–6

Slootmaker A, Herman J and Habing H J 1985 *Astron. Astrophys. Suppl.* **59** 465–83

Snyder L E and Buhl D 1974 *Astrophys. J.* **189** L31–3

Snyder L E, Jewell P R, Dinger A S, Dickinson D F and Buhl D 1986 *Astron. J.* **92** 416–30

Spencer J H, Johnston K J, Moran J M, Reid M J and Walker R C 1979 *Astrophys. J.* **230** 449–55

Staveley-Smith L, Allen D A, Chapman J M, Norris R P and Whiteoak J B 1989 *Nature* **337** 625–7

Staveley-Smith L, Cohen R J, Chapman J M, Pointon L and Unger S W 1987 *Mon. Not. R. Astron. Soc.* **226** 689–701

Sternberg A 1986 *PhD Thesis* Columbia University, NY

Strelnitskij V S 1984 *Mon. Not. R. Astron. Soc.* **207** 339–54

Sullivan W T 1973 *Astrophys. J. Suppl.* **25** 393–432

Sullivan W T and Kerstholt J H 1976 *Astron. Astrophys.* **51** 427–50

Sutton E C and Herbst E 1988 *Astrophys. J.* **333** 359–65

Tarter J C and Welch W J 1986 *Astrophys. J.* **305** 467–83

te Lintel Hekkert P, Habing H J, Caswell J L, Norris R P and Haynes R F 1988 *Astron. Astrophys.* **202** L19–22

Thomasson P 1986 *Quart. J. R. Astron. Soc.* **27** 413–31

Thompson A R, Clark B G, Wade C M and Napier P J 1980 *Astrophys. J. Suppl.* **44** 151–67

Thompson A R, Moran J M and Swenson G N 1986 *Interferometry and Synthesis in Radio Astronomy* (New York: Wiley)

Troland T H, Heiles C, Johnson D R and Clark F O 1979 *Astrophys. J.* **232** 143–57

Trumpler R J and Weaver H F 1953 *Statistical Astronomy* (Berkeley: University of California Press) pp 336–412

Turner B E 1970 *J. R. Astron. Soc. Canada* **64** 221–37

—— 1979 *Astron. Astrophys. Suppl.* **37** 1–332

—— 1982 *Regions of Recent Star Formation* (Dordrecht: Reidel) pp 425–32

—— 1985 *Astrophys. J.* **299** 312–33

Turner B E, Palmer P and Zuckerman B 1970 *Astrophys. J.* **160** L125–9

Turner J L and Welch W J 1984 *Astrophys. J.* **287** L81–4

Ukita N, Hasegawa, Kaifu N, Morita K I, Okumura S, Suziki H, Ohishi M and Hayshi M 1987 *IAU Symp. No 115* (Dordrecht: Reidel) p 178

Unger S W, Chapman J M, Cohen R J, Hawarden T G and Mountain C M 1986 *Mon. Not. R. Astron. Soc.* **220** 1P–6P

van Langevelde H J and Diamond P J 1991 *Mon. Not. R. Astron. Soc.* **249** 7P–10P

Vardya M S 1987 *Astron. Astrophys.* **182** 75–9

Walker R C 1984 *Astrophys. J.* **280** 618–28

Walker R C, Matsakis D N and Garcia-Barreto J A 1982 *Astrophys. J.* **255** 128–42

Weaver H, Dieter N H and Williams D R W 1968 *Astrophys. J. Suppl.* **16** 219–74

Weaver H, Williams D R W, Dieter N H and Lum W T 1965 *Nature* **208** 29–31

Welch W J and Marr J 1987 *Astrophys. J.* **317** L21–3

Weliachew L 1971 *Astrophys. J.* **167** L47–52

Weliachew L, Fomalont E B and Greisen E W 1984 *Astron. Astrophys.* **137** 335–42

Weliachew L N and Lucas R 1988 *Rep. Prog. Phys.* **51** 605–45

Werner M W, Beckwith S, Gatley I, Sellgren K, Berriman G and Whiting D L 1980 *Astrophys. J.* **239** 540–8

Western L R and Watson W D 1983 *Astrophys. J.* **275** 195–200

—— 1984 *Astrophys. J.* **285** 158–73

White G J 1979 *Mon. Not. R. Astron. Soc.* **186** 377–81

Whiteoak J B and Gardner F F 1973 *Astrophys. Lett.* **15** 211–5

—— 1983 *Mon. Not. R. Astron. Soc.* **205** 27P–32P

—— 1986 *Mon. Not. R. Astron. Soc.* **222** 513–23

Wilking B A and Claussen M J 1987 *Astrophys. J.* **320** L133–7

Wilson T L, Walmsley C M, Menten K M and Hermsen W 1985 *Astron. Astrophys.* **147** L19–22

Wilson T L, Walmsley C M, Snyder L E and Jewell P R 1984 *Astron. Astrophys.* **134** L7–10

Wilson W J and Barrett A H 1968 *Science* **161** 778–9

—— 1972 *Astron. Astrophys.* **17** 385–402

Wilson W J, Barrett A H and Moran J M 1970 *Astrophys. J.* **160** 545–71

Winnberg A, Baud B, Habing H J and Olnon F M 1985 *Astrophys.* **291** L45–50

Winnberg A, Graham D, Walmsley C M and Booth R S 1981 *Astron. Astrophys.* **93** 79–84

Wood P R, Bessel M S and Whiteoak J B 1986 *Astrophys. J.* **306** L81–4

Wouterlout J G A, Brand J and Henkel C 1988 *Astron. Astrophys.* **191** 323–40

Wouterlout J G A, Habing H J and Herman J 1980 *Astron. Astrophys.* **81** L11–2

Wouterlout J G A and Walmsley C M 1986 *Astron. Astrophys.* **168** 237–47

Wright M C H and Plambeck R L 1983 *Astrophys. J.* **267** L115–20

Yorke H W 1986 *Ann. Rev. Astron. Astrophys.* **24** 49–87

Zuckerman B, Palmer P, Penfield H and Lilley A E 1968 *Astrophys. J.* **153** L69–76

Zuckerman B, Yen J L, Gottlieb C A and Palmer P 1972 *Astrophys. J.* **177** 59–78

Rep. Prog. Phys. **54** (1991) 1069–1122. Printed in the UK

Intergalactic matter

A C Fabian† and X Barcons‡

† Institute of Astronomy, Madingley Road, Cambridge CB3 0HA, UK
‡ Departamento de Física Moderna, Universidad de Cantabria, 39005 Santander, Spain

Abstract

Intergalactic matter is best seen in clusters of galaxies, where the mass of the diffuse intracluster medium exceeds that of the galaxies. Apart from some clouds observed in absorption against background quasars, there are few signs elsewhere of matter lying between galaxies, althouth it is most unlikely that galaxy formation was so efficient that all the baryons were turned into galaxies or that galaxies once formed did not eject any gas into intergalactic space. The major reason that so little gas is seen is that even large amounts of it are difficult to observe.

The review begins with a brief discussion of the state of our own interstellar medium and follows with a detailed study of the diffuse medium in clusters of galaxies. The distribution of the intracluster medium and the role of cooling in the cores of many clusters are given special attention. The similar properties of the hot gas in massive early-type galaxies are also discussed. Clusters and massive early-type galaxies contain the best-studied bodies of diffuse gas known. Observations made over a very wide range of wavelengths, from the radio to x-ray bands, show that the intracluster medium is complex and multiphase, consisting mainly of hot diffuse gas, but with cold embedded clouds in the cluster core.

The possible properties of a diffuse intergalactic medium are then reviewed, including heat sources and the allowed degree of clumping. The most crucial observations are the absence of any measurable distortion to the microwave background spectrum and the lack of an absorption trough shortward of the wavelength of Lyman α in the spectra of quasars. There is as yet no clear evidence for any diffuse intergalactic medium, although it is widely predicted to exist and could be the dominant baryonic component in the Universe.

Finally, the intergalactic absorption-line clouds, both those forming the 'Lyman α forest' in the spectra of distant quasars and the metal-rich systems, are discussed in detail.

This review was received in its present form in April 1991.

133

Contents

1. Introduction

1.1. Prologue

Diffuse gas is found throughout the Universe. Although it is probably the dominant baryonic constituent, astronomers took a long time to recognize its existence. Interstellar and intracluster gas are now observed routinely but truly intergalactic gas is elusive because of its low density. An understanding of such gas is very much dependent upon what can be seen, so some appreciation of the processes by which the gas is detected is needed.

The history of the development of our understanding of intracluster and intergalactic gas has, to some extent, parallelled that of interstellar gas in our own galaxy. Basic detection by ground-based optical and radio, or space-based x-ray, techniques was followed first by the mapping of the distribution of the gas and then by a theoretical grasp of the role of heating and cooling processes. Later it was realized that in most interesting cases the gas is multiphase, turbulent, and generally complicated. A good example of this is the discovery and study of the interstellar medium in early-type galaxies. All luminous objects presumably formed from gas and its study is of great importance for our understanding of those formation processes.

The physical relation between gas inside and outside galaxies is, generally speaking, poorly understood, although it is likely to be important. Only in clusters of galaxies are imprints of intergalactic gas detected in the interstellar medium of the member galaxies. For example, spiral galaxies in rich clusters are usually deficient in cold neutral gas, presumably due to the presence of the high-pressure intracluster medium (ICM). Also, in most clusters of galaxies, cooling of intracluster gas in their cores leads to massive inflows of cooled gas onto the central dominant galaxy at rates of several hundreds of solar masses per year. Thereby the central galaxy, typically the largest galaxy in the cluster, continues to form.

The first step towards the detection and understanding of gas between galaxies is provided by an insight into our own interstellar medium (ISM). We review the current understanding in section 1. The multiphase structure of this gas shows that different observational techniques have to be used to detect cosmic gas. There are, however, differences between the physical conditions in a spiral galaxy, where centrifugal forces support a widespread distribution of gas, and those in spherical, pressure-supported, systems such as early-type galaxies or clusters. In order that gas is widely distributed in those cases, its sound speed must be comparable with the local velocity dispersion, i.e. speeds of a few hundred $km\,s^{-1}$ and temperatures of 10^6–10^7 K in galaxies or 500–1200 $km\,s^{-1}$ and 10^7–10^8 K in clusters. This means that hot gas, emitting mainly at x-ray wavelengths, has to dominate in spherical systems. The extent and behaviour of gas in spherical systems—clusters of galaxies and early-type galaxies—is discussed in section 2. Whereas the mass of gas in a normal spiral galaxy represents less than about ten per cent of the mass in stars, the fraction becomes unity or greater for clusters of galaxies. We observe here the accumulated stellar mass-loss and infall from several hundred galaxies. The ICM exerts a measurable influence on the galaxies and may prove to be the best probe of the underlying dark matter.

Comparatively little is known about truly intergalactic gas, which is discussed in section 3. There are strong arguments for its existence, including the simple argument that it would be surprising if galaxy formation were 100 per cent efficient. Indeed, most of the baryons in the Universe probably reside in intergalactic gas. Where intergalactic gas is most clearly observed, it is in clouds which absorb the light of distant QSO. Not all of these absorption clouds are obviously associated with galaxies and they probably represent a small fraction of the intergalactic gas that is present. We show that most of the intergalactic medium (IGM) is highly ionized and is now in the temperature range $10^4 \, \text{K} < T_{IGM} < 10^6 \, \text{K}$. Finally, we identify some problem areas and consider some likely fields of observational progress.

In all the cosmological arguments, a standard Friedmann matter-dominated universe is assumed with $H_0 = 50 \, \text{km s}^{-1} \, \text{Mpc}^{-1}$ and $q_0 = 1/2$.

1.2. Observations of interstellar gas

Interstellar gas is simply the gas between the stars in a galaxy. In this chapter, we discuss mainly the gas in our own galaxy (see Thronson and Shull 1990, for discussions of other galaxies). It is concentrated in the galactic plane with a mean density of $\sim 1 \, \text{cm}^{-3}$ and a mean thickness of a few 100 pc. The abundances of the elements are roughly similar to those in the Sun, with hydrogen dominant. About 20 per cent of the mass (or more) is in molecular form and about 1 per cent is in dust grains. As we discuss, the ISM is very inhomogeneous and turbulent.

The detailed properties of the ISM are well-studied by absorption techniques, the depth and profiles of absorption lines being measured in the spectra of distant stars. The atoms of most interstellar gas reside in their ground state and so resonance absorption lines (such as the hydrogen Lyman series) from that state are the strongest lines. For the commonest elements, most of these lines lie in the ultraviolet and x-ray wavebands. Detectable optical absorption lines occur from abundant species that have an excited state a few eV about the ground state. This means that the strongest optical lines are from Na I, Ca II (the H + K lines), Ca I, K I, Ti II and Fe I. These lines are generally very narrow and are useful for determining velocity structures.

A particularly important line for ground-based observations of neutral hydrogen is the radio 21 cm line, which is due to a spin flip of the electron and proton at a rate of $3 \times 10^{-15} \, \text{s}^{-1}$. A typical hydrogen atom then takes about one tenth of a galactic rotation to flip. Both absorption and emission are commonly observed by this transition and it has proved to be an extremely powerful tracer of H I. The differential rotation of the galaxy allows, through the Doppler effect, the distribution of neutral hydrogen (H I) to be mapped. Molecular lines (e.g. OH, NH and CO) occur in the radio and submillimetre bands from rotational transitions and are used to probe dense clouds.

Photoelectric absorption is, in principle, an excellent method for determining the column density of most ions along the line of sight to some background source. Detailed measurements must await sensitive high spectral-resolution x-ray detectors (McCammon and Sanders 1990).

Optical recombination lines, in which an electron is captured into an excited state and then decays back to the ground state are common in diffuse nebulae. Most electron–proton recombinations lead to the emission of an Hα (λ6563) photon which, with other lines, makes the appearance of many nebulae characteristically pink. The measurement of diffuse Hα radiation from the galactic plane can be used to infer the electron density. Ionized gas such as H II is readily detected through bremsstrahlung.

Absorption at radio frequencies may also occur. The frequency dependence of pulse arrival times from pulsars (dispersion) is a further probe of such gas.

The methods for detection of interstellar, intracluster and intergalactic gas are summarized in table 1.

Table 1. Detection of diffuse gas.

Waveband	Absorption	Emission	Other
Radio	Free–free 21 cm	21 cm Bremsstrahlung Recombination lines	Pulsar dispersion Faraday rotation Distortion of sources
Infrared		Masers Molecules Dust	
Optical	Dust Ca II, Na I etc	Recombination radiation	
Ultraviolet	Dust Resonance lines	Resonance lines	
X-ray	Photoelectric absorption	Bremsstrahlung lines	
γ-ray			Probe molecular clouds

1.3. The state of interstellar gas

1.3.1. A multiphase medium.
We begin by assuming that the ISM is isobaric, i.e., at constant pressure $P = nT$ (cm^{-3} K). This can be understood as a special case of hydrostatic equilibrium where the gravitational potential is constant. The specific cooling rate (i.e., the energy lost by cooling processes per unit time and unit volume) is $q = n^2\Lambda(T)$, where n is the gas density and $\Lambda(T)$ the cooling function which acounts for the different physical processes that cause the cooling (see next section for details). The gas is thermally unstable if $q \propto T^\gamma$ with $\gamma < 0$, or, in other words, if the cooling function $\Lambda(T)$ does not grow as fast as T^2. This happens in most of the regimes. Note that all gas is cooling, but gas which is slightly denser finally cools much more rapidly than the rest.

Let there now be some heat source present. We assume that it is distributed and has a specific rate $n\Gamma$ erg^{-3}s^{-1} (most heating rates are proportional to density), where Γ is independent of T. Then an equilibrium, in which heating and cooling balance, occurs where

$$n\Gamma = n^2\Lambda \tag{1.1}$$

i.e.

$$\Lambda = \frac{\Gamma}{n} \propto \frac{T\,\Gamma}{P}. \tag{1.2}$$

An equilibrium of this kind can often be found around 10^4 K due to the steep increase of cooling as hydrogen (and helium) are ionized. At high heating rates and low densities the gas tends to heat up and, if gravity is the sole confining agent, will expand out of the system and form coronal gas. At low heating rates and high densities the gas can cool and condense into molecular clouds (and stars). Three phases are thus common (cold, $T < 100$ K; warm, $T \sim 10^4$ K and hot, $T > 10^6$ K).

1.3.2. The cosmic cooling curve. The sum of all the emission processes (bremsstrahlung, recombination radiation, etc) leads to the cosmic cooling curve $\Lambda(T)$ (figure 1) which is a measure of the volume emissivity per unit particle density (Λ in units of $erg^3 s^{-1}$) as a function of temperature. The shape of this curve is dominated by bremsstrahlung at the highest temperatures ($> 10^7$ K for a cosmic plasma) since most of the common elements are completely ionized. Then as lower temperatures are reached ($< 10^7$ K) more elements recombine and their recombination and line radiation dominates. Finally, around $\sim 10^4$ K, the hydrogen recombines and electron collisional excitation and other processes become important. A very rough approximation for Λ in the temperature range $10^5 < kT < 10^9$ K is

$$\Lambda \simeq 10^{-19}T^{-0.5} + 10^{-27}T^{0.5} \, erg \, cm^3 \, s^{-1}. \tag{1.3}$$

The cosmic cooling curve represents the radiative energy loss of optically-thin (no self-absorption) gas in collisional ionization equilibrium.

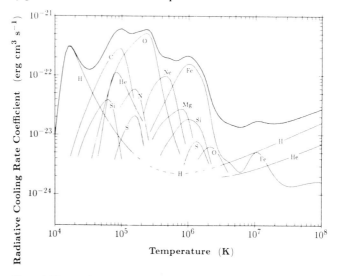

Figure 1. The cosmic cooling curve for an optically thin gas of solar abundances (kindly provided by H Bohringer). The contributions of the major elements is indicated.

Rapid cooling (see e.g. Shapiro and Moore 1976) or a strong photoionizing flux (see e.g. Krolik *et al* 1981) can change the cooling rate considerably. Much of the observed low density interstellar gas is in photoionization equilibrium where the rate of photoionizations is balanced by recombinations

$$n_{HI} \int U_\nu \sigma_{pe} \, d\nu = n_e n_i \alpha_r, \tag{1.4}$$

where n_{HI}, n_e and n_i are the neutral hydrogen, electron and ion densities, respectively and U_ν is the number density of ionizing photons ($L/4\pi R^2 h\nu$ for a point source), σ_{pe} is the photoelectric absorption cross-section and α_r the recombination rate coefficient. The ratio of ionizations per unit volume ($\propto nL/R^2$) to recombinations ($\propto n^2$) is proportional to $\xi = L/nR^2$, so that ξ can be a useful measure of the ionization conditions for a particular ionizing spectrum. Curiously, the ionization state of the cool, diffuse ISM is similar to that of the photosphere of the Sun (i.e. ξ is similar).

The cosmic cooling curve at temperatures below 10^3 K is dominated by the collisional excitation of low lying levels in ions such as C II and Si II. These elements have ionization potentials below 13.6 eV, the ionization energy of hydrogen. Most of the excitation is provided by electrons and the fraction of electrons present in the gas is crucial. Much of the gas can then be optically thick. Molecular line radiation and energy exchange with dust are often very important at low temperatures ($T < 200$ K).

1.3.3. Heat sources. Heat sources commonly considered are cosmic rays and soft x-ray ionization (Field *et al* 1969), mechanical heating such as supernovae and stellar winds (Cox and Smith 1974; Castor *et al* 1975; McKee and Ostriker 1977) and photoionization from hot stars. Only the first two are likely to be distributed. Photoionization means that Γ is a function of the ionization state and thus of T and P. Consequently, the situation is more complicated than outlined in 1.3.2. Different heat sources apply in different regions and for different temperature gas. Also, flows change the energy equation (e.g. by adding adiabatic cooling or shock heating). This means that the concepts of the 'violent interstellar medium' (McCray and Snow 1979) and the 'galactic fountain' (Shapiro and Field 1976; Bregman 1980) are quite appropriate.

Massive stars put much energy into the interstellar medium through ionization fronts (totalling 5×10^{39} erg s^{-1}) stellar winds (10^{41}–10^{42} erg s^{-1}) and supernovae (10^{42} erg s^{-1}). They do occur preferentially in regions of denser gas clouds close to where they were formed and so it is not clear that this energy is deposited directly into a typical part of the interstellar medium. Many of the changes in our understanding of the ISM over the past two decades have parallelled the appreciation of the importance of the various heat sources.

1.3.4. The importance of supernova remnants. At the present time there is some debate as to the role of supernova heating and the volume filling factor of the hot phase. The maximum radius, R_{max}, that a supernova remnant expands to is roughly obtained when there is pressure equilibrium of the remnant and the external ISM pressure. Typically $R_{max} \approx 100$ pc if $nT \sim 10^{4-3}$ K. Such a remnant will last at least one sound crossing time before interstellar 'noise' etc disrupts it. At a rate of 1 supernova per 100 yr, each of which blows a hole of area πR_{max}^2, the total fraction of the galactic disc covered in holes exceeds unity if $R_{max} > 30$ pc. We therefore see that much of the galaxy could be filled with the hot phase so that the surface filling factor $f_A > 0.5$. Cox and Smith (1974) suggested that a network of tunnels of hot gas would occur as the supernova remnants overlap. They model the ISM with a volume filling factor $f_V \sim 0.5$.

Thermal conduction is neglected in Cox's model. Given that the multiphase medium assumes a mix of cold and hot gas then conduction could be important. The timescale, t_{evap}, on which a cold spherical cloud of radius R embedded in hot gas would be evaporated by electron conduction is obtained by considering Spitzer conductivity. Then

$$t_{evap} \approx 3 \times 10^5 n \left(\frac{R}{pc} \right)^2 \left(\frac{T}{10^6 \, K} \right)^{-5/2} \text{yr.} \tag{1.5}$$

(The above formula should be modified if the electron mean free path is greater than R; Cowie and McKee 1977.)

McKee and Ostriker (1977) have built a detailed model of the ISM in which evaporating clouds are an integral part. Partially ionized edges to these clouds explain the observation of intervening O VI absorption found in stellar spectra. They consider that the hot phase has $f_V \sim 0.75$.

1.3.5. The galactic halo.

A hot galactic halo was predicted by Spitzer (1956) and discovered observationally by Savage and de Boer (1981). In between these dates some 21 cm gas was found beyond the galactic disc. Much of this gas is in the form of high velocity clouds, many of which are falling onto the disc. The hotter gas was found through its Si IV and C IV absorption along lines-of-sight to the Magellanic Clouds. The ratio of their column densities indicate that the gas temperature is 8×10^4 K. This gas is near the top of the cooling curve and extends out to a few kpc. There is no evidence yet for gas with temperatures greater than 4×10^5 K at higher distances from the galactic disc, although a small halo at $\sim 3 \times 10^6$ K may exist around the galactic Centre (Garmire, quoted by York 1982). The total mass of gas in the halo is less than $5 \times 10^8 \, M_\odot$. The data are consistent with the halo corotating with the disc (indeed this is an important assumption in deducing the location of the gas).

1.4. Current understanding of our ISM

1.4.1. Distribution of gas and heat sources.

Recent summaries of the ISM are given by Cowie (1986), Heiles and Kulkarni (1987), Heiles (1987), Gusten and Mezger (1982) and in the conference proceedings edited by Hollenbach and Thronson (1987). The picture is that of a dynamic medium consisting of dense cold clouds embedded in a warm intercloud medium. Pervading that is a hot phase of debatable filling factor. The medium is stratified in the perpendicular distance from the plane with coldest and densest material having the smallest scale height.

The mean gas density at the centre of the plane, $\bar{n} = 1.2^{-3}$ and the total mass of H I gas is $5 \times 10^9 \, M_\odot$. The mass of $H_2 \sim 1$ to $5 \times 10^9 \, M_\odot$ (it is extrapolated from measurements of the less abundant but more readily detected CO molecule) and a further $10^9 \, M_\odot$ may be in the warm intercloud medium. The surface density of H I is roughly constant over galactic radii of 4–20 kpc with a constant scale height of 135 pc. The scale height increases rapidly (with a warp) beyond 20 kpc. The pressure $\sim 4000^{-3}$ K. It is common to assume that most of the ISM is in, or close to, pressure equilibrium. The electron density in the ionized gas, $\bar{n}_e \approx 0.05^{-3}$ with a scale height 260 pc. Much of the halo gas appears to be infalling (Albert 1983). There is little evidence for higher halo gas and the existence of a million degree corona is controversial. Whether the total mass of the ISM is roughly constant with time, increasing due to accretion or decreasing because of star formation, is not yet known.

The primary energy sources for the ISM are the massive stars formed in giant molecular clouds (GMC). They provide photoelectric heating for the cold gas and mechanical heating for the warm.

The filling factor in hot gas is uncertain. McKee and Ostriker (1977) predict a large filling factor and Heiles (1987) estimates $f_V \sim 90$ per cent from the 'standard' supernova rate. However, estimates of the filling factor of cold gas lie between 40 and 80 per cent (Heiles 1987). These estimates are fairly secure. The most likely resolution of this problem lies in the estimate of the heating rate by massive stars. The 'standard' supernova rate may have been overestimated and, more important, the massive stars tend to form in associations (OB associations). Consequently, most of the supernova and winds act at a few localized 'hot spots' along the spiral arms (see e.g. McCray and Snow 1979). Much of their energy can then be radiated away or vented mechanically into the corona.

A further uncertainty lies in the dust content of the galaxy (dust absorbs the ultraviolet light and then reradiates the energy in the infrared). Recent interpretation of the appearance of spiral galaxies lying at a variety of inclination angles suggests that they may be optically thick at visible wavelengths (Valentijn 1990).

Most theoretical descriptions of the ISM assume that the cold matter is in clouds that are roughly spherical. Observations do not support this and indicate a wide range of structures, few of which are spherical. Folded sheets and ropes are more plausible. The full role of magnetic fields and interstellar turbulence has yet to be understood (see Scalo 1990 for an interesting discussion).

1.4.2. Star formation. Little, if any, of our ISM is primordial gas. It has been processed through stars and chemically enriched. The source of the enriched gas is stellar winds, planetary nebulae and supernovae. The major sinks of gas are star formation and galactic mass loss.

Star formation is a complicated business, the details of which are not well known. The estimates of the numbers of stars formed in a given mass range, the initial-mass-function (IMF), are questionable. It is a general, and strongly held, assertion that the IMF changes little from place to place (the 'universal' IMF). There is no strong justification for this as the observed IMF is an average over many regions that could differ considerably. A good discussion of all aspects of the IMF is given by Scalo (1986).

In order for a star to form, a region of gas has to become gravitationally bound. This requires the mass of the cloud to exceed the Jeans mass for that temperature, which means that the cloud lies within its own accretion radius;

$$R_a = \frac{GM_J}{c_s^2} \tag{1.6}$$

where c_s is the sound speed in the gas. Then

$$M_J \approx \left(\frac{kT}{G\mu m} \right)^{1/2} \frac{1}{\sqrt{\rho}}. \tag{1.7}$$

If the gas radiates during its collapse then it can remain isothermal and M_J reduces as the density ρ increases. Smaller lumps within the initial cloud are then Jeans unstable and the cloud can fragment (Hoyle 1953). This process stops when the sub-clouds are opaque and cannot remain isothermal. The mass of the fragments is then $\sim 0.01 M_\odot$ (Rees 1977). To obtain the stars observed then requires subsequent coalescence or a lack of fragmentation (see e.g. Silk 1980).

Recent theories of star formation (see Shu *et al* 1987) doubt the importance of fragmentation, concentrating instead on the formation of a stellar core and the subsequent accretion of surrounding gas. The final mass of a star is in part determined by the amount of such gas accreted, which takes place through a disc owing to the angular momentum of the initial gas cloud. The disposal of angular momentum and of magnetic fields, both of which act to oppose gravitational collapse, are crucial elements of these theories.

The star formation rate in the 4–8 kpc ring where most of the molecular gas (e.g. CO) lies is $\sim 3 \, M_\odot \, \text{yr}^{-1} \, \text{kpc}^{-1}$ if an IMF extending over all masses (0.16 to $50 \, M_\odot$) is used. However, the total mass of stars observed in that ring requires a rate of only $\sim 0.5 \, M_\odot \, \text{yr}^{-1} \, \text{kpc}^{-1}$ over a galactic age. Agreement can be obtained only if there is a high ($> 1 \, M_\odot$), low-mass cut-off to the IMF. This problem also implies that the star formation rate in our galaxy has not necessarily decreased with time.

1.4.3. A tentative summary. The interstellar medium in the Milky Way is dynamic. A possible overall steady-state occurs with $\sim 1 \, M_\odot \, \text{yr}^{-1}$ of (mainly) low-mass star formation proceeding throughout the disc fairly uniformly. The mean filling factor of the cold and warm gas in the plane of the disc is at least 50 per cent. Some of the dense clumps of gas coagulate into the GMC within which massive star formation spasmodically breaks out.

One implication from this is that the existence of massive stars may be a poor diagnostic of the overall star formation rate of a region. The formation of massive stars may be a 'disease' associated with the GMC.

A fairly clear idea emerges from the study of the ISM: environmental effects (heating sources, etc) are important for the state of the gas and *vice versa*. This is also true in the intracluster gas and to some unknown extent in the diffuse intergalactic medium.

2. Intracluster gas

2.1. General properties

Vast regions of intergalactic gas occur in clusters and groups of galaxies. The sound speed of the gas is similar to the velocity dispersion of the cluster which is typically 500 to 1200 km s^{-1} and so the gas predominantly radiates x-rays. This emission was discovered in the early 1970s with the Uhuru satellite (Gursky *et al* 1971; although M87 had been detected earlier, Byram *et al* 1966). Thermal bremsstrahlung and line radiation were confirmed as the source of the emission by the discovery of a strong iron emission line due to Fe XXV and Fe XXVI (Mitchell *et al* 1976, see figure 2). Diffuse x-radiation is the principal source of information on the intracluster medium (ICM). There is indirect evidence for the gas in 'head–tail' radio sources and from theories of the propagation of double-lobe radio sources. A comprehensive review of the properties of the ICM is given by Sarazin (1986).

Much intracluster gas has an electron density, n_e, in the range of 10^{-4}–10^{-2} cm^{-3} and a temperature $T \sim 10^7$–10^8 K, and is observed over regions of 1 to 2 Mpc. The total bremsstrahlung luminosity, L_x, is $\sim 10^{43} - 3 \times 10^{45} \, \text{erg} \, \text{s}^{-1}$. The 6.7 keV iron emission line is observed in all clusters that are bright enough for a detection to be made (see e.g. Rothenflug and Arnaud 1986) and the gas has 0.2–0.3 times solar abundance in iron (Edge 1989). The work of Canizares *et al* (1982) and Mushotzky *et al* (1981) on cooling flow clusters shows O, Ne, Si and S also, with O probably overabundant compared to solar material.

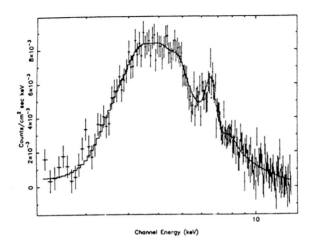

Figure 2. The raw x-ray spectrum of the Perseus cluster of galaxies obtained with the EXOSAT gas scintillation proportional counter. Note the strong emission line due to iron at 6.7 keV.

Some cluster properties show correlations which are fairly strong if just x-ray quantities are used (e.g. x-ray luminosity L_x versus gas temperature T_x; figure 3). The temperature correlates with optically-determined velocity dispersion (which is a notoriously uncertain quantity) and with the central galaxy number density. Generally, the deeper the potential well, the more gas and galaxies it contains and the more luminous it is in x-rays.

Evolution of the x-ray luminosity of clusters has recently been discovered (Edge *et al* 1990; Gioia *et al* 1990). The evolution is in the sense that the most luminous clusters (those with $L_x > 10^{45} \, \mathrm{erg \, s^{-1}}$) have the greatest space density at the current epoch and were less numerous at earlier times, even as recent as redshifts, z, of 0.15. Edge *et al* (1990) interpret this as due to the mergers of smaller subclusters to form larger clusters, a process supported by the observation that clusters are themselves strongly clustered (Bahcall and Soneira 1983; Lahav *et al* 1989) and also contain much substructure (Geller 1984). This means that the luminous clusters are each assembled from up to 10 separate subclusters. Mergers have important consequences for the state of the ICM, which will not then be completely hydrostatic, uniform and homogeneous.

2.2. The gas distribution

The intracluster gas acts as a fluid on galactic scales. The electron–electron coupling time, $t_{e-e} \approx 2 \, 10^5 T_8^{3/2} n_{-3}^{-1} \, \mathrm{yr}$, where $T = 10^8 T_8 \, \mathrm{K}$ and $n = 10^{-3} n_{-3}$, and the

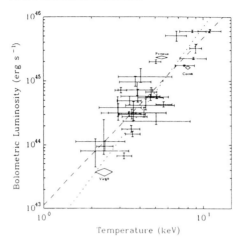

Figure 3. The correlation between bolometric x-ray luminosity and the measured x-ray temperature of the ICM (from Edge 1989).

electron–ion coupling time is 1840 times larger. The crossing time of the cluster exceeds this,

$$t_{\text{cross}} = \frac{R}{\langle v^2 \rangle^{1/2}}$$
$$\approx 10^9 R_{\text{Mpc}} v_8^{-1} \text{ yr}. \tag{2.1}$$

where $v = v_8 10^8 \text{ cm s}^{-1}$ is the velocity dispersion of the cluster galaxies.

The sound crossing time of the gas,

$$t_s = \frac{R}{c_s} \approx t_{\text{cross}} = 10^9 R_{\text{Mpc}} T_8^{-1/2} \text{ yr} \tag{2.2}$$

where the size of the region is R_{Mpc} Mpc. This is much less than the age of the cluster (presumably H_0^{-1}) provided that $R \lesssim 5 \text{Mpc}$. Any radial flows of the gas must therefore be subsonic otherwise impossibly large mass flow rates are implied ($\gg 10^4 \text{ M}_\odot \text{ yr}^{-1}$). Intracluster gas is then close to hydrostatic support so that

$$\frac{\mathrm{d}P_{\text{gas}}}{\mathrm{d}r} = -\rho_{\text{gas}} \frac{\mathrm{d}\phi}{\mathrm{d}r} = -\rho_{\text{gas}} g. \tag{2.3}$$

This means that measurements of P_{gas} and ρ_{gas}, the gas pressure and density as a function of radius r (i.e. $n_e(r)$ and $T_e(r)$), allows $\phi(r)$ and so $M_{\text{cluster}}(r)$ to be determined. Much of the mass of a cluster, M_{cluster} is dark matter and x-ray observations can probe its distribution. The gas will arrange itself (convect) so that isobaric surfaces are on equipotentials which will be roughly spherical, even for quite a lumpy or flattened mass distribution. Analysis therefore assumes spherical equipotentials, or at most two sets of spherical equipotentials (e.g. A754; Fabricant *et al* 1986). We note, however, that the continual infall of subclusters into a cluster will keep the gas stirred and turbulent such that there can be significant subsonic motions (see Fabian 1990, Evrard 1990). Such motions mean that a turbulent pressure term should be added to (2.3) and that x-ray-determined estimates of M_{cluster} from the distribution of n_e and T_e alone will be lower limits on the mass. The addition of gas from subclusters may also make the ICM inhomogeneous, since it is most unlikely from both theory and observations that the different gases mix into a uniform gas instantaneously (or at all).

We now ignore these cautionary remarks and proceed under the assumption that the gas is completely hydrostatic and uniform. This is a useful approximation for the bulk of the gas in an average sense. If the gas is isothermal (which it probably is not) then (2.3) implies

$$kT\frac{dn_e}{dr} = -n_e\mu m\frac{GM(r)}{r^2} \tag{2.4}$$

and

$$\rho_{\text{gas}} \propto \exp\left(-\frac{\phi(r)}{(kT/\mu m)}\right) \tag{2.5}$$

where μ is the mean molecular mass. Also, if the galaxies have an isothermal velocity distribution (also unlikely) then

$$\rho_{\text{gal}} \propto \exp\left(-\frac{\phi(r)}{\sigma_{\text{los}}^2}\right) \tag{2.6}$$

where σ_{los} is the line-of-sight velocity disperion of the cluster. Then

$$\rho_{\text{gas}} \propto (\rho_{\text{gal}})^\beta \tag{2.7}$$

where

$$\beta = \frac{\mu m\sigma_{\text{los}}^2}{kT} \tag{2.8}$$

(introduced by Cavaliere and Fusco-Femiano 1976 as τ). We might expect $\beta \sim 1$ so $c_s^2 \approx \sigma_{\text{los}}^2$. However, its application indicates otherwise. The above isothermal–isothermal model suggests the use of King's (1966) approximation of an isothermal distribution where

$$\rho_{\text{gal}}(r) = \rho_{\text{gal}}(0)(1 + (r/a)^2)^{-3/2} \tag{2.9}$$

so that

$$\rho_{\text{gas}}(r) = \rho_{\text{gas}}(0)(1 + (r/a)^2)^{-3\beta/2} \tag{2.10}$$

where a is the core radius. This can then be fitted to x-ray images of clusters (emissivity$\propto \rho_{\text{gas}}^2$). Jones and Forman (1984) obtain

$$\langle \beta_{\text{image}} \rangle = 0.65 \tag{2.11}$$

so

$$\rho_{gas} \approx \rho_{gas}(0)(1 + (r/a)^2)^{-1}. \tag{2.12}$$

On the other hand, x-ray spectral and optical velocity dispersion measurements of T and σ_{los} sometimes give

$$\langle \beta_{spec} \rangle \approx 1.2 \approx 2 \langle \beta_{image} \rangle. \tag{2.13}$$

The reasons for this discrepancy are not yet clear, although it is unlikely that the gas and galaxies are isothermal. It many cases there is good agreement (Edge 1989) but for the well-studied Coma and Perseus clusters, for example, where the ratio of temperatures exceeds unity and the ratio of velocity dispersions is less than unity, respectively, there is clearly a discrepancy. Two explanations have been made, the first assumes that there is considerable velocity anisotropy in the galaxy distribution (Solanes and Salvador Solé 1990) and the second that there is considerable turbulent pressure in the gas, perhaps because of a recent subcluster merger (Fabian 1990).

Earlier attempts to overcome a lack of knowledge of the equation of state of the cluster gas assumed that it is polytropic i.e. $P \propto \rho^\gamma$. This does not necessarily mean that γ is the ratio of specific heats, and it is little more than a mathematical expediency. Using it in the equation of hydrostatic support yields

$$\frac{\gamma}{(\gamma - 1)} \frac{k}{\mu m} \frac{dT}{dr} = -\frac{d\phi}{dr} \tag{2.14}$$

so that

$$T = T_c + \frac{(\gamma - 1)}{\gamma} \frac{\mu m}{k} (\phi_c - \phi) \tag{2.15}$$

and

$$\frac{\rho}{\rho_c} = \left(1 + \frac{(\gamma - 1)}{\gamma} \frac{(\phi_c - \phi)}{(kT_c/m)} \right)^{1/(\gamma - 1)}. \tag{2.16}$$

The subscripts refer to values at the centre. The density equation limits to the exponential isothermal form as $\gamma \to 1$.

The polytropic approach is still in common use, especially when needing to extrapolate to large radii. There is no particular reason, however, why γ has to have a single fixed value throughout a cluster. The core radius is also poorly defined. Most estimates of the total mass of gas in a cluster rely on some large extrapolation assuming γ is constant. This is the main reason why there are conflicting results in the literature (see e.g. Cowie *et al* 1987). In the best determination of the overall properties of the ICM in the Coma cluster, Hughes *et al* (1988) find some small departure from isothermality and discuss the limits of polytropic models in more detail. Hughes (1989) later extends the study beyond a Mpc to obtain good limits on the total mass of the cluster and of the ICM within about 3 Mpc. He shows that the ICM may become a large fraction of the cluster mass (20–50 per cent) at these radii, compared with being about 10 per cent within the inner 0.5 Mpc.

It is possible to obtain gas density and temperature profiles without assuming an equation of state (Fabian *et al* 1981). The x-ray surface brightness profile can be deprojected, assuming some geometry (e.g. spherical) and a distance to the cluster, to yield count emissivities as a function of radius. The emissivity depends upon n_e and T_e as well as the detector response and the effects of intervening photoelectric absorption, the last two of which are assumed known. A further relationship between n_e and T_e is obtained from the pressure via the equation of hydrostatic equilibrium. The densities obtained in this way are usually determined to better than 10 per cent, whereas the temperatures are somewhat dependent upon the assumed value of g (usually estimated from σ_{los}) that is used in the hydrostatic equilibrium equation. One pressure, typically the outer pressure, is required to start the solution and this is usually adjusted so that most of the cluster gas has a temperature consistent with x-ray spectral measurements. These spectra have usually been obtained with wide field-of-view experiments. When spatially resolved spectra become available from x-ray satellites such as ROSAT, ASTRO-D, AXAF and XMM, then we shall be able to solve directly for n_e and T_e without requiring g (which can then be measured, of course).

2.3. The Sunyaev–Zeldovich effect

X-ray emission is not the only means of observing the hot intracluster medium. The low energy photons of the cosmic microwave background can be Compton-scattered on passing through the gas and experience an energy shift, $\Delta\epsilon$, given by

$$\left\langle \frac{\Delta\epsilon}{\epsilon} \right\rangle \approx \frac{4kT_e}{m_e c^2} \tag{2.17}$$

where ϵ is the initial photon energy. The proportion of photons scattered is given by the Thomson depth,

$$\tau_T = \sigma_T \int n_e \, dl. \tag{2.18}$$

This leads to a microwave dip in the direction of the cluster in the Rayleigh–Jeans part of the cosmic blackbody spectrum

$$\frac{\Delta T}{T} = -\frac{2k\sigma_T}{m_e c^2} \int n_e T_e \, dl. \tag{2.19}$$

This is the Sunyaev–Zeldovich (1969) effect and measures the integral of the pressure along the line of sight. It has now been detected in several clusters (Birkinshaw *et al* 1984). Since the x-ray luminosity is proportional to $\int n_e^2 T_e^{1/2} \, dl$, a combination of x-ray and microwave measurements can lead to an independent estimate of cluster distance. To understand this, consider a spherical region of gas of radius R at distance D, subtending an angle θ at the observer. The x-ray luminosity

$$L_x \propto n_e^2 T_e^{1/2} R^3 \tag{2.20}$$

and

$$\frac{\Delta T}{T} \propto n_e T_e R. \tag{2.21}$$

The observer measures the flux

$$F_x \propto \frac{L_x}{D^2} \propto n_e^2 T_e^{1/2} \theta^3 D \tag{2.22}$$

so, substituting for n_e,

$$F_x \propto \left(\frac{\Delta T}{T}\right)^2 \frac{\theta}{T_e^{3/2}D}. \tag{2.23}$$

Measurements of F_x, $\Delta T/T$, and T_e then lead to an estimate of D. If the redshift of the cluster is known, say from optical measurements, then the Hubble constant is obtained from one cluster and q_0 from two! Of course in practice, the cluster is not a uniform isothermal sphere of gas and a statistical approach has to be made. Van Speybroeck (1987) estimates that H_0 will be determined to an accuracy of 10 per cent with AXAF, a factor of ten improvement on the current uncertainty. The present best measurements of H_0 by this method, using the clusters A665 (Birkenshaw *et al* 1991) and A2218 (McHardy *et al* 1990), give values of $48 \, \mathrm{km \, s^{-1} \, Mpc^{-1}}$ and $24 \pm 10 \, \mathrm{km \, s^{-1} \, Mpc^{-1}}$, respectively. Unlike the second estimate, the first uses a consistent model for both the x-ray and radio measurements and gives good agreement with a commonly adopted value for H_0.

2.4. The origin of the ICM and galaxy stripping

The intracluster medium is hot and chemically enriched. The mass of gas is comparable to that in stars (i.e. excluding dark matter) and so current stellar mass-loss is relatively unimportant. The origin of its heat is simple to explain since it just has the same specific kinetic energy as the galaxies (i.e. $\beta \sim 1$); it is gravitational in the same way that the motions of the galaxies are gravitational in origin. There may, of course be some extra heat from supernovae and outflows from member galaxies and also some widespread cooling could have occurred in the past.

The origin of the metal enrichment is explained by supernovae during the formation of the stellar populations in the member galaxies and was predicted before the iron-emission line in the ICM was discovered (Larson and Dinerstein 1975). Following the discussion of cluster enrichment by Thomas and Fabian (1990), we note that supernovae from 'normal' star formation (i.e. with a galactic IMF) enriches and heats above the escape temperature most of the gas in galaxies with shallow potential wells (say $T_{\mathrm{virial}} <$ few $\times 10^6 \, \mathrm{K}$). Much of the enriched gas is thereby lost from that galaxy and because only a small fraction of the gas mass need form stars before the gas is expelled, the metallicity is low. Only in massive galaxies is much gas retained and even there at least half the gas is expelled with a metallicity of 0.1–0.3 times the solar value. The observed cluster abundance of ~ 0.3 solar is explained if the last phase of star formation was associated with potential wells with $T_{\mathrm{virial}} \sim 10^7 \, \mathrm{K}$ i.e. the brighter member galaxies. This picture is consistent with the hierarchical formation of clusters.

Gas remaining bound to an individual galaxy may be liberated in a galaxy–galaxy collision (Spitzer and Baade 1951) or by ram-pressure stripping in the ICM. This can be understood by considering a spherical mass of gas of density ρ_i and radius R in a galaxy of mass M and velocity dispersion σ falling through a medium of density ρ_0 at velocity v. The ram force due to the ICM

$$F_{\mathrm{ram}} \approx \rho_0 v^2 \pi R^2 \tag{2.24}$$

and the maximum gravitational restoring force is

$$F_{\mathrm{grav}} \approx \rho_i \frac{4}{3}\pi R^3 \frac{GM}{R^2}. \tag{2.25}$$

The gas is then pushed out if

$$\rho_0 v^2 > \frac{4\rho_i GM}{3R} \tag{2.26}$$

so for stripping, $\rho_0 v^2 \gtrsim 3\rho_i\sigma^2$. This criterion is for sudden stripping of gas that has previously accumulated in the galaxy. If there is continuous mass loss at a rate per unit volume ($\alpha \sim \rho_*/\tau$, where ρ_* is the stellar density and τ is the mass loss time, $\sim 10^{12}\,\mathrm{yr}$) then the condition for continuous stripping is

$$\tfrac{1}{2}\rho_0 v^3 \gtrsim \int \alpha|\phi|\,dz \tag{2.27}$$

where ϕ is the gravitational potential. This is derived from the kinetic energy flux into the gas (Takeda *et al* 1984) and reduces to

$$\rho_0 v^3 \gtrsim 16\Sigma\sigma^2\tau \tag{2.28}$$

where Σ is the column density of stars.

Gas pushed out of a galaxy will be shredded, dispersed and decelerated rapidly on a time

$$t \sim 3\left(\frac{R}{v}\right)\left(\frac{\rho_i}{\rho_0}\right)^{1/2}. \tag{2.29}$$

When a galaxy moves subsonically through the ICM, its ISM may be stripped by viscous processes, turbulence and the effects of the Kelvin–Helmholtz instability (Nulsen 1982). The rate of mass loss is given by

$$\dot{M} \simeq \pi R^2 \rho v \left(\frac{12}{R_e}\right)^{1/2} \tag{2.30}$$

where the Reynold's number $R_e = 3(R/\lambda)(v/c_s)$. If conduction is uninhibited then gas may be evaporated from a galaxy at a rate

$$\dot{M} \approx 8\rho c_s \pi R^2 \tag{2.31}$$

(Cowie and Songaila 1977).

The action of ram-pressure stripping should in part be responsible for the clear lack of spiral galaxies in rich clusters. Edge (1989) finds that the anticorrelation between L_x and the spiral fraction in a cluster is one of the strongest observed in clusters (figure 4).

2.5. Cooling flows

The gas density in the core of many clusters is sufficiently high that the radiative cooling time is less than the age of the cluster (i.e. $t_{cool} < H_0^{-1}$). The weight of the overlying gas then causes the gas density to rise and establishes a cooling flow (see Fabian *et al* 1991 for a recent review). The inflow velocities are highly subsonic. X-ray observations of this phenomenon indicate that the mass flow rates can be substantial

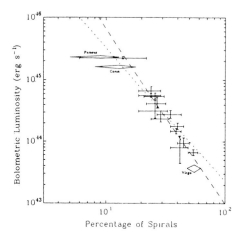

Figure 4. The correlation between bolometric x-ray luminosity of clusters and the fraction of member galaxies which are spirals (from Edge 1989).

and in the range of hundreds of solar masses per year. It appears that we are observing the continued formation of central cluster galaxies.

The cooling time of the gas can be obtained from density and temperature profiles estimated by the method given at the end of section 2.2. Where $t_{cool} < H_0^{-1}$, the rate at which mass is deposited through cooling, $\dot{M}$, is given by

$$L_{cool} \approx \frac{5}{2} \frac{kT}{\mu m} \dot{M}, \tag{2.32}$$

where L_{cool} is the total luminosity within that region. The factor of 5/2 represents the enthalpy of the gas (i.e. the sum of thermal energy and PdV work done). Short cooling times are associated with high gas densities and bright regions and so x-ray images of cooling flows show a highly peaked surface brightness profile. This is common in 30–50 per cent of rich clusters (Stewart *et al* 1984b), in some poor clusters (Schwartz *et al* 1980; Canizares *et al* 1983) and in elliptical galaxies (Nulsen *et al* 1984). Recent studies (Pesce *et al* 1990; Stewart *et al* 1990) show that the fraction exceeds 50 per cent. Most nearby clusters (Virgo, Centaurus, Hydra, Perseus etc) contain cooling flows and it is only those with two dominant central galaxies such as Coma that do not show a sharp x-ray peak. Constraints on more distant clusters are not very strong due to the relatively poor spatial resolution of the commonly used imaging proportional counter (IPC) on the Einstein Observatory, but will be improved with ROSAT.

Unambiguous evidence for cooling is provided by the low x-ray temperature spectral components observed in high resolution spectra. Mushotzky *et al* (1981), Mushotzky and Szymkowiak (1988), Canizares *et al* (1982) and Canizares *et al* (1988) find emission lines characteristic of gas in the 5×10^6 to 2×10^7 K range in the core of the Perseus cluster, where the mean gas temperature is nearer 8×10^7 K. Lines are also observed in M87 (Lea *et al* 1982) and in A496 (Nulsen *et al* 1982). The emission measures ($\int n_e^2 \, dV$) of gas at the lower temperatures agree well with a cooling interpretation. Ulmer *et al* (1987) find evidence for a temperature decrease in the core of the Perseus cluster from a strip scan measurement. Much work has been carried out on the Perseus cluster as it is the brightest x-ray cluster.

2.5.1. Mass deposition in cooling flows. As the surface brightness profile gives $L(r)$ we can determine $\dot{M}(r)$, the mass deposition profile. The x-ray images are not as peaked as they could be if all the matter flowed to the centre (Fabian *et al* 1984). We generally find that

$$\dot{M}(r) \propto r. \tag{2.33}$$

Distributed mass deposition is presumably due to a range of densities present at all radii. The gas is inhomogeneous. Nevertheless, it is instructive to consider the equations of a homogeneous flow, which roughly represent the mean conditions of an inhomogeneous flow. We have the equation of continuity;

$$\dot{M} = 4\pi r^2 \rho v \tag{2.34}$$

the pressure equation (ignoring highly subsonic flow terms)

$$\frac{dP}{dr} = -\rho \frac{d\phi}{dr} \tag{2.35}$$

and an energy equation,

$$\rho v \frac{d}{dr} \left(\frac{5}{2} \frac{kT}{\mu m} + \phi \right) = n^2 \Lambda. \tag{2.36}$$

If the cooling region (where $t_{cool} < H_0^{-1}$) is at constant pressure ($d\phi/dr = 0$), then $n \propto T^{-1}$ and

$$\rho v \frac{d}{dr} \left(\frac{5}{2} \frac{kT}{\mu m} \right) = n^2 \Lambda \tag{2.37}$$

and if

$$\Lambda \propto T^\alpha \tag{2.38}$$

then

$$nv \frac{dn^{-1}}{dr} \propto n^{2-\alpha} \tag{2.39}$$

so

$$v \frac{dn}{dr} \propto n^{3-\alpha}. \tag{2.40}$$

From continuity, if $\dot{M}$ is constant,

$$v \propto n^{-1} r^{-2} \tag{2.41}$$

so

$$\int_{\infty}^{n} \frac{dn}{n^{4-\alpha}} \propto \int_{0}^{R} r^2\, dr \tag{2.42}$$

and

$$n \propto R^{-3/(3-\alpha)}. \tag{2.43}$$

This is proportional to $R^{-6/5}$ for bremsstrahlung. The density rises inward as the temperature falls. Constant pressure is a fair approximation to the core region of a cluster. Gravity is not particularly important, except perhaps for focussing the flow, until the gas has cooled to about the virial temperature of the central galaxy. Then the gas heats up as it flows in further and the pressure rises (Fabian and Nulsen 1977). The flow velocity $v \simeq r/t_{cool}$, which is highly subsonic.

When the flow is inhomogeneous we can estimate $\dot{M}(r)$ by assuming that the gas is composed of a number of phases, the densest of which cools out of the flow at the radius under consideration, or by model fitting (White and Sarazin 1987). In the first approach, the cooling region is divided into a number of concentric shells of size compatible with the instrumental resolution. The luminosity, δL_i of the ith shell can then be considered to be the sum of the cooling luminosity of the gas cooling out at that radius from the mean temperature T_i at rate $\delta \dot{M}$ and the luminosity of gas flowing across the shell experiencing temperature and potential changes ΔT_i and $\Delta \phi_i$;

$$L_i = \delta \dot{M}_i \frac{5}{2} \frac{kT_i}{\mu m} + \left(\frac{5}{2} \frac{k \Delta T_i}{\mu m} + \Delta \phi_i \right). \tag{2.44}$$

In a most detailed approach (Thomas *et al* 1987), allowance has been made for as many phases at a radius as there are shells within that radius and the cooling function and spectrum have been integrated carefully. A typical mass deposition profile is shown in figure 5. It agrees fairly well with that obtained by assuming that the gas is homogeneous, principally because most of the energy is lost on cooling from the average cluster temperature $T_{cluster}$ at temperatures close to $T_{cluster}$. This new approach does allow us to measure the spread of densities in the gas at any radius. It is this which determines the manner in which mass is deposited (Nulsen 1986). It is inferred that the intracluster gas must contain a density spread of at least a factor of two. This may not be surprising when it is recalled that it has been enriched in metals which must have mixed different gases together and that there is a continuing infall of gas-rich subclusters, which may themselves contain cooling gas and anyway contain a wide range of gas densities.

The result that $\dot{M}(r) \propto r$ means that the deposited matter has $\rho \propto r^{-2}$ which is essentially an isothermal halo such as inferred for the dark matter around galaxies. It is assumed that whatever condenses out of the cooled gas orbits about, or through, the central galaxy such that its mean radius is similar to that where it was formed.

In many cases, the rate of mass deposition is large, with 50-500$M_\odot$ yr^{-1} being common for a cluster of galaxies. An outstanding problem now presents itself, namely what does the cooled gas form? If stars, then the IMF must be non-standard or the central galaxies would be much bluer and brighter than is observed.

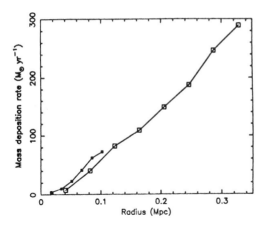

Figure 5. The mass deposition profile of the cooling flow in the cluster A2199 inferred by Thomas *et al* (1987).

2.5.2. Alternatives to cooling. The problem of what forms from a cooling flow has caused the assumptions behind the phenomena of cooling flows to be questioned. A heat source, for example, could offset the cooling so that little gas is actually deposited. Heat sources, or fluxes, that have been considered are cosmic rays (Tucker and Rosner 1982; Rephaeli 1987; Schwarz *et al* 1990), conduction (Bertschinger and Meiksin 1986; Rosner and Tucker 1989), galaxy motions (Miller 1986; Pringle 1989) and supernovae (Silk *et al* 1986). None of these proposals successfully confronts the evidence from spectroscopic x-ray line measurements of low temperature components that are entirely consistent with a cooling interpretation. It would indeed be a strange heat source that allowed the gas to lose 90 per cent of its thermal energy in cooling from 80 million K down to 8 million K but then prevented it cooling further. There is, of course, no evidence for large stockpiles of gas at some intermediate temperature. The x-ray spectroscopic data also indicate that cooling flows are long lived since gases with a wide range of cooling times give the same mass cooling rate.

To consider heat sources in more detail, we note that it is generally difficult to keep the gas stable whilst heating it. Most heat sources, such as cosmic rays, heat at a rate proportional to the gas density, whereas the cooling varies as the density squared. Heating may then cause the gas to become more unstable by tending to increase the temperature and pressure of the lowest density phases but allowing the denser gas to carry on cooling. If conduction occurs unimpeded then cooling of the core gas can set up a temperature gradient that is offset by a conductive heat flux. Whilst such a situation can occur, it is restricted to only a small part of parameter (initial density and temperature) space. The energy equation (at constant pressure) is

$$\frac{\rho v}{\mu m} \frac{5}{2} k \frac{dT}{dr} = n^2 \Lambda - \frac{1}{r^2} \frac{d}{dr} \left(r^2 \kappa \frac{dT}{dr} \right). \tag{2.45}$$

The first term on the RHS is the cooling, which at constant pressure and for bremsstrahlung, varies as $T^{-3/2}$. The second term is the conductive heat flux which varies as $T^{7/2}$ (it does not depend on density). Their widely different temperature dependences makes it difficult to allow a balance where one term does not dominate. Conduction dominates where the temperature is high and cooling where it is low. Generally, cooling dominates at the centre of a flow and conduction can be important further out (Nulsen *et al* 1982; Bregman and David 1988). If conduction is dominant, then it tends to make the gas almost isothermal, in disagreement with observations. The x-ray spectroscopic observations then demonstrate that conduction is inhibited, probably by tangled magnetic fields which greatly increase the electron mean free path.

Supernova heating from stars formed from cooled gas (Canizares *et al* 1982; Lea *et al* 1982; Silk *et al* 1986) can at most change the mass deposition estimates by a factor of two. This is because the energy from a supernova can only heat a mass of gas sufficient to form another supernova progenitor (and an IMF's worth of lower-mass stars) to 8×10^6 K. There is no evidence for supernovae around the central galaxies in cooling flow clusters (Caldwell and Oemler 1981).

Finally on the topic of heating, it is worth remembering that cooling flows occur in a wide variety of clusters, both with and without strong radio sources (e.g. Cyg A versus AWM4) and in deep and shallow potential wells (e.g. Perseus versus Hydra). Any heat source necessary to counteract the radiative cooling would represent a major heat flow, of $\gtrsim 10^{62}$ erg per cluster. In our view, the current lack of understanding of star formation in our own galaxy means that it is not a simple business to extrapolate to other situations. Star formation can proceed at hundreds of solar masses per year without necessarily being evident optically, provided that only low mass stars are formed. There is no problem with the total mass deposited, which is distributed out to 100–300 kpc from the central galaxy and does not pile up within its centre. Clusters are full of dark matter and there can be no problem with some (or even all) of it being baryonic.

2.5.3. The evolution of the cooling gas and star formation. In the limit of zero viscosity, the gas in a cooling flow should be homogeneous and cool into a central singularity. This is because of gravity causing any denser gas to fall ahead and join gas with similar properties (the gas is not unstable to infinitesimal perturbations; see Malagoli *et al* 1987 and in particular Balbus and Soker 1989). Only near the centre of the flow, where $v \sim v_{\rm ff}$, the free-fall velocity would the flow become inhomogeneous. However a real flow does have viscosity, is turbulent and contains magnetic fields. The behaviour of gas blobs in a flow has been explored by Nulsen (1986) and more recently by Loewenstein (1989), Hattori and Habe (1990) and Tribble (1991). A large gas blob of size r and overdensity $\delta\rho$ will try to move ahead of the mean flow and reaches a terminal velocity

$$v_{\rm T} \simeq v_{\rm Kepler} \sqrt{\left(\frac{\delta\rho}{\rho_0}\frac{r}{R}\right)}. \tag{2.46}$$

This relative motion will then cause the blob to spread out and fragment (Nittman *et al* 1982). r/R is reduced and the relative velocity of the overdense gas, $v_{\rm T}$, is reduced. Magnetic fields can help to pin the gas to the mean flow so that it comoves. The net result is that large, slightly overdense blobs at large radii from the centre of the flow are turned into an emulsion of smaller and very overdense blobs at smaller radii. The densest gas will cool out of the flow (i.e. $T \to 0$ K) at intermediate radii. The density distribution of gas at a given radius will tend to evolve a 'cooling tail' (volume filling fraction $f \propto \rho^{-(4-\alpha)}$, where α is the exponent of the cooling function; Nulsen 1986; Thomas *et al* 1987). This allows mass to be deposited in a distributed manner. If a spread of densities exists throughout the cluster, then gas may be deposited by cooling well beyond the radius where the mean cooling time is H_0^{-1}. It is also possible that the ICM consists of a relatively uniform hottest phase with an embedded inhomogeneous, cooler phase that forms the cooling flow in the densest regions around the central cluster galaxy (Johnstone *et al* 1990). This cooler phase (with a highest temperature about half that of the hottest phase) is the remnant of the denser parts of infalling subclusters.

Gas blobs can drop out of pressure equilibrium on cooling if $t_{cool} < t_{cross}$, the sound crossing time;

$$\frac{t_{cool}}{t_{cross}} \propto \frac{T^3}{R} \tag{2.47}$$

using the simple approximation to the cooling function below 10^7 K (1.3). This just means that the gas is cooling faster than sound waves can maintain constant pressure. For typical cluster pressures ($nT \sim 10^5$–10^6 $^{-3}$ K) then the cooling becomes isochoric (constant density rather than constant pressure) at temperatures $T \leqslant 3\,10^6 r_{kpc}^{0.32}$ K (Cowie *et al* 1980), where $r = r_{kpc}$ kpc. Even parsec size blobs will become isochoric around the peak of the cooling curve. The cooling time is still initially rapid below 10^4 K and the gas may cool to 100 K before being repressurized by shocks.

Such shocks will lead to optical radiation which could explain some of the optical and ultraviolet line emission observed at the centres of *some* cooling flows (see e.g. Kent and Sargent 1979; Heckman 1981; Hu *et al* 1983; Johnstone *et al* 1987; Heckman *et al* 1989). It is not enough, however, as the number of Balmer photons per hydrogen atom passing through the shock, H_{rec}, is only ~ 1 and the observed line luminosities require $H_{rec} \sim 100$ to 1000 (Johnstone *et al* 1987).

$$L(H\beta) \approx 10^{39} H_{rec} \left(\frac{\dot{M}}{100\,M_\odot\,yr^{-1}} \right) erg\,s^{-1} \tag{2.48}$$

is unobservable unless $H_{rec} \geqslant 10$ (depending upon the distance to the cluster and the area of the emission region).

There must be some extra source of ionization which is unlikely to be photoionization by an active nucleus or hot stars (Johnstone and Fabian 1988). The source has not been identified at the moment but there is a current suspicion that turbulence in the gas is involved (Heckman *et al* 1989; Loewenstein and Fabian 1990), perhaps through mixing layers (Begelman and Fabian 1990). Turbulence is inferred from the large observed Doppler widths of the optical emision lines (Heckman *et al* 1989). Note that there is a wide dispersion in the luminosity of optical emission lines from cooling flows and some large flows identified from their x-ray properties (e.g. A2029) have no detected optical flux. This can be explained by different levels of chaotic motions in different clusters (perhaps related to their recent merger history).

The colours and magnitudes of the central galaxies in cooling flows indicate that only a small fraction of the x-ray inferred mass deposition rate, $\dot{M}_x$, forms stars with a disc-galaxy IMF (see e.g. O'Connell 1987). Most disc galaxies such as our own are bright and blue through forming stars at $\leqslant 5\,M_\odot\,yr^{-1}$, whereas a large cooling flow such as that around NGC1275 in the Perseus cluster is depositing several hundred solar masses per year. It is presumed that most of this matter condenses into low mass stars (Fabian *et al* 1982; Sarazin and O'Connell 1983). There are many differences between the conditions in a disc galaxy and those in a cooling flow. The pressure of the gas is 100 to 1000 times higher, which can reduce the Jeans mass by a factor of 10 to 30 (Jura 1977). Dust is likely to be absent in the hot gas, due to sputtering (Draine and Salpeter 1979), and unlikely to form before the gas has cooled and condensed (although there does appear to be dust in large quantities in the cores of some clusters, Bregman *et al* 1990). This may in turn inhibit the formation of GMC, which seem to be necessary for massive star formation in our galaxy. In particular, the disruption of infalling gas blobs means that they must be much smaller than the gas clouds giving rise to most of the massive star formation in spiral galaxies. Perhaps we are chauvinists with respect to star formation and only recognize star formation if it resembles that which is obvious in our galaxy.

There is definitely a small amount of visible star formation in some central galaxies. A stars are apparent in NGC1275 (Rubin *et al* 1977) to an extent such that the 'normal' star formation rate, $\dot{M}_{*}, \sim 2$ per cent of $\dot{M}_x$ (Fabian *et al* 1984; Gear *et al* 1985). The central galaxy in A1795 (O'Connell 1987), PKS 0745-191 (Fabian *et al* 1985), NGC6166 and M87 (Bertola *et al* 1987) all show evidence for a few per cent of $\dot{M}_x$ passing into 'normal' stars. More widespread evidence for star formation is obtained from the $4000\,\text{Å}$ break, D, in the continuum spectra of the central galaxies. This break is characteristic of late-type stars and is filled in (weakened) by the presence of young stars. Johnstone *et al* (1987) find that D correlates with $\dot{M}$. Up to a few per cent of $\dot{M}_x$ appears to pass into 'massive' stars (mostly A stars of a few $M_{\odot}$). Only in PKS 0745-191 (where $\dot{M}_x$ is 500–1000 $M_{\odot}\,\text{yr}^{-1}$ (Arnaud *et al* 1987) is $\dot{M}_{*} \sim 30$–100 $M_{\odot}\,\text{yr}^{-1}$. D is only weakened in those objects with observed emission lines and so there is some correlation between massive star formation and optical line emission. Hintzen and Romanishin (1988) have mapped the excess blue light in a number of cooling flows and find it strongly peaked toward the centres.

The distribution of low-mass stars in central cooling flow cluster galaxies has been the subject of a recent study by Johnstone and Fabian (1989) in the case of 1E111.9–3754 which has $\dot{M} \sim 200\,M_{\odot}\,\text{yr}^{-1}$. Imaging in the I band by Maccagni *et al* (1988) shows that the surface brightness profile matches that of the mass-deposition profile inferred from x-ray observations extending much further than the visible component. This suggests that low-mass stars are indeed present and formed by the cooling flow process.

Roughly it can be stated that

$$\frac{\dot{M}(\text{normal IMF})}{\dot{M}(\text{low mass stars})} \approx \frac{M(\text{visible})}{M(\text{dark})} \tag{2.49}$$

for the underlying galaxy. This, together with the density profile of distributed gas, suggest that cooling flows may be related to galaxy formation.

A recent development is the discovery of large amounts of cold neutral H I gas in the cores of some cooling flows by Jaffe *et al* (1990) and Bregman *et al* (1990). This, together with the discovery of dust already mentioned and of CO emission in the central galaxy in the cooling flow in the Perseus cluster, NGC1275, by Lazareff *et al* (1989) and Mirabel *et al* (1989), shows that the inner regions of cooling flows are exceedingly complex and highly multiphase. Components have now been identified in that flow (of $\sim 200\,M_{\odot}\,\text{yr}^{-1}$, Fabian *et al* 1981) at several $10^7\,\text{K}$ and a few $10^6\,\text{K}$

(x-ray spectra), $10^4 - 10^5$ K (Lyα and optical emission lines) and now $< 10^3$ K (CO and H$\textsc{i}$ gas). The Doppler widths of the optical lines indicate that the warm gas (at least) has chaotic velocities of a few hundred km s^{-1}. Diffuse radio emission is also observed from this region. Jaffe (1990) has very recently mapped the H$\textsc{i}$ absorption against the diffuse radio source and finds about 10^{10} M$_\odot$ of cold gas within a radius of about 10 kpc. Daines *et al* (1991) use this result to build a model of colliding clumps of cold gas in the core of a turbulent cooling flow. The clumps accrete other clouds until they exceed the Jeans mass and then form stars. Some of the optical line emission can then be due to shocks in the cloud–cloud collisions, which is an efficient way of tapping the kinetic energy there. Extrapolating the H$\textsc{i}$ mass to larger radii suggests that there may be $> 10^{11}$ M$_\odot$ of cold gas in that cluster or at least 10 per cent of the accumulated mass deposition from the cooling flow.

Large quantities of cold gas have also now been discovered through x-ray absorption. White *et al* (1991) have analysed the x-ray spectra from 21 clusters of galaxies observed with the *Einstein Observatory* solid states spectrometer (SSS). Simple model spectra appropriate for isothermal gas appear to be adequate to fit the observed spectra in the few cases in which a cooling flow is not inferred from x-ray imaging data (e.g. the Coma cluster, A1367 and A2256). The majority of the clusters, for which the images suggest the presence of a cooling flow, are not well-fitted by the simple model and require both the increased low energy line and continuum emission of a cooling flow and excess absorption (above that expected from line-of-sight absorption in our galaxy). This analysis simultaneously confirms the mass deposition rates obtained from the x-ray images and reveals a new component in the intracluster medium. The excess absorption corresponds to a column density of about 10^{21-2}. Since the field of view of the SSS corresponds to at least about 200 kpc, the mass of absorbing gas is at least 5×10^{11} M$_\odot$, if of solar abundance. The mass can, of course, be reduced by a factor of two to three if the dominant absorber (oxygen) is overabundant by the same factor (as suggested by some line spectra (Canizares *et al* 1988)).

The cold gas represents a significant new component of the intracluster medium and means that, within the inner few 100 kpc of a cooling flow cluster, the masses of cold and hot gas are roughly equal and that both exceed the visible stellar mass. The gas must exist in cold clouds and may just be undetectable in the non-cooling flow clusters because the covering fraction is lower there. Further work using soft x-ray spectral images, such as are being obtained from ROSAT, will measure the covering fraction of the cold gas and determine its mass and distribution in a much wider selection of clusters.

2.5.4. Distant cooling flows and galaxy formation. We have already noted that cooling flows may have some relationship to galaxy formation. If a substantial fraction of the mass of a protogalactic cloud was virialized so that $H^{-1} > t_{\text{cool}} > t_{\text{ff}}$, then the gas can cool into the galaxy as a cooling flow. The densest (probably central) gas blobs may collapse without being virialized and form massive stars that enrich and mix the remaining gas. Of further interest in this mechanism is the possibility that much of the dark matter is baryonic, i.e. low mass stars formed in the flow. Such a scenario has been discussed by Thomas and Fabian (1990). Ashman and Carr (1988, 1991) have also considered possible ways in which cooling flows could have formed widespread dark matter, including 'dark clusters' of low-mass stars.

The obvious way to search for distant cooling flows is to make soft x-ray observations with ROSAT, and later with AXAF and XMM. Results from ROSAT should be available in the near future. In the meantime we must make use of indirect methods. Extended optical emission lines are relatively common around distant radio galaxies and quasars. The line emission from some 3CR radio galaxies is remarkably similar (if more luminous) to that of nearby cooling flows such as that around NGC1275. If there is a surrounding flow then the optical gas must be at high pressure and confined by hotter gas. Evidence for high pressure gas surrounding radio-loud quasars has been obtained from a photoionization calculation applied to the observed [O III]/[O II] line ratio of the gas at tens of kpc from the nucleus of a sample of quasars (Fabian *et al* 1987; Crawford and Fabian 1989; Forbes *et al* 1990). The ionization parameter $\xi = L/nR^2$ is adjusted to give the observed line ratio at the observed gas radius, R, where the ionizing luminosity of the quasar, L, is obtained from its UV and x-ray spectrum. The gas density, n, inferred there is then so high so that the pressure of the confining ICM implies mass deposition rates of many $100 \, M_\odot \, yr^{-1}$. If the gas is unconfined it would disperse very rapidly and we would be most unlikely to observe it, unless there were an unreasonably large reservoir of cold gas waiting to be ionized. There is evidence accumulating that many radio-loud quasars are in poor clusters (e.g. Yee and Green 1984; 1987). Hintzen and Romanishin (1986) have also suggested that 3C275.1 lies in a cluster cooling flow from the appearance of its extended [O III] gas. Forbes *et al* (1990) find that the pressure of the confining gas increases with quasar redshift up to $z \sim 1$. Heckman *et al* (1989) find that this trend increases out to $z \sim 3$.

One picture for the evolution of cooling flows (Fabian *et al* 1986) suggests that they were common in poor clusters (e.g. MKW3s). Rich clusters are then assembled from the mergers of these subclusters and the merger process can then disrupt the individual flows. A large flow may survive relatively undisturbed if a small subcluster collides with a larger one, but the merger of two large subclusters in the formation of a Coma-like cluster could well spread the cooling gas throughout the core. Thus an unfocussed flow could persist even in Coma. The merging of subclusters will also make the intracluster medium inhomogeneous, as required. This picture is consistent with the evolution of clusters reported in section 2.1.

Only a small fraction of the cooling gas is needed at the centre of a galaxy to power even a quasar. Activity, and in particular radio emission (Valentijn and Bijleveld 1983; Jones and Forman 1984; Burns 1990) is common in cooling flows. The recent decrease in the numbers of very powerful quasars can then be related to the assembly of clusters from quasar-active subclusters which are subsequently disrupted (Fabian and Crawford 1990). Cooling-flow-deposited dark matter is spread throughout the cluster and so much cluster dark-matter may be baryonic (Thomas and Fabian 1990).

2.5.5. Summary. Cooling flows are common. As more observations are made of these regions they are found to be increasingly complex, with a wide range of phases detected and apparently large chaotic velocity fields. Tangled magnetic fields suppress conduction and large quantities of cold gas accumulate. Star formation presumably takes place at high rates, particularly of low-mass stars. Star formation must then be environment-dependent. This means that at least some dark matter is baryonic and is forming now from x-ray emitting gas.

2.6. The interstellar medium of early-type galaxies

2.6.1. Introduction. It was thought for many years that early-type galaxies—ellipticals and S0s—were mostly devoid of gas. After all, little gas was observed. Some optical emission lines from, at most, $10^8 M_\odot$ of gas were seen in active objects, but the average 'normal' elliptical galaxy looked pretty boring. Radio H I measurements revealed little more gas and a general picture emerged in which early-type galaxies were kept swept clean by supernova-driven galactic winds. We now know this picture to be wrong and that early-type galaxies contain large amounts of gas, of order 10^9-$10^{10} M_\odot$ which is not very dissimilar from the mass of gas in spirals.

The reason that the existence of such a large amount of gas was largely unsuspected is that it is very hot and so only readily detectable with x-ray instruments. This is one further part of the revolution in our understanding of elliptical galaxies that has taken place over the past ten years. First it was shown that their shape is not primarily due to rotation (Bertola and Capaccioli 1975) and lastly there is much evidence accumulating that their stellar population is not particularly old (Pickles 1985). Now, with a complex interstellar medium and often with shells (Malin and Carter 1980), early-type galaxies seem to be just as complicated as spiral ones.

Carefully reasoned arguments against any substantial interstellar medium in ellipticals were presented in 1976 by Faber and Gallagher. They emphasize that some mechanism must act to remove gas if it is not present. This is because stars lose mass as they evolve at a rate, which for a relatively old population is

$$\alpha_\star = \frac{\rho_\star}{\tau} \approx 10^{-12} \rho_\star \, \text{yr}^{-1} \tag{2.50}$$

where $\rho_\star$ is the mass density of stars and τ is the mass-loss timescale (10^{12} yr). This rate is fairly well known from studies of the mean white dwarf mass in our galaxy. This is $0.8 M_\odot$ which, if the average progenitor is $\sim 1 M_\odot$, gives $\sim 0.2 M_\odot$ per star that evolves through the red giant phase. The less numerous massive stars lose proportionally more but it is the lower mass stars that dominate in total mass. A large elliptical galaxy ($L > \bar{L}$, say) has a total stellar mass $\sim 10^{12} M_\odot$ and so produces $\sim 1 M_\odot \, \text{yr}^{-1}$. If this were to accumulate over a Hubble time (H_0^{-1}) then it would total $\geqslant 10^{10} M_\odot$. Of course, star formation could consume this gas, but a star formation rate with an IMF similar to a normal spiral (which produces stars at a rate of a few $M_\odot \, \text{yr}^{-1}$) would make an elliptical brighter and bluer than it obviously is. Jura (1977) pointed out that the higher pressure of gas in an elliptical would lower the Jeans mass and perhaps cause the IMF to skew towards lower-mass stars. This is probably the correct answer, although high pressure alone may not be sufficient.

Most discussions centred on the galactic winds proposed and calculated by Johnson and Axford (1971) and Mathews and Baker (1971). In these models, the stellar mass loss is heated by supernovae such that it becomes unbound and flows out of the galaxy. Computations show that the gas would go supersonic close to the core ($\lesssim 1$ kpc) and from there outward possess a density profile $\rho \propto r^{-2}$. Although the gas would be hot in the core ($T \gtrsim 10^7$ K) it would cool adiabatically as it flows and have a low density and so be undetectable by x-ray instruments flown so far (White and Chevalier 1984). Faber and Gallagher (1976) did briefly consider a static hot atmosphere but assumed that the gas density followed the star density into the central stellar cusp. The hot gas would there be dense enough to be detectable by its $H\beta$

emission, which is not generally observed. In reality, the gas does not have such a steep density profile in the core, which incidentally would imply a prodigous cooling rate!

X-ray emission from early-type galaxies was first detected in the Einstein Observatory images of the Virgo cluster (Forman *et al* 1979). These showed diffuse patches of emission centred on the larger galaxies, together with more widespread and intense emission on M87. A plume of x-ray emission was found to the north west of M86, which is plunging through the Virgo cluster at $1500\,\mathrm{km\,s^{-1}}$. This plume provides strong evidence for a gaseous origin for the x-ray emission. Neither M dwarfs nor binary x-ray sources (possible sources that might appear diffuse in great numbers) would be preferentially swept out of a galaxy. Further evidence came from observations of the nearby radio galaxy, Cen A, which showed a diffuse component that also could be plausibly explained as due to hot gas (Feigelson *et al* 1981). Long and Van Speybroeck (1983) amassed x-ray data on a wide variety of galaxies (late- and early-type) and showed that

$$L_\mathrm{x} \sim 10^{-3} \qquad \text{to} \qquad 10^{-4} L_\mathrm{opt}. \tag{2.51}$$

All galaxies appear to have substantial x-ray luminosities. Biermann and Kronberg (1983) then looked at some ellipticals in weak groups and discussed their x-ray emission in terms of hot gas, which could be cooling at the centre. Nulsen *et al* (1984) extended this work to isolated galaxies, showing that galactic winds do not operate and that cooling, presumably with associated star formation, occurs instead.

2.6.2. The x-ray properties of early-type galaxies. Large and detailed compilations of the x-ray emission from early-type galaxies have been made by Forman *et al* 1985; Trinchieri and Fabbiano (1985); Trinchieri *et al* (1986) and Canizares *et al* (1987) (see also Fabbiano 1989). There are correlations between the x-ray luminosity and both optical and radio luminosities.

Generally,

$$L_\mathrm{x} \propto L_v^\alpha \qquad 1.5 < \alpha < 2.3 \tag{2.52}$$

with the most recent estimates emphasizing the steeper slope (Sarazin 1990). The total observed mass of x-ray emitting gas is $\sim 10^9$ to $10^{10}\,\mathrm{M_\odot}$ with the gas density rising as $r^{-\gamma}$ ($1.2 < \gamma < 1.6$). It is debatable as to whether the emission follows the stellar profile well. Any outer radius of the x-ray emitting gas is certainly $> 30\,\mathrm{kpc}$ and poorly determined. It is not clear that one could be defined. The influence of surrounding gas and stripping etc is unknown. Where measured (NGC4472, which lies in a southern clump in the Virgo cluster) the temperature of the gas increases with radius. A mean temperature has been established for the gas in some more galaxies at $1.5 \times 10^7\,\mathrm{K}$.

The thermal pressure of the gas is much higher (a factor ~ 100) than that in spiral galaxies and the cooling time of much of the gas (most within 30 kpc radius) $\leqslant H_0^{-1}$. The immediate implications of the x-ray observations are that there are no steady-state winds and so no large heat input from supernovae. Furthermore early-type galaxies are very massive and so possess extensive dark halos. There is widespread cooling and, presumably, continuous low-mass star formation. The interstellar medium in elliptical and S0 galaxies is continuously being recycled and redistributed. How these conclusions are reached is outlined in the next few sections.

2.6.3. Winds and heat sources. The arguments against any galactic winds or strong supernova heating are simply described by considering a relatively isolated elliptical galaxy, NGC1395 (Nulsen *et al* 1984). An x-ray luminosity of $8 \times 10^{40} \, \mathrm{erg \, s^{-1}}$ peaks from a region of radius 10 kpc within which there is $5 \times 10^{8} \, \mathrm{M_\odot}$ of gas at $10^{7} \, \mathrm{K}$. Any wind would escape at the speed of sound, which gives a mass loss rate;

$$\dot{M} \approx \frac{M_{gas}}{R/c_s} \approx 13 \, \mathrm{M_\odot \, yr^{-1}} \qquad (2.53)$$

which is an order of magnitude higher that could be provided by stellar mass loss. In fact, a wind would go supersonic at a smaller radius so the above is only a lower limit to $\dot{M}$. The widespread observation that early-type galaxies have similar x-ray emission (we are not aware of any examples outside the cores of rich clusters where x-ray emission should have been detected and was not) demonstrates that this is not a time-dependent result (e.g. 'galactic gusts', Sanders 1981; MacDonald and Bailey 1981). On the other hand a cooling flow gives

$$\dot{M} \approx \frac{M_{gas}}{t_{cool}} \approx 1.3 \, \mathrm{M_\odot \, yr^{-1}} \qquad (2.54)$$

since $t_{cool} \sim 4 \times 10^{8} \, \mathrm{yr}$, which is acceptable.

The gravitational binding energy of the gas, assuming that the galaxy has a mass $M_{gal} \simeq 10^{12} \, \mathrm{M_\odot}$; $GM_{gas}M_{gal}/R \simeq 5 \times 10^{57} \, \mathrm{erg}$ is similar to the thermal energy,

$$\frac{3M_{gas}kT}{\mu m} \approx 5 \times 10^{57} \, \mathrm{erg}. \qquad (2.55)$$

This already hints at a massive halo. The total kinetic energy liberated by supernovae at a rate of 1 per $100 \, \mathrm{yr^{-1}}$ and $10^{51} \, \mathrm{erg}$ per supernova, is

$$E_{sn} \approx 2 \times 10^{59} \, \mathrm{erg} \qquad (2.56)$$

over H_0^{-1} (or $\sim 10^{58} \, \mathrm{erg}$ in a cooling time). As the total x-ray emission over a Hubble time is $5 \times 10^{58} \, \mathrm{erg}$, the supernova rate must be lower than one per 400 yr.

It might be argued that the kinetic energy from supernovae is radiated before reaching the hot gas or is transported through the hot gas without heating it. Both are implausible since there is no extra luminosity (unless perhaps in the infrared) and viscosity would surely prevent very efficient transport in the hot gas.

2.6.4. Cooling and mass deposition. It has already been mentioned that the cooling time of much of the observed gas in early-type galaxies is relatively short ($\ll H_0^{-1}$). In the absence of heating (see section 2.5.2), the gas will then cool at a rate

$$\dot{M} = L/H \qquad (2.57)$$

where H is the enthalpy of the gas ($\frac{5}{2}kT/\mu m$). Unlike clusters of galaxies where any mass injection is now negligible, the stellar mass loss is likely to dominate the gas within the galaxy.

The mass deposition profile can then be estimated (Thomas *et al* 1986) if there are no radial flows (matter injected at radius r cools out there) from

$$\Delta \dot{M} = \frac{\Delta L(r)}{H(r)} \tag{2.58}$$

where $\Delta L(r)$ is the incremental cooling luminosity from r to $r + \Delta r$. Then

$$\dot{M}(r) = \sum_n \Delta \dot{M}(r). \tag{2.59}$$

Such a steady-state would predict a density profile of a form similar to that of the stars, but is observed to be slightly flatter. A cooling inflow can also be estimated as for clusters (section 2.5). The estimates generally show that the mass cooling rate is less than the expected mass injection rate unless the galaxy is at the centre of a large group or weak cluster (e.g. NGC1399, NGC4472). The discrepancy is even worse if the rates within the de Vaucouleurs effective radius (r_e) are considered. Half the mass should be injected with this radius.

A more complicated flow with several phases and heat and mass injection is perhaps most appropriate, but the present x-ray data do not warrant too much sophistication in analysis. Thomas *et al* (1986) find that reasonable results can be accommodated with two phases and some heat injection (e.g. from some low supernova rate). This suggests that the gas in ellipticals is multiphase and complex, perhaps with some hot phases flowing outward and cooler ones flowing in. Much of the stellar mass loss may actually not mix in with the x-ray detected hot phase but collapse to low temperatures rapidly (White and Chevalier 1984; Thomas 1986). To see this, consider two red giants losing mass and approaching each other at some impact parameter b. If there is no interstellar medium then the mass loss can expand to a radius greater than b (if there is time) and shock on colliding. The temperature of the gas from many such collisions will be such (few 10^6 K) that the sound speed is comparable with the stellar velocity dispersion. If however there is already an ISM, then this can pressure confine the mass loss to a radius $\ll b$ and maintain it at a high density where it can cool and prevent it from being shocked. Such mass loss may lead to a 'rain' or 'fog' of shredded mass loss within an elliptical galaxy. This gas can contain dust as it has not been exposed to the 10^7 K gas—a point relevant to IRAS observations of dust emission from ellipticals (Jura 1986; Tytler 1987c). Just how, and what, stars form under these conditions is obscure, but 21 cm observations show that the gas cannot remain long without condensing.

There are strong similarities between the hot ISM in elliptical galaxies and the cooling flows in cluster cores, except that in ellipticals there is the added complication of the continuous injection of stellar mass loss at rates similar to those by which cooling depletes the atmosphere.

The evolution of the hot gas in early-type galaxies depends on how they formed, on their early environment and on the IMF and supernova rate when young. Models for evolution of the gas, in which winds play a role at early epochs, are discussed by Mathews and Loewenstein (1986), Mathews (1989) and David *et al* (1990).

2.6.5. Gravitational masses and mass profiles. X-ray measurements of the hot gas in elliptical galaxies and clusters will eventually provide a powerful means for determining their gravitational mass profiles (see e.g. Mushotzky 1987). The equation of hydrostatic equilibrium may be written as

$$\frac{d\phi}{dr} = -\left(\frac{kT_{gas}}{\mu m}\right)\left(\frac{d\ln\rho_{gas}}{dr} + \frac{d\ln T_{gas}}{dr}\right). \tag{2.60}$$

$$\phi = \int \frac{GM}{r^2}\,dr \tag{2.61}$$

is thus obtained from the gas pressure and density profiles. Although the density profiles measured so far are robust, the temperatures carry large uncertainties. Some progress has been made using Einstein Observatory IPC spectra of the cluster emission around M87 (Fabricant and Gorenstein 1983) and Focal Plane Crystal Spectrometer spectra of the inner regions (Stewart *et al* 1984a). These measurements all indicate that M87 is surrounded by an extensive dark halo. Confirming evidence is obtained from velocity measurements of the globular clusters surrounding that galaxy (Mould *et al* 1987; Huchra and Brodie 1987). Turbulence pressure in the hot gas is usually ignored (see section 2.2).

In the absence of any other spectral data, we can estimate a lower limit to the gravitational binding mass of an early-type galaxy from observing gas at radius r_o with a pressure P_o and temperature T_o (Fabian *et al* 1986). The method is analogous to those used on stars to obtain limits on central conditions from surface properties, excepting this time we observe the centre and do not know how far away the 'outside' is. All that is assumed is that the gas is pressure-confined by an 'unseen' hydrostatic, convectively stable atmosphere in which the pressure decreases outward to some external pressure P_∞. Remember that the central pressure is much higher than that in a spiral and so P_o is likely to far exceed P_∞. If the gas is not confined then it will expand and the wind arguments already mentioned rule that out. The key point about this outer atmosphere is that the total binding mass is minimized when its temperature gradient (which decreases outward) is steepest. This occurs for an atmosphere which is marginally convectively stable and so follows an adiabat.

Consider the outer confining atmosphere extending from r_o to some outer radius r_∞ where $P = P_\infty$. Under the assumption of convective stability the minimum mass is

$$M_T \geqslant \frac{5}{2}\frac{kT_o r_o}{G\mu m}\frac{[1 - (P_\infty/P_o)^{2/5}]}{(1 - r_o/r_\infty)}. \tag{2.62}$$

This limit applies to all systems in which gas is in hydrostatic support (e.g. galaxies, groups and clusters). Dimensionally, it resembles a virial mass (as any such combination should). It is the numerical coefficient that is important. A safe conclusion for many elliptical galaxies is that, $M_T \geqslant 5 \times 10^{12}\,M_\odot$ and probably $> 10^{13}\,M_\odot$; the mass to (blue) light ratio $(M/L_b) \geqslant 74$, in solar units, and probably > 150.

2.6.6. The interstellar medium and the propagation of jets. One of the major energy transport processes in the Universe is the jet. Dissipation of the energy at the end of a jet is most often observed at radio wavelengths in the various types of extended radio sources (see e.g. Bridle and Perley 1984; Begelman *et al* 1984). The collimation of jets appears to occur on very small scales close to the central 'engine'. Thereafter, the jet must propagate through the interstellar medium of the galaxy and perhaps through the surrounding intracluster, intragroup or intergalactic medium. Most powerful radio sources are at the centres of elliptical galaxies and so their jets pass through a hot surrounding medium, the thermal pressure of which may help to confine them (Bicknell 1986). The pressure profile of the interstellar gas, $p \propto r^{-1.2}$ (Thomas *et al* 1986), then determines the shape of the jet from scales of a few parsecs to hundreds of kpc.

Intracluster and intragroup gas is the 'working surface' for double-lobed radio sources, which would otherwise expand relativistically. In the case of the well-known powerful source Cygnus A studied by Alexander *et al* (1984), the balance of ram pressure from the cluster gas and the energy in the radio lobes cause the hot spots to expand at a velocity $v \sim 0.126c$. The intracluster medium may provide the sideways confinement of the lobes (Arnaud *et al* 1984).

Motion of twin-jetted source relative to the intracluster medium causes a 'head–tail' radio source to form. A good example of one of these is NGC1265, which is falling into the Perseus cluster at a velocity of at least $2000 \, \text{km s}^{-1}$. Studies of such sources can give information on the intracluster gas distribution and on the orbits of galaxies (see e.g. O'Dea *et al* 1987). Less extreme examples of bent jets are seen in wide-angle-tail sources which may represent lower velocity galaxies or perhaps are the result of an abrupt motion of the intracluster medium as would occur in a subcluster collision. VLBI observations of the superluminal radio-source 4C39.25 (Marcaide *et al* 1989) also suggest the presence of a high-pressure confining medium.

Interactions between jets and the density inhomogeneities expected in the interstellar and intracluster medium should lead to knots in the jets (such as seen in M87) and emission lines if the blobs are cool (such as observed in A194; van Breugel *et al* 1985). This area has been little explored but should reveal the properties of the density distribution of the gas.

Another fact that emphasizes the interaction between jets and the surrounding gas is the correlation between the size of extragalactic radio-sources and the redshift (Rosen and Wiita 1988). This actually suggests a hot intergalactic gas whose pressure rises very steeply with redshift.

2.6.7. Relationship of an active nucleus to the interstellar medium. A powerful active nucleus can photoionize, heat and push out much of the gas in a spiral galaxy (Begelman 1985). As the interstellar medium in an elliptical galaxy is fairly ionized already, only Compton heating or cooling is important. This is generally unimportant on radii greater than one kiloparsec unless the luminosity of the nucleus exceeds several 10^{46} erg (e.g. a powerful quasar).

Very close to the active nucleus (i.e. within $\sim 1 \, \text{pc}$, the gas may form a two (or more) phase medium consisting of cool ($\sim 10^4 \, \text{K}$) dense ($\sim 10^{10-3}$) clouds in pressure equilibrium with a Compton-heated hot ($\sim 10^8 \, \text{K}$) intercloud gas. Most of the mass is in the intercloud medium. The interstellar medium is at least one source of material to be accreted into this region. As most of the interstellar medium in an elliptical (and perhaps also in a spiral) galaxy is hot ($\geqslant 10^6 \, \text{K}$) then the accretion process may resemble Bondi-flow (Nulsen *et al* 1984; Begelman 1986). This process provides a direct coupling between the active nucleus and its surroundings. Fabian and Crawford (1990) show that Compton cooling of a surrounding cooling flow by a central quasar prevents distributed mass deposition in the inner parts of the flow (few 100 pc) and for all of it then to be accreted by the nucleus, thereby providing feedback such that the quasar is self-fuelled. If the galaxy is stripped or heavily disturbed, say, by a collision with another galaxy, or at the centre of a group or cluster, by collision with a subcluster, then the accretion flow can be much reduced and activity cease.

2.6.8. *Colliding galaxies.* Spitzer and Baade (1953) showed that some galaxies are likely to collide. The large cross-sections and relative velocities of galaxies mean that some collisions are inevitable. The extent and importance of this phenomenon at the present time is not clear. It is often invoked to explain 'starburst' galaxies and other active galaxies.

The cool gas in colliding galaxies will be strongly shocked. Grains can survive the shock and molecular hydrogen forms in the high-density post-shock gas (it is assumed that the gas cools). Large masses of molecular gas can then form. Total H_2 masses inferred from CO observations of 'colliding' galaxies (e.g. NGC6240, Joseph *et al* 1984; Rieke *et al* 1985) can be in the range $4 \times 10^9 - 4 \times 10^{10} \, M_\odot$; i.e. most of the interstellar medium is converted to a molecular form.

Massive star formation then occurs in the dense molecular gas, which resembles a gigantic molecular cloud. Infrared observations are important and suggest $\sim 10^{12} L_\odot$ and star formation rates of $100 \, M_\odot \, yr^{-1}$ in some cases. Given the total mass of gas, such a rate can only last for $\sim 10^8 \, yr$. Such events must presumably lead to many supernovae and a galactic wind and are one source of an enriched intergalactic medium.

2.7. Intrasupercluster gas

Clusters of galaxies are not uniformly distributed in the Universe. The most widely used measure of the lumpiness of their distribution is the correlation function, which measures the 'excess' probability of finding a cluster at a certain distance r from a previously found one. The cluster correlation function fits a power law

$$\xi(r) = (r/r_0)^{-1.8} \tag{2.63}$$

where r_0 ranges from 24 Mpc (Sutherland 1988) to 50 Mpc (Bahcall and Soneira 1983; Lahav *et al* 1989). This indicates that clusters tend to be grouped in greater structures called superclusters.

Some superclusters can be optically identified in the sky (Bahcall 1988). Their geometrical shape is often highly non-spherical, with typical sizes as high as 100 Mpc, suggesting that they are not dynamically relaxed. The question here is whether or not superclusters contain significant amounts of gas. Such a gas could have been accreted from a diffuse IGM and (possibly) virialized into the supercluster potential well. At low redshifts, Persic *et al* (1988) have searched for any excess in the (2–10 keV) x-ray emission in optically-identified superclusters. No evidence is seen for such an emission. Jahoda and Mushotzky (1989) have scanned the 'Great-Attractor' region in x-rays and find excess emission. If extragalactic, it could be due to either hot intrasupercluster gas or to the enhanced galaxy density.

One good candidate object in which to search for an intrasupercluster medium (ISCM) is the Shapley Supercluster, which contains 6 of the 50 x-ray brightest clusters in a very massive region ($\hat{\Omega} > 1$) of radius at least 75 Mpc (Raychaudhury *et al* 1991). However, a scan of it by the x-ray satellite *Ginga* shows no evidence for any ISCM (Day *et al* 1991).

The conclusion is that there is no compelling evidence, at present, for the existence of significant amounts of intrasupercluster gas, although the limits are not conclusive.

3. The intergalactic medium

3.1. Introduction

Apart from the gas contained in galaxies and clusters, there is the possibility that the space between them is not empty, i.e., an intergalactic medium (IGM) is present in the Universe. Although its nature and origin are not known, it would be very surprising if the process of galaxy formation had been so efficient that all of the baryonic content of the Universe resides in galaxies and clusters. Actually, current theories of galaxy formation, which invoke biasing (i.e., luminous structures only form in very high density regions, Kaiser 1984) imply the existence of large amounts of uncollapsed baryonic dark matter filling intergalactic space.

In spite of these facts, no compelling observational evidence for the existence of a *diffuse* IGM exists at present. This does not mean that the intergalactic space is empty, because localized structures (like the Lyman α clouds discussed in section 3.3) do indeed exist. Depending on its features, a diffuse IGM should be indirectly detectable (see Shaver 1987 for a review), which allows us to, at least, place some limits on its physical parameters. This is explained in section 3.2.

Observations of high-redshift QSO, reveal the presence of multiple absorption features in their spectra (see, e.g., Weymann *et al* 1981 for a review). These absorption features are usually divided between broad absorption lines and narrow absorption lines, depending on their velocity width. The first of them are usually understood as arising in clouds of material associated with or ejected by the QSO that is being observed (see Turnshek 1988 for a recent review). The origin of the narrow absorption lines is, however, more controversial. The importance of the study of these lines towards our understanding of the IGM will become evident in the next sections.

Some of these narrow lines can be identified as being created by heavy elements (in which case they are called metal absorption lines), but the vast majority of them occur shortward of the Lyman α emission line in QSO, and are naturally associated with Lyman α absorption features. A few of these Lyman α absorption lines have prominent damping wings, corresponding to column densities $N_{HI} \geqslant 10^{19} \, cm^{-2}$ and they are called damped Lyman α lines, while the remainder are simply called Lyman α lines, constituting the so-called Lyman α forest for each QSO.

In section 3.3 we review some of the features concerning these non-damped Lyman α lines, which are very likely to constitute a different population from the other absorption systems. However, Tytler (1987a) pointed out that there is an unexpected continuity between the distribution in column density N_{HI} for the Lyman α lines, the metal-rich absorption lines discussed in section 3.4 and the damped Lyman α systems studied in section 3.5 (but see Bechtold 1988). This could be an indication that all narrow absorption systems belong to the same population. This subject is still controversial, but the evolution, clustering and other properties of the three families of narrow absorption systems are so different that it is quite likely that they arise in different classes of objects (see, e.g., Sargent 1988).

3.2. A diffuse intergalactic medium

3.2.1. The state of the IGM. It seems likely that a diffuse IGM exists and that it contains most of the baryons in the Universe. Just how many of them and in which physical state is not known. There are, however, some relevant constraints on its properties that are worth exploring.

The standard theory of primordial nucleosynthesis together with observations of present-day Deuterium abundance (Yang *et al* 1984, Boesgaard and Steigman 1987) places an absolute upper limit of about 0.1 for the fraction contributed by baryons to the closure density of the Universe (Ω_{baryon}), although even this limit has been questioned in the framework of inhomogeneous nucleosynthesis models (Malaney and Fowler 1988). A lower limit to the IGM density must exist because galaxies eject some material through their life (via supernovae and winds). Standard primordial nucleosynthesis also sets a lower limit to Ω_{baryon} from $^3He + D$ abundance at 0.04. Since luminous matter in galaxies already contributes $\Omega_{baryon} \sim 0.01$, a similar amount is contributed by intracluster gas and there could be some other baryonic dark matter in galaxies, we set a conservative lower limit of $\Omega_{IGM} > 0.01$. Figure 6 shows these upper and lower limits to Ω_{IGM}.

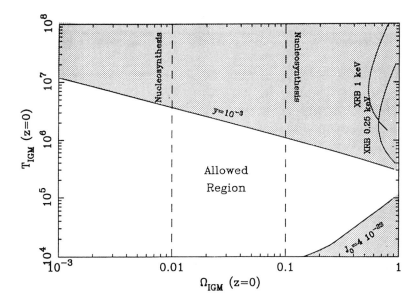

Figure 6. Constraints on the fraction of the closure density in the Universe in a diffuse IGM (Ω_{IGM}), from Barcons *et al* (1991). The separate lines show the limits from nucleosynthesis, from the soft x-ray Background, from the Gunn–Peterson test and from the *y*-parameter from COBE.

As a further comment on the physical state of the IGM, we note that it is unlikely that it contains any dust. This would produce a systematic reddening of distant galaxies and QSO which is not supported by observations. Ostriker and Heisler (1985) suggested that even the QSO redshift cut-off could be due to dust, although it has been shown that high-redshift absorbing objects do not contain enough dust to produce this obscuration (Fall and Pei 1989; see, however, Fall *et al* 1989). In what follows, we shall assume that the IGM is made only of primordial gas. This is probably an over simplification, since the material that galaxies eject is metal rich. If, for example, the ratio of gas to galaxy density stays fairly constant, independent of galaxy concentration, the intracluster metal abundance of ~ 0.3 could well be a general feature of the IGM.

Of course, this should not be taken as a strict conclusion, since clusters are more abundant in ellipticals than the field. However, the metallicity of the IGM is unlikely to be small and that will certainly have to be explored.

A very stringent limit on the amount of diffuse neutral hydrogen at high redshift can be placed from what is known as the Gunn–Peterson (1965) test. A diffuse distribution of H I should produce a trough in the spectra of QSO shortward of the 21 cm and Lyman α lines (in the QSO rest frame). No such troughs are observed. A tight limit comes from the work of Steidel and Sargent (1987), who were able to set an upper limit for $n_{\mathrm{H\,I}}$ at $z \sim 2.64$ of $\sim 10^{-12}\,\mathrm{cm}^{-3}$. For a given amount of diffuse hydrogen, this sets a lower limit to its temperature. Figure 6 shows these limits (extrapolated to $z = 0$ with simple adiabatic cooling) assuming ionization equilibrium (see section 3.3) and including both collisional excitation and photoionization by a UV background of $\sim 4 \times 10^{-23}\,\mathrm{erg\,s}^{-1}\,\mathrm{cm}^{-2}\,\mathrm{A}^{-1}$ in the Lyman limit at $z = 0$ (see Barcons *et al* (1991) for further details). An interesting constraint at low redshifts was obtained from the lack of any absorption trough in the UV spectrum of the nearby QSO, 3C273 (Davidsen 1977).

The cosmic background explorer (COBE) has already produced some interesting constraints on the amount of ionized gas in the Universe (Mather *et al* 1990). The lack of distortion in the submillimeter part of the microwave background radiation (MBR) sets an upper limit to the possible Compton scattering of microwave photons by hot electrons (cf section 2.3). This is measured via the y parameter

$$y = \sigma_{\mathrm{T}} c H_0^{-1} \int_0^{z_{\mathrm{max}}} \mathrm{d}z\, \frac{kT_e(z)}{m_e c^2} n_e(z)(1+z)^{-5/2} \tag{3.1}$$

where T_e and n_e are the electron temperature and density, σ_{T} is the Thomson cross-section and z_{max} is the redshift at which the IGM is reheated. A preliminary and conservative upper limit for y is $\sim 10^{-3}$. Assuming $n_e(z) \propto (1+z)^3$ and adiabatic cooling ($T_e(z) \propto (1+z)^2$), upper limits for the IGM temperature at $z = 0$ can be found for each value of the density Ω_{IGM}. This is also shown in figure 6 for $z_{\mathrm{max}} = 3.5$ and $z_{\mathrm{max}} = 6$. These limits rule out the possibility that the x-ray background (XRB) arises as thermal bremsstrahlung of a diffuse and very hot IGM ($T_{\mathrm{IGM}} > 10^8\,\mathrm{K}$; see below).

One last constraint on the $(\Omega_{\mathrm{IGM}}, T_{\mathrm{IGM}})$ plane comes from the upper limit to the extragalactic flux at $0.25\,\mathrm{keV}$ of $\sim 60\,\mathrm{keV\,s}^{-1}\,\mathrm{cm}^{-2}\,\mathrm{sr}^{-1}\,\mathrm{keV}^{-1}$ (MacCammon and Sanders 1990) and at $1\,\mathrm{keV}$ of $\sim 20\,\mathrm{keV\,s}^{-1}\,\mathrm{cm}^{-2}\,\mathrm{sr}^{-1}\,\mathrm{keV}^{-1}$ (Wu *et al* 1991). Any ionized IGM would contribute to this flux via bremsstrahlung radiation. Sherman (1982) first studied the constraints on T_{IGM} for $\Omega_{\mathrm{IGM}} = 1$. Here (figure 6) we show the permitted values of T_{IGM} and Ω_{IGM} by this upper limit, for $z_{\mathrm{max}} = 6$.

Now, in view of figure 6, we can conclude that there is some IGM contributing $0.01 < \Omega_{\mathrm{IGM}} < 0.1$ to the closure density of the Universe. The hydrogen in the diffuse IGM must be ionized because of the absence of any Gunn–Peterson effect. Its temperature must exceed $10^5\,\mathrm{K}$ at $z \sim 4$ even if $\Omega_{\mathrm{IGM}} \sim 0.04$. It is however likely that it is cooler than the typical intracluster medium, unless reheating has occurred very recently ($z_{\mathrm{max}} < 6$). The energy density in the IGM is of the same order of magnitude as the binding energy in galaxies.

3.2.2. Reheating. In the standard Big-Bang theory, after matter becomes transparent to radiation ($z \approx 1500$), the atoms cool faster than the radiation, which has a temperature of $2.7\,\mathrm{K}$ today. Some departure from this picture must have occured in the post-recombination Universe in order to efficiently reheat the IGM to a temperature many orders of magnitude greater than the nominal Friedmann one. Independent of the details, only in a highly nonlinear phase of the Universe could such a large amount of energy have been deposited such that the IGM has attained a minimum temperature of 10^5 K now.

The first obvious attempt to explain the high degree of ionization of the IGM, in relation with the absence of any Gunn–Peterson effect, is QSO photoionization. As our knowledge of the QSO population has improved and the limits on the density of neutral hydrogen have become more severe, this interpretation has been challenged (Shapiro and Giroux 1987, Giroux and Shapiro 1990). The point is that with the apparent decline of the QSO number density at redshift $\geqslant 3$ and with the known high-redshift QSO luminosity functions, the 'Strömgren' regions around QSO do not overlap, leaving a large fraction of the IGM not photoionized. Contrary to this, Donahue and Shull (1987) pointed out that for a low enough density IGM ($\Omega_{IGM} \leqslant 0.1$), the filling factor of ionized regions reaches unity by a redshift 2.75, and could be as high as 3.5 if $\Omega_{IGM} \leqslant 0.04$. These authors also argue that QSO luminosity functions and ionizing fluxes are not well enough known to require an extra source of ionization for the IGM. The conclusion is that, if QSO provide the only ionizing flux for the IGM, the Gunn–Peterson test could be marginally met if $\Omega_{IGM} < 0.04$, although progress on the study of the high-redshift QSO population could well change this result.

It is also possible that the IGM was reheated by primeval galaxies (Bechtold *et al* 1987, Shapiro 1990). If galaxies and massive stars formed early, they could have provided 10 more times of ionizing flux than QSO (in the absence of dust). Very early metal enrichment of the Universe is then expected (Shapiro 1990).

Another model for the reheating of the IGM arises in the explosion scenario for galaxy formation (Ostriker and Cowie 1981). According to this picture, the energy produced by the death of stars in galaxies could propagate in an explosive form through the IGM. The blast wave that develops around this primordial galaxy will then reheat the surrounding IGM. Temperatures of $T_{IGM} \approx 10^7 - 10^8$ K are easily reachable as the 'bubbles' overlap. This conflicts with COBE limits on the *y* parameter when considered together with upper limits on the temperature fluctuations in the MBR (Barcons *et al* 1991). A further consequence of this model is the dust and metal enrichment of the IGM, because explosions would have been generated with galactic material.

Within the explosion model, and for reasonable energy release per galaxy ($\approx 10^{61}$ erg), it is not difficult to produce structure on scales up to $\approx$ 5Mpc, corresponding to the observed galaxy–galaxy correlations (Ostriker 1988). The generation of power on larger scales requires, however, more energetic events. Superconducting cosmic strings have been suggested as possible mechanisms for the generation of very energetic perturbations in the IGM (Ostriker *et al* 1986). The temperature of the IGM will be even greater in this case and in direct conflict with the absence of distortions in the MBR spectrum.

Other models for the reheating of the IGM based upon more localized phenomena have also been proposed, the physics being basically gravitational reheating by large structures. The main problems are again the Gunn–Peterson test which requires a uniformly ionized IGM and the lack of fluctuations in the MBR.

3.2.3. The intergalactic medium and the cosmic background radiation. If, as suggested by the Gunn–Peterson test, the IGM is highly ionized, protons and electrons will emit bremsstrahlung radiation. Then, a uniform ionized IGM will necessarily produce a contribution to the cosmic background radiation (CBR). The wavelengths where this contribution appears depend on the thermal history of the IGM and the detectability of this background will depend upon the IGM density and the foreground radiation created by discrete sources at the corresponding wavelengths. Also, the presence of hot electrons in the intergalactic space, will necessarily produce a distortion in the microwave background radiation (MBR) which has been already discussed in 3.2.1.

It was once thought that the extragalactic cosmic x-ray background (XRB) discovered by Giacconi *et al* (1962) could arise as thermal bremsstrahlung in the IGM. The main reasons for this were that the XRB spectrum in the 3–100 keV band is consistent with 40 keV thermal bremsstrahlung (Marshall *et al* 1980) and that the brightest x-ray sources have steeper spectra than the XRB (see Boldt 1987 for a review). This model was first suggested by Hoyle (1963) (and soon rejected by Gould and Burbidge 1963) and considerably improved later (Cowsik and Kobetich 1972, Field and Perrenod 1977, Guilbert and Fabian 1986, Barcons 1987). It basically requires $T_{IGM}(z = 0) \sim 10^8$ K and $\Omega_{IGM} \approx 0.25$. It is therefore quite clear from figure 6 that the lack of distortion in the MBR spectrum does not allow a uniform IGM to produce the XRB.

Since the volume emissivity of the gas is $\propto n_{IGM}^2$ and the Compton y parameter $\propto n_{IGM}$, clumping of the gas could in principle reconcile this model with the observations. The required clumpiness parameter of the IGM should then be

$$C = \sqrt{\frac{\langle n_{IGM}^2 \rangle - \langle n_{IGM} \rangle^2}{\langle n_{IGM} \rangle^2}} \sim 25 \tag{3.2}$$

For illustrative purposes, we assume the IGM is confined in spherical regions of radius R with a density of such blobs n_b. These regions could be either pressure or gravity confined. Pressure confinement would require a yet hotter tenuous general medium (Guilbert and Fabian 1986, Barcons and Fabian 1988). Relativistic effects then come into play in the Compton distortion of the MBR which make the y parameter unacceptably high. Gravity confinement, assuming that the blobs are virialized, will require $R \sim 10^2 \Delta^{-1/2}$ Mpc and $n_b \sim 10^{-9} \Delta^{3/2}$ Mpc^{-3}, where $\Delta = (\delta\rho/\rho)$ is the mass overdensity in the blobs. For values of $\Delta < 100$ this will produce more fluctuations and anisotropies both in the XRB and in the MBR than what is actually observed. Much greater values of Δ would correspond to galaxy-sized sources. It can be then concluded that a hot IGM cannot give rise to the XRB, even if it is clumpy.

There are, however, other effects of the IGM on the cosmic background radiation, and in particular in the MBR through Compton scattering. MBR fluctuations are very sensitive to a possible reionization of the Universe (Vishniac 1987, Efstathiou 1988). The basic idea is that primordial fluctuations present in the radiation field of the Universe at $z \sim 1500$ (the so-called recombination epoch) can be washed out, at least on scales of less than a few degrees, whilst completely new fluctuations appear. There is also the possibility that the Universe has never really recombined.

The consequences on the MBR of a re-ionization depend very much on the thermal history of the IGM. If the IGM is ionized at $z \geqslant 100$, then all previous matter-coupled fluctuations are erased, and a high density baryonic Universe ($\Omega_{baryon} \approx 1$) does not conflict with current upper limits on anisotropies (Efstathiou 1988).

3.3. The Lyman α forest

3.3.1. Main features. Narrow absorption lines are usually parametrized by their equivalent width (W) defined as $W = \int d\lambda (f_c - f_{obs})/f_c$, where f_c is the continuum flux and f_{obs} the observed flux. This quantity is given (in the optical) in Å and it roughly represents the fraction of photons that have been absorbed times the width of the line. However, for cosmological purposes it is more convenient to define the rest-frame equivalent width $W_r = W/(1 + z)$ as the equivalent width seen by an observer located in the absorption system reference frame. Only high resolution spectroscopy allows direct fitting of two physically interesting parameters: the column density N_{HI} and the velocity dispersion parameter b ($b = \sqrt{2}\sigma_v$).

For intermediate resolution studies (≈ 1 Å) of the Lyman α forest, a cut-off in the rest equivalent width of about 0.36 Å is generally assumed, because weaker lines could be confused with noise. When higher resolution is available, and profile fitting can be performed, a lower limit for N_{HI} is also taken for the same purposes.

In general, the distribution of rest equivalent widths is quite well fitted (fairly independently of resolution) with a single exponential

$$P(W_r) \propto \exp(-W_r/W_*) \tag{3.3}$$

where $W_* \approx 0.5$ Å (Murdoch *et al* 1986). No serious variations with the redshift are seen in this distribution.

It has been widely believed until recently from ~ 1 Å resolution studies that the parameter b was concentrated around ~ 30 km s^{-1} (Carswell *et al* 1987). However, recent very high-resolution spectroscopic studies at ~ 0.02 Å resolution (Pettini *et al* 1990) have shown the existence of systems with b as low as the resolution ~ 6 km s^{-1} and a correlation between b and the column density N_{HI} whose implications will be discussed in 3.2.3.

The distribution of column densities goes as follows. Low redshift Lyman α lines seem to fit a single power law

$$p(N_{HI}) \propto N_{HI}^{-\beta} \tag{3.4}$$

where β is about 2 (Atwood *et al* 1985). This power law extends from about $N_{HI} = 10^{13.5}$ cm^{-2} to at least 10^{16} cm^{-2}. At higher redshifts (Carswell *et al* 1987) this power law seems to break below $N_{HI} = 10^{14.5}$ cm^{-2} where $\beta \approx 1$.

Lyman α lines are thought to arise in clouds (the Lyman α clouds) containing some neutral hydrogen. The physical state of these clouds is not very well undestood at present, although some models do exist (see 3.3.3). This fact arises from the lack of determination of the background ionizing field, and other parameters. For example, Sargent *et al* (1982) were able to put an upper limit on the size of Lyman α clouds of about 1 Mpc from the absence of correlation in the redshift of Lyman α lines towards a double QSO. On the other hand, Foltz *et al* (1984) found that 13 out of 16 lines in both images of the gravitationally lensed QSO 2354+007 are common in both lines of sight, which suggests a minimum size of about 5–25 kpc for the Lyman α clouds. However, this observation was at a very low resolution and the lines detected were consequently very strong. In addition, if Lyman α clouds cluster on these scales, we could be just seeing the effect of clustering along both lines of sight.

If the size and the temperature of the Lyman α clouds and the intensity of the ionizing background could be efficiently estimated, the number density of neutral and ionized hydrogen in the Lyman α clouds could be known (under the assumption of an approximate spherical symmetry). Unfortunately, these points are still controversial as we shall see below.

3.3.2. Redshift distribution and clustering. The fact that Lyman α clouds are thought to be intervening objects, rather than clouds of gas associated with the QSO is mostly based on their redshift distribution. Comparison of the number of Lyman α lines in a high redshift QSO with the number of lines of a lower redshift QSO reveals a very strong increase of their number with increasing redshift. It is customary to parametrize the number of Lyman α lines per unit redshift with a single power law

$$\frac{\mathrm{d}\mathcal{N}}{\mathrm{d}z} \propto (1+z)^{\gamma} \tag{3.5}$$

where the slope γ changes somehow from QSO to QSO (and even with the method used, see section 3.3.4), but it has a value of about two (Murdoch *et al* 1986, Carswell *et al* 1987, Bajtlik *et al* 1988).

A comoving population of fixed-size absorbers would give a number of Lyman α lines per unit redshift (see, e.g., Weymann *et al* 1981)

$$\frac{\mathrm{d}\mathcal{N}}{\mathrm{d}z} \propto (1+z)(1+2q_0z)^{-1/2} \tag{3.6}$$

which means an exponent γ between 0.5 and 1 depending on the cosmological model. This suggests strong evolution of the Lyman α cloud population: clouds either disappear or reduce their cross-section with cosmic time.

Once this general trend in the number of lines is allowed for, the distribution of clouds is basically Poissonian (Sargent *et al* 1980). This clearly argues in favour of the intervening hypothesis—the Lyman α clouds are uniformly distributed in space and deviations from a Poisson distribution, if any, only occur on very small scales.

In principle, with intermediate resolution data (e.g., the data used by Sargent *et al* 1980) only velocity splittings above some $300 \, \mathrm{km \, s^{-1}}$ can be tested. Therefore, in order to search for clustering on smaller scales higher resolution is needed. Webb (1988) found that Lyman α lines weakly cluster on scales of about $50 \, \mathrm{km \, s^{-1}}$, the redshift correlation function being negligible on scales $\geqslant 300 \, \mathrm{km \, s^{-1}}$, in agreement with Sargent *et al* (1980). There has been, however, a recent claim by Ostriker *et al* (1988) that there is an excess of small scale ($\approx 500 \, \mathrm{km \, s^{-1}}$) splittings in a sample made up of (mostly) intermediate resolution data. This has been checked against randomly simulated spectra (Webb and Barcons 1991) and it has been shown to be completely consistent with a Poissonian distribution. As a conclusion, Lyman α lines seem to obey Poisson statistics on scales $\geqslant 1h^{-1} \mathrm{Mpc}$, although some clustering (only detectable with high resolution data) is present on smaller scales.

There have also been other searches for inhomogeneities in the Lyman α forest. Carswell and Rees (1986) found no significant modulations in the number of lines on scales $\approx 5000 \, \mathrm{km \, s^{-1}}$ (the expected size of a void) in the high resolution spectra of Q0420-338 and PKS2000-330. This seems to contradict the finding of a large gap in the distribution of Lyman α lines in the spectrum of Q0420-338 (Crotts 1987). However, such large gaps are not so unexpected even in Poisson statistics (Ostriker *et al* 1988). In any case, a void structure, like the one observed in the distribution of nearby galaxies (de Lapparent *et al* 1986) does not map into the Lyman α forest at high redshift (Pierre *et al* 1988; Duncan *et al* 1989). Another way to search for inhomogeneities in the Lyman α forest (Webb and Barcons 1991) is to look for correlations in the rest equivalent width. It has also failed to detect any structure beyond $300 \, \text{km s}^{-1}$.

3.3.3. Physical models. As discussed above, the physical nature of the Lyman α clouds is still unknown. There are, however, two classes of models for the physical state of the clouds, namely, pressure-confined clouds and gravity-confined clouds.

In the pressure-confined model, Lyman α clouds are (probably primordial) clouds of material in the intergalactic space, confined by the pressure of a general tenuous IGM (Sargent *et al* 1980, Ikeuchi and Ostriker 1983, Ostriker and Ikeuchi 1983). The typical cloud temperature is a few times $10^4 \, \text{K}$ if the line widths are thermal and $b \sim 30 \, \text{km s}^{-1}$, but it will certainly be $< 10^4 \, \text{K}$ if the results of Pettini *et al* (1990) are confirmed. The hydrogen of the cloud is ionized by the general UV background and by collisions of the atoms among themselves. Assuming ionization equilibrium, the equations governing the structure of the clouds are (see, e.g., Sargent *et al* 1980, Black 1981)

$$(n_{\text{H I}} + n_{\text{H II}})T_{\text{cloud}} = n_{\text{IGM}}T_{\text{IGM}} \tag{3.7}$$

$$\frac{n_{\text{H II}}}{n_{\text{H I}}} = 7.7 \times 10^3 \left(\frac{T_{\text{cloud}}}{10^4 \, \text{K}}\right)^{3/4} \left(\frac{n_{\text{H II}}}{10^{-3} \, \text{cm}^{-3}}\right)^{-1} \left(\frac{I}{10^{-21} \, \text{erg cm}^{-2} \, \text{s}^{-1} \, \text{sr}^{-1}}\right)$$
$$+ 4.0 \times 10^4 \left(\frac{T_{\text{cloud}}}{10^4 \, \text{K}}\right)^{5/4} \exp\left[-15.8\left(\frac{T_{\text{cloud}}}{10^4 \, \text{K}}\right)^{-1}\right] \tag{3.8}$$

where the influence of helium and heavier elements has been neglected, $n_{\text{H I}}$ and $n_{\text{H II}}$ represent the densities of neutral and ionized hydrogen, and I is the UV ionizing flux in the Lyman limit. If the pressure of the IGM, the ionizing flux I, and the cloud temeperature are known, the density and ionization fraction of a Lyman α cloud can be inferred.

If $T_{\text{cloud}} > 10^4 \, \text{K}$ (Sargent *et al* 1980, Ostriker and Ikeuchi 1983) the state of the cloud is dominated by photoionization, collisions being unimportant. The total particle density of the cloud is then low ($n_{\text{cloud}} \approx 10^{-5} \, \text{cm}^{-3}$) and the ionization fraction $n_{\text{H II}}/n_{\text{H I}}$ high ($\geq 10^4$). Clouds then have a size (for a typical $N_{\text{H I}} \approx 10^{14} \, \text{cm}^{-2}$) of the order of 10–100 kpc, in agreement with the size inferred from the work of Foltz *et al* (1984).

If, on the contrary, $T \ll 10^4 \, \text{K}$ and the observed line widths are not thermal photoionization is unimportant for any reasonable value of the UV flux, and the clouds are dense and almost neutral. These clouds should be then aspherical and/or small in order to give rise to an observed column density of $10^{14} \, \text{cm}^{-2}$ (Barcons and Fabian 1987a, Pettini *et al* 1990). In addition, highly-flattened clouds could explain, rather naturally, the distribution of column densities seen in the Lyman α forest (Barcons and Fabian 1987a, Milgrom 1988). In this class of models, clustering of clouds on scales $\leqslant 100 \, \text{kpc}$ should be present, which is indicated by the work of Webb (1988).

Pressure-confined Lyman α clouds have been claimed to give a natural explanation for evolution (Ostriker and Ikeuchi 1983). Adiabatic expansion implies $P_{\text{IGM}} \propto (1 +$

$z)^5$, which for a fairly constant value of T_{cloud} would imply $\gamma \approx 2.5$–3 (see section 3.3.2). However, recent work by Carswell *et al* (1987) has shown that the cloud density n_{cloud} could even *decrease* with redshift, whilst pressure-confinement implies a rather strong increase of n_{cloud} with redshift.

A completely different class of models for the physics of the Lyman α clouds involving gravity-confinement was first introduced by Black (1981). Rees (1986) suggested that in a cold dark matter scenario for galaxy formation, the formation of dark minihalos giving rise to virial temperatures of about 10^4 K for the baryonic component, is an unavoidable prediction of theory. These minihalos could then host the Lyman α clouds (gas being mostly ionized by the UV background). Evolution in the number of lines per unit redshift will, in this case, arise as an effect of merging of the minihalos with cosmic time, giving rise to galaxies. Similar models have been discussed by Ikeuchi and Norman (1986) and by Bond *et al* (1987). The distribution of column densities spanning several orders of magnitude could in this case be explained mostly as an 'impact parameter' effect, since gas density inside the clouds would be highly inhomogeneous (Ikeuchi *et al* 1989).

Another important issue about the physics of Lyman α clouds is their origin and possible relation to galaxy formation. Bond *et al* (1987) discuss in detail the possibility that clouds lie in high density regions of the underlying cold dark matter field, with a size of about 100 kpc. This model gives about the right number of clouds per unit volume, evolution and distribution of column density, but depends sensitively on the free expansion model taken for the evolution of individual clouds. Ikeuchi and Norman (1986) present the same idea, but with some hot dark matter in order to account for very-large scale structure in the Universe. Barcons and Webb (1990) have recently shown that Lyman α clouds do not significantly cluster around high redshift galaxies (assumed to be represented by heavy-element absorption systems). This argues against Lyman α clouds being primeval objects formed in a biased scenario, although evolutionary effects could be relevant (i.e. clouds near galaxies could more easily disappear).

Vishniac and Bust (1987) have considered the possibility that Lyman α clouds are created by shocks in the IGM, triggered by explosions (*à la* Ostriker and Cowie 1981). The weak correlation function of Lyman α lines requires that clouds are finally destroyed in the very high density regions where galaxies form after explosions. This model could have some observational support from the fact that a doublet (separated by a velocity splitting of 114 km s^{-1}) of 5 lines in the Lyman series was seen towards the QSO S50014+81 (Chaffee *et al* 1985). A possible interpretation could be that we are seeing the edge of a cosmic bubble—perhaps the remnant of an explosion (Ozernoy 1987).

A completely different model was suggested by Hogan (1987) where Lyman α clouds are very small objects (see also Tytler 1987a, Barcons and Fabian 1987a, 1987b) arising inside galaxies (from shocks for example). Lyman α absorption would then be correlated on the size of a galaxy (according to the observation by Foltz *et al* 1984). The major point against this interpretation is that it looks difficult to produce column densities $N_{HI} \approx 10^{14}$ cm^{-2} in a galaxy where the total column density is about 10^{20} cm^{-2} (Rees 1988).

Further progress on the problem of the origin and physical state of Lyman α clouds requires more reliable high-resolution data samples. Also the UV flux at high redshift and the true size of the Lyman α clouds need to be known. Such studies deserve much effort.

3.3.4. The inverse effect. This is a point, in the study of the distribution of Lyman α lines, that has raised a lot of controversy in the last few years. The fact is that, apart from the general increase of the number of lines with redshift (see section 3.3.2), there is a relative lack of lines with absorption redshift increasingly approaching the QSO emission redshift. This was first pointed out by Weymann *et al* (1981).

One of the consequences of the inverse effect has been that different values of γ (see (3.5)) found in different samples (and even by different methods) (Sargent *et al* 1980, Carswell *et al* 1982). This situation has been emphasized by Tytler (1987b) and Murdoch *et al* (1986). In an analysis of a heterogeneous data sample (including high- and low-resolution data) Tytler (1987b) concluded that this lack of lines starts far away from the QSO emission redshift. Instead of (3.5), this author introduced a new parameter δ in the number of lines found per unit redshift

$$\frac{d\mathcal{N}}{dz_{abs}} \propto (1 + z_{abs})^\gamma \left(\frac{\lambda_r}{1215\text{Å}}\right)^\delta \tag{3.9}$$

where $\lambda_r = 1215\text{Å}(1 + z_{abs})/(1 + z_{em})$ is the absorption wavelength in the QSO rest-frame. Tytler (1987b) found a very strong dependence on λ_r ($\delta \approx 4 \pm 1$), using those Lyman α lines falling between the Lyman β and the Lyman α QSO emission lines.

On the other hand, Murdoch *et al* (1986), with a more restricted and homogeneous (intermediate resolution) sample found that this lack of lines is restricted to a wavelength range within some 5 Mpc of the QSO. A similar conclusion was reached by Bajtlik *et al* (1987) who renamed the lack of lines the proximity effect.

Webb and Larsen (1988) fitted a value for δ separately for high and intermediate resolution data, the main conclusion being, in both cases, that the inverse effect is only significant in a region close to the QSO and does not extend down to the Lyman β emission line as previously claimed by Tytler (1987b). Another important point raised by Webb and Larsen (1988) is that the inverse effect does not seem to be correlated with the QSO luminosity, or with the emission redshift.

With the largest intermediate resolution data sample Lu *et al* (1991) have very recently confirmed the existence of the inverse effect. They also show that it appears to be restricted to a ~ 16 Mpc region around the QSO (although an extension to greater distances cannot be excluded) and that is not correlated with the luminosity or the radio properties of the QSO. This last feature is important since radio-loud QSO are thought to be associated with clusters of galaxies (Yee and Green 1984, 1987) and in those objects Lyman α clouds could be destroyed by the ICM. There is, however, no evidence for this.

Current possible explanations for the inverse effect are: (i) Lyman α clouds are very small and those near the QSO fail to completely occult the QSO continuum emitting region, and (ii) QSO ionize their neighbourhood and so clouds near the QSO are more highly ionized and undetectable as absorption systems. The first possibility (Tytler 1987b, Barcons and Fabian 1987b) requires some fine tuning of the Lyman α cloud size and the QSO continuum emitting region size, and the inverse effect would probably extend to regions well away from the QSO. In addition some of the lines near to the QSO would not completely absorb the QSO light at the corresponding wavelength (which is observed in some cases). Clouds will then be almost neutral and small, as in the model of Hogan (1987).

The second explanation, would require a rather low UV background (of the order of the integrated UV background from QSO) in order that photoionization near the

QSO is not dominated by the background (Bajtlik *et al* 1987). As explained in 3.2.2 this could conflict with the absence of any Gunn–Peterson effect. In fact, if this model is correct, the inverse effect could be used to estimate the UV background, for a known QSO luminosity. In any case, clouds should be large and highly photoionized in this case. Moreover, when a QSO falls near the line of sight towards another QSO, a 'hole' in the Lyman α forest should appear in the spectrum of the far QSO due to ionization by the nearest QSO. The observational status of this effect is, at present, far from definitive, but deserves some effort in the future, given its importance.

3.4. Metal absorption lines

One of the most convincing arguments against the Lyman α forest lines arising in galaxies is their poor metal content (Williger *et al* 1989). Generally speaking, only for those Lyman α systems whose column density N_{HI} exceeds 10^{17} cm^{-2} (the Lyman limit absorbers), are associated metal lines detected (see, e.g. Robertson *et al* 1983). The usual interpretation is that the most abundant low-column density Lyman α clouds have primordial element abundances (mainly H and He), while metal absorption features should arise after stellar processing has started (in galaxies). Only if Lyman α clouds are almost neutral (Tytler 1987a) and/or highly flattened (Chaffee, quoted in Black 1981), are metal lines associated with the Lyman α forest not expected to be seen, assuming that Lyman α clouds do contain metals. As these hypotheses are not very fashionable, and some photoionization is not dominated by the background (Bajtlik *et al* 1988) and the properties of Lyman α systems and metal absorption systems are seen to be very different (see below), they are usually assumed to belong to different populations (see, however, Tytler 1987a).

3.4.1. Main features and classification. Metal lines are, by far, much less numerous than Lyman α lines (at a ratio of about 1:100). In order to avoid contamination with the Lyman α forest, they are usually identified longwards of the QSO Lyman α emission line. Some of those lines have redshifts $z_{abs} \approx z_{em}$ and are believed to be produced by the galaxy hosting the QSO, and usually excluded from any statistical analysis.

More or less unbiased samples of heavy-element absorption lines can be found in the literature. For low-redshift QSO, the Mg II doublet is the standard identifier (Young *et al* 1982, Tytler *et al* 1987 and others), whilst for higher redshifts CIV samples are used (Bergeron and Boissé 1984, Sargent *et al* 1988).

As mentioned before, heavy-element absorption systems are thought to arise in galaxies, that have undergone some stellar evolution (only stars can produce heavy-elements from primordial material). The expected number of metal absorption lines per unit redshift along a line of sight can be written as

$$\frac{d\mathcal{N}}{dz} = \frac{c}{H_0} n_{gal}(z = 0)\pi R_{gal}^2(z = 0)(1 + z)(1 + 2q_0 z)^{-1/2} \tag{3.10}$$

where n_{gal} is the (average) number of galaxies per unit volume, and R_{gal} the effective gas-rich (average) radius of a galaxy (assumed spherical). No evolution has been assumed. For reasonable values of these two parameters it is found that the expected number of metal lines is more than one order of magnitude below the observed one (Bergeron 1988a, 1988b). This is known as the *cross-section* problem. There are two obvious ways to explain this discrepancy.

The first possibility is that there are many faint unobserved galaxies which actually produce most of the metal absorption features. This possibility is, however, very unlikely, because in this case it would be practically impossible to detect any emission feature from the metal absorption systems, and some have been certainly detected (see Bergeron 1988c and section 3.4.3).

The other obvious possibility is that the galaxies responsible for absorption lines have much larger metal-rich gaseous envelopes than it is usually assumed. In this sense, the detection of some emission lines at the absorption redshift and quite far away from the line of sight in the case of some Mg II systems (as in the case of Cristiani 1987), supports this hypothesis. One explanation related to earlier parts of this review is that of Crawford *et al* (1987) in which it is proposed that all galaxies had massive cooling flows in the past ($\dot{M} \approx 100 M_\odot \mathrm{yr}^{-1}$). The radius of the optically active metal-enriched region of such massive cooling flows is of the order of 100 kpc, which gives enough covering fraction in the sky to meet the cross-section problem.

In conclusion, the number of heavy-element absorption lines seen towards QSO suggests early ($z \geqslant 2$) galaxy formation, and that those galaxies have a metal-rich extended halo significantly greater than it is observed in our environment. This could be very important for understanding the details of the late stages of galaxy formation. The effective value of the closure density in clouds of column density less than 10^{19-2} is about 8×10^{-3}.

3.4.2. Evolution and clustering. As in the case of Lyman α lines, it is customary to parametrize the number of absorption systems per unit redshift (3.5) in a single power law with an exponent γ. For the Mg II systems, an evolution in the same sense as in the Lyman α forest is found, with $\gamma \approx 2.4 \pm 0.8$ (Lanzetta *et al* 1987), although there is evidence that this steep evolution is restricted to the stronger lines (Sargent, private communication). For CIV sytems, however, the situation is different. Values of γ compatible with zero (Boissé 1988) or even negative (Sargent 1988) are found (see also Bechtold 1988). These systems, being at higher redshifts, could be showing the onset of the metal enrichment of the galactic gas due to the combined effect of stellar nucleosynthesis and supernovae (Steidel *et al* 1988, Sargent 1988). Again, the strength of the evolution depends on the strength of the lines, the strong ones only being numerous at lower redshifts. Part of the evolution may be due to changes in the metal abundance of the gas, rather than in the density of absorbers.

Clustering analyses of the metal absorption line population have only been done in the case of the CIV systems (Sargent 1988). There is a clear excess of velocity splittings up to 500 km s^{-1}. It is very unlikely that this clustering is due to galaxy motions within clusters. It rather seems that the true galaxy–galaxy correlation function is being seen at redshifts unavailable by direct observation. In addition, a simple hierarchical model for the evolution of the correlation function predicts roughly the correlation function of metal absorption systems from the observed galaxy–galaxy clustering at $z = 0$.

It is also worth mentioning the possible detection of an especially aligned super-cluster detected at $z \approx 2$ in the region around Tol 1037-27 (Jakobsen *et al* 1986, Sargent and Steidel 1987). An unusually high number of lines was found in a region of size $50 h^{-1}$ Mpc along the line of sight in the spectra of two QSO separated by a distance of $\approx 5 h^{-1}$ Mpc, suggesting the existence of a large-scale non-spherical structure at high redshift.

It must be emphasized that the clustering amplitude for metal lines is much larger than the one for Lyman α lines. This is used to argue against the interpretation, by Tytler (1987a) that all absorption lines belong to the same population (Sargent 1988). Currently fashionable ideas on galaxy formation predict, however, a clustering amplitude which increases with mass. If mass has any relation with column density (and it probably has), then it is not surprising that Lyman α clouds, being less massive, are less clustered than metal absorption systems (usually associated with large N_{HI}).

3.4.3. Identification of metal absorption systems. In the last few years, some effort has been devoted to the search for emission features from metal absorption systems. Imaging and spectroscopy of the regions around QSO with strong Mg II absorption systems has been performed. Only in a few cases (Blades 1988) has this search been unsuccessful. The majority of systems detected have strong [O II] and/or [O III] emission lines, and have relatively high luminosities (Bergeron 1988a, Miller *et al* 1987, Cristiani 1987, Yanni *et al* 1987, 1989; see Bergeron 1988a, 1988b, 1988c for reviews). These intervening galaxies are usually well separated from the line of sight to the QSO (29 arcsec in one case, Yanni *et al* 1987) therefore supporting the idea of very extended gaseous halos. All of the detected galaxies have a strong blue continuum and evidence for star formation.

It has also been found by Lanzetta and Bowen (1990) that for those Mg II systems that have been seen in emission, there is an anticorrelation between the absorption equivalent width and the impact parameter.

3.5. Damped Lyman α systems

A recent study by Wolfe *et al* (1986) has revealed the existence of some very strong damped Lyman α absorption lines shortward of the QSO Lyman α emission line. These systems, although not very numerous (15 in 68 QSO in the above sample) could be due to thick galaxy discs at high redshift. Column densities of these objects are of the same order (or even greater) than the column density in our galaxy. The velocity dispersion of the damped Lyman α systems is not easily measured from these studies, because of the damping wings and of the low-resolution used, but 21 cm absorption seems to indicate a rather low level of turbulence ($\sigma_v \approx 10\,\mathrm{km\,s^{-1}}$, Briggs and Wolfe 1983) although a second component with $\sigma_v \approx 30\,\mathrm{km\,s^{-1}}$ could also be present (Wolfe 1986). In a new sample, Wolfe (private communication) finds no obvious evolution and measures a column density distribution above an extrapolation of the Lyman α forest distribution.

The idea that these systems could be due to disc galaxies, comes from a few facts (Wolfe 1986): (i) the contribution of these systems to the density parameter of the Universe is of the order of the density contributed by present-day galaxies (i.e. the effective $\Omega \approx 0.005$); (ii) the distribution in column density is well-fitted if a population of randomly oriented discs is assumed (as in the case of the Lyman α clouds, Barcons and Fabian 1987a). The *cross-section* problem already mentioned reappears here, since a gaseous envelope of the order of 100 kpc is needed for each high-redshift galaxy (assuming a comoving population) to give rise to the right number of damped Lyman α systems observed. If galaxies at high redshift have gaseous discs with a size of about 100 kpc, the spherical collapse picture for galaxy formation should be modified. Collapse will then proceed in two phases: first, a highly anisotropic collapse produces a disc, and second the disc collapses radially to produce present-day galaxies.

The importance of the study of these systems has been enhaced by the recent discovery of a very high-redshift one ($z_{abs} = 3.392$) towards Q0000-26 (Webb *et al*

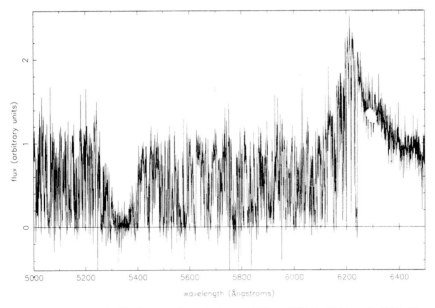

Figure 7. The Lyman α forest in the $z = 4.11$ QSO 0000-26 (Webb *et al* 1988). The Lyman α emission line in the QSO peaks at about 6200 Å. The noise level in the data can be seen longward of that line. All of the large features shortward of the emission line are absorption lines, most of which are due to Lyman α in intervening clouds. The large absorption line at about 5330 Å is a damped Lyman α system.

1988) (see figure 7). As argued by Wolfe (1986), this could be a new way to make a survey for galaxies at high redshift.

Some recent attemps to detect Lyman α emission from these disc systems have renewed the interest in their study. Hunstead and Pettini (1989) report a possible detection for the system at $z_{abs} = 2.465$ on the QSO Q0836+113. From the Lyman α luminosity of this object they infer that $\approx 1 \, M_\odot \, yr^{-1}$ of stars are being formed. A systematic search for such a type of emission associated with four damped Lyman α systems by Smith *et al* (1989) has given no positive result, although the inferred upper limit for the star formation rate is about 2–$4 \, M_\odot \, yr^{-1}$. All of these results suggest that unless dust is strongly present in those systems, disc galaxies form early but rapid star formation does not take place until late. Recent work by Fall *et al* (1989) has shown that these systems do contain some dust (dust to gas ratio $\sim 1/4$ of our galaxy). This will be enough to prevent Lyman α emission being observed in most cases.

4. Epilogue

The study of intergalactic gas consists of observing the observable and later trying to understand what has been seen. The densest and/or brightest clumps tend to show up first and the diffuse confining medium last. Confinement by pressure equilibrium, modified by gravitation and dynamic forces does appear to apply generally. Broadly speaking, the role of magnetic forces is unknown and is usually neglected.

Some problem areas where progress is needed are:

- When did reheating occur and what were the dominant heating mechanisms? Was it related to galaxy formation and was there any feedback between the two processes?

- Is the IGM basically of primordial composition, or does it have a metal content approaching that of clusters?

- Is the IGM very inhomogeneous (see e.g. Cen *et al* 1990 for some indications of how inhomogeneous it might be in a cold dark matter Universe)? Is Ω_{baryon} roughly constant throughout the Universe or does $\Omega_{galaxies}/\Omega_{baryon}$ vary? Are the voids in the galaxy distribution full of gas?

- What are the Lyα clouds and what objects should they be identified with at the present epoch? What happens to them at $z < 2$? What is their size and ionization state and has a single cloud been resolved spectroscopically yet?

- How well mixed is the ICM and what are the scale sizes and relative densities of the inhomogeneities?

- How important is the role of magnetic fields and turbulence in the ICM? Are the properties intermittent (Ruzmaikin *et al* 1989)?

- Is star formation different in different environments? Can there be a predominantly very low-mass mode?

- How did the gas in galactic halos evolve? Did it expand or accrete?

- What fraction of the required dark matter in galaxies and clusters is baryonic?

Finally, we note that existing and proposed telescopes and instrumentation promise exciting new discoveries and progress on the observational front. For example:

- The Cosmic Background Explorer, COBE, will soon improve on the limits on the Compton y-parameter of the diffuse IGM and will measure (or limit) fluctuations in the microwave background from reheating.

- UV spectra from the Hubble Space Telescope, HST, should improve on the Gunn–Peterson limits at low redshift and help with the identification of Lyα clouds. It may also improve on the limits on Ω_{baryon} from measurements of the deuterium abundance.

- The x-ray satellite, ROSAT, is mapping the diffuse hot ISM in our galaxy and will obtain powerful limits on any extragalactic soft x-ray background. It is also mapping the soft x-ray emission from the ICM in a variety of clusters and early-type galaxies.

- The broad-band x-ray telescope, BBXRT, the Japanese satellite ASTRO-D and later missions such as AXAF and XMM will measure good x-ray spectra of the ICM and map the distribution of hot gas in many extragalactic objects in detail.

- The Infrared Space Observatory, ISO, will map regions of star formation, probe galaxy formation through measurements of diffuse light and identify the role of dust in many diffuse gases. It might also detect low mass star formation.

- Improved ground-based optical (and infrared) spectrographs and larger telescopes mean that the Lyα forest and other cooler components in the IGM and in the ICM will be studied in detail.
We anticipate that studies of cosmic gas will continue to be a very active field.

Acknowledgments

This review is a much-updated and improved version of a series of Lectures given by ACF at the Saas-Fee school organized and published by the Swiss Astronomical Society. We thank Alastair Edge, Helen Parnell and Hans Bohringer for copies of figures and acknowledge partial support from NATO Collaborative Research Grant 0250/88.

References

Albert C E 1983 *Astrophys. J.* **272** 509
Alexander P, Brown M T and Scott P F 1984 *Mon. Not. R. Astron. Soc.* **208** 751
Arnaud K A, Fabian A C, Eales S A, Jones C and Forman W 1984 *Mon. Not. R. Astron. Soc* **211** 981
Arnaud K A, Johnstone R M, Fabian A C, Crawford A C, Nulsen P E J, Shafer R A and Mushotzky R F 1987 *Mon. Not. R. Astron. Soc.* **227** 241
Ashman K M and Carr B J 1988 *Mon. Not. R. Astron. Soc.* **234** 219
—— 1991 *Mon. Not. R. Astron. Soc.* **249** 13
Atwood B, Baldwin J A and Carswell R F 1985 *Astrophys. J.* **292** 58
Bahcall N A 1988 *IAU Symp. 130: Large Scale Structures of the Universe* ed J Audouze, M-C Pelletan and A S Szalay (Dordrecht: Kluwer)
Bahcall N A and Soneira R E 1983 *Astrophys. J.* **270** 20
Bajtlik S, Duncan R C and Ostriker J P 1988 *Astrophys. J.* **327** 570
Balbus S and Soker N 1989 *Astrophys. J.* **341** 611
Barcons X 1987 *Astrophys. J.* **313** 54
Barcons X and Fabian A C 1987a *Mon. Not. R. Astron. Soc.* **224** 675
—— 1987b *Mon. Not. R. Astron. Soc.* **229** 157
—— 1988 *Mon. Not. R. Astron. Soc.* **230** 189
Barcons X, Fabian A C and Rees M J 1991 *Nature* **350** 685
Barcons X and Webb J K 1990 *Mon. Not. R. Astron. Soc.* **244** 30P
Bechtold J 1988 *High Redshift and Primeval Galaxies* ed D Kunth, J Bergeron and B Rocca-Volmerange (Editions Frontières)
Bechtold J, Weymann R J, Lin Z and Malkan M A 1987 *Astrophys. J.* **315** 180
Begelman M C 1985 *Astrophys. J.* **297** 429
—— 1986 *Nature* **322** 614
Begelman M C, Blandford R D and Rees M J 1984 *Rev. Mod. Phys.* **56** 255
Begelman M C and Fabian A C 1990 *Mon. Not. R. Astron. Soc.* **244** 26P
Bergeron J 1988a *IAU Symp. No. 130: Large Scale Structures of the Universe* ed J Audouze, M–C Pelletan and A S Szalay (Dordrecht: Kluwer)
—— 1988b *The Post Recombination Universe* ed N Kaiser and A N Lasenby (Dordrecht: Kluwer)
—— 1988c *QSO Absorption Systems: Probing the Universe* ed J C Blades, C A Norman and D Turnshek (Cambridge: Cambridge University Press)
Bergeron J and Boissé P 1984 *Astr. Astrophys.* **133** 374
Bertola F and Capaccioli M 1975 *Astrophys. J.* **200** 439
Bertola F, Gregg M D, Gunn J E and Oemler A 1987 *Astrophys. J.* **303** 624
Bertshinger E and Meiksin A 1986 *Astrophys. J.* **306** L1
Bicknell G V 1986 *Astrophys. J.* **305** 109
Biermann P and Kronberg P P 1983 *Astrophys. J.* **268** L69
Birkinshaw M, Gull S F and Hardebeck H 1984 *Nature* **309** 34
Birkinshaw M, Hughes J and Arnaud K A 1991 *Astrophys. J.* submitted
Black J H 1981 *Mon. Not. R. Astron. Soc.* **197** 553
Blades J C 1988 *QSO Absorption Lines: Probing the Universe* ed J C Blades, C A Norman and D Turnshek (Cambridge: Cambridge University Press)
Boesgaard A M and Steigman G 1987 *Ann. Rev. Astron. Astrophys.* **23** 319
Boissé P 1988 *High Redshift and Primeval Galaxies* Ed J Bergeron, D Kunth and B Rocca-Volmerange (Editions Frontières)
Boldt E 1987 *Phys. Rep.* **146** 215

Bond J R, Szalay A S and Silk J 1987 *Astrophys. J.* **324** 627

Bregman J D 1980 *Astrophys. J.* **236** 577

Bregman J D and David L P 1988 *Astrophys. J.* **326** 639

Bregman J D, McNamara B R and O'Connell R W 1990 *Astrophys. J.* **351** 406

Briggs F H and Wolfe A M 1983 *Astrophys. J.* **268** 76

Bridle A H and Perley R A 1984 *Ann. Rev. Astron. Astrophys.* **22** 319

Burns J O 1990 *Astron. J.* **99** 14

Byram E T, Chubb A and Friedman H 1966 *Science* **152** 66

Caldwell C N and Oemler A 1981 *Astron. J.* **86** 1424

Canizares C R, Clark G W, Jernigan J G and Markert T H 1982 *Astrophys. J.* **262** L33

Canizares C R, Fabbiano G and Trinchieri G 1987 *Astrophys. J.* **312** 503

Canizares C R, Markert T H and Donahue M E 1988 *Cooling Flows in Clusters and Galaxies* ed Fabian A C (Dordrecht: Kluwer) p 63

Canizares C R, Stewart G C and Fabian A C 1983 *Astrophys. J.* **272** 449

Carswell R F and Rees M J 1986 *Mon. Not. R. Astron. Soc.* **223** 18P

Carswell R F, Webb J K, Baldwin J A and Atwood B 1987 *Astrophys. J.* **319** 709

Carswell R F, Whelan J A J, Smith M G, Boksenberg A and Tytler D 1982 *Mon. Not. R. Astron. Soc.* **198** 91

Castor J, McCray R and Weaver R 1975 *Astrophys. J.* **200** L107

Cavaliere A and Fusco-Femiano R 1976 *Astron. Astrophys.* **49** 137

Cen R Y, Jameson A, Liu F and Ostriker J P 1990 *Preprint*

Chaffee F H, Foltz C B, Röser H-J, Weymann R J and Latham D W 1985 *Astrophys. J.* **292** 362

Cowie L L 1986 *Ann. Rev. Astron. Astrophys.* **24** 499

Cowie L L, Fabian A C and Nulsen P E J 1980 *Mon. Not. R. Astron. Soc.* **191** 399

Cowie L L, Henriksen M J and Mushotzky R F 1987 *Astrophys. J.* **312** 593

Cowie L L and McKee C F 1977 *Astrophys. J.* **211** 135

Cowie L L and Songaila A 1977 *Nature* **266** 501

Cowsik R and Kobetich E J 1972 *Astrophys. J.* **177** 585

Cox D P and Smith B W 1974 *Astrophys. J.* **189** L105

Crawford C S, Crehan D A, Fabian A C and Johnstone R M 1987 *Mon. Not. R. Astron. Soc.* **224** 1007

Crawford C S and Fabian A C 1989 *Mon. Not. R. Astron. Soc.* **239** 219

Cristiani S 1987 *Astron. Astrophys.* **175** L1

Crotts A P S 1987 *Mon. Not. R. Astron. Soc.* **230** 48

Daines S J, Fabian A C, Thomas P A and Johnstone R M 1990 *Mon. Not. R. Astron. Soc.* submitted

David L P, Forman W and Jones C 1990 *Astrophys. J.* **359** 29

Davidsen A F, Harting G F and Fastie W G 1977 *Nature* **269** 203

Day C S R, Fabian A C, Edge A C and Raychaudhury S 1991 *Mon. Not. R. Astron. Soc.* in press

de Lapparent V, Geller M J and Huchra J P 1986 *Astrophys. J.* **302** L1

Donahue C M and Shull J M 1987 *Astrophys. J.* **323** L13

Draine B T and Salpeter E 1979 *Astrophys. J.* **231** 77

Duncan R C, Ostriker J P and Bajtlik S 1989 *Astrophys. J.* **345** 39

Edge A C 1989 *PhD* University of Leicester

Edge A C, Stewart G C, Fabian A C and Arnaud K A 1990 *Mon. Not. R. Astron. Soc.* **245** 559

Efstathiou G 1988 *Large Scale Motions in the Universe* ed V C Rubin and G V Coyne (Princeton, NJ: Princeton University Press) p 299

Evrard A E 1990 *Astrophys. J.* **363** 349

Fabbiano G 1989 *Ann. Rev. Astron. Astrophys.* **27** 87

Faber S M and Gallagher J S 1976 *Astrophys. J.* **2104** 365

Fabian A C 1990 *Physical Processes in Hot Cosmic Plasmas* ed W Brinkmann, A C Fabian and F Giovanelli (Dordrecht: Kluwer) p 271

Fabian A C, Arnaud K A, Nulsen P E J and Mushotzky R F 1986 *Astrophys. J.* **305** 9

Fabian A C and Crawford C S 1990 *Mon. Not. R. Astron. Soc.* **247** 439

Fabian A C, Crawford C S, Johnstone R M and Thomas P A 1987 *Mon. Not. R. Astron. Soc.* **228** 963

Fabian A C, Hu E M, Cowie L L and Grindlay J 1981 *Astrophys. J.* **248** 47

Fabian A C and Nulsen P E J 1977 *Astrophys. J. Mon. Not. R. Astron. Soc.* 479

Fabian A C, Nulsen P E J and Arnaud K A 1984 *Mon. Not. R. Astron. Soc.* **208** 179

Fabian A C, Nulsen P E J and Canizares C R 1982 *Mon. Not. R. Astron. Soc.* **201** 933

—— 1984 *Nature* **311** 733

Fabian A C, Nulsen P E J and Canizares C R 1991 *Astr. Astrophys. Rev.* **2** 191

Fabian A C *et al* 1985 *Mon. Not. R. Astron. Soc.* **216** 923

Fabricant D, Beers T C, Geller M J, Gorenstein P, Huchra J P and Kurtz M J 1986 *Astrophys. J.* **308** 580

Fabricant D and Gorenstein P 1983 *Astrophys. J.* **267** 535

Fall S M and Pei Y C 1989 *Astrophys. J.* **337** 7

Fall S M, Pei Y C and McMahon R G 1990 *Astrophys. J.* **341** L5

Feigelson E D, Schreier E J, Delvaille J P, Giacconi R, Grindlay J E and Lightman A P 1981 *Astrophys. J.* **251** 31

Field G B, Goldsmith D BV and Habing H J 1969 *Astrophys. J.* **155** L149

Field G B and Perrenod S C 1977 *Astrophys. J.* **215** 717

Foltz C B, Weymann R J, Roser J-J and Chaffee F H 1984 *Astrophys. J.* **281** L1

Forbes D A, Fabian A C, Crawford C S and Johnstone R M 1990 *Mon. Not. R. Astron. Soc.* **244** 680

Forman W, Jones C and Tucker W 1985 *Astrophys. J.* **293** 102

Forman W, Schwarz J, Jones C, Liller W and Fabian A C 1979 *Astrophys. J.* **234** L27

Gear W K, Gee G, Robson E I and Nott I G 1985 *Mon. Not. R. Astron. Soc.* **217** 281

Geller M J 1984 *Commun. Astron. Space Sci.* **10** 47

Giacconi R, Gursky H, Paolini F and Rossi B 1962 *Phys. Rev. Lett.* **9** 439

Gioia I M, Henry J P, Maccacara T, Morris S L, Stocke J T and Wolter A 1990 *Astrophys. J.* **356** L35

Giroux M and Shapiro P R 1990 *Physical processes in fragmentation and star formation* ed R Capuzzo Delcetta, C Chiosi and A DiFazio (Dordrecht: Kluwer) in press

Gould R J and Burbidge G R 1963 *Astrophys. J.* **138** 969

Guilbert P W and Fabian A C 1986 *Mon. Not. R. Astron. Soc.* **220** 439

Gunn J E and Peterson B A 1965 *Astrophys. J.* **142** 1653

Gursky H E, Kellogg E, Leong C, Tananbaum H and Giacconi R 1971 *Astrophys. J.* **167** L81

Gusten R and Mezger P 1982 *Vistas in Astron.* **26** 159

Hattori M and Habe A 1990 *Mon. Not. R. Astron. Soc.* **242** 399

Heckman T M 1981 *Astrophys. J.* **250** L59

Heckman T M, Baum S A, van Breugel W J M and McCarthy P 1989 *Astrophys. J.* **338** 48

Heiles C 1987 *Astrophys. J.* **315** 555

Heiles C and Kulkarni S R 1987 *Proc NATO ASI on Physical Processes in Interstellar Clouds* ed G E Morfill and M Scholer (Dordrecht: Reidel)

Hintzen P and Romanishin W 1986 *Astrophys. J.* **311** L1

—— 1988 *Astrophys. J.* **327** L17

Hogan C 1987 *Astrophys. J.* **316** L59

Hollenbach D J and Thronson H A 1987 *Interstellar Processes* (Dordrecht: Kluwer)

Hoyle F 1953 *Astrophys. J.* **118** 513

—— 1963 *Astrophys. J.* **137** 993

Huchra J P and Brodie J 1987 *Astron. J.* **93** 779

Hu E M, Cowie L L and Wang Z 1985 *Astrophysics. J. Suppl.* **59** 447

Hughes J P 1989 *Astrophys. J.* **337** 21

Hughes J P, Yamashita K, Okumura Y, Tsunemi H and Matsuoka M 1988 *Astrophys. J.* **327** 615

Hunstead R W and Pettini M 1989 *The Epoch of Galaxy Formation* ed C S Frenk *et al* (Dordrecht: Kluwer)

Ikeuchi S, Murakami I and Rees M J 1989 *Mon. Not. R. Astron. Soc.* **236** 21P

Ikeuchi S and Norman C A 1986 *Astrophys. J.* **312** 485

Ikeuchi S and Ostriker J P 1986 *Astrophys. J.* **301** 522

Jaffe W 1990 *Astron. Astrophys.* **240** 254

Jaffe W, de Bruyn A G and Sijbreng D 1988 *Cooling Flows in Clusters and Galaxies* ed A C Fabian (Dordrecht: Kluwer) p 145

Jahoda K and Mushotzky R F 1989 *Astrophys. J.* **346** 638

Jakobsen P, Perrymam M A C, Ulrich M H, Macchetto F and di Serego Alighieri S 1986 *Astrophys. J.* **303** L27

Johnson H E and Axford W I 1971 *Astrophys. J.* **165** 381

Johnstone R M and Fabian A C 1988 *Mon. Not. R. Astron. Soc.* **233** 581

—— 1989 *Mon. Not. R. Astron. Soc.* **224** 75

Johnstone R M, Fabian A C, Edge A C and Thomas P A 1990 *Preprint*

Johnstone R M, Fabian A C and Nulsen P E J 1987 *Mon. Not. R. Astron. Soc.* **224** 75

Jones C and Forman W 1984 *Astrophys. J.* **276** 38

Joseph R D, Meikle W P S, Robertson N A and Wright G S 1984 *Mon. Not. R. Astron. Soc.* **209** 111

Jura M 1977 *Astrophys. J.* **129** 268

Kaiser N 1984 *Mon. Not. R. Astron. Soc.* **284** L9

Kent S M and Sargent W L W 1979 *Astrophys. J.* **230** 667

King I R 1966 *Astron. J.* **71** 64

Krolik J H, McKee C F and Tarter C B 1981 *Astrophys. J.* **249** 422

Lahav O, Edge A C, Fabian A C and Putney A 1989 *Mon. Not. R. Astron. Soc.* **238** 881

Lanzetta K M and Bowen D 1990 *Astrophys. J.* **357** 321

Lanzetta K M, Turnshek D A and Wolf A M 1987 *Astrophys. J.* **322** 739

Larson R B anbd Dinerstein H L 1975 *Publ. Astron. Soc. Pacific* **87** 911

Lazareff B, Castets A, Kim D-W and Jura M 1989 *Astrophys. J.* **336** L13

Lea S M, Mushotzky R F and Holt S S 1982 *Astrophys. J.* **262** 24

Loewenstein M 1989 *Mon. Not. R. Astron. Soc.* **238** 15

Loewenstein M and Fabian A C 1990 *Mon. Not. R. Astron. Soc.* **242** 120

Long K S and Van Speybroeck L P 1983 *Accretion Driven Stellar X-ray Sources* ed W Lewin and E P J Van Den Heuvel (Cambridge: Cambridge University Press)

Lu L, Wolfe A M and Turnshek D A 1991 *Astrophys. J.* **367** 19

Maccagni D, Garilli B, Gioia I M, Maccacaro T, Bettolani G and Wolter A 1988 *Astrophys. J.* **334** L1

MacCammon D and Sanders W T 1990 *A. Rev. Astron. Astrophys.* **28** 657

MacDonald J and Bailey M E 1981 *Mon. Not. R. Astron. Soc.* **197** 995

Malagoli A, Rosner R and Bodo G 1987 *Astrophys. J.* **319** 632

Malaney R A and Fowler W A 1988 *Astrophys. J.* **333** 632

Malin D and Carter D 1980 *Nature* **280** 643

Marcaide J M, Alberdi A, Elósegui P, Schalinski C J, Jackson N and Witzel A 1989 *Astron. Astrophys.* **211** L23

Marshall F E, Boldt E A, Holt S S, Miller R, Mushotzky R F, Rose L A, Rothschild R and Seremitsos P J 1980 *Astrophys. J.* **235** 4

Mather J C *et al* 1990 *Astrophys. J.* **354** L37

Mathews W G 1989 *Astron. J.* **97** 42

Mathews W G and Baker J C 1971 *Astrophys. J.* **170** 241

Mathews W G and Loewenstein M 1986 *Astrophys. J.* **306** L7

McCray R and Snow T P 1979 *A. Rev. Astron. Astrophys.* **17** 213

McHardy I M, Stewart G C, Edge A C, Cooke B, Yamashita K and Hatsukade I 1990 *Mon. Not. R. Astron. Soc.* **242** 215

McKee C F and Ostriker J P 1977 *Astrophys. J.* **218** 148

Milgrom M 1988 *Astron. Astrophys.* **202** L9

Miller J S, Goodrich R W and Stephens S A 1987 *Astrophys. J.* **94** 633

Miller L 1986 *Mon. Not. R. Astron. Soc.* **220** 713

Mirabel I F, Sanders D B and Kazes I 1989 *Astrophys. J.* **340** L9

Mitchell R J, Culhane J L, Davison P J and Ives J C 1976 *Mon. Not. R. Astron. Soc.* **176** 29P

Murdoch H S, Hunstead R W, Pettini M and Blades J C 1986 *Astrophys. J.* **309** 19

Mould J R, Oke J P and Memec J M 1987 *Astron. J.* **92** 53

Mushotzky R F 1987 *Astrophys. Lett.* **26** 43

Mushotzky R F, Holt S S, Smith B W, Boldt E A and Serlemitsos P J 1981 *Astrophys. J.* **244** L47

Mushotzky R F and Szymkowiak A E 1988 *Cooling Flows in Clusters and Galaxies* ed A C Fabian (Dordrecht: Kluwer) p 47

Nittmann J, Falle S A E G and Gaskell P H 1982 *Mon. Not. R. Astron. Soc.* **201** 833

Nulsen P E J 1982 *Mon. Not. R. Astron. Soc.* **198** 1007

—— 1986 *Mon. Not. R. Astron. Soc.* **221** 377

Nulsen P E J, Stewart G C and Fabian A C 1984 *Mon. Not. R. Astron. Soc.* **208** 185

Nulsen P E J, Stewart G C, Fabian A C, Mushotzky R F, Holt S S, Ku W H M and Malin D F 1982 *Mon. Not. R. Astron. Soc.* **199** 1089

O'Connell R W 1987 *Proc. IAU Symp. 127* ed T De Zeeuw (Dordrecht: Reidel) p 167

O'Dea C P, Sarazin C L and Owen F N 1987 *Astrophys. J.* **316** 113

Ostriker J P 1988 *IAU Symp. 130: Large Scale Structures of the Universe* ed J Audouze, M-C Pelletan and A S Szalay (Dordrecht: Kluwer)

Ostriker J P, Bajtlik S and Duncan R C 1988 *Astrophys. J.* **327** L35

Ostriker J P and Cowie L L 1981 *Astrophys. J.* **243** L127

Ostriker J P and Heisler J 1985 *Astrophys. J.* **278** 1

Ostriker J P and Ikeuchi S 1983 *Astrophys. J.* **268** 163

Ostriker J P, Thompson C and Witten E 1986 *Phys. Lett.* **280B** 231

Ozernoy L 1987 *QSO Absoption Systems: Probing the Universe* ed J C Blades, C A Norman and D Turnshek (Space Telescope Science Institute Publications)

Persic M, Rephaeli Y and Boldt E 1988 *Astrophys. J.* **327** L1

Pesce J E, Edge A C Fabian A C and Johnstone R M 1990 *Mon. Not. R. Astron. Soc.* **244** 58

Pettini M, Hunstead R W, Smith L J and Mar D P 1990 *Mon. Not. R. Astron. Soc.* **246** 545

Pickles A J 1985 *Astrophys. J.* **296** 340

Pierre M, Iovino A and Shaver P A 1988 *Astron. Astrophys.* **197** L3

Pringle J E 1989 *Mon. Not. R. Astron. Soc.* **239** 479

Raychaudhury S, Fabian A C, Edge A C, Jones C and Forman W 1991 *Mon. Not. R. Astron. Soc.* **248** 101

Rees M J 1977 *Mon. Not. R. Astron. Soc.* **176** 483

—— 1986 *Mon. Not. R. Astron. Soc.* **218** 25P

—— 1988 *QSO Absorption Lines: Probing the Universe* ed J C Blades, C A Norman and D Turnshek (Cambridge: Cambridge University Press)

Rephaeli Y 1987 *Mon. Not. R. Astron. Soc.* **225** 851

Rieke G *et al* 1985 *Astrophys. J.* **290** P116

Robertson J G, Shaver P A and Carswell R F 1983 *XXIV Colloque International de Liège* ed J P Swings

Rosen A and Wiita P J 1988 *Astrophys. J.* **330** 16

Rosner R and Tucker W H 1989 *Astrophys. J.* **338** 761

Rothenflug R and Arnaud M 1986 *Astron. Astrophys.* **147** 337

Rubin V C, Ford W K, Peterson C H and Oort J H 1977 *Astrophys. J.* **211** 693

Ruzmaikin A, Sokoloff D and Shukurov A 1989 *Mon. Not. R. Astron. Soc.* **241** 1

Sanders R H 1981 *Astrophys. J.* **244** 820

Sarazin C L 1983 *The Interstellar Medium in Galaxies* ed H A Thronson and J M Shull (Dordrecht: Kluwer)

—— 1986 *Rev. Mod. Phys.* **58** 1

Sarazin C L and O'Connell R W 1983 *Astrophys. J.* **258** 552

Sargent W L W 1988 *IAU Symp. 130: Large Scale Structures of the Universe* ed J Audouze, M-C Pelletan and A S Szalay (Kluwer: Dordrecht)

Sargent W L W, Boksenberg A and Steidel C 1988 *Astrophys. J. Suppl.* **68** 539

Sargent W L W and Steidel C 1987 *Astrophys. J.* **322** 142

Sargent W L W, Young P J, Boksenberg A and Tytler D 1980 *Astrophys. J. Suppl.* **42** 41

Sargent W L W, Young P J and Schneider D P 1982 *Astrophys. J.* **256** 374

Savage B D and de Boer K S 1981 *Astrophys. J.* **243** 460

Scalo J 1986 *Fund. Cos. Phys.* **11** 1

—— 1990 *Physical Processes in Fragmentation and Star Formation* ed R Capuzzo-Delcetta, C Chiosi and A DiFazio (Dordrecht: Kluwer) in press

Schwartz D A, Schwarz J and Tucker W H 1980 *Astrophys. J.* **238** L59

Schwarz R, Bohringer H and Morfill G 1990 *Preprint*

Shapiro P R 1990 *Ann. NY Acad. Sci.* **571** 128

Shapiro P R and Field G B 1976 *Astrophys. J.* **205** 762

Shaprio P R and Giroux M L 1987 *Astrophys. J.* **321** L107

Shapiro P R and Moore R T 1976 *Astrophys. J.* **207** 460

Shaver P 1987 *High Redshift and Primeval Galaxies* ed D Kunth, J Bergeron and B Rocca-Volmerange (Editions Frontières)

Sherman R D 1982 *Astrophys. J.* **256** 370

Shu F H, Adams F C and Lizano S 1987 *Ann. Rev. Astron. Astrophys.* **25** 23

Silk J 1980 *Saas-Fee Lecture Course* Geneva Observatory

Silk J, Djorgovski G, Wyse R F G and Bruzual G A 1986 *Astrophys. J.* **307** 415

Solanes J M and Savador-Solé E 1990 *Preprint*

Smith H E, Cohen R D, Burns J E, Moore D J and Uchida B A 1989 *Astrophys. J.* **347** 87

Spitzer L 1956 *Astrophys. J.* **124** 20

Steidel S and Sargent W L W 1987 *Astrophys. J.* **318** L11

Stewart G C, Canizares C R, Fabian A C and Nulsen P E J 1984a *Astrophys. J.* **278** 536

Stewart G C, Edge A C and Fabian A C 1990 *Preprint*

Stewart G C, Fabian A C, Jones C and Forman W 1984b *Astrophys. J.* **285** 1

Sutherland W J 1988 *Mon. Not. R. Astron. Soc.* **239** 159

Takeda H, Nulsen P E J and Fabian A C 1984 *Mon. Not. R. Astron. Soc.* **208** 261

Thomas P A 1986 *Mon. Not. R. Astron. Soc.* **220** 949

Thomas P A and Fabian A C 1990 *Mon. Not. R. Astron. Soc.* **246** 156

Thomas P A, Fabian A C, Arnaud K A, Forman W and Jones C 1986 *Mon. Not. R. Astron. Soc.* **222** 655

Thomas P A, Fabian A C and Nulsen P E J 1987 *Mon. Not. R. Astron. Soc.* **228** 973

Thronson H A and Shull J M 1990 *The Interstellar Medium in Galaxies* (Dordrecht: Kluwer)

Tribble P 1991 *Mon. Not. R. Astron. Soc.* **248** 741

Trinchieri G and Fabbiano G 1985 *Astrophys. J.* **296** 447

Trinchieri G, Fabbiano G and Canizares C R 1986 *Astrophys. J.* **310** 637

Tucker W H and Rosner R 1982 *Astrophys. J.* **296** 447

Turnshek D 1988 *QSO Absorption Lines: Probing the Universe* ed J C Blades, C Norman and D Turnshek (Cambridge: Cambridge University Press)

Tytler D 1987a *Astrophys. J.* **321** 49

—— 1987b *Astrophys. J.* **321** 69

—— 1987c *Preprint*

Tytler D, Broksenberg A, Sargent W L W, Young P J and Kunth D 1987 *Astrophys. J. Suppl.* **64** 667

Ulmer M, Cruddace R G, Fennimore E E, Fritz G G and Snyder W A 1987 *Astrophys. J.* **319** 118

Valentijn E A 1990 *Nature* **346** 153

Valentijn E A and Bijleveld W 1983 *Astron. Astrophys.* **125** 223

Van Breugel W, Filippenko A V, Heckman T and Miley G 1985 *Astrophys. J.* **293** 83

Van Speybroeck L 1987 *Astrophys. J.* **322** 597

Vishniac E T 1987 *Astrophys. J.* **322** 597

Vishniac E T and Bust G S 1987 *Astrophys. J.* **319** 14

Webb J K 1988 *IAU Symp. 124: Observational Cosmology* ed A Hewitt, G Burbidge and L-Z Fang (Dordrecht: Kluwer)

Webb J K and Barcons X 1991 *Mon. Not. R. Astron. Soc.* **250** 270

Webb J K and Larsen I P 1988 *High Redshift and Primeval Galaxies* ed J Bergeron, Kunth D and B Rocca-Volmerange (Editions Frontières)

Webb J K, Parnell H C, Carswell R F, McMahon R G, Irwin M J, Hazard C, Ferlet R and Vidal-Madjar A 1988 *ESO Messenger* **51** 15

Weymann R J, Carswell R F and Smith M G 1981 *Ann. Rev. Astron. Astrophys.* **19** 41

White D A, Fabian A C, Johnstone R M, Mushotzky R F and Arnaud K A 1991 *Mon. Not. R. Astron. Soc.* in press

White R E and Chevalier R A 1984 *Astrophys. J.* **280** 561

White R E and Sarazin C L 1987 *Astrophys. J.* **318** 612, 621, 629

Williger G M, Carswell R F, Webb J K, Boksenberg A and Smith M G 1989 *Mon. Not. R. Astron. Soc.* **237** 635

Wolfe A M 1986 *Phil. Trans. R. Soc. Lond.* A **320** 503

Wolfe A M, Turnshek D A, Smith H E and Cohen R D 1986 *Astrophys. J. Suppl.* **61** 49

Wu X-Y, Hamilton T, Helfand D J and Wang Q 1991 *Astrophys. J.* in press

Yang J, Turner M S, Steigman G, Schramm D N and Olive K A 1984 *Astrophys. J.* **281** 493

Yanny B, Hamilton D, Schommer R A, Williams T B and York D G 1987 *Astrophys. J.* **323** L19

Yanny B, York D G and Gallagher J S 1989 *Astrophys. J.* **338** 735

Yee H K C and Green R F 1984 *Astrophys. J.* **280** 79

—— 1987 *Astrophys. J.* **319** 28

York D G 1982 *A. Rev. Astron. Astrophys.* **20** 221

Young P J, Sargent W L W and Boksenberg A 1982 *Astrophys. J. Suppl.* **48** 455

Zeldovich Ya B and Sunyaev R A 1969 *Astrophys. Space Sci.* **4** 301

Rep. Prog. Phys. **54** (1991) 579–633. Printed in the UK

Active galactic nuclei

D E Osterbrock

Institute for Advanced Study, Princeton, New Jersey 08540, USA

Abstract

Our current knowledge of active nuclei is reviewed. The importance of observational data taken over a wide range of frequencies, from radio and infrared through optical and ultraviolet to x-rays and γ-rays, is emphasized. Important overall principles include the continuity from quasars and QSOs through Seyfert and radio galaxies to low-luminosity LINERs, the importance of considering roughly cylindrically symmetric (rather than spherically symmetric) structures, and that the various regions generally have different axes and planes of symmetry, and are often warped.

This review was received in June 1990.

0034-4885/91/040579+55$14.00 © 1991 IOP Publishing Ltd

Contents

1. Historical introduction

The study of active galactic nuclei is one of the most rapidly growing subjects in present-day astrophysics. As the sites of the release of energy on the most powerful sustained rates and compact scales we know, they clearly are of great intrinsic interest. As the most luminous objects we know, they are the best markers we have of the distant reaches of the universe. From both points of view, understanding their physical nature and structure, and thus how to use them to measure cosmological distances and times are two of the most important aims of current astrophysical research. The greatest difficulty in attempting to review work in this field is the huge number of papers, covering so many different methods of research, that are being published. An excellent recent book on the subject is *Quasar Astronomy* by Weedman (1986).

The observational study of active galactic nuclei (AGNs) began with Fath (1909), who noted in a spectroscopic survey of the brightest spiral 'nebulae' that although most of them had absorption-line spectra, which he correctly interpreted as the integrated light of large numbers of stars, one, NGC 1068, also had six emission lines in its spectrum. He recognized them all as characteristic emission lines of planetary nebulae; today we know them as [O II] $\lambda 3727$, [Ne III] $\lambda 3869$, Hβ, and [O III] $\lambda\lambda 4363$, 4959, 5007. Other astronomers, especially Slipher (1917), soon obtained much better spectra of NGC 1068 and of the somewhat similar NGC 4151. Hubble (1926), in his epoch-making paper on the 'extragalactic nebulae', emphasized the planetary-nebula-like emission-line spectra of three AGNs (as we call them today), NGC 1068, 4051 and 4151. Then seventeen years later Seyfert (1943) studied these and other galaxies, and isolated the small fraction of them which show many relatively high-ionization emission lines in their nuclear spectra. These nuclei are invariably highly luminous, and their emission lines are wider than the lower-ionization emission lines that occur in the spectra of the nuclei of many 'normal' galaxies, he reported. These properties, broad emission lines covering a wide range of ionization, arising in a small, bright ('semi-stellar') nucleus, became the defining characteristics of the class of objects we call Seyfert galaxies, the most numerous known type of AGNs.

They were very little studied, however, until after the optical identification of several of the strongest radio sources with galaxies. One of the first of these was Cyg A = 3C 405, identified by Baade and Minkowski (1954) with a cD galaxy at redshift $z = 0.057$. Its rich emission-line spectrum was observed to be very similar, in line widths and high levels of ionization, to the spectra of Seyfert galaxies. Soon these characteristic features in the spectrum became the recognized signature by which many (but not all) radio galaxies could be identified. These are the next most numerous type of AGNs we know, after the Seyfert galaxies.

In addition to the radio galaxies, a fraction of the early identified radio sources appeared to be stars, with no trace of a galaxy or nebula in their images. Their spectra were continuous, without absorption lines, but with broad emission lines which could not be identified. Many attempts were made to understand these 'stellar radio sources' as peculiar stars, perhaps white dwarfs or subdwarfs with highly deviant abundances of the elements, but no physically consistent interpretation was found along these lines. Then Schmidt (1963) solved the puzzle, identifying several well-known nebular emission lines with the then unusually large redshift $z = 0.158$ in the 'stellar' radio source 3C 273. Immediately after this, Greenstein and Matthews (1963) identified several similar emission lines in 3C 48, with redshift $z = 0.367$. This was larger than the redshift of any galaxy known at the time, but 3C 48 appeared to be a 14th magnitude star. It was immediately clear that both these objects are highly luminous, and could be observed to very great distances. They are not stars, but quasi-stellar radio sources, usually referred to as 'quasars'. They are now understood to be AGNs, so luminous and so distant that the galaxy in which they are cannot (or could not) be detected on available photographic images. Now with CCDs and other high-quantum efficiency, linear two-dimensional imaging detectors, the quasi-stellar nucleus can be subtracted with good accuracy, revealing the galaxies around many of the nearer quasars.

Corresponding radio-quiet high-luminosity stellar-appearing objects were found soon afterwards. At first they were called quasi-stellar objects, or QSOs, but gradually the distinction faded out, and they are now referred to simply as quasars by most research workers in the field. In this review we shall often preserve the distinction. As a result of many systematic discovery programs, at present we know many quasars and QSOs with redshifts up to $z \approx 4$, but with a fairly strong cutoff around $z \approx 3.5$, so that at the present writing only ten are known with $z > 4$ (Schneider *et al* 1989). They are the most distant objects we know in the universe, but there appears to be a limit to the distance, or light-travel time, at which we can observe them. They have a wide range of luminosities. One of the chief aims of AGN research is to understand them physically so that the luminosity of an individual object can be estimated from its spectrum with sufficient accuracy to calculate its distance.

Table 1. Approximate space densities here and now (Osterbrock 1982).

Type	Number Mpc^{-3}
Field galaxies	10^{-1}
Luminous spirals	10^{-2}
Seyfert galaxies	10^{-4}
Radio galaxies	10^{-6}
QSOs	10^{-7}
Quasars	10^{-9}

For rough orientation purposes table 1 lists the approximate space densities here and now of the various types of AGNs we have discussed, in comparison with representative densities of two classes of normal galaxies (Osterbrock 1982). The quasars and QSOs are the rarest but most luminous types of AGNs; radio and Seyfert galaxies are more numerous in space but intrinsically less luminous. Observation and theory alike have made it more and more apparent that QSOs and Seyfert galaxies are not different types of objects, but rather names we use for AGNs at the high- and low-luminosity ends of a continuous sequence of physically similar objects. In fact Schmidt and Green (1983) have introduced the terminology that AGNs in galaxies with total absolute magnitude more luminous than $M_B = -23$ are called quasars (or QSOs), while those less luminous than $M_B = -23$ are called Seyfert galaxies. Note, however, that to say QSOs and Seyfert galaxy nuclei belong to one physically continuous sequence does not imply that they are identical except in luminosity, size and other scales, any more than to say O stars and M dwarfs belong to one physically continuous main sequence implies that they are identical except in luminosity, size and other scales. The luminosities of AGNs are very high; a nucleus with $M_B = -23$ has, in order of magnitude, $L \approx 10^{12} L_\odot \approx 10^{45.6}$ erg s^{-1} integrated over the ultraviolet, optical and infrared spectral regions 0.1 μm $\leq \lambda \leq$ 100 μm (Edelson and Malkan 1986), or $L \approx 5 \times 10^{12} L_{\odot} \approx 2 \times 10^{46}$ erg s^{-1} integrated over the entire 'observed' range (including interpolations and sketches below upper limits) 10^{10} Hz $\leq \nu \leq 10^{25}$ Hz (Ramaty and Lingenfelter 1982, Urry 1990). This is released within a very small volume, typically with a dimension of less than a few tens of light days, as will be discussed quantitatively in the next section. The energy release is far larger than that which can be derived from stars we know, or any stellar-like objects operating on thermonuclear reactions. The only energy-release process that seems possible is the liberation of gravitational energy. The one that seems most plausible by far is the gravitational release of energy in a rotating accretion disk around a massive black hole, as has been well reviewed by Rees (1977, 1978, 1984).

No matter what the energy-liberation process is, radiation pressure will be important, and any spherically symmetric object which is stable must be gravitationally bound against being blown apart by it. This is the Eddington condition

$$L \leq L_{\mathrm{E}} = 4\pi c G m_{\mathrm{H}} M / \sigma_T = 1.3 \times 10^{38} (M/M_{\odot}) \text{ erg s}^{-1}.$$

Here σ_T is the electron-scattering cross section, and the condition has been written for a pure hydrogen, completely ionized object; any larger opacity would correspond to a smaller limit on the luminosity. For an AGN with $L \approx 10^{46}$ erg s^{-1}, this corresponds to $M \gtrsim 10^8 \ M_{\odot}$, the minimum mass for a spherically symmetric situation. In more complicated geometries, for instance a roughly cylindrically symmetric structure, it is still useful as an order of magnitude estimate. Furthermore, if enough fuel is available, the luminosity tends to build up to close to this self-limiting maximum value.

The luminosity gives the rate at which mass is converted to energy

$$\dot{M}/M_{\odot} \approx (0.1/\epsilon) \left(L/5.7 \times 10^{45} \text{ erg s}^{-1} \right)$$

in solar masses per year, with ϵ the efficiency or fraction of the energy which escapes disappearing into the black hole and is radiated. For $\epsilon \approx 0.1$, probably a high estimate of the efficiency, $\dot{M} \approx 1.8 \ M_{\odot}$ yr^{-1} for an AGN with $L = 10^{46}$ erg s^{-1}.

AGNs radiate their energy over a wide range of energies, from γ-ray and x-rays through the ultraviolet, optical and infrared spectral regions to the far-infrared and radio-frequency regions. This is the reason the luminosity quoted above is so much higher integrated over the entire energy range. Thus observations at all frequencies yield important information on the structure and nature of AGNs.

Because the AGNs are so small in angular scale as to be unresolved except for the very outermost parts of a few of the very nearest examples, we can observe only the integrated radiation of the entire object. Thus interpreting the measured data to deduce the structure of the AGN is far from simple. Nebulae, nova shells and supernova remnants are valuable 'laboratories' from which spatially resolved information is available, to suggest and also to test physical ideas that may be important in understanding AGNs.

2. Optical spectra

2.1. Classification

The emission-line spectra of Seyfert galaxies have been very well studied in the optical spectral region. They are similar to the spectra of gaseous nebulae, but cover a wider range of ionization. In the simplest classification scheme, they may be divided into two types, Seyfert 1 and Seyfert 2. The Seyfert 2 nuclei have emission-line spectra with line widths typically 350 km s^{-1} full width at half maximum (FWHM), ranging from 200 to 700 km s^{-1} in different objects (Dahari and De Robertis 1988a). These lines include both permitted lines of H I, He I and He II, and forbidden lines of which the strongest are [O III] $\lambda\lambda4959, 5007$, [N II] $\lambda\lambda6548, 6583$; other lines include [O I] $\lambda\lambda6300, 6364$, [S II] $\lambda\lambda6716, 6731$, often [Fe VII] $\lambda6087$, and in many cases [Fe X] $\lambda6375$. The Seyfert 1 nuclei spectra include all these 'narrow' broad emission lines, plus much broader emission lines of H I, He I, He II and Fe II, typically with FWHM 3000 km s^{-1}, ranging from 500 to 7000 km s^{-1} in different objects. The full widths at nearly zero intensity (FWOI) of these same lines, so well as they can be defined, range from 5000 to 30 000 km s^{-1} (Osterbrock and Shuder 1982).

It is evident from these line spectra that in AGNs, as in nearly all other astronomical objects we know, H is by far the most abundant element, He next, O, Ne, N, C next, etc. The simplest interpretation is that the narrow lines are emitted in a 'narrow-line region' (NLR) in which the velocity field ranges up to a few hundreds of km s^{-1}, and the broad lines in a 'broad-line region' (BLR) in which the velocity field ranges up to as high as 10 or 15 km s^{-1}. The (nearly) complete absence of forbidden line emission from the BLR can only mean that in it they are *all* collisionally de-excited; this implies that the electron density throughout this region is much higher than the *critical densities* for collisional de-excitation of *all* the strong forbidden lines observed from the NLR and in gaseous nebulae (see, e.g., Osterbrock 1989). From the known transition probabilities and collision strengths for collisional excitation of the forbidden lines of the expected more abundant ions, this limit is roughly $N_e \gtrsim 10^8$ cm^{-3}.

An upper limit to the electron density in the BLRs may be set from the observed presence in many Seyfert 1 and QSO spectra of semi-forbidden C III] $\lambda1909$ emission, with a FWHM comparable to those of the permitted lines. Its transition probability, $A = 96$ s^{-1}, intermediate between those of typical permitted and forbidden lines, and its critical electron density $N_{ec} \approx 3 \times 10^9$ cm^{-3}. The density in the regions of the BLR in which C III] $\lambda1909$ is emitted with appreciable strength cannot be much higher than this, and an intermediate density $N_e \approx 10^9$ cm^{-3} has frequently been taken as roughly representative of the BLR. However, recent evidence discussed in section 2.7 argues for higher densities, and thus for appreciable collisional de-excitation.

The interpretation then of a Seyfert 1 AGN is that it contains a BLR and a NLR; of a Seyfert 2, that it contains only a NLR, or as we shall discuss in some detail in section 6, that it contains a NLR and a 'hidden' BLR that we do not directly observe, because of intervening material that cuts off the direct radiation from it toward us. Intermediate cases can also be recognized. Seyfert 1.5 is the classification type generally used for objects whose spectra show H I emission lines with strong broad and strong narrow components, while Seyfert 1.8 and 1.9 are the types used for objects in which the broad component of Hβ is very weak or undetectable, respectively (Osterbrock 1984).

2.2. Diagnostics

Diagnostic information is available on the physical conditions within the NLRs. From the measured ratios of the strengths of emission lines with different upper levels, such as [O III] $[I(\lambda4959) + I(\lambda5007)]/I(\lambda4363)$ and [S II] $I(\lambda6716)/I(\lambda6731)$, a representative 'mean' electron temperature and density may be derived for each observed NLR. Typical values are $T \approx 15\,000$ K and $N_e \approx 3 \times 10^3$ cm^{-3}. There is no direct diagnostic that gives a representative mean temperature for the BLR, but on general grounds an estimate $T \approx 15\,000$ K also is plausible.

Of course these are highly simplified representations of what must be in reality an extremely complicated situation. In the somewhat analogous case of a stellar atmosphere we know that the density and temperature increase inward, but for the roughest semi-quantitative description we may represent it by a single 'effective' temperature and a single 'mean' density. A better representation is to take two representative points, still better, three, and the only true description is the detailed run of mean temperature and density with depth, with the fluctuations around these mean values in space and time. In the case of a Seyfert 1 galaxy a description in terms of one representative point is too simplified to be useful; we must specify at least two points, 'the BLR' and 'the NLR'. In fact there must be a continuous transition between these two idealized 'regions', and large variations in the physical conditions within each of them. Certainly diagnostic line ratios of different ions give different mean temperatures and/or densities, indicating real variations in T and N_e within the NLR; the quoted values are the best overall fits to the various different determinations. In all gaseous nebulae we know, ionized gas tends to be clumped in condensations on all scales down to the smallest observable, so the mean density indicated by emission-line processes, dominated by the denser regions of the condensations, is higher than the mean density given by the total number of particles within the whole volume of the nebula. This can be represented by a 'filling factor' f, giving the fraction of the volume occupied by dense condensations. It corresponds to describing the nebula as consisting of two phases: gas, with electron density N_e and relative volume f, and vacuum or a much hotter, lower-density gas with the same pressure and relative volume $1 - f$. For typical nebulae $f \approx 10^{-2}$.

From this picture the approximate size of a typical NLR in a Seyfert galaxy nucleus can be estimated from its luminosity in a particular emission line, for instance Hβ. It is well understood as arising largely from recombination of protons and electrons to levels $n \geq 4$, followed by downward radiative transitions, and the luminosity in Hβ can be written

$$L(\mathrm{H}\beta) = N_e \, N_p \, \alpha_{\mathrm{H}\beta} h\nu_{\mathrm{H}\beta} f \, V.$$

Here $\alpha_{\mathrm{H}\beta}$ is the effective recombination coefficient for emission of an Hβ photon, $h\nu$ is its energy, and the product of the first four factors is the emission coefficient per unit volume, while V is the total volume of the NLR. To a sufficiently good approximation $N_e = N_p + 1.5 N_{\mathrm{He}}$ (H completely ionized to protons, and He half to He$^+$ and half to He^{2+}). The recombination coefficient $\alpha_{\mathrm{H}\beta}$ varies only slowly with T, and if to be specific we assume $T = 10^4$ K and a spherical volume,

$$V = \frac{4\pi R^3}{3}$$

will give the radius R. The most luminous NLRs of Seyfert galaxies have $L(\mathrm{H}\beta) \approx 2 \times 10^8 L_\odot$, which gives $M \approx 7 \times 10^5 \left(10^4/N_e\right) M_\odot$ and $R \approx 20 f^{-1/3} \left(10^4/N_e\right)^{2/3}$ pc. Such an NLR with $N_e = 10^4$ cm^{-3} will therefore have a mass of ionized gas $M \approx 10^6 M_\odot$ and, for an assumed filling factor $f = 10^{-2}$, $R \approx 90$ pc. In fact a few of the nearest Seyfert 2 NLRs are resolved on direct images, and have apparent diameters of the order 100 pc or so.

Typical BLRs are much smaller. The observed Balmer decrements, or ratios of intensities of H I lines $I(\mathrm{H}\alpha) : I(\mathrm{H}\beta) : I(\mathrm{H}\gamma)$ etc. show that other processes in addition to recombination contribute to the H I line emission in these denser objects. However, for a rough estimate we can ignore them and use the same approximate calculation. The most luminous Seyfert 1 AGNs typically have $L(\mathrm{H}\beta) \approx 10^9 L_\odot$, which gives $M \approx 36 M_\odot \left(10^9/N_e\right)$ and $R \approx 0.015 f^{-1/3} \left(10^9/N_e\right)^{-2/3}$ pc. Thus for a representative density $N_e \approx 10^9$ cm^{-3}, the mass of ionized gas in the BLR is only $M \approx 40 M_\odot$, and for an assumed $f \approx 10^{-2}$, $R \approx 0.07$ pc ≈ 0.2 light year. This is far too small to hope to resolve, even in the nearest Seyfert 1s, and to date none has been resolved. Figure 1 is a highly schematic drawing (not necessarily all at the same scale) of these various regions, with the central black hole, its BLR shown as cylindrically symmetric, and the NLR as spherically symmetric.

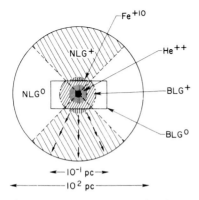

Figure 1. Schematic representation of structure of an active galactic nucleus, including central black hole and accretion disk (black), cylindrically symmetric broad-line region (BLR) containing ionized (BLG$^+$) and neutral (BLG^0) gas, with highest stages of ionization such as He^{2+} concentrated closest to ionization source. Narrow-line region (NLR) taken here to be spherically symmetric, but ionized (NLG$^+$) only in the core in which ionizing radiation from central source can escape or penetrate through the NLR. The figure is necessarily not to scale. (Osterbrock 1978a)

2.3. Photoionization

Next let us discuss the energy-input mechanism for the ionized gas, first in the NLR. The relatively low electron temperature, $T \approx 15\,000$ K, together with the ionization extending to such high stages as O^{2+} and Ne^{2+}, shows that the main energy-input mechanism must be photoionization. The only other forms of energy input we know, conversion of kinetic energy into heat either through shock waves ('cloud–cloud collisions') or beams of particles being stopped, require much higher temperatures for the observed degree of ionization.

Furthermore AGNs have a strong continuous spectrum. Presumably it arises from the accretion disk. In typical objects, in the optical and observable ultraviolet spectral regions, it follows approximately a power-law form

$$L_\nu = C\nu^{-n}$$

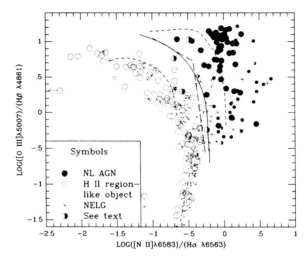

Figure 2. Reddening-corrected [O III] $\lambda5007/\mathrm{H}\beta$ vs [N II] $\lambda6583/\mathrm{H}\alpha$ intensity ratios. Symbols for various objects are as shown in the key in the lower left corner of the figure. Short broken curves represent calculated ratios from AGN models with power-law input spectrum as described in the text, and solar and 0.1 solar abundances (upper and lower respectively). The long broken curve is the prediction from the composite model described in text. The chain curve is the prediction of the shock wave model. The full curve divides AGNs from H II-region like objects (Veilleux and Osterbrock 1987).

with $n \approx 1.2$. This spectrum, extrapolated further into the ultraviolet and x-ray regions, apparently is the source of the ionizing photons. This is indicated by the fact that the equivalent width of the observed $\mathrm{H}\beta$ emission, that is

$$\Delta\nu_{\mathrm{H}\beta} = \frac{L(\mathrm{H}\beta)}{L_\nu}$$

is approximately constant for most Seyfert 2 NLRs. This is exactly what is expected under photoionization by a spectrum of fixed form, such as a power law, if the total number of ionizing photons is balanced by the total number of recombinations

$$\int_{\nu_0}^{\infty} \frac{L_\nu}{h\nu}\, \mathrm{d}\nu = N_e N_p \alpha_B f V$$

where α_B is the effective recombination coefficient while the luminosity in $\mathrm{H}\beta$ comes from a definite fraction of those recombinations

$$L(\mathrm{H}\beta) = N_e N_p \alpha_{\mathrm{H}\beta} h\nu f V.$$

A power-law form for the photoionizing spectrum is harder than any O star spectrum, which always falls off roughly as $\exp(-h\nu/kT_*)$. This explains both the higher stages of ionization in the AGN spectra than that in H II regions, and the great strength of the lines of low ionization, such as [O I] and [S II], which arise in the long partly ionized region maintained by the high-energy, penetrating photons.

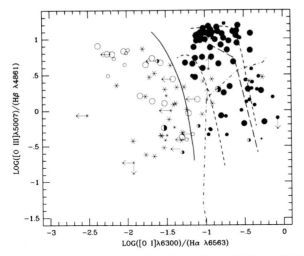

Figure 3. Reddening-corrected [O III] $\lambda5007$/Hβ vs [O I] $\lambda6300$/Hα intensity ratios. Symbols and curves as in figure 2. (Veilleux and Osterbrock 1987)

These qualitative conclusions are confirmed by quantitative models, calculated using various power-law input spectra, chosen to match approximately the observed luminosities and frequency dependence of the observed NLRs. Some of the most complete of such calculated models are those by Aldrovandi and Contini (1984, 1985), Contini and Aldrovandi (1983), Ferland and Netzer (1983), Halpern and Steiner (1983), Stasinska (1984a, 1984b), and Binette (1985). For instance, figures 2, 3 and 4 show the line ratios I ([N II] $\lambda6583$)/$I(H\alpha)$, I ([O I] $\lambda6300$)/$I(H\alpha)$, and I ([S II] $\lambda6716 + \lambda6731$)/$I(H\alpha)$ plotted against $I([O\ III]\ \lambda5007)/I(H\beta)$. These are the best diagnostic ratios to distinguish between AGN spectra and H II region or starburst-galaxy spectra of those originally proposed by Baldwin *et al* (1981). The black circles represent measured ratios for the AGNs, and the open circles are the measured ratios for H II regions in other galaxies, starburst, and H II region galaxies, all known to be photoionized by O stars. The full line on each diagram is the empirical division between AGNs and H II regions. The short broken lines represent sequences of AGN models, calculated with input power-law spectra with $n = 1.5, N_e \approx 10^3$ cm^{-3}, and either essentially solar abundances of the elements (upper right curves on all three diagrams) or with the abundances of all the heavy elements reduced by a factor 10 relative to H and He (lower left curves). Along each of these curves the ionization parameter

$$\Gamma = \frac{1}{4\pi r^2 c N_e} \int_{v_0}^{\infty} \frac{L_\nu}{h\nu}\,\mathrm{d}\nu = \frac{Q}{4\pi r^2 c N_e}$$

essentially the ratio of density of ionizing photons to density of free electrons, decreases from $\Gamma = 10^{-1.5}$ at the upper left end to $\Gamma = 10^{-4}$ at the lower right. The long broken lines represent three models calculated with the same input spectrum and solar abundances, but containing two types of clouds, with $N_e = 10^2$ and 10^6 cm^{-3}, to mimic roughly the effects of density variations within the object and collisional de-excitation.

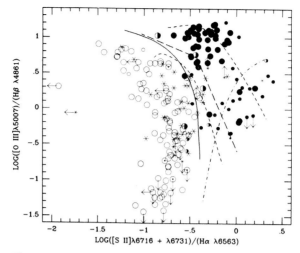

Figure 4. Reddening-corrected [O III] $\lambda5007$/Hβ vs [S II] ($\lambda6716 + \lambda6731$)/H$\alpha$ intensity ratios. Symbols and curves as in figure 2. (Veilleux and Osterbrock 1987)

It can be seen that the calculated models with power-law input spectra match the observed line ratios for AGNs reasonably well, and do not agree at all with the observed ratios for H II regions and galaxies. These latter, however, are well represented by models with O star input photoionizing spectra (Veilleux and Osterbrock 1987). Note that for the composite density and hence presumably the most realistic AGN models, the [O I] and [S II] data are matched well with 'normal' solar abundances, but for [N II] an increase in the abundance (in the first order, of N alone) by roughly a factor of three is indicated by the observed ratios.

The only justification for assuming a power-law form of spectrum is that it is simple. A broken power law, more nearly flat at high energies, may be the next best approximation to reality (see section 3.3). Much more complicated forms, based on observational data and reasonable extrapolations, have been used by Mathews and Ferland (1987), Ferland and Persson (1989), and others. Some other types of input spectra, all sharing the property of including photoionizing spectra over a wide range of energies, have been investigated by Binette *et al* (1988).

It should be noted that, in addition to recombination, collisional excitation also makes an appreciable contribution to the strength of the H I lines, particularly of Lα and Hα. The reason is that the gas clouds in AGNs, photoionized by a 'hard' spectrum (containing relatively many high-energy photons in comparison with O-star spectra) have large partly ionized zones in which both H^0 and free electrons exist, in contrast to gaseous nebulae, in which the ionization of H and the free electron density drop very abruptly together at the edge or boundary of the neutral region. The NLR models typically give calculated intensity ratios $I(H\alpha)/I(H\beta) = 3.1$, which must be used to determine the reddening by dust observationally, rather than 2.85 as in pure recombination (Gaskell and Ferland 1984).

2.4. Broad-line region

It is not so clear that photoionization is the main energy-input mechanism to the ionized gas in the BLR, but this interpretation seems most likely. The main diagnostic evidence is that the equivalent widths of the H I emission lines, specifically either Hα or Hβ, expressed in terms of the featureless continuum, are more or less the same for Seyfert 1 and 2 galaxies, QSOs, and radio galaxies, and fit the predicted relationship for a power law with $n \approx 1.2$ (Yee 1980, Shuder 1981). This is the expected result of photoionization, as discussed in section 2.3, but it is not unique; if some other energy-input mechanism, say injection of fast particles or high-kinetic-energy clouds, were closely proportional to the luminosity in the featureless continuum, the observational result would be the same. No such process has been suggested. In addition, the observed tight correlation between continuum and broad emission-line variability makes it difficult to consider any other source. Furthermore, detailed photoionization models can be adjusted to fit approximately the observed spectra. However, it is quite possible that a non-negligible fraction of the heating in some or all BLRs results from dissipation of mechanical energy, as well as from photoionization.

Calculating the expected emission-line spectra of model photoionized BLRs is more complicated than for gaseous nebulae and NLRs. In the latter objects the density is low enough so that the fundamental nebular approximation, that essentially all ions and atoms are in their lowest energy levels and that processes involving ions in excited levels can be neglected, except for emission of photons, is generally valid. Furthermore the optical depths in nearly all the lines are small, and for other lines, such as the Lyman series of H, it is a good approximation to consider them infinitely thick optically. Very complete calculations are available for the H I and He II recombination spectra in this nebular approximation (Hummer and Storey 1987). They take into account all the radiative and collisional processes that are relevant up to densities $N_e \approx 10^7$ cm^{-3} for the H I spectrum and 10^9 cm^{-3} for He II. The energy-level diagram of He I is more complicated, but calculations on a reasonably good approximation are also available for it (Brocklehurst 1972, Almog and Netzer 1989). They also include all relevant collisional processes (but the various cross sections are not as accurately known for this two-electron atom), and in addition approximately include radiative-transfer effects in a uniform finite-thickness slab approximation. As a result these He I calculations are stated to be correct (within the limitations of the highly simplified model geometry and the available values of the collision cross sections) up to $N_e \approx 10^{14}$ cm^{-3}.

Most of the earlier theoretical work on BLRs was done on the H I line spectrum of a dense, optically thick region (Kwan and Krolik 1981). Generally these papers assumed uniform density and temperature, but went beyond the low-density nebular approximations in including collisional and radiative excitation and ionization from excited levels, and also line radiative-transfer effects in a simplified approximation. One of the striking early observational discoveries in the spectra of redshifted quasars is that H I Lα is relatively weak with respect to the Balmer lines, so that the intensity ratio $I(L\alpha)/I(H\beta) \approx 10$ typically, rather than 30 as expected in nebular-type (low-density) spectra. The reason is that in dense BLRs Lα photons are not simply scattered, but are 'destroyed' by Balmer-line absorption processes and collision processes which remove H^0 atoms from the excited 2^2P level before they emit a Lα photon. Dust extinction also plays a role, but the discrepancy does not result from it alone.

The best recent papers take into account heating by photoionization, include other ions in addition to H and He, determine the equilibrium temperature at each point from the balance with radiative cooling, and calculate the emergent spectrum (Kwan 1984). The relevant parameters, in addition to the form of the photoionizing spectrum, are the input ionization parameter Γ, the assumed constant density N_0 and the total optical depth of the model 'cloud' (semi-infinite slab) at the Lyman limit. In a general way, models with ionization parameters derived from the luminosities and reasonable interpolations and extrapolations of the observed spectra of AGNs into the photoionizing region, $N_e \approx 10^9$ cm^{-3} and sizes of BLRs previously mentioned approximately fit their observed emission-line spectra.

One very interesting but complicated set of diagnostics is the strength of the Fe II features in the optical region. These are blends of large numbers of individual emission lines belonging to several multiplets especially concentrated in the regions around $\lambda\lambda 4570$, 5120, 5320. They make up a significant fraction of the flux in the optical part of the spectrum in many Seyfert 1 galaxies (Wills *et al* 1985).

All the observed lines arise from upper levels connected with the ground configuration of Fe II by permitted ultraviolet resonance transitions. They are evidently excited by collisional and radiative fluorescence processes, and the optical depths in the resonance lines are clearly large ($\tau_0 \sim 10^4$ in the stronger transitions). Since Fe$^+$ is a relatively low stage of ionization (the ionization potential of Fe^0 is 7.9 eV), the Fe II lines are emitted quite strongly in the large partly ionized zone, or 'transition zone', in which the ionization of H is dropping from nearly completely H$^+$ to nearly completely H^0 (Netzer and Wills 1983). In this region the bulk of the ionization occurs by highly penetrating x-ray photons. The Mg II emission $\lambda\lambda 2798$, 2803 lines observed in many (redshifted) QSOs and quasars, and Ca II $\lambda\lambda 8498$, 8542, 8662, observed in many (low-redshift) Seyfert 1s (Persson and McGregor 1985, Persson 1988) also arise in this zone, as does a significant fraction of the H I line emission (Collin-Souffrin *et al* 1982, Collin-Souffrin and Dumont 1986).

Quite recently evidence from time variations of the strengths and profiles of broad emission lines in some AGNs has suggested that their BLRs may be considerably smaller than previously estimated (see section 2.7). This would imply that their mean densities are considerably higher than 10^9 cm^{-3}. Calculated models are therefore required with densities up to $N_e \approx 10^{13}$ cm^{-3}. In this regime the physical conditions are more nearly similar to those in stellar atmospheres then in typical gaseous nebulae. Three-body recombination, large optical depths in many lines, and even Stark broadening (which affects the optical depths) must be taken into account. The models calculated to date with mean densities $N_e \approx 10^{13}$ cm^{-3} do not match the observed line spectra and line profiles of BLRs nearly as well as the models with characteristic mean densities $N_e \approx 10^9$ cm^{-3}, but the models with mean densities up to $N_e \approx 10^{10.5}$ cm^{-3} are quite acceptable (Rees *et al* 1989). All the published models assume simple density laws, but undoubtedly the actual BLRs are much more complicated.

One problem in any photoionization model, whether of an NLR or a BLR, is the physical origin of the small filling factor, that is, the clumping as gas into density condensations or 'clouds'. If not confined they should quickly dissipate, essentially at the speed of sound (≈ 10 km s^{-1}) and after the 'sound crossing time', approximately 10^4 y for a BLR with $R = 0.07$ pc, should no longer exist. Perhaps they are constantly regenerated by turbulent processes; perhaps they are confined by a hot, low-density invisible gas with which they are in pressure equilibrium. The most widely accepted picture is that this intercloud gas is at $T \gtrsim 10^8$ K (Krolik *et al* 1981). However, recent theoretical studies by Fabian *et al* (1986) and Mathews and Ferland (1987) show that such gas could not escape detection somewhere in the x-ray, ultraviolet or infrared continua of AGNs, as it has to date. Thus whether it exists, and its nature if it does, are uncertain. Rees (1987) has suggested that magnetic fields may provide the confinement mechanism. The models previously described simply adopt various power-law forms for the pressure of the assumed confining medium without specifying their physical cause.

So many energy levels and so many transitions are involved in the Fe II emission-line spectrum that it is difficult to model with currently available computers and calculated atomic parameters. Ca II, with only five relevant levels, is much more straightforward. The model calculations show that its emission lines must arise in slightly ionized regions at very large optical depths for ionizing radiation from the central source, that is, large column densities, $N_{\rm H} \gtrsim 10^{24.5}$ cm^{-2}. At these large densities and large optical depths the radiative heating and cooling processes familiar in most nebular contexts are ineffective, and processes such as Compton scattering of high-energy photons by bound electrons, photoionization of C^0 and H^- by photons with $h\nu < h\nu_0$, and photoionization of Fe^0 K and L electrons by x-rays all come into play (Collin-Souffrin *et al* 1988, Ferland and Persson 1989). Even small amounts of dissipation of kinetic energy by heat would also be important (Collin-Souffrin *et al* 1986).

2.5. Accretion disk continuum

The observed continuous spectra of AGNs are very complicated mixtures of several components. One is a power law extending from the infrared to soft x-ray spectral regions, as previously described, but superimposed on it is a broad continuum which extends over the range roughly $10^{14.5}$ Hz $< \nu < 10^{16.5}$ Hz, with a peak somewhere between. This is generally called the 'big blue bump' (meaning on a plot of flux F_ν, or even better νF_ν against frequency), and it is further contaminated by the 'little blue bump', in the near ultraviolet, consisting of Balmer continuum emission and many unresolved Fe II emission multiplets (Wills *et al* 1985). The big blue bump is generally attributed to dissipation of energy in an accretion disk near the central black hole (Shields 1978). Various types of accretion disks might exist, for instance thin disks, radiation tori and ion tori, differing in the physical processes that dominate in fixing their structure (Begelman 1985). Which actually occurs, or which occur in which specific objects, depends on details of the release of energy in the accretion disk that are not yet understood. The simplest and most straightforward case to calculate is the (geometrically) thin disk, which is optically thick. Most published interpretations of observational data have been made in terms of this model. For the high-luminosity QSOs, in which the underlying integrated-stellar galaxy continuum spectrum makes the smallest contribution and the fit to the blue bump is best determined, an accretion-disk continuum provides a significantly better match to the observed spectrum than a blackbody does. Both have two parameters, M and $\dot{M}$ for the accretion disk, or L and T for the blackbody. The calculated spectra of accretion disks about either a Schwarzschild (non-rotating) or Kerr (rotating) black hole fit equally well; the indicated mass of the Schwarzschild black hole is about two to three times that of the corresponding Kerr black hole. The calculated spectrum of an accretion disk depends on the angle i between its normal and the line of sight, but this angle cannot be determined from the fitted data; to the accuracy of the data an increase in i can be compensated for in the emergent spectrum by an increase in M and $\dot{M}$ (Malkan 1983).

Observations over a wide frequency range are necessary to eliminate or fit the power-law continuum, the Balmer continuum, the Fe II features, and the underlying integrated-stellar galaxy continuum. The energy dissipation in the accretion disk increases inward toward the black hole, as a result of the Keplerian velocity field in the disk. This leads to an increase of effective temperature inward, with the highest energy photons coming from the inner edge of the accretion disk. The emergent spectrum at each radius has been calculated for various theoretical models in various approximations (Czerny and Elvis 1987, Wandel and Petrosian 1988). The most recent published treatment takes into account detailed calculations based on the Kerr metric of a rotating black hole spun up to the maximum angular momentum-to-mass ratio (Sun and Malkan 1989). On physical grounds it should be the best approximation to the actual situation in nature (Thorne 1974). There are relativistic effects especially on the radiation from near the inner edge of the accretion disks. The emergent spectrum depends on the mass of the black hole M, the accretion rate $\dot{M}$, and the inclination of the disk to the line of sight. All the cited papers agree that their respective models and fitting procedures give AGN black holes masses ranging from $10^{7.5}$ to $10^{9.5} M_\odot$, the higher range of masses $M = 10^8$ to $10^{9.5} M_\odot$ corresponding to QSOs and the lower range $10^{7.5}$ to $10^{8.5} M_\odot$ corresponding to Seyfert 1 nuclei. Furthermore, the most luminous QSOs tend to have mass accretion rates and luminosities nearly as large as the Eddington luminosities L_E corresponding to the derived masses of their black holes, while the less luminous Seyfert 1 nuclei typically have luminosities only a few percent of their Eddington luminosities. The most recent, and to date most physically complete model calculations, are those of Laor and Netzer (1989). They found that the upper limit to the luminosity for a *thin* disk is $L < 0.3L_E$, and that all the AGN-model thin disks are dominated by radiation pressure.

However, it must be recognized that real problems remain in fitting the observed spectra with thin accretion disk models. Earlier calculations predicted a large absorption discontinuity at the Lyman limit ($\lambda 912$), while the Laor and Netzer (1989) calculations predict a large emission discontinuity there. The observations show neither, and furthermore the polarization predictions are difficult to reconcile with the observational data (Antonucci *et al* 1989). The case is not closed.

2.6. Ultraviolet line spectra

The ultraviolet line spectrum of AGNs is simply an extension of the optical line spectrum to the region beyond the atmospheric cutoff imposed by the Earth's atmosphere. In low-redshift Seyfert nuclei the ultraviolet region is not observable from the ground; in high-redshift QSOs it is redshifted into the optical region and becomes observable, but correspondingly the optical region is redshifted toward the infrared and out of the region in which optical detectors work. For instance in a QSO with $z = 1.7$, Lα is shifted to $\sim \lambda 3280$, barely observable with any efficiency at a ground based observatory, but Hβ is shifted to $\sim \lambda 13125$, beyond the range of sensitivity of a CCD. As a result, reasonably complete data are available on the emission lines over a wide range in wavelengths only for fairly bright Seyfert galaxies, observable in the ultraviolet with the IUE satellite. They show that the ultraviolet spectra of Seyfert 1 galaxies are quite similar to those of QSOs (Wu *et al* 1983, Clavel and Joly 1984).

In QSOs with emission lines, H I Lα, if in the observable range, is nearly always the strongest line. Very frequently C IV $\lambda 1549$, a close doublet which is unresolved in all broad-line objects, is also quite strong. Thus in objective prism or grism surveys aimed at finding more QSOs, if only a single emission line is seen in a spectrum, the best working hypothesis is that it is Lα, or if not Lα, λ1549.

The best tests of any model for a particular AGN or a particular class of AGNs are comparison of the predicted spectrum with observational data over the entire ultraviolet, optical and infrared regions, to as long a wavelength as possible (Ferland and Osterbrock 1986, Ferland and Osterbrock 1987, Oliva and Moorwood 1990).

2.7. Temporal variations

A considerable fraction of all observed Seyfert 1 nuclei, and many QSOs as well, have been observed to vary in light. Among those which vary with reasonably large amplitude for which time-resolved spectral data are available, the strengths of the broad emission lines and their profiles have also been observed to vary. From the physical picture it is seen that the gas in the BLR is photoionized by continuum radiation from the central source, and that the photoionizing continuum varies together with the optical continuum, these variations must give information on the size and geometry of the BLR. This subject has been very well summarized recently by Peterson (1988).

The first observational recognition of these time-variable broad-line profiles occurred from comparisons of spectra of the same object taken at times separated by months or even years. Systematic study of the phenomenon requires observations at much more frequent intervals. These are difficult to arrange, except with a telescope, spectral scanner and a team of observers dedicated to the project. The most complete data now available have observations of NGC 5548 with average sampling intervals of 3 to 4 nights. In general, the response of a finite BLR to a pulse of photoionizing continuum radiation is expected to occur on a time scale $\tau_{LT} = R/c$, the light-travel time across the region. (The recombination time, on which the gas responds to the pulse on its arrival, is short compared to typical light-travel times.) Cross-correlating the continuum and individual broad emission-line light curves (for instance H I Lα, Mg II λ2798, C III] λ1909, C IV λ1549, or Hβ) for different assumed lags, and finding the lag for which the cross-correlation is a maximum is the most objective method of determining the light-travel time. This phase shift depends on the geometrical structure of the object and can be calculated for simple models; it is straightforward to show that, for instance, for a thin spherical shell of gas centred on the nucleus, the phase shift is exactly τ_{LT} as defined previously (Gaskell and Sparke 1986).

In these space-based observations of NGC 5548, different time delays have been found for different lines, namely $R \approx 4$ to 10 light days for the highest ionization He II λ1640 and N V λ1240 emission lines, ~ 12 light days for H I Lα, 8 to 16 for C IV λ1549, and 26 to 32 for C III] λ1909, while for Mg II λ2798 the variations are small and the phase lag is not well determined but is even longer (Clavel *et al* 1991). These results agree with the photoionization model predictions that the degree of ionization decreases outward from the central source (unless the density decreases very strongly) and the usual assumption that the density probably also decreases outward, but not faster than R^{-2}. Ground based measurements on the variations of Hα and Hβ in NGC 5548 by Netzer *et al* (1990) over a 5-month interval in 1988 give a phase lag corresponding to $R = 7$ light days. These authors assumed various simplified models for the BLR structure, used the observed continuum light curve as the input photoionizing spectrum, calculated the response of each model, and varied the parameters to get the best fit. For a spherical thin-shell model they found $R = 7\pm3$ light days. (This agrees with the IUE results for Lα.) Thin shells with $R \geq 14$ light days could be excluded at the 95% confidence level, likewise thick shells with inner radius $R_i = 4$ light days and outer radius $R_o > 40$ light days, also thin disks inclined by 60° to the line of sight with $R_i = 7$ light days and $R_o \geq 30$ light days. On the other hand a large consortium of observers, observing the same nucleus over 10 months in 1988–89, with what appeared to be better signal-to-noise-ratio data and at a time of larger variation, found a phase lag $R \sim 20$ light days (Peterson *et al* 1991). Their analysis is still in progress, but there appears to be a clear contradiction between the two sets of data on the phase lag of the Hβ variation. Note the difference with the phase lag ~ 12 days found for Lα for essentially the *same* time interval (Clavel *et al* 1991). It is not theoretically understood at this time.

For this object the standard photoionization model which best fits the relative emission-line strengths, $\Gamma = 10^{-2}$ (Mushotzky and Ferland 1984), has, for mean $N_e = 10^{10}$ cm^{-3}, a radius $R = 150$ light days, larger by a factor 20 than that implied by the time-variation measurements interpreted by the spherical thin-shell models of Netzer *et al* (1990).

Two Seyfert AGNs extensively studied for variation are NGC 4151 and Akn 120. For NGC 4151 the early data gave a wide range of values of τ_{LT} (or R, expressed in light days) for different lines, but closer analysis shows the differences are dominated by observational uncertainties resulting from inadequate time sampling and insufficient signal-to-noise-ratio data. All the measurements agree with $R \sim 7$ light days, with an uncertainty by a factor two. This does not disagree significantly with the value from the static photoionization model that best fits the observed mean broad-line spectrum, $R \sim 16$ light days.

The recent optical variation measurements for NGC 4151 are among the best available for any AGN, because this low-luminosity Seyfert 1 is close and relatively bright. Hence it was measured with a very good signal-to-noise ratio (Maoz *et al* 1991). The cross-correlation technique, applied to the broad Hα and Hβ emission lines, give $R = 9 \pm 2$ days, in reasonable agreement with the previous measurements and with the photoionization models. However, with the recent best data it proved possible to investigate the 'transfer function', or response of several possible simplified BLR models to the light variation of the central continuum source. The models with a large ratio of the outer to inner radii (~ 10) and with the line emission either constant or decreasing outward from the centre fit the observed variations best.

On the other hand for Akn 120 the cross-correlation technique gives $R \lesssim 30$ light days, but the ionization parameter Γ that best fits the mean observed spectrum then demands $N_e \gtrsim 10^{11}$ cm^{-3}. Thus for some AGNs, including NGC 4151, the best BLR time-variation data agree with the photoionization models for mean densities $N_e \approx 10^{10}$ cm^{-3}, but for several others including NGC 5548, Akn 120 and Mrk 279 (Maoz *et al* 1990) the sizes, at least for Hβ, are smaller than expected at these densities. For this reason the models with mean densities as high as $N_e = 10^{13}$ previously mentioned are being investigated. Comparisons of their predictions with the observed spectra seem to show that most of the line emissions comes from lower densities, with $N_e = 10^{10}$ cm^{-3} probably a more representative mean density, but up to $N_e = 10^{11}$ cm^{-3} is possible (Rees *et al* 1989, Ferland and Persson 1989). Some of these BLR models assume a wide range of densities, with a power-law decrease outward. A crucial test is the C III $\lambda977$/C III] $\lambda1909$ line ratio, which is predicted to be about 4 at $N_e = 10^{11}$ cm^{-3}, but about 0.3 if $N_e = 10^{10}$ cm^{-3}, both at $T = 15\,000$ K, a reasonable temperature estimate (Nussbaumer and Schild 1979). There are few published observations of C III $\lambda977$, partly because of its short wavelength and the related problem that it lies in the Lα 'forest' of absorption lines in the spectra of distant QSOs. However, in the IUE observations of six AGNs with the IUE by Green *et al* (1980), four show C III] $\lambda1909$ emission, and two of these are listed as having measured C III $\lambda977$, with intensity ratios $I(\text{C III }\lambda977)/I(\text{C III] }\lambda1909) = 2$ for PKS 1302–102 and 0.9 for PG 1247 + 268. These would indicate mean densities (as measured by the C III lines) around 3×10^{10} cm^{-3} for these two AGNs. However, later, better data suggest that the purported measurement of $\lambda977$ for PKS 1302–102 was not correct. For several other QSOs observed with the IUE, C III $\lambda977$ has not been detected (Gondhalekar 1990a, Kinney *et al* 1991). For these objects a very rough upper limit appears to be $I(\lambda977)/I(\lambda1909) < 0.5$, corresponding to mean densities on any models published to date $N_e < 3 \times 10^{10}$ cm^{-3}. All these QSOs, however, are considerably more luminous than the Seyfert 1 galaxies observed for variability, and the question of the viability of standard photoionization models is definitely an open one at present.

An alternative interpretation to the smaller size and consequent higher densities derived from time variations is that the geometry is anisotropic. If the ray from varying nucleus to the gas which responds to the variation makes an angle θ to the ray from the nucleus to the observer, then the time lag of the variation by this element of gas is $r(1 - \cos\theta)/c$. If the gas which is observed to vary lies within a cone of small half-angle θ, the phase lag

$$\langle r \rangle (1 - \langle \cos\theta \rangle)/c$$

may be much less than R/c. This type of model, with a cylindrical geometry, a small opening angle, and the rear side occulted was proposed for 3C 446 by Bregman *et al* (1986). It is the preferred interpretation of Perez *et al* (1989a, 1989b) for their time variation data for many AGNs. It would seem to require that only a small fraction of AGNs be observed to vary. However, if the radiation is seen down the axis of a torus, which is opaque in directions near its equatorial plane, so the BLR itself is not visible from those directions, all observed BLRs could be variable. These cylindrically symmetric type models are discussed in section 4.

Some data are also available on the variability of the broad emission lines Lα and C IV $\lambda1549$ in high-luminosity QSOs, from multiple IUE observations (Gondhalekar 1990b). Since the observed emission-line spectra are essentially independent of luminosity, the ionization parameter Γ and electron density N_e are supposed to be independent of L and hence the time scale for variation

$$\tau_{\text{LT}} \propto R \propto L^{1/2}.$$

However, the available observational data do not show this dependence, but in fact suggest that the time scale for variation is more nearly independent of luminosity. Thus again, more sophisticated models, probably with variable density and non-spherical geometry, will be necessary to fit the observed AGNs (Gondhalekar 1990b).

Variations are not expected in the 'ordinary' narrow emission lines in times less than hundreds of years, and none have been observed. However, variations have been observed in [Fe X] $\lambda6375$ in a few high-ionization AGNs, most certainly in NGC 5548, on time scales of a few years. This is expected from photoionization models, as such high stages of ionization as Fe^{+9} are predicted to be at distances of 0.1 to 1 pc from the ionization source (Veilleux 1988).

2.8. Completeness of Seyfert galaxy samples

As mentioned in section 1, the first known Seyfert galaxies were recognized and classified on the basis of their emission lines on slit spectrograms. As only a few per cent of luminous galaxies are Seyferts, this is a relatively inefficient method of finding new ones. Since many Seyfert 1 and some Seyfert 2 nuclei have strong blue and ultra-violet continua, it is possible to identify candidates on this basis from low-dispersion objective-prism photographic spectra surveys. This programme was very successfully carried out at Byurakan Observatory by the late B. E. Markarian and his collabora-tors, and is now generally referred to as the First Byurakan Survey or FBS (Lipovetsky *et al* 1987). About 10% of the galaxies isolated by their ultraviolet continuum in this survey turned out to be Seyfert galaxies, on the basis of slit spectra; the other 90% are mostly star-burst galaxies in which the blue continuum comes from the hot stars rather than the accretion disk. More recently Markarian began a second survey (SBS), which has been carried on since his death by his collaborators, using fine-grain plates, objec-tive prisms of various dispersions, and multiple limiting exposures with the prisms in different orientations (Markarian *et al* 1987). With these techniques they have been able to find much fainter Seyfert galaxies than had been previously recognized. There are many examples as faint as apparent magnitude $m_B = 17$ or 18. In many cases slit spectra are obtained with the 6-m telescope before the identification is published. Nearly every galaxy listed in this SBS has been confirmed as a Seyfert galaxy. How-ever, the level of completeness, that is what fraction of the Seyferts in the field have been found, is not at all well known.

Many of the known Seyfert galaxies were first identified as candidates in the FBS. But many Seyfert 2 galaxies were *not* found, because their blue continua are too faint to be picked up. Programs for obtaining slit spectra of *all* emission-line galaxies, or all galaxies down to a given magnitude, such as the Center for Astrophysics (CfA) redshift survey, have shown this by turning up many 'new' Seyfert galaxies (Huchra *et al* 1982, Phillips *et al* 1983). By comparison with the CfA survey, it is known that among the brighter galaxies the FBS is 67% complete for Seyfert 1s, but only 44% for Seyfert 2s (Lipovetsky *et al* 1987).

Other, more recent, objective-prism surveys have aimed at finding Seyfert-galaxy candidates primarily by their emission lines, especially [O III] $\lambda\lambda4959, 5007$, rather than by the blue continuum. The largest area is covered by the University of Michigan survey (MacAlpine and Williams 1981). This method was applied, using a relatively high-dispersion objective prism to detect relatively weak emission lines, to a large field surrounding the North Galactic Pole (Wasilewski 1983). The limit of completeness of such a survey clearly cannot really be given simply in terms of a limiting apparent magnitude; instead, it depends on a combination of the magnitude and the strength of the emission line or lines detected. Faint galaxies with strong emission lines can be detected as well as brighter galaxies with weaker emission lines (Salzer 1989).

Thus a true complete magnitude-limited sample is difficult to obtain, or even to define. Earlier attempts at such samples were seriously deficient in lower-luminosity Seyfert 2 galaxies. The best approximation to a complete, magnitude-limited sample is one drawn from a slit-spectrum survey like the CfA, and taking the sample only down to a magnitude limit well above that of the catalogue. This leaves only a few objects, but is relatively well defined physically (Edelson 1987). Even so, it implies some ill-defined limit to the strength of the emission lines. It is clearly quite possible that many galaxies that are not called Seyfert galaxies have the same or similar phenomenon going on in their nuclei, but the characteristic emission lines are too faint to be detected. Some of the LINERs discussed in section 6.4 fit into this category, but 'less active', less luminous, nuclei may well remain undetected.

3. Other wavelength regions

3.1. Radio-frequency spectral region

Quasars and radio galaxies are the most luminous radio sources known. The emission process is well known to be synchrotron emission by relativistic electrons in a magnetic field, leading to power-law spectra. The most luminous of all are the Fanaroff–Riley 'class II' radio sources, double-lobed structures with the strongest emission at the outer edges of the lobes. Long linear structures lead out to the lobes from the nuclei and are evidently the channels along which the energy flows out to where the plasma is braked and radiates. The linear structures can be traced as well aligned down to very small linear structures near the nucleus, generally called jets. The subject is a very extensive one, treated in whole books and detailed review articles, for example by Miley (1980), Bridle and Perley (1984), and Begelman *et al* (1984). We cannot hope to go into it in depth here. However, it should be noted that many of these galaxies are broad-line radio galaxies which, like the quasars, have optical emission-line spectra similar to Seyfert 1 galaxies. Others are narrow-line radio galaxies, with similar optical spectra to Seyfert 2s. The broad-line radio galaxies are invariably cD, D or E galaxies in form, rather than spirals, chiefly Sa and Sb, as the Seyferts are. It should also be noted that the entire radio luminosity, and the energy that goes into the jets is only of the order of 1% of the total luminosity of the AGNs in these objects (Caganoff 1989). Finally, these class II radio galaxies have a linear relationship between [O III] $\lambda5007$ luminosity in their nuclei and total radio power, as do Seyfert galaxies (Rawlings *et al* 1989).

Seyfert galaxies are much weaker radio sources, but within recent years much observational data have been obtained on them (Wilson and Ulvestad 1982a, b, Ulvestad and Wilson 1984a, b). From many early measurements it appeared that Seyfert 2 galaxies are on the average stronger radio sources than Seyfert 1s, and Ulvestad and Wilson (1984a) confirmed this for their sample. They also found that the Seyfert 2 radio sources in their sample on the average are larger (typical diameter ~ 600 pc) than Seyfert 1 radio sources (typical diameter ~ 100–200 pc). In all cases the high-resolution VLA radio maps show that the nucleus of the radio source coincides with the optical nucleus, and in most of them the remaining structure is linear and jet-like. Thus it seems that the same type of processes which generate two-sided plasma jets in radio galaxies do so, perhaps in modified form, in Seyfert galaxies also, but interactions with interstellar matter in these spirals stop the jets in relatively short distances. Probably the smaller typical sizes of the jets in Seyfert 1s result from the interaction of the jet material with the denser BLR gas, as highly simplified calculations suggest. There is also no correlation of the direction of the axis of the jet with the directions of either the long or short axes of the optical galaxy, as projected on the sky, indicating that the jets are neither parallel to nor perpendicular to the plane of the main body of the galaxy (Ulvestad and Wilson 1984b).

In particular, in the nearby, bright Seyfert 2 galaxy NGC 1068, a jet is observed, centred on the nucleus, and about 1 kpc (13″) long. It lies in position angle 33°. There is also radio emission on a larger scale in this same direction over a distance of at least 9 kpc (2′), probably mostly associated with the disk of the galaxy. Several small, bright ionized-gas condensations, discovered by Walker (1968), lie very close to the jet, and strengthen the connection between the plasma (radio-emitting) structure and the ionized-gas (optical-line emitting) structure (Wilson and Ulvestad 1982b). A similar, highly elongated radio jet in NGC 4151, in position angle 77°, is also closely connected with the optical emission lines. These are but two examples of cylindrically symmetric structures which so many of the observations indicate.

More recently, radio measurements have been made of a complete, unbiased sample of bright spectroscopically selected Seyfert galaxies at 1.5, 6 and 20 cm (Edelson 1987). All of them have $m_{pg} = 14.5$ or brighter, and the number of Seyfert 1 and Seyfert 2 galaxies are very nearly equal in this sample. All of them were detected as radio sources. The Seyfert 2 galaxies tend to have radio spectra $F_\nu \propto \nu^{-n}$ with $n = 0.7$ over the range 6 to 20 cm. The Seyfert 1 galaxies have nearly the same index (0.66 $\pm$ 0.28 dispersion compared with 0.71 $\pm$ 0.23 for the Seyfert 2s), but with a slight tendency to have flatter power laws (5 = 25% of the Seyfert 1s have $n < 0.5$ but only 2 = 10% of the Seyfert 2s do). Edelson has concluded that at high frequencies the Seyfert 1s and 2s appear to differ; about 25% of the Seyfert 1 galaxies have curved spectra which tend to flatten out at high frequencies while none of the Seyfert 2s do. Thus the Seyfert 2 galaxies have radio spectra similar to those of other 'normal' optically thin synchrotron sources, clustered around $n = 0.7$ to 0.8. The Seyfert 1 galaxies have a broader range of radio spectra, with flatter spectra and more tendency to flatten at high frequency. This may result from a flat-spectrum, optically thick core in some Seyfert 1s, becoming visible near 1.5 cm.

The mean 6-cm radio luminosities of Seyfert 1s and 2s from this sample are indistinguishable. This is probably correct, for here the sample is complete (to $m_{pg} = 14.5$ apparent magnitude), while previous samples were largely based on the Markarian objective-prism survey, which selected galaxies by their ultraviolet continuous spectra. It was thus biased against low-luminosity Seyfert 2 galaxies (which have only weak ultraviolet continua), as the subsequent CfA optical spectra survey proved by finding many new low-luminosity Seyfert 2s not included in the Markarian lists. Since the optical and radio luminosities are correlated, the ultraviolet-selected objects are thus not only biased against low optical luminosity but also against low radio luminosity. In the magnitude-limited sample the average ratio of radio-to-optical luminosity is a factor 1.9 larger for Seyfert 2 galaxies than for Seyfert 1s (Edelson 1987).

The most recent collection of radio data (21-cm continuum) of a larger sample of bright radio galaxies, spectroscopically selected, shows little clear evidence of a difference in average radio properties between Seyfert 1 and 2 galaxies (Giuricin *et al* 1990). Likewise, Ulvestad and Wilson (1989) have extended their 6 and 20 cm survey to include all 57 Seyfert galaxies known with $z < 0.0153$ that are accessible to the VLA. In this distance-limited survey they find only small differences between the radio properties of Seyfert 1 and 2 galaxies, which are but marginally significant. Thus all the recent radio-frequency data confirm that the previously reported differences were biased by the exclusion of lower-luminosity, redder Seyfert 2s, not found by the Markarian survey.

3.2. Infrared continuum

It has long been well known that Seyfert galaxies and QSOs emit much of their radiation in the infrared continuum. In particular NGC 1068 had been the first galaxy to be observed with almost every new infrared detector or infrared telescope. The main mechanism of infrared continuum emission in Seyfert 2s has long been understood as resulting from thermal reradiation by dust particles, heated by the central source. Extinction by dust is very important in QSOs and Seyfert galaxy nuclei (MacAlpine 1985). NGC 1068 and a few other well observed Seyfert 2 galaxies show a silicate absorption feature at $\lambda = 10.4$ μm, and a silicate emission feature at $\lambda = 19$ μm, similar to those that would and do result from dust in our Galaxy. In NGC 1068 the infrared luminosity is approximately 3×10^{11} $L_\odot$, approximately half of it coming from a. warm, compact source with measured diameter $1''$ (≈ 80 pc), the other half from a more extended, cooler component, the dust in the galaxy. In Seyfert 1 galaxies, quasars and QSOs some of the infrared radiation also comes from heated dust, but in addition some has long been thought to arise in a power-law source (Rieke and Lebovsky 1979).

The recent IRAS long-wavelength measurements have added greatly to our knowledge of all classes of AGNs and their host galaxies. Indeed, these observations showed that it is possible, from the IRAS measurements of known Seyfert galaxies, to develop criteria in a far infrared colour–colour plot for finding Seyfert galaxy candidates (Miley *et al* 1985). Subsequent follow-up optical spectra showed a fairly large fraction of these candidates to be actual AGNs, and deep-pointed IRAS measurements to low flux levels thus the most efficient known method for finding faint, distant Seyfert galaxies (Keel *et al* 1988). However, as both these papers stated, and as Salzer and MacAlpine (1988) confirmed with further observational data, the IRAS colours do not find *all* Seyfert galaxies, and *do* include among the candidates many starburst and H II-region galaxies, which turn out from slit spectra *not* to be Seyferts.

Somewhat contradictory results as to the nature of the infrared continuum have been obtained from the IRAS measurements. One study is based on a sample of 29 of the brightest AGNs for which there are good measurements from the IRAS regions (100–12 μm) through the ground-based infrared and optical to the IUE ultraviolet spectral region (Edelson and Malkan 1986). After removing the underlying galaxy spectrum and the H I and Fe II emission features, the remainder of the continuum from 10^{-1}–10^2 μm was fitted by a combination of a power law ν^{-n} (with strength and index n to be determined), an ultraviolet blackbody (with strength and temperature to be determined—this component could equally well have been fitted by an accretion disk), an optically thin Balmer continuum component and a '5 μm excess', fitted by a parabola centred at 5.2 μm. The three Seyfert 2 galaxies in this sample (including NGC 1068), plus three Seyfert 1 galaxies and one radio galaxy, 3C 84 = NGC 1275, could not be fitted by this combination. Their infrared radiation is clearly thermal emission from dust covering a range of temperatures. Their derived dust temperatures are in the range 40–80 K, and their dust masses ≈ 100 $M_\odot$, corresponding to gas masses of approximately 10^4–10^5 $M_\odot$, a reasonable estimate for an NLR. The great importance of radiation by dust is well brought out by the analysis of Edelson *et al* (1987).

For the relatively dust-free Seyfert 1 galaxies such as II Zw 136, NGC 3516, Mrk 335 and others, the power-law component with $n \approx 1.3$ does fit and the infrared continuum is therefore interpreted as primarily synchrotron emission. These objects have relatively low internal reddening, as estimated from their $\lambda 2175$ (graphite dust) absorption. In the far infrared the fluxes generally peak near 80 μm, and a 'turnover' (decrease in flux at longer wavelengths) due to synchrotron self-absorption is also included among the fitting parameters (Edelson and Malkan 1986). Strong arguments that the non-stellar infrared continua out to 10 μm are primarily power-law, synchrotron emission had previously been given by Malkan and Filippenko (1983). They argued that, if the source were hot thermal dust, the infrared flux should fall steeply (have a 'cutoff') at wavelengths shorter than 2 μm. In this picture the 5 μm excess *can* best be fitted by thermal emission by dust near the nucleus, heated by the optical and ultraviolet radiation of the accretion disk. The short wavelength fall-off is defined by the 1500 K evaporation temperature of graphite grains, and the long wavelength fall-off by complete absorption of the radiation over the limited solid angle subtended by the dust clouds at the nucleus. A torus with the dust between 1–2 pc (inner edge) distance and 200–500 pc (outer edge, $T \approx 100$–300 K) will fit the observed 5 μm excess (Barvainis 1987). Another good analysis of the continuum radiation in these terms is due to Carleton *et al* (1987).

However, a large sample of bright QSOs and quasars for which IRAS data are available was analysed by another group with a different interpretation (Sanders *et al* 1989). This sample is restricted to the bright QSOs from the PG survey, and does not include any Seyfert 2s. The main observational result is that the gross shape of the energy distributions over the range $\nu = 10^{12}$–10^{17} Hz ($\lambda = 3 \times (10^5$–$10^{-1})$nm) is similar for all these objects, and can be fitted plausibly by two broad components of thermal emission. One is the 'big blue bump', the radiation of the accretion disk. The other is the 'infrared bump' between $\lambda = 2$ and 10^3 μm. The latter is interpreted as entirely resulting from dust, at distances from the central source from 0.1 pc (the absorbing torus) out to more than 1 kpc (in the disk). It includes *both* the power-law *and* the 5 μm 'excess'. Quantitative calculations show that the amount of dust expected to be present is sufficient to produce the required infrared radiation. Optically-thin emission from dust near the source contributes to the radiation in the region around $\nu = 3 \times 10^{14}$ Hz ($\lambda = 1$ μm) between the two bumps. It is supposed to come from the most refractory graphite or graphite-coated particles, with sublimation temperature 2000 K. (The theoretical analysis is 'in preparation'.) Thus these authors see no convincing evidence for significant power-law radiation in the 10^{12}–10^{17} Hz region in these objects.

Sanders *et al* (1989) argue that the dust and gas clouds are distributed in a *warped* disk, in agreement with section 4.1 of this review and that, as a result of this warp, the covering factor for dust to be heated by the central source is $\Omega \approx 0.1$. As they discuss, there is good evidence for a 'tilted disk' (or warp) near the centre of our Galaxy, and further out in many Seyfert galaxies.

Sanders *et al* (1989) emphasize that evidence for power-law (synchrotron) emission in the far infrared spectra of AGNs has come primarily from blazars (BL Lac objects and optically violently variable quasars) and flat-spectrum radio-loud quasars, of which 3C 273 is a prime example. These types of objects are relatively rare. Blazars, with their relatively smooth, near power-law spectra ($n \approx -1$), high polarization and rapid variability, are best interpreted as synchrotron emission from a relativistic jet over the entire radio-frequency to ultraviolet continuum (Saikia and Salter 1988). 3C 273 does vary in the optical and near-infrared on time scales as short as 1 day, and on time scales of months at $\lambda = 10$ μm. Thus there is good evidence for non-thermal, power-law synchrotron radiation in it. It is an exceptional object, and the process is exceptional according to the interpretation of Sanders *et al* (1989).

A strong argument that there is a power-law continuum has been that the infrared (3 μm) and x-ray (2 keV) fluxes of AGNs are correlated. This argues that these fluxes are produced by the same process, and therefore that the infrared is not thermal emission from dust (Elvis *et al* 1986). However, more recently McAlary and Rieke (1988), using more recent data and more nearly complete samples, have found no such correlation for the Seyfert 1 galaxies, and only a weak correlation for PG quasars. Sanders *et al* (1989) likewise find no strong evidence for a correlation between the entire infrared (2–100 μm) flux the x-ray (2–10 keV) flux, and conclude that the data do not require a common origin or a power law linking these two regions. Thus at present there is no consensus on the origin of the infrared radiation in QSOs and most Seyfert 1 galaxies. Very probably *both* thermal dust emission *and* the power law contribute, to different relative amounts in different objects.

3.3. X-ray region

From the early days of x-ray astronomy, Seyfert 1 galaxies and QSOs were recognized as luminous objects in this energy range (Wilson 1979). The $H\alpha$ (broad emission-line) flux was discovered to be proportional to the x-ray flux, and a striking result was the prediction on this basis of the x-ray flux in the LINER M 81, and its confirmation, as described in section 6.4. The surveys in the 2–20 keV range found AGNs to have power law spectra with a narrow range around $n = 0.7$ (Mushotzky 1984). A later survey of more objects at lower energy (0.2–4 keV) gave somewhat larger scatter, with a mean $n \approx 1.0$ for the radio-quiet objects, mostly QSOs but including some Seyfert 1s, and $n \approx 0.5$ for the fewer quasars and radio galaxies. Furthermore, these data showed an increase above the power law at low-energies, or 'soft spectral excess' in many of these objects (Wilkes and Elvis 1987).

The most recent survey, covering an even wider energy range (0.1–10 keV), shows still more complexity. The objects observed are largely x-ray flux-limited samples, mostly Seyfert 1 (and 1.5) with a few 'narrow emission-line galaxies' (Seyfert 1.9 or 1.99 in the sense of section 6.3), one quasar (3C 273), and one Seyfert 2 (NGC 1068). The general result is that in the 2–10 keV range a power law with $n = 0.7 \pm 0.2$ (average deviation) fits well. At lower energies many objects have complex spectra. Some have soft x-ray excesses. In addition to the power law, the absorbing column along the line of sight can be determined from the amount of absorption at low energy. Typical values range from 10^{21} to 10^{23} H atoms cm^{-2} (the absorption is actually primarily by elements heavier than H and He and an assumption about abundances is involved in the calibration). As stated in section 6.1 there is a tendency for more of the low luminosity nuclei to have large absorbing columns, suggesting the high luminosity objects may have blown away and dispersed the clouds around them. The soft excesses can be modelled either by a second component with a steeper power law, or by partial absorption by material in clumps, allowing some of the radiation to pass through between them, unattenuated. In most but not all cases, the two-component model fits the data better than the 'leaky-absorber' model (Turner and Pounds 1989).

X-ray emission in AGNs with a power-law spectrum with $n = 0.7$ can be associated with the infrared power-law continuum with $n = 1.2$ discussed in the previous section, if the latter is actually present. The most plausible process is synchrotron emission in the infrared (low-energy) region, and Compton upscattering of these photons plus synchrotron emission in the x-ray (high-energy) region. Models based on these ideas are called synchrotron self-Compton models and require a broken power-law of electron energies. Fairly natural assumptions give the break in flux power laws (n) by about 0.5 at energy ~ 1 keV. These models are well summarized by Zdziarski (1986) and by O'Dell *et al* (1987).

The more complicated x-ray spectra resulting from the recent, most accurate data complicate this picture (Turner and Pounds 1989, Kruper *et al* 1990). The several observed spectral components imply several synchrotron self-Compton regions, or perhaps a combination of one with a high-temperature blackbody spectrum, or perhaps other, non-power-law electron energy spectra (Urry 1990, Urry *et al* 1990). Very complex models, bridging the range from x-ray to infrared, have been calculated (Band 1987) and fitted to observed AGN spectra (Band and Malkan 1989).

Finally, x-ray variations have been observed in some AGNs on a time scale of hours. A recent, quite well determined case is in Mrk 335, where large amplitude variations have been detected on times of ~ 6 h. This is consistent with variations in the accretion disk at 5 Schwarzschild radii about a black hole with mass $10^7 \, M_\odot$ (Turner and Pounds 1988, who give references to earlier similar observations). X-ray variations at the 10% level, which are not simple to interpret theoretically, have been observed with doubling times from 10^2 s (Seyfert 1s) to 10^6 s (QSOs) (Barr and Mushotzky 1986).

3.4. γ-ray radiation

X-rays have been observed from several AGNs to energies above 0.1 MeV, which might be called γ-rays, and 3C 273 has been measured to over 100 MeV. The flux is not much below an extension of the x-ray spectrum to this energy. Hence a significant amount of energy is emitted in these very high-energy photons. Little is definitely known as to the mechanism, but an extension of the processes responsible for the x-ray emission is the most obvious possibility. Others include pair production and annihilation in a hot, Doppler-broadened region and Penrose–Compton scattering. However, 3C 273 may not be representative of the general class of AGNs in its γ-ray properties (Ramaty and Lingenfelter 1982).

4. Torus model

4.1. Cylindrical symmetry

There are many reasons for supposing the structure of AGNs to be more nearly cylindrically symmetric about an axis than spherically symmetric about a point. One is the theoretical idea of an accretion disk, the other is the observation in many objects of radio-frequency jets, which appear to be axes of the structure (Shields 1978, Osterbrock 1978). There is no strong observational reason to suppose that the axis (angular momentum vector) of the AGN coincides in direction with the axis of the galaxy in which it is, and comparison of the directions of jets with images of their host galaxies suggest they are often not in the same direction (Tohline and Osterbrock 1982, Ulvestad and Wilson 1984b). Thus the highly schematic AGN of figure 1, in which the BLR is disk-shaped, should be understood as having its axis in an arbitrary direction. In this picture ionizing photons can escape into the NLR along the axis, but not along the equatorial plane of the BLR where the optical depth is greater. There are many indications that the lower-density NLR merges continuously into gas in the disk of the host galaxy. Hence a schematic picture like figure 5 might be more appropriate, in which the BLR is tipped with respect to the NLR, also shown as a disk, but with the density condensations represented as spherical clouds. In reality no doubt there is a continuous transition from BLR to NLR both in density and in axis or plane; thus the NLR would be warped, as is observed for gas near the centre of our galaxy and in many other spirals in which the interstellar H I distribution has been mapped.

There are very few nearly edge-on Seyfert galaxies, as detected by their optical emission lines. (NGC 4388 is one of the rare known exceptions.) No doubt this is due to extinction by dust in the planes of their host galaxies (Keel 1980). However, hard x-rays can escape freely, and indeed there are Seyfert 1.8 and 1.9 galaxies with luminosities in this band comparable to Seyfert 1 galaxies, but with much weaker broad Hα emission. Some of them are nearly edge-on, but others among them are

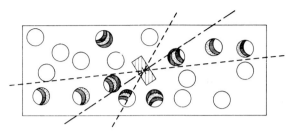

Figure 5. Schematic AGN model, showing tipped BLR disk in larger NLR disk. Most of the ionizing photons escape from the BLR disk in the cone about its axis, but a few 'tunnels' may exist in other directions. Highest degree of ionization, indicated by close cross-hatching, occurs on faces of clouds closest to the central accretion disk. Actually the BLR probably merges and warps continuously into the NLR. (Osterbrock 1978b)

not, suggesting that considerable extinction also occurs in the plane of the BLR, and that these objects are seen with their BLR edge-on to us. These and other correlations of x-ray luminosity, path length along the line of sight to the nucleus as shown by soft x-ray absorption, and narrow-line luminosity led to the suggestion that many if not all Seyfert 2 galaxies contain BLR regions, which however are obscured by the dust surrounding them, more or less in the neutral regions marked BLG^0 on figure 1 (Lawrence and Elvis 1982).

4.2. Seyfert 2s with 'hidden' BLRs

The earlier interpretation of the differences between Seyfert 1, 1.5 and 2 galaxies was in the amount of dense gas in the BLR, and the relative numbers of ionizing photons absorbed in it and in the NLR. However, that this is not correct in all cases was clearly shown by spectropolarimetric measurements of the Seyfert 2 AGN, NGC 1068 (Antonucci and Miller 1985, Miller 1988). These data show that the forbidden lines are only slightly polarized, by approximately 1%. However, after removal of the galaxy absorption-line continuum, the featureless continuum (presumably radiated by the nucleus or accretion disk itself) is much more strongly plane-polarized. The degree of polarization is approximately 16%, independent of wavelength and in a constant direction. The direction of polarization is with the E vector approximately perpendicular to the axis of the radio jet in this AGN. Furthermore, in the polarized light spectrum, weak *broad* emission lines of H I and Fe II, characteristic of a Seyfert 1 spectrum, can be seen. Their FWOI are about 7500 km s^{-1}, much wider than in the Seyfert 2 spectrum seen in natural light. More recent spectropolarimetric measurements of eight other Seyfert 2 nuclei with high polarization show that at least four of them have similar 'hidden Seyfert 1' spectra, broad H I and in some cases Fe II emission lines, visible only in the plane-polarized spectrum (Miller and Goodrich 1990).

A general property of these and still more Seyfert 2 nuclei observed with broadband polarimeters is that their continua are plane-polarized perpendicular to the jet or other radio structure. On the other hand, Seyfert 1 nuclei have much weaker polarization, and in the opposite sense, namely parallel to the radio jet or linear structure (Antonucci 1983). This same direction also occurs in quasars and QSOs. In all the objects for which there are spectropolarimetric measurements, the degree of polarization is independent of wavelength over the observed range, suggesting that electron scattering is the mechanism. The suggested interpretation is that in Seyfert 1 nuclei a thin disk is seen directly, more or less face on, and the polarization is weak but parallel to the axis because it is averaged over the disk. On the other hand in the Seyfert 2s the disk is not seen directly, but radiation from the central source and the BLR which escapes along the axis (of the cone marked NLG$^+$ in figure 1) is seen only if it is scattered toward us. Hence it is polarized on the average perpendicular to the axis of the disk. A schematic thick-disk model of this type is shown in figure 6. From the widths of the plane-polarized H I emission lines in NGC 1068, an upper limit to the temperature of the scattering electrons is $T < 10^6$ K. In NGC 1068 Miller (1988) has also made spectropolarimetric measurements of regions near but outside the nucleus, namely a strip more or less along the direction of the jet, and a bright knot about 5 in ($\approx$ 400 pc) northeast of the nucleus. These areas are also strongly polarized and show the broad H I and Fe II emission lines in polarized light, in each case with plane perpendicular to the direction of the nucleus. In the knot the H I lines are not as broad as in the nuclear polarized-light spectrum, indicating that dust scattering rather than electron scattering is dominant in the knot. This interpretation is borne out by the differing wavelength dependence of the polarized light fluxes from the knot and the nucleus. From the temperature limit previously stated, the scattering electrons appear to be part of an outward flowing wind, Compton heated by the direct ultraviolet and x-ray radiation of the central source, and cooled by expansion. The total optical depth is $\tau \approx 0.1$. A predicted signature of this temperature, density and radiation field is Fe K x-ray line and continuum emission (Krolik and Kallman 1987, Krolik and Begelman 1988). This feature has been observed, with a strength that agrees with the hidden Seyfert 1 picture of NGC 1068 (Koyama 1989).

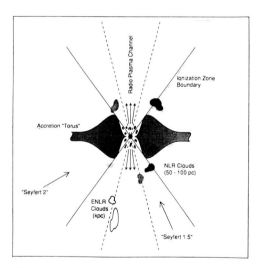

Figure 6. Schematic drawing of central source of AGN, surrounded by thick accretion torus, which channels radio plasma and ionizing radiation along the axis. Directions from which the object's spectrum is Seyfert 1, 1.5 and 2 are marked (Pogge 1988c).

NGC 1068 is a unique Seyfert 2 in several respects, as previously mentioned. Hence the observations of 'hidden BLRs' by the same spectropolarimetric technique used on the other Seyfert 2 AGNs, as mentioned before, is important in establishing that it occurs in more than this one object. Whether *all* Seyferts have this same structure is another question. Half the eight Seyfert 2 galaxies observed spectropolarimetrically to date show the presence of a 'hidden' Seyfert 1 AGN, but this sample of eight was selected as objects of known high polarization. If *all* Seyfert galaxies were physically the same type of objects, seen from different orientations, many more highly polarized Seyfert 2s would be expected than are observed (Miller and Goodrich 1990). As described in the previous section, the infrared and radio properties of Seyfert 1 and 2 galaxies are now understood to be relatively similar. Thus the picture is an attractive one, but far from proven at present.

4.3. Ionization cones

The nearest Seyfert galaxy NLRs can be resolved. One of the most studied is again the bright Seyfert 2 NGC 1068. As previously mentioned, it is far from typical of this class, but its nearness makes it a rewarding object for study. Very complete long-slit data are available (Baldwin *et al* 1987), and also Fabry–Perot imaging data (Cecil *et al* 1990). These two papers are the most recent of a continuing series of improvements in detector technology, going back to the photographic spectra of Walker (1968), who first mapped the velocity field and homogeneities in this object.

The long-slit spectra cover the central 10 kpc × 2 kpc ($140'' \times 30''$) with good wavelength and spatial resolution. They show that in the outer part of the galaxy ($r > 1000$ pc, based on $H = 75$ km s^{-1} Mpc^{-1}) the gas lies in a fairly normal disk, with a flat rotation curve. Within this distance the velocity field is more complicated, containing several condensations or high-velocity clouds with velocities up to several hundred km s^{-1}. In this inner region these optical condensations tend to lie along the axis of the radio jet, as recognized by Wilson and Ulvestad (1983).

The velocity field will be discussed in the next section, but the emission-line spectra themselves show a mixture of two disk components projected on top of each other. They can be separated from each other by the diagrams of figures 2, 3 and 4. One component consists of normal H II regions, photoionized by hot stars. They are particularly strong in a ring of about $15''$ (1 kpc) radius, and within it. The other component consists of gas photoionized by the central source within the AGN, as shown by the fact that measured line-ratios for it cluster close to the AGN lines in the diagnostic diagrams. This component is located in space along the axis of the jet and its extension. It can best be seen in ionization maps based on continuum-subtracted images taken with interference filters centred on [O III] $\lambda5007$ and on Hα + [N II] $\lambda\lambda6548$, 6583 in the rest system of the Seyfert galaxy. Hα alone would be preferable, but the somewhat wider filters which include the [N II] lines are also easier to obtain and use. Dividing the [O III] image by the Hα + [N II] image, after first cutting out the lowest levels of each (which are dominated by noise), and then plotting the contours only of high [O III] / (Hα +[N II]), essentially the same as high [O III] / Hα, produces a qualitative 'ionization map'. The regions it shows are those photoionized by the hard spectrum of the AGN nucleus, while the H II regions photoionized by hot stars drop out (Pogge 1988a). In NGC 1068 the region photoionized by the nuclear continuum is triangular-shaped, (evidently a cone in projection) centred on the axis of the jet and its extension. It disappears where it crosses the ring of H II regions (which are brighter than it), and then reappears further out. It can be followed to a distance of about $25''$ (nearly 2000 pc). The observed structure is quite similar to that predicted for one side of the ionized 'narrow-line gas' in models such as those shown in figures 2, 3 and 4. Evidently the other side of the cone lies below the central plane of this galaxy, rich with dust, and is therefore not seen.

Similar ionization maps for approximately 20 other nearby, non-interacting Seyfert 1 and Seyfert 2 galaxies have been obtained (Pogge 1989a, 1989b). Of the eleven Seyfert 2 galaxies, eight show extended ionized gas, with strong [O III] $\lambda\lambda 4959$, 5007, and emission-line ratios consistent with photoionization by a hard source. The [O III] / (Hα + [N II]) maps reveal distinctly conical high-ionization regions, with the nucleus at the apex, in four of these eight Seyfert 2 galaxies. In the objects for which high-quality radio continuum maps are available, the cones are aligned with the axes of the cones. Although the number of objects is still small, this is strong evidence for the ionization-cone picture. The opening angles of the cones range from 30° to 90°, with a mean of perhaps 50°. Only three of the nine Seyfert 1 galaxies show extended high-ionization emission. In none of them does the ionization map show either a cone, or a halo (as a cone seen approximately end on, from a direction within it might appear). These data thus do *not* show the difference between the presumed 'bare nucleus' Seyfert 1s and 'hidden nucleus' Seyfert 2s that would be expected from this picture. None of these Seyfert 1s appear to be 'pole-on' versions of the Seyfert 2s with extended conical photoionized regions. Although there is much good evidence that some if not all Seyfert 2 galaxies have a hidden BLR, the statement that Seyfert 1 nuclei have the same structures seen face-on is not explicitly confirmed.

In an independent study, ten nearby Seyfert galaxies with known linear radio structure were imaged in [O III] and Hα (Haniff *et al* 1988). Most of the objects are Seyfert 2s, but two, Mrk 6 and Mrk 79 are Seyfert 1.5s. In all ten cases the [O III] emission outside the nucleus is aligned with the radio structure to the accuracy of the data (about 5°). There is also good agreement between the Hα and radio structure, but not as much as with the [O III] images, indicating the relatively greater contribution of H II regions to the Hα image. Furthermore, these same authors have found evidence for anisotropic emission of the ionizing radiation from the nucleus in at least two of these Seyfert 2 galaxies, Mrk 3 and Mrk 78 (Wilson *et al* 1988). In each object they estimated the available number of ionizing photons from the nucleus from its optical, ultraviolet or x-ray brightness, a plausible extrapolation of its spectrum to $h\nu > h\nu_0 = 13.6$ eV by a power law, and an estimate of the solid angle subtended at the nucleus by the photoionized gas. The number of photoionizations required is given directly by the integrated flux of the extended emission in Hα or Hβ. For both Mrk 3 and Mrk 78 the discrepancy is at least a factor of ten, probably indicating that the ionizing flux which escapes to the extended emission in the direction of the axis of the jet is significantly greater than expected from the continuum flux which escapes in our direction. The connection with the torus model is clear; it is sometimes referred to as the 'beaming' of ionizing radiation. However, it is also possible that the ionizing flux is larger than the extrapolation of the power law would suggest.

5. Velocity fields

5.1. Narrow-line profiles

AGNs are unique among galaxies in the property that defines them, namely having 'wide' line profiles with a larger velocity spread than in gaseous nebulae and stars. The only interpretation that seems physically possible is that the profiles result from the internal velocity fields in the objects. Since the nuclei are mostly unresolved, the information the profiles bear are integrated and not straightforward to interpret, but they are the best information we have. Much of this work has been summarized by Osterbrock and Mathews (1986). The highest spectral resolution data, including those of Whittle (1985a,b,c) with resolution $\Delta\lambda \approx 1.1$ Å ≈ 65 km s^{-1}, Vrtilek and Carleton (1985) with $\Delta\lambda \approx 0.4$ Å ≈ 25 km s^{-1}, and Veilleux (1989) with $\Delta\lambda \approx 0.17$ Å ≈ 10 km s^{-1} have been especially important. Correlations between line-profile parameters and extinction have been studied in detail by Dahari and De Robertis (1988b).

In most respects the narrow lines in Seyfert 1 and Seyfert 2 galaxies have similar properties, and they may be discussed together. [O III] $\lambda\lambda4959, 5007$ are the strongest unblended narrow lines and therefore the most studied. In Seyfert 1 and 2 galaxies their full widths at half maximum (FWHMs) have a broad distribution, centred perhaps at 350 to 400 km s^{-1}, and extending from about 200 to about 900 km s^{-1}. NGC 1068, the very well studied Seyfert 2 galaxy, has unusually wide 'narrow' lines, with FWHM of [O III] 1200 km s^{-1}. The [O III] FWHMs are correlated (with considerable scatter) with [O III] luminosity and with the radio-frequency luminosity of the nucleus, indicating that the velocities in the NLR are driven by or associated with the energy input to this region. There are also weaker, but significant, correlations of the narrow-line profiles with the morphological type of the host galaxy, and possibly with the luminosity of its bulge. This indicates that the gravitational effects of the galaxy affect the velocity field in the NLR (Whittle 1989, Veilleux 1991a).

In many Seyfert 1s and 2s the narrow-line profiles can be observed to be asymmetric. Almost invariably this is in the sense that the profile has a longer tail or wing extending to the shorter wavelength side and a more abrupt fall-off toward longer wavelength. No purely rotational velocity field will produce such profiles. The simplest possible interpretation is that the asymmetry results from radial flow combined with extinction by dust. Various combinations are possible; the sense of the flow is not definite because the physical structure of the emitting region is not certain. If the flow is radially outward, dust distributed throughout the AGN will, on the average, cut down the emission lines from the far (receding) side more than from the near (approaching) side, producing the observed asymmetry. On the other hand, if the emission comes from small, optically thick (to ionizing radiation) clouds which contain dust, the emission will come primarily from the sides of the clouds which face the ionizing source. The radiation from clouds on the far side of the source will then come directly to the observer without as much extinction as radiation from clouds on the near side, which must escape through the cloud. Hence, in this picture, if the clouds are falling radially inward, the line profiles will have the observed sense of asymmetry.

The most physically attractive picture of the NLR is that the flow is outward, with a high-temperature, low-density, invisible plasma wind carrying the low-temperature, high-density, emitting clouds within it. Such models, among others, fit the observed profiles well (Vrtilek 1985). All these models have dust distributed in the same regions as the gas. However, there are some observational indications that distributed dust is not the only factor responsible for the line asymmetry. Considerable scatter occurs in the relation between asymmetry and extinction (measured by the Hα/Hβ intensity radio). Cylindrical rather than spherical symmetry is indicated by many observed features of NLRs, and an occulting or optically thick central disk, cutting off radiation from the far side of the NLR, may be the cause of the asymmetry in at least some galaxies (Veilleux 1991b).

In a fair number of bright Seyfert 1 and 2 galaxies, profiles of many narrow emission lines of different ions have been measured. In each NLR, the profiles have the same general form (degree of asymmetry). In many objects there is a correlation between the FWHM and critical electron density for de-excitation, in the sense that the lines that can be emitted at higher densities have larger FWHMs. An example is Mrk 1066, in which the FWHM ranges from 300 km s^{-1} for [S II] with $N_{ec} \approx 10^{3.5}$ cm^{-3} to 400 km s^{-1} for [O III] with $N_{ec} \approx 10^{5.8}$ cm^{-3}. In 70% of the AGNs for which [O III] $\lambda4363$, with $N_{ec} \approx 10^{7.5}$ cm^{-3}, profiles have been measured, they are broader than those for [O III] $\lambda\lambda4959$, 5007, with $N_{ec} \approx 10^{3.8}$ cm^{-3}. In many NLRs there is also a correlation between line width and ionization potential, in the sense that the ions of highest ionization potential have the largest FWHMs. Since the velocity decreases outward from the central ionizing source in practically every model, these results indicate that the ionization decreases outward in typical NLRs, and that the mean electron density also decreases outward, if fitted by a power law approximately as $N_e \propto r^{-n}$ with $0 < n < 2$. [O I] $\lambda6300$, a low-ionization line with a relatively high critical density $\left(N_{ec} \approx 2 \times 10^6 \text{ cm}^{-3}\right)$, typically has relatively broad wings, indicating that the emitting clouds are optically thick to ionizing radiation and contain, in their 'transition zone', some O^0 atoms.

De Robertis and Shaw (1990) have attempted to resolve observationally the question of whether the velocity field is primarily radially outward or inward in the NLR, by studying the asymmetry as a function of ionization and critical electron density for collisional de-excitation. Their general result, obtained by comparing high-quality observational data with simplified, spherically symmetric kinematic models, is that the dust extinction is within the emitting clouds and the direction of flow is therefore inward. However, as many other possible types of models, specifically cylindrically rather than spherically symmetric, were *not* investigated, the conclusion is still open to question.

In addition the narrow line profiles often show some substructure, in some cases strong enough to show 'shoulders' or secondary peaks. This is especially clear with high resolution data (Veilleux 1989, 1991c). No doubt these effects result from non-homogeneous gas distribution. In some of the nearest AGNs, NGC 1068 (Baldwin *et al* 1987) and NGC 4151 (Schulz 1987) for example, such inhomogeneities can be directly resolved. Several more are seen in other AGNs in the spectra of Veilleux (1989, 1991c). The NLR is certainly a complicated structure. Radial motions, probably outflow, modified by extinction and/or obscuration, photoionization from a central source, and inhomogeneities all form parts of its description. We will return to a more nearly complete discussion of the velocity field in the NLR in section 5.3.

5.2. Extended emission-line region

Long-slit spectra provided the first convincing evidence for the ionization cones described in section 4.3. Unger *et al* (1987) took spectra of seven Seyfert galaxies, known to have jets or linear radio structure, with the slit aligned parallel and perpendicular to the jets. Those with the slit parallel to the linear structure showed the gas photoionized by the hard spectrum extending far from the nucleus, as much as 20 kpc in some cases, but with very little extent in the perpendicular direction. Unger *et al* called the ionized gas far from the nucleus the extended narrow-line region (ENLR). Though the interference-filter images map the entire cone, the spectra give radial-velocity information. These spectra show that the velocity field in the ENLR is chiefly the smooth rotation field of the galaxy in which they lie. The lines are much narrower than in the NLR, typically with FW0I $\lesssim$ 50 km s^{-1} rather than around 300 km s^{-1}.

Likewise Baldwin *et al* (1987) mapped the velocity field in NGC 1068 in great detail, with long-slit spectra taken at several different positions. Again, they show that outside a radius of about 10″ (≈ 750 pc) the gas photoionized by the nucleus has the smooth galactic rotation curve characteristic of the disk. Inside this distance the field is more complicated but appears to correspond to rotation with a smoothly varying line of nodes. One possible interpretation is that in this region the midplane of the gas is warping between the plane of the galaxy and that of the NLR (perpendicular to the axis of the radio jet). In addition, in the outer region a few isolated clumps have velocities significantly different from the rotational velocity field. Their line ratios show they are all photoionized by the AGN. All of the clumps that are seen projected against dusty regions have redshifts with respect to the rotation curve, that is are falling toward the disk. Near the nucleus several clumps are seen with a range of velocities; these are the condensations associated with the jet. An even more detailed and complete analysis of the velocity field in the NLR region of NGC 1068 has been given by Cecil *et al* (1990) with similar general conclusions. Likewise in the Seyfert 1 galaxy NGC 4151 Schulz (1988) found similar patterns of the photoionization by the AGN along the extended axis of the jet.

5.3. Overall NLR velocity field

The observational data previously summarized clearly show that the velocity field in the inner part of the NLR is not rotational, but radial (if spherical) or in the ρ and z direction (if cylindrically symmetric). Its spatial structure is clearly more nearly cylindrically symmetric and associated with the axis of the jet instead of the axis of the main body of the host galaxy. It is not clear whether this velocity field is directed outward or inward. Some observational evidence favours inward motion, but it is highly model dependent. Theoretical expectations favour outward motion. The motion of the radio plasma in the jet is clearly outward, and since the NLR is so closely associated with it, this again would seem to favour outward motion for it. The extension of the NLR is probably continuous, through the ENLR to the gas in the disk of the galaxy. Its velocity field is rotational. The spatial structure warps from perpendicular to the axis of the jet to perpendicular to the axis of the galaxy.

The ionizing radiation from the nucleus is anisotropic, perhaps collimated or 'beamed', and these ionization cones in many Seyfert 2 galaxies include segments of the main body of the disk. The evidence for anisotropy of the AGN radiation is not very marked for Seyfert 1s. They may in fact have thin disks and 'cones' with opening angles of nearly 180°.

Far from the nucleus and from the extended axis of the jet, isolated clouds or elements of gas seem to be falling toward the disk in NGC 1068, the one best studied Seyfert 2. Although it cannot be (or has not been) observed, probably the mainly rotational velocity field in the innermost parts of the ENLR contain a slight inward component, which probably also continues in the central plane of the NLR, bringing material in to the accretion disk. This overall pattern is schematically represented in figure 7, which is probably the best present working hypothesis for the velocity field in the NLR and ENLR.

5.4. BLR velocity field

The BLR is completely unresolved for all observed AGNs, as expected from the calculated sizes. It is expected to remain even so with the much better angular resolution

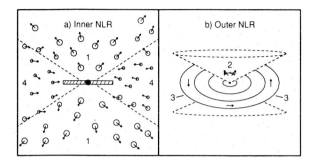

Figure 7. Schematic representation of the velocity field in the inner ($r \lesssim 50$ pc) and outer ($r \gtrsim 50$ pc) regions of the NLR. The line-emitting gas in zones 1, 2, 3 and 4 corresponds to the wind-accelerated, jet-associated, and inner and outer gravity-dominated components, respectively. Note that the rotation axis of the outer, gravity-dominated component (which merges into the interstellar matter in the galaxy) is not necessarily perpendicular to the axis of the double cone associated with the jet. Clouds in zone 5 probably contain substantial amounts of neutral gas. (Veilleux 1989)

of the Hubble Space Telescope ($\sim 0.1''$). This prediction of course should, however, be checked observationally. Thus the only observed information on the velocity fields in the BLRs is in the form of the broad-line profiles. An atlas containing Hα, Hβ, He I λ5876 and He II in 19 Seyfert 1 galaxies has been published by Osterbrock and Shuder (1982). As previously stated, the H I lines have a wide range of FWHMs, from approximately 500 km s^{-1} in Mrk 359, NGC 4051 and Akn 564 up to approximately 7000 km s^{-1} in Mrk 279, Mrk 876 and Mrk 926. The FWOIs, as well as they can be defined, range from approximately 4×10^3 km s^{-1} in Mrk 359 to approximately 3×10^4 km s^{-1} in Mrk 876. Most Seyfert 1s thus have much wider lines than Seyfert 2s, though a small group of 'narrow-line Seyfert 1s', of which Mrk 359 is the prototype, have only slightly larger H I FWHMs than these.

There is no common pattern for the broad-line profiles, as there is for the narrow-line profiles. Approximately half the well observed objects have essentially symmetric broad-line profiles; the remainder with asymmetric profiles in some cases extend further to the blue, in others, to the red. Thus radial flow combined with extinction is not the main velocity field in the BLR.

There are real differences among the broad-line profiles in a single Seyfert 1 galaxy. Usually Hβ is somewhat broader than Hα (1.16 ± 0.05 average deviation is the ratio in the sample mentioned), He I still broader (1.36 ± 0.05), and He II probably still broader, though its profile is generally badly blended with neighbouring Fe II features and is less well defined. All calculations of the H I spectrum under BLR conditions show that the Hβ/Hα intensity ratio increases with an increasing N_e and with increasing ionization parameter Γ. Thus, from the observed profiles, the velocity dispersion in the BLR must increase with either one or other of these parameters. Thus if for instance N_e = constant, the velocity dispersion must increase inward with Γ, which increases as R^{-2}; or if Γ = constant, the velocity dispersion must increase inward with N_e, which increases as R^{-2} for constant Γ. Likewise, calculations show that the intensity ratio He I λ5876/Hβ increases with N_e and Γ, and the same argument applies again. Thus the profiles show that the velocity dispersion increases inward in the BLR. In addition the Fe II emission lines have FWHMs typically somewhat smaller than the broad H I lines. The exact value is difficult to quantify, because all the Fe II multiplets are badly blended, but ratios of FWHMs of Fe II to Hβ between 0.8 and 1.0 will satisfy the observational data. This again indicates that the velocities decrease outward in the BLR.

A rotational velocity field about a central black hole has this property and it is natural to assume that it is responsible for the observed profiles. Calculations have been made which show that simple models combining rotation with outward directed flow, perpendicular to the disk, will produce profiles that match the best-quality observed ones. Likewise, however, assumed velocity fields with only the outward flow confined to a cone whose axis is normal to the disk, assumed density distributions, cutoffs in the flow velocity, or cone opening angles will fit almost all of the observed profiles, if the asymmetries are ignored. These are kinematical models only, but they indicate at least some of the directions to explore for complete physical models (Osterbrock and Mathews 1986).

Much the same is true of the broad emission-line profiles in QSOs. Different broad lines tend to have similar, but not identical profiles. The lower-ionization lines probably have somewhat smaller FWHMs than the higher-ionization lines, as in Seyfert 1 nuclei. Lα, the strongest emission line, does *not* have a particularly asymmetric profile, as would be expected for such an optically thick resonance line, if the velocity field was primarily radial inflow or outflow. The FWHMs of C IV λ1549, a strong and quite unblended line, in a large sample of quasars (primarily) and QSOs range from about 2000 to 7000 km s^{-1}, with a mean of 4500 km s^{-1}, somewhat higher than in Seyfert 1 galaxies, in which the mean is more like 3000 km s^{-1} (Wilkes 1984, 1986).

The earliest physical models of the velocity field in the BLR considered radiation-pressure-driven outflow. This predicts a logarithmic form for the line profile, which approximately fits many of the observed profiles. Additional acceleration by a hot wind helps to increase the stability of the clouds, and essentially preserves the same form of line profile. However, as previously mentioned, essentially all such wind media that would lead to stability have been ruled out on the grounds that they should be observable in one spectral region or another, but have not been observed (Mathews and Ferland 1987, Mathews and Doane 1989).

If the observed broad-line profiles represent outflow, the rate of mass loss ranges from 0.1 to 10 $M_\odot$ y^{-1}. This is of the order of the mass accretion rate. If we imagine that an AGN's lifetime is of the order of a few percent of the Hubble time, corresponding to the fraction of spirals that have AGNs, the integrated mass-loss rate is comparable with the mass in the black hole, which does not seem unreasonable.

If the velocity field is largely rotational flow, the mass-loss rate is much smaller. In addition, the magnitudes of the velocities deduced from the line widths comparable with those expected under gravitational forces. This has been particularly emphasized by Wandel (1989) who found that the velocity dispersions from the FW0Is lead, through the relation $v^2 = GM/R$, to a dynamical mass for the black hole that agrees fairly well with the value for the mass found from fitting the ultraviolet continuum with an accretion disk model. However, there are many only poorly-known parameters and the specific numerical values are uncertain. The most recent and complete data on dynamical masses found in this way are in Padovani *et al* (1990).

The best information on the nature of the BLR velocity fields should come from the temporal-variation measurements summarized in section 2.7. By determining the time-lag between variations in the continuum and various *parts* of the profile, for instance the red side or the blue side, it is in principle possible to localize the motions to some extent. Gaskell (1988), using IUE spectra of NGC 4151, found that the variations in the blue wing of C IV λ1549 lagged the red one by 3.4 ± 3.4 day, and the same for Mg II λ2798 by 4.5 ± 3.1 day. He interpreted this as evidence for inflow (the nearer side of the object has a redshift with respect to the centre), but the effect is too small to be convincing. Much better ground-based data for this same object give a lag of only 1 day for the blue wing of Hα, considerably smaller than the probable error. Purely radial motion, either inflow or outflow, is ruled out by these data. Combinations of chaotic and orbital motion are allowed (Maoz *et al* 1991).

On the other hand for the considerably more luminous (and thus larger) QSO Fairall 9, Koratkar and Gaskell (1989), again using IUE spectra taken over several years, found $R = 124 \pm 39$ light day as the time lag for C IV λ1549 with respect to the continuum, and the direction of motion preferentially inward. The chance of purely random or chaotic motion was given as only 10%. Mg II λ2798 gave a somewhat larger mean size, 167 ± 61 light day, and a chaotic (which includes orbital in all directions) velocity field.

Finally, using the IUE spectra of six objects taken from 1978 to 1988, Koratkar and Gaskell (1991) found that both pure outflow and inflow were excluded at the 99% level. Purely chaotic or circular motion is not ruled out, though these authors prefer the disk or isotropic orbital interpretations. However, much observational research is going on in this field at present, and none of these results should be regarded as final. The recent large study of regular, frequently spaced spectra of NGC 5548 should give much more definite information but the results for the velocity field were not available at the time of writing (Clavel *et al* 1991, Peterson *et al* 1991).

The BLR, close to the nucleus, is the most interesting but least well understood part of active galactic nuclei. The velocity field is particularly uncertain: taken to be outflow in most theories, demanding a confining medium which does not seem to be there, and with the observational evidence favouring chaotic or orbital motions, but still highly uncertain.

One very interesting idea, recently put forward or actually revivified, is that the 'clouds' in the BLR are actually the atmospheres of stars, held together by gravity rather than by an external medium. One version has them as red-giant stars, another version as main-sequence stars 'bloated' by the intense radiation field near the central source. In any case they are photoionized and heated by the central source; their surfaces are continuously flowing off as a wind and being ablated. The calculations, all quite recent and rather simplified, suggest that the observed emission-line spectrum and profiles can be produced by such models. One problem lies in replenishing the supply of stars, for their rate of mass loss is high (Voit and Shull 1988, Norman and Scoville 1988, Penston 1988, Kazanas 1989).

5.5. Possible magnetic fields

There is certainly ionized gas in AGNs, and almost certainly rotation. Under these circumstances, magnetic fields may naturally arise. The observed radio emission is synchrotron radiation, the result of the acceleration of relativistic electrons in a magnetic field. At the centre of our own Galaxy, long filaments of radio emitting plasma are observed, clearly associated with a poloidal type magnetic field. Thus very probably magnetic fields are prevalent in AGNs, and especially in the BLRs close to their centres. Some of their effects on the radiation and structure have been discussed (Rees

1984). Otherwise they have not been very much considered in discussions of the velocity fields, but it is clear that they may be important in some situations (Greyber 1989).

6. Continuity of AGN properties

6.1. QSOs and Seyfert 1 galaxies

Seyfert galaxies were first recognized as galaxies, then as galaxies with unusually luminous nuclei with strong, broad emission lines. Quasars were first identified as quasistellar radio sources with similar strong, broad emission lines. All the evidence since then has tended to indicate these are but two somewhat distance-dependent names for similar physical entities. Many of the most luminous radio galaxies are morphologically classified as N galaxies, meaning bright, near-quasi-stellar objects, with only a faint, barely detectable extended component or 'fuzz'. Many objects such as I Zw 1, II Zw 1, Mrk 876, and 3C 120 have appeared *both* as Seyfert galaxies and QSOs in various catalogues. This continuity of physical appearance has been particularly emphasized by Morgan and Dreiser (1983). In recent years, using high-quality digital detectors and subtraction techniques, various authors have been able to detect faint galaxy images about the AGNs that had previously been described as quasars or QSOs. Out to about $z \lesssim 0.5$, nearly every such object has yielded to these techniques. The emission-line spectra show no consistent differences. Furthermore, high-quality spectra show an underlying integrated stellar absorption-line spectrum in many objects, in addition to the strong featureless continuum of the central source. In many cases these spectra are earlier in spectral type than normal spirals, indicating the results of recent star formation in the active galaxies (see also section 7.5). Two very good reviews of this material, with many references to the original papers, are by Hutchings (1983) and Miller (1985).

As mentioned in section 2.4, the observed ratio of Hα luminosity to continuum luminosity, $L(\mathrm{H}\alpha)/L_C$ is constant over a wide range of luminosity from Seyfert 1 galaxies through QSOs, suggesting that the form of the spectrum and the amount of the luminosity absorbed (covering factor $\Omega/4\pi$) is also constant (Yee 1980, Shuder 1981). Attempts have been made to measure the covering factor ($\Omega/4\pi$ subtended at the source by the photoionized clouds) from the observed ratio of soft x-ray flux (which is absorbed in the photoionization processes) to the hard x-ray flux (which may pass through unabsorbed if the column density is not too large). These suggest that the covering factors are larger for low-luminosity objects than for higher-luminosity ones (Lawrence and Elvis 1982, Reichert *et al* 1985). However, more recent, higher-resolution measurements of more objects show that the x-ray situation is quite complicated. Nevertheless they tend to support this decrease in covering factor with luminosity, but with considerable scatter (Mushotzky 1988, Turner and Pounds 1989). Similarly, comparison of the equivalent widths of both Lα and C IV λ1549 in high-luminosity QSOs with those in Seyfert 1 galaxies, show a decrease (by about a factor two) in covering factor over this range of luminosity (Wu *et al* 1983). Again, the scatter is large (Wilkes 1986).

6.2. Quasars and radio galaxies

Quasars and broad-line radio galaxies appear to be related much as QSOs and Seyfert 1 galaxies are. The broad-line radio galaxies are almost invariably N, cD, D or E galaxies rather than spirals, and are rare among the E galaxies. There are few if any published surveys of D, cD or N galaxies that are *not* radio galaxies, but practically no radio-quiet Seyfert galaxies of these morphological types are known. Most probably the radio plasma that escapes from the nucleus along the jets in relatively gas-free cD, D and E galaxies is stopped by interaction with interstellar gas in Seyfert galaxies.

The outstanding spectral difference between the radio-quiet and radio-loud objects is that many Seyfert 1 galaxies and QSOs have strong, broad Fe II emission features in the optical region, while few quasars and broad-line radio galaxies do. This subject has been reviewed by Osterbrock (1985), who gives many references to the original papers. The ultraviolet Fe II multiplets arising from the same upper levels as the observed optical features are very strong in all Seyfert 1 galaxies (Veron-Cetty *et al* 1983). They are observed in the broad-line radio galaxies as well, indicating that Fe^+ *is* present in their nuclei, but in lower abundance or less highly excited (Wills *et al* 1985). The physical interpretation is not known at present. Models which schematically take into account increased heating by relativistic electrons seem to reproduce some but not all of the observed effects: they predict weakening of the Fe II optical multiplets but do not give the observed H I relative line strengths (Ferland and Mushotzky 1984, Cesar *et al* 1985).

6.3. Few high-luminosity Seyfert 2s

Although Seyfert 1 nuclei and QSOs form a continuous sequence, there are no, or only very few, Seyfert 2 nuclei with luminosities as bright as $M_B = -23$. Only one 'Seyfert 2 QSO' has been reported, E 0449-184, originally discovered as an x-ray object. Its redshift is $z = 0.338$ and its absolute magnitude $M_B = -23.2$ (assuming $H_0 = 50$ km s^{-1} Mpc^{-1} as the value of the Hubble constant). There is no sign of any broad emission lines in its spectrum, but it has strong, typical Seyfert 2 narrow emission lines, with FWHM ≈ 600 km s^{-1} (Stocke *et al* 1982). This is larger than average for a Seyfert 2, but fits well on the correlation (with considerable scatter) between L([O III]) and FWHM.

The fact that high-luminosity Seyfert 2 AGNs are so rare must be incorporated into any overall model of AGNs. On the traditional picture it means that a high-luminosity, strong continuum from the central source, and an extensive BLR always go together. On the 'hidden-BLR' picture, it means that the occulting torus becomes very weak or disappears around $M_B \approx -23$ (Miller and Goodrich 1990).

6.4. LINERs

Although QSOs and Seyfert galaxy nuclei make up most of the *known* AGNs, lower-luminosity ones certainly also exist. One is the nucleus of M 81, in which very weak, broad Hα emission was detected by Peimbert and Torres-Peimbert (1981) and confirmed by Shuder and Osterbrock (1981). Based on the correlation between broad Hα and x-ray luminosity, Elvis and Van Speybroeck (1982) predicted that it should be detectable as an x-ray source and did find it. Further x-ray data have shown that the nuclear source is relatively soft, with a power-law form spectrum ν^{-n} with $n \geq 2$. This is steeper than the characteristic value $n \approx 0.7$ for AGNs, but within the range of a few others. It can be very approximately fitted by an accretion-disk model corresponding to black-hole mass $M \leq 10^4$–10^5 $M_\odot$ and $\dot{M} \leq 10^{-4}$–10^{-3} $M_\odot$ y^{-1}, in which the x-ray emission arises in the disk itself (Fabbiano 1988). Long-slit spectra show narrow emission lines of [N I], [O III] and [Fe VII] (only a small range in wavelength was observed) in the NLR close to the nucleus, and give only an upper limit to the black-hole mass $M \lesssim 10^7 \, M_\odot$ (Keel 1989).

Many such 'Low-Ionization Nuclear Emission-Line Regions' or LINERs have been identified. They are nuclei with emission-line spectra showing a generally lower-level of ionization than Seyfert 2s, but with the characteristic relatively strong [O I], [S II] and [N II] lines of an AGN. Many LINERs have relatively weak emission lines, and to detect them and to measure their strengths with any precision it is necessary to subtract the underlying galaxy absorption-line spectrum fairly accurately.

When the class of LINERs was first isolated by Heckman (1980), it was suggested that they might be objects in which shock-wave heating is the main energy-input mechanism, rather than photoionization by a hard spectrum. At relatively low levels of ionization (below log [O III]/Hβ = +0.5) the diagnostic ratios of figures 2, 3 and 4 do not distinguish between these two possibilities. However, it now seems much more likely that LINERs are simply the extension of Seyfert 2 nuclei to lower luminosities, smaller ionization parameters, and somewhat larger exponents n in the representative power-law photoionizing spectrum. Figures 1, 2 and 3 extend smoothly down to LINERs, and several nuclei with ionization near the Seyfert 2-LINER 'boundary' log [O III]/H$\beta \approx$ +0.5 show weak He II $\lambda4686$ emission. It can only be understood as resulting from photoionization by a hard spectrum. The models calculated on this basis fit LINERs well (Ferland and Netzer 1983, Halpern and Steiner 1983, Keel 1983a, Stasinska 1984a, Binette 1985).

Very strong confirming evidence of photoionization is available from IUE observations of one LINER, NGC 4579. They show that there is a point-like ultraviolet source at the nucleus of this galaxy, with an observed flux which, with assumed power-law index $n = 1.4$, fits the measured x-ray flux and provides the required number of ionizing photons to power the observed Hβ flux (Goodrich and Keel 1986).

Many LINERs have been detected in spiral galaxies. For instance in a well defined sample, observed and with the continuum carefully subtracted to reveal the presence of very weak nuclear emission lines, *all* showed Hα and [N II]. Of these nuclei 5% were Seyferts, and of the remainder approximately 80% of the Sa and Sb nuclei showed LINER-type spectra. This percentage dropped steeply through Sc, to essentially zero at Scd, probably at least in part because of the increased strength of the H II region type spectra in the nuclei of these objects, resulting from recent star formation. Furthermore, approximately 10% of these LINERs in this survey show weak, *broad* Hα lines (FWOI $\gtrsim$ 3000 km s^{-1}). They might be classified as Seyfert 1.99 (a very weak, broad component of Hα barely detectable!) on the traditional scheme. They must be an extension of 'typical' (previously known) AGNs to smaller black-hole masses and lower accretion rates (Stauffer 1982, Keel 1983b).

In addition to the spirals, the nuclei of a few elliptical galaxies have LINER spectra. Two bright, well studied examples are NGC 1052 and NGC 4278. In NGC 1052 a weak, broad Hα emission component has been detected, again indicating an AGN similar in general type to those in Seyfert 1 galaxies (Filippenko and Sargent 1985). Many more similar cases have been found by careful analysis of high signal-to-noise-ratio spectra (Filippenko 1985, Filippenko 1989). Presumably there are 'mini-AGNs' with smaller masses and mass accretion rates than Seyfert 1s and QSOs. Probably the values previously cited for the nucleus of M 81 are roughly applicable. There are probably wide ranges in both mass and mass accretion rate, for the luminosity ($\dot{M}$) and FWOI (determined by M and R) are not well correlated and the objects do not form a one-parameter sequence in these observational parameters (Keel 1983a).

The lowest-luminosity AGN currently known is the nucleus of the late-type, nearby, dwarf spiral NGC 4395 (Filippenko and Sargent 1989). Its absolute magnitude is only $M_B \approx -10$ (in the continuum). Its narrow emission lines have FWHMs < 60 km s^{-1}, but are otherwise characteristic of a Seyfert 2 nucleus, with [O III] $\lambda 5007$/H$\beta \approx 7$, strong [O I] $\lambda 6300$ and [S II] $\lambda\lambda 6716$, 6731, and measurable He II $\lambda 4686$, [Ne V] $\lambda 3426$ and [Fe X] $\lambda 6375$. In addition Hα and Hβ have weak broad components with FWoI ~ 4000 and 7000 km s^{-1} respectively. Thus this nucleus fits in most ways the definition of a very low-luminosity Seyfert 1.8 nucleus with unusually narrow lines. The luminosity in the *broad* Hα emission component is L(Hα) $\approx 1.2 \times 10^{38}$ erg s^{-1}, down by about a factor 10 from M 81. This broad component would probably not be detectable in an earlier-type spiral galaxy with a correspondingly more luminous *stellar* nucleus.

Note however that *all* LINER-type spectra are *not* the results of photoionization. Morphologically, a certain number of LINERs are extended, nebulous objects in which the form strongly suggests that shock-wave heating is in progress. None of these show He II $\lambda 4686$. Clearly LINERs include objects of quite different physical nature; but those in the nuclei of galaxies are AGNs photoionized by a hard but weak spectrum (Heckman 1987, Filippenko 1989).

Finally we may briefly discuss the question as to whether our Galaxy has an AGN at its nucleus. The optical extinction to the centre ($A_V \approx 35$ magnitudes) is so high that there is no chance to observe the nucleus directly in the optical region. There is a compact non-thermal radio source, Sgr A*, at the centre, with loops and filaments of non-thermal radio emission apparently emerging from it. Far infrared and radio-frequency velocity measurements suggest but do not prove the presence of a black hole with mass $\approx 5 \times 10^6$ $M_\odot$. There may also be a small, highly variable x-ray source at the nucleus. However, its x-ray luminosity, about 2×10^{38} erg s^{-1} is smaller by a factor of approximately 100 than that of the nucleus of M 81. This is probably the best estimate or upper limit to the activity of a possible AGN in our Galaxy (Genzel and Townes 1987, Fabbiano 1988).

7. Fuelling

7.1. Introduction

One of the most interesting subjects in the study of AGNs is the question of the supply of mass that fuels them, thus 'feeding the monster' (Gunn 1979). For instance, a black hole with mass $10^{8.5}$ $M_\odot$, with luminosity 0.1 L_E, and if its efficiency of mass conversion is 0.1, consumes about 0.7 $M_\odot$ y^{-1}. If the mass supply is available and the two factors remain constant, its mass will increase exponentially on a time scale of 4.4×10^8 y. However, the mass supply must be available at the accretion disk, with a radius perhaps 100 times its Schwarzschild radius, that is about 5×10^{15} cm $\approx 10^{-3}$ pc. This is practically zero on the scale of the galaxy, and to fall that close to the centre, the mass must lose practically all of the angular momentum it had on the scale of the galaxy. Thus the problem of fuelling AGNs is the problem of how mass arrives at the accretion disk, close to the black hole, with low angular momentum. Many different mechanisms have been studied (Rees 1984). In recent years considerable observational evidence has tended to favour interactions between galaxies and mergers as the main fuelling mechanism. We will discuss this evidence, and theoretical calculations related to it in this section, after reviewing the observational data on the luminosity function of AGNs (relative numbers here and now), the presence of black holes in inactive galactic nuclei, and the statistics of AGNs in various environments.

7.2. Luminosity functions of AGNs

The luminosity function of Seyfert galaxies means the number per unit luminosity interval (or optical absolute magnitude interval) per unit volume. Even though Seyfert galaxies are quite luminous, the quantitative information on the luminosity function is limited to 'here and now', actually to about redshift $z \approx 0.03$–0.05. The luminosity function can be derived from a sample that is complete down to a given apparent magnitude, basically from a census of the objects. For each galaxy the redshift gives the distance and hence absolute magnitude, and the limiting magnitude of the sample then gives the volume out to which that galaxy could be observed in the sample. Its contribution to the luminosity function is one object in that volume; the sum of all the densities derived in this way is the luminosity function.

A good, more recent determination of the luminosity function by this method is by Meurs and Wilson (1984). Their result is that for absolute magnitude $M_B = -21$ and fainter, about 10% of all field galaxies are Markarian galaxies, and approximately 8% of Markarian galaxies between $M_B = -20$ and -21 are Seyferts. But at higher luminosities the fraction that are Seyferts increases rapidly. More luminous than about $M_B = -22.5$, nearly all field galaxies are Markarian galaxies and almost all of them are Seyfert galaxies! (These absolute magnitudes are based on an assumed Hubble constant $H_0 = 50$ km s^{-1} Mpc^{-1}; for any other value they can be scaled accordingly.) Integrated over their whole range of absolute magnitude, $-24 < M_B < -19$, Seyfert galaxies are about 1.1% of all galaxies in that interval. This luminosity function converges smoothly with the luminosity function of QSOs at brighter absolute magnitude.

Furthermore, the separate luminosity functions of Seyfert 1 and 2 galaxies can be found in the same way. The Meurs and Wilson (1984) work shows the Seyfert 1s to have a weak maximum at $M_B = -21$, and the Seyfert 2s at $M_B = -20$. Essentially *all* the high-luminosity objects are Seyfert 1s. According to these luminosity functions the numbers of Seyfert 1s and 2s *per unit volume* are the same, but *to a given apparent magnitude* the number of Seyfert 1s is about twice the number of Seyfert 2s.

However, as noted in section 2.8, the Markarian survey, from which all of Meurs and Wilson's (1984) sample was drawn, is seriously incomplete in Seyfert 2 galaxies. Hence the relative number of Seyfert 2s, and their luminosity function, are underestimated. Probably the best area for Seyfert galaxy statistics is the Wasilewski (1983) field, covering 825 square degrees at the North Galactic Pole, searched for emission-line galaxies down to $m_B \approx 17$, and is believed to be complete to $m_B = 15.7$. Osterbrock and Shaw (1988) obtained slit spectra of all the Seyfert-galaxy candidates, as well as many of the other emission-line galaxies, and found the *observed* numbers of Seyfert $1 + 1.5$ to Seyfert $1.8 + 1.9$ to Seyfert 2 to be 4; 2; and 9 respectively. Correcting for the higher luminosity of the Seyfert 1s, this corresponds to relative space densities (luminosity functions integrated over absolute magnitude) $0.12 : 0.10 : 0.78$, with large uncertainties because of the small number of objects. Probably all the Seyfert 1.5s and half the Seyfert 1.8s and 1.9s would be classified Seyfert 1, the rest of the Seyfert 1.8s and 1.9s as Seyfert 2s, if only these two types were used. Thus a large upward correction to the number density of Seyfert 2 galaxies from the Meurs and Wilson luminosity function is required, to approximately five times the number of Seyfert 1s. This ratio has been only slightly changed by the discovery of one more Seyfert 1 and one more Seyfert 2 in the Wasilewski field by Bothun *et al* (1989). A similar ratio, 5.3 ± 2.4, has been derived by Salzer (1989) from a complete (magnitude-limited) sample of emission-line galaxies in a larger field from the University of Michigan (UM) survey. Edelson (1987), in a complete magnitude-limited sample ($m_B \leq 14.5$) based on the CfA survey, found 25 Seyfert 1 galaxies and 23 Seyfert 2, corresponding to a ratio of space densities of Seyfert 2s to 1s of about 2.6, rather than 4.9 as in the Wasilewski field. In either case, since the Seyfert 1s are much more nearly complete in the Meurs and Wilson sample, the fraction of field galaxies brighter than $M_B = -19$ which are Seyfert 1 is about 1%, and Seyfert 2, about 3 to 5%.

In a very interesting recent study, Spinoglio and Malkan (1989) have used the 12 μm IRAS measurements to form a complete sample of AGNs. The flux at this wavelength, they argue, is least affected by dust absorption and emission, and thus is best correlated with the bolometric luminosity of the AGN. The resulting luminosity functions for Seyfert 1 and 2 galaxies agree well with those derived from the CfA survey. Approximately 20% of the galaxies in the flux-limited 12 μm sample contain active nuclei. Thus this appears to be an extremely efficient technique for discovering AGNs.

On any evolutionary picture in which Seyfert 1s evolve into Seyfert 2s, or *vice versa* (or both, which has been observed nearly to happen, on scales of a few months, in NGC 4151 by Penston and Perez (1984)) the relative fractions are proportional to the relative times spent in each form. If *all* galaxies went through the Seyfert stages, they would spend 1% of their (observable) lives as Seyfert 1s, say 4% as Seyfert 2s, and the rest as field galaxies. If only *some* galaxies did so, the fraction of their lifetimes they spent as 'field galaxies' would be smaller, and as Seyferts correspondingly larger.

On the other hand, if all Seyfert 2s had hidden BLRs, visible only from a cone centred about an axis, the half angle of this cone would be about 40°, its whole opening 80°, for a representative AGN.

More recently Cheng *et al* (1985) have discussed the luminosity function of Seyfert 1 (and 1.5) *nuclei*, which is more significant than that of the galaxies with the nuclei included. They tried to correct for the starlight of the galaxy itself, using available colour and galaxy surface-brightness information. The derived luminosity function agrees reasonably well with that of Meurs and Wilson, when the latter is adjusted to remove approximately the galaxy contribution to the total magnitude. The luminosity function for QSOs ($M_B < -23$) has been determined by Schmidt and Green (1985), based on a colour survey of the entire sky accessible from Palomar (the Palomar–Green or PG survey). The luminosity function for Seyfert 1 nuclei matches up quite well with it around absolute magnitude -23. For rough orientation, the luminosity function is about 10^{-6} AGNs Mpc^{-3} mag^{-1} at $M_B = -20.5$, 10^{-7} at $M_B = -22.0$, 10^{-8} at $M_B = -23.5$ and 10^{-9} at $M_B = -25.2$. For the most luminous absolute magnitude, the space density is very low, the volume surveyed and the light-travel time correspondingly large. Hence assumptions must be made about the evolution of AGNs. This is a fascinating subject, but not at all well determined physically. In fact, understanding QSOs physically so that their rate of formation and evolution can be calculated is the ultimate aim of AGN research.

7.3. Black holes in normal galaxies

To detect a black hole directly is difficult, because its only interaction with the outside world is gravitational. Among stars, the technique is to find a single-lined spectroscopic binary in which the dark companion has a mass greater than the upper limit to a neutron star, and a strong upper limit to its luminosity. In galaxies the procedure is to measure velocities close to the nucleus which require a central point-like mass, rather than the integrated mass density of the stars.

As stated in section 6.4, far-infrared and high radio-frequency measurements of gas emission-line velocities less than 1 pc from the centre of our Galaxy suggest the presence of a black hole of mass about $5 \times 10^6 \ M_\odot$. This interpretation is not certain, however, for it depends on assumptions which, though reasonable, are not unique as to the unobserved velocity components normal to the line of sight (Genzel and Townes 1987).

In other galaxies the technique is to measure the line widths and projected velocity field in the integrated stellar absorption-line spectrum very near the nucleus. Only the nearest galaxies can be investigated, and the angular resolution set by seeing is the chief observational limitation. Results to date include probable detection of a black hole of mass $\sim 5 \times 10^7 \ M_\odot$ in M 31, and $\sim 8 \times 10^6 \ M_\odot$ in M 32. The inner parts of the nuclei of both these nuclei show rapid rotation, and an increase in velocity dispersion at their centres. Again, the interpretation of the observed data is not certain, but the presence of black holes in both these inactive galactic nuclei seems quite likely (Dressler and Richstone 1988, Kormendy 1988).

7.4. Clusters, groups, and neighbour galaxies

It has long been known that the relative number of AGNs (with respect to 'normal' galaxies) is much smaller in rich clusters of galaxies than in the general field. A very good summary is given by Dressler *et al* (1985). From a relatively homogeneous body of spectra of galaxies in 14 rich clusters (average $z = 0.04$) and of field galaxies, they found the frequency of AGNs to be only approximately 1% in the cluster sample, compared with approximately 5% in the field galaxies. The clusters on the average are well known to have earlier-type (gas-poor) galaxies (typically E and SO) than the field and, as previously noted, Seyfert galaxies are more frequent among Sa and Sb spirals. However, this only partly accounts for the difference in frequency of AGNs between clusters of galaxies and the field; much of this difference must result from the differing environments. However, a relatively few clusters contain a higher frequency of AGNs, the best example is 3C 295 in which about 10% of the most luminous galaxies (top three magnitudes) are AGNs (Dressler and Gunn 1983).

These statistics strongly suggest that the gravitational interactions of galaxies in clusters, and their interaction with the intergalactic medium in clusters, does not deliver mass to the nuclei. Probably the intergalactic medium sweeps some interstellar gas out of the cluster galaxies, reducing the amount available as fuel (Gisler 1978). The large relative velocities of galaxies in clusters also reduce the strength of their interactions.

From counts of galaxies in the fields of Seyfert galaxies, Petrosian (1982) found that Seyferts do occur in looser clusters, in the same proportion as in the general field. However they tend to avoid the denser central regions of such clusters. In addition Seyfert 2s tend to occur more frequently in loose clusters than Seyfert 1s.

At the other extreme, it seems very well established that AGNs occur much more frequently in galaxies which have a nearby 'companion' galaxy. (The word in this context is not meant to imply that the 'neighbour' galaxy is in a long-term bound orbit, as in the case of stars, but that it is close enough to interact gravitationally. If this is the case, and the relative motion is slow enough, the orbit tends to decay in only a few periods and the two galaxies merge or interact strongly.) The tendency of Seyferts to occur in pairs or interacting galaxies has long been known (Adams 1977, Vorontsov-Velyaminov 1977). Dahari (1984) investigated this hypothesis quantitatively in a survey of a well-defined redshifted-limited sample of Seyfert galaxies. With a quantitative definition of a close companion in terms of projected distance and redshift, he found the fraction of Seyferts with companion galaxies to be about 15%, while the upper limit in a comparison sample of field galaxies was only 3% with companions. The overabundance of *very close* companions of *comparable* size to the Seyferts was even greater. Fuentes-Williams and Stocke (1988) did not confirm this result, finding only very marginal evidence that Seyferts have an excess of 'companions' of comparable size. The results of both studies are quite sensitive to the correction for background projected galaxies ('optical companions' in the terminology of double-star astronomy). Dahari used a procedure based on the *local* background density, while Fuentes-Williams and Stocke used *average* background corrections. Dahari's sample contained about twice as many Seyfert galaxies as Fuentes-Williams and Stocke's, and also has a smaller redshift limit, $z \leq 0.03$ compared with $z \leq 0.05$.

The most recent survey, by MacKenty (1989) based on a sample of 51 Seyfert galaxies with $z < 0.043$ and an approximately equal number of non-Seyfert Markarian galaxies, confirms Dahari's result. Seyferts are more likely than field galaxies to have close 'companions'. However, the non-Seyfert Markarian galaxies, basically starburst galaxies or galaxies in which star formation is now going on strongly, are equally likely as Seyferts to have companions. Many Seyfert galaxies also have star formation going on in them, others do not. Evidently there is a strong connection between these two processes, which strengthens the idea that fuelling by perturbations of interstellar gas, which is known to be important in star formation, is also involved in AGNs.

One reason for the discrepancy with the results of Fuentes-Williams and Stocke is that their sample of Seyfert galaxies is drawn from relatively early published lists of these objects, which, as explained in section 2.8, are relatively deficient in Seyfert 2s. Dahari's and MacKenty's samples are drawn from more recent catalogues, whose contents are more nearly representative of the total population of Seyfert galaxies. Dahari did not treat Seyfert 1s and 2s separately in his statistics, but MacKenty (1989) did, and found the Seyfert 2s more likely to have companions than Seyfert 1s. Despite the small sizes of the samples, the result appears to be significant. This same result, that Seyfert 2 galaxies are more likely to have companions than Seyfert 1s, was first found by Petrosian (1982). This, and their greater prevalence in loose clusters, are two of the few morphological differences known between these spectroscopically-defined classes, and thus one of the few counterarguments to the hypothesis that both are identical types of objects seen from different orientations. Thus the sample Fuentes-Williams and Stocke used, being deficient in Seyfert 2s, was also deficient in companions.

It has also long been known that many Seyfert AGNs are in interacting systems, galaxies in which the gravitational interaction with their companions is visible by their perturbed structure. Adams (1977) and Vorontsov-Velyaminov (1977) called attention to this observational result, and many quantitative statistical studies since then have confirmed it (Kennicutt and Keel 1984, Keel *et al* 1985, Dahari 1985, MacKenty 1990). In general, all these studies show that an excess of interacting galaxies have AGNs, but *strongly* interacting systems (such as the extreme Vorontsov-Velyaminov and Arp objects) tend *not* to have AGNs. These strongly interacting systems have strong star formation going on, as do many of the less strongly interacting ones.

Many morphological studies of low-redshift QSOs also show direct evidence of interaction in the morphology of the 'fuzz' or host galaxy in which they are found

(Hutchings 1983, Hutchings *et al* 1984).

Finally, several so-called 'multiple-nuclei' Seyfert galaxies are known, which appear to be close collisions in progress. Some of them may be actual mergers. Some of them show apparent tidal tails. Other objects, in which only an apparently single nucleus is seen, have strong tidal tails and apparently contain two nuclei too close to be resolved. The bright, peculiar Seyfert 1 galaxy Mrk 231, appears spectroscopically and morphologically to be a recently merged system. These and other observational evidence for the interaction interpretation of AGNs are well discussed by Fricke and Kollatschny (1989).

7.5. Interactions and mergers

The observational data of the last section strongly suggest that interactions of galaxies are an important mechanism for providing fuel to the central accretion disk in AGNs. Any interaction will introduce a non-axisymmetric perturbation. These perturbations, by breaking down conservation of angular momentum for individual orbits, can allow interstellar gas either originally present in the galaxy or acquired in the interaction, to fall inward to distances of order 1 kpc from the nucleus (Norman and Silk 1983). The fact that most AGNs are in spirals, gas-rich objects, suggests that interstellar gas does provide the fuel. As gas accumulates at this distance scale, presumably in the form of molecular clouds, instabilities in this self-gravitating disk, with cloud–cloud (dissipative) collisions can lead to an instability that allows the gas to fall inward on a relatively short time scale. Numerical examples worked out by Lin *et al* (1988) suggest time scales of order several $\times 10^8$ y for black holes of mass 10^7–10^8 $M_\odot$. Although these authors consider that this instability can take the gas all the way in to distances of order 1 pc, a more attractive possibility has been put forward by Hernquist (1989a). It is that on the scale of 100 pc the increased density in the gas can trigger a burst of star formation. Observationally, such starbursts are seen in many AGNs, as previously noted, as has been particularly emphasized by Weedman (1983). From this stage the remaining gas will contract to the centre more rapidly as it becomes fully self-gravitating. At a scale of order 1 pc, if a black hole is present viscosity will take over and get the gas into the accretion disk. If not, possibly a dense star cluster will form, and from it a black hole.

An overall hierarchical set of processes, with several alternative paths from the scale of the galaxy down to the central accretion disk has thus been suggested by Hernquist (1989a). It is shown in figure 8, which can be consulted for a fuller explanation and estimates of many of the relevant time scales. Many details remain to be worked out, but the general ideas that interactions are involved, that several stages, and several alternate processes within each of them are involved, and that star formation is often an important aspect of the AGN phenomenon, are all in good accord with the available observational results. Another discussion emphasizing the importance of bars and non-axisymmetric perturbations is the review by Shlosman *et al* (1990).

Following up an earlier morphological study by Simkin *et al* (1980), MacKenty (1990) confirmed the widespread prevalence in the Seyfert class of disturbed, interacting or peculiar galaxies. Many of the remainder have bars or rings. In addition, approximately 20 to 25% of the sample he studied had amorphous (or 'difficult to classify') structures. Their colours do not agree with those of elliptical galaxies, and they can most simply be interpreted as the remnants of interactions or interactions in the recent past.

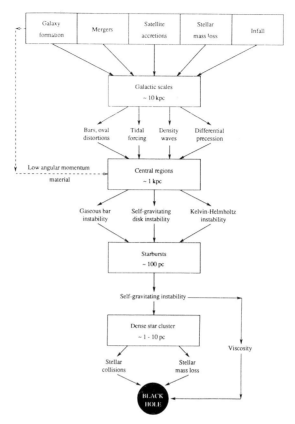

Figure 8. Simple hierarchical model for the origin and fuelling of activity in galactic nuclei. High-angular momentum gas deposited on galactic scales (top of figure) is driven to the central ~ 1 kpc by intermediate processes. Low-angular momentum material settles there directly. Once sufficient gas accumulates there, instabilities can drive it to the inner ~ 100 pc, triggering a starburst. Continued collapse, as the gas becomes fully self-gravitating, may then form and can fuel a central black hole. (Hernquist 1989)

Galaxy interactions are complicated to model, and progress in understanding them can only be made with large computers and sophisticated programs. The early calculations of Toomre and Toomre (1972) showed that strongly perturbed systems, similar to those observed, are indeed expected to arise in gravitational interactions between galaxies. The idea that such interactions can deliver mass to the nucleus was made in this paper, in the context of stars rather than gas. Detailed n-body calculations, aimed at interpreting observational data on the statistics of Seyferts with companions, have been made by Byrd *et al* (1986, 1987). These calculations reproduce semi-quantitatively the association of the strength of the tidal perturbation $(M_{\mathrm{p}}/M_{\mathrm{T}})/(p/a)^{3}$, where M_{p} is the mass of the perturbing galaxy, M_{T} the disk plus halo mass of the perturbed galaxy, p the perigalactic distance of the perturber and *a* the galaxy disk radius, with the presence or absence of nuclear activity. By taking account of projection effects, and possible 'companion' galaxies too faint to have been observed, they show that tidal interactions might account for nearly *all* AGNs. All these calculations are based on an *n*-body code in which only gravitational interactions are taken into account.

The most recent calculations of Hernquist (1989a, 1989b) treat stars and gas separately, using a hybrid *n*-body-hydrodynamics code. As these calculations are continued, and made even more sophisticated, it will be possible to test the overall picture described here more thoroughly.

8. Evolution

8.1. Fuelling black holes

According to the ideas of the previous sections, an AGN requires a black hole plus fuel delivered at it. Some galaxies, such as M 31, M 32 and our Galaxy, have small black holes but are not active. There may be inactive galaxies with more massive black holes; they would be difficult to detect at larger distances. It seems likely that the black hole initially formed in an AGN episode, but this is not necessary. Once a black hole exhausts its fuel, it becomes quiet. However the *n*-body calculations of Byrd *et al* (1986, 1987) suggest that this may occur in an episodic manner, with 'latency times' of a few times 10^8 y before the fuelling begins, and between episodes of fuelling. This behaviour is related to the free-fall time and gravitational instability in the disk. An existing black hole can be reactivated by further interactions which refuel it. The evolution of the mass of an individual black hole must be to increase (if it is fuelled) from Seyfert galaxy to QSO, according to the derived masses of section 2. This complex type of luminosity evolution of an individual AGN is quite different from the smoothly decreasing (with cosmic time) luminosity evolution generally assumed in fitting QSO magnitude and redshift counts, as for instance by Schmidt and Green (1985). From the scanty observational evidence, it is certainly conceivable that a large fraction of 'normal' galaxies contain black holes ready and available for refuelling.

8.2. Formation rate

From the interaction picture, the rate of formation of AGNs increases rapidly with the number density of galaxies. Thus the observed increase in number density of luminous QSOs with redshift z can be qualitatively understood. Simplified calculations on this basis by De Robertis (1985) show that the density-evolution rates required by the counts can be approximately matched, with a maximum number density (in comoving coordinates) around $z = 2$–3. Further, as the expansion continues, if the more recently formed black holes have smaller mean masses, the mean luminosity of the QSOs will also decrease. This can mimic the luminosity evolution required by the counts, by a proper choice of assumed conditions. However, there is such a wide range of possible parameters, and the details of the interaction process are so simplified, that it is difficult to evaluate the significance of this result.

8.3. QSOs in formation

In the IRAS survey, a significant number of galaxies were discovered with the bulk of their radiation in the infrared spectral region. A significant number of them, 10 of the 324 galaxies in the IRAS bright galaxy survey, have luminosity $L > 10^{12}\ L_\odot$, that is in the QSO range. Their mean redshift is $z = 0.055$ and the maximum is $z = 0.08$. Two of them are previously known Seyfert galaxies, Mrk 231, a Seyfert 1, and Mrk 273, a Seyfert 2. Optical spectra show that one other is a Seyfert 1.5, another a Seyfert 1.8, and the rest, except for one with an H II region spectrum, are Seyfert 2s. All but one of them have extended optical images, and seven of the 10 appear to be 'peculiar', being either mergers (as Mrk 231 was previously described) or distorted systems. Seven of them have been observed in the mm-wavelength region for the CO $(1 \rightarrow 0)$ emission line, and all of them rank among the most luminous CO sources observed. Sanders *et al* (1988b) identify these ultraluminous infrared galaxies as QSOs in the process of formation. They are still highly distorted, and shrouded with dust, which is apparent not only from the CO but from the strong reddening of their emission-line spectra. As the dust is blown away by the AGN radiation, supernovae, and stellar winds, Sanders *et al* (1988b) hypothesize, and the nucleus becomes more directly apparent, there will be a marked shift in the spectral energy distribution to shorter wavelengths and the characteristic AGN features will appear. The number of bright QSOs in the same redshift range with $L > 10^{12}\ L_\odot$ is five, comparable with the number of ultraluminous infrared galaxies. Thus it seems quite likely that these are indeed the first stage in the QSO process.

9. Miscellaneous

9.1. Molecular gas

Seyfert galaxies are rich in interstellar dust, and as expected from this, also contain large amounts of molecular gas. CO, one of the most abundant molecules in molecular clouds in our Galaxy, and the most readily detectable, has in fact been observed in many Seyferts (Bieging *et al* 1981, Heckman *et al* 1989). The observational data seem to show that the Seyfert 2 galaxies have larger molecular gas contents than the Seyfert 1s. This is one of the few remaining observational differences which appear to conflict with the idea that Seyfert 1 and 2 galaxies are essentially similar systems seen from different orientations. To date no broad-line CO has been detected; the line profiles of the observed CO are most nearly similar to those of the H I $\lambda = 21$ cm lines in the same object. This suggests that the molecular gas is associated more with the disk of the galaxy than with the NLR (Heckman *et al* 1989). Molecular H_2 has only forbidden emission lines in the infrared, and is much more difficult to detect. However, three low-excitation lines of its $(1, 0)$ vibrational band have been measured in NGC 1068, with relatively narrow widths (Oliva and Moorwood 1990). There is no reason to suppose that the abundance ratio CO/H_2 is appreciably different from that in our Galaxy.

Very recently high-resolution interferometric measurements have been made of CO emission in two Seyfert galaxies, NGC 3227 and NGC 7469. At $6''$ resolution the latter shows a concentration of molecular gas with a mass $\sim 4 \times 10^9\ M_\odot$ and diameter ~ 1.5 kpc, centred on the nucleus (Sanders *et al* 1988a). The higher resolution (2–$3''$) observations show that in both these galaxies a large fraction of the molecular gas is concentrated within a few hundred parsecs of the nucleus. In NGC 3227 the central structure is partly resolved, and suggests that a significant amount of the CO lies in a roughly toroidal region, centred on the nucleus. Its velocity field includes a rotational component, with apparently other complications as well. In NGC 7469 the molecular gas distribution is more complicated, but seems to have a similar overall pattern. On the other hand in the third Seyfert galaxy of which interferometer measurements have been reported to date, NGC 5033, the CO emission is more diffuse, spread over several kiloparsecs (Meixner *et al* 1990). Clearly, high resolution maps of the CO emission in more Seyfert galaxies, compared with atomic-line and dust continuum maps, will be very useful for understanding the structure of AGNs and their surroundings.

9.2. Warmers

As stated in section 2.3, all the observational evidence agrees that in AGNs photoionization by hard photons from the central source is the dominant mechanism of energy input to the ionized gas. These high-energy photons are generally considered to be generated by non-thermal processes such as relativistic synchrotron radiation and Compton scattering, or in the inner parts of the accretion disk itself, as described in section 3.3. However, an alternate view is that in Seyfert 2 galaxies and LINERs, the photoionizing radiation may come from stars, rather than from non-thermal sources. The most recent form of this idea is the suggestion by Terlevich and Melnick (1985) that the ionization is produced by high-energy photons from postulated 'extremely hot and luminous stars', which they have named 'Warmers'. They are supposed to be 'bare core Wolf–Rayet stars,' the descendents of massive progenitors ($M > 60 \ M_\odot$), with temperatures up to 2×10^5 K. Observational evidence that any such stars exist is scanty indeed, and theoretical justification that such high effective temperatures actually occur is very weak. However, if sufficiently many stars of arbitrarily high temperature are postulated, clearly any type of power-law, broken power-law or more complicated form of photoionizing spectrum can clearly be matched by some population of such objects. Of course many starburst galaxies are known to exist, with emission-line spectra like H II regions, photoionized by OB stars. They are the objects marked by circles and asterisks in figures 2, 3 and 4. The AGNs, to be understood on this basis, must have much more luminous and hotter stars, and many more of them.

The Warmer picture of AGNs has been discussed and defended by Terlevich *et al* (1987). They have attempted to explain the x-ray and radio-frequency properties of AGNs, and some of the emission-line broadening, to supernovae produced from the massive, rapidly evolving stars. However, the rapid optical and especially x-ray variations in many AGNs are extremely difficult to explain by this picture. Likewise jets, polarization of the optical continuum on Seyfert 2 galaxies and the apparent dichotomies of figures 2, 3 and 4 are not, at present, understood on this basis. Neither is the apparent continuity in the physical properties of LINERs, Seyfert 2 and Seyfert 1 nuclei explained by it.

10. Conclusion

Although we do not fully understand AGNs in the way in which we understand stars, many of the main ideas of their nature seem to be coming into focus. Perhaps our knowledge of AGNs in 1990 is roughly equivalent to our predecessors' knowledge of the nature of stars in 1935 or 1940. They had been known since antiquity, and had been observed astrophysically with telescopes since the latter part of the last century; Seyfert galaxies were 'discovered' in 1943 and QSOs twenty years later. Our rate of learning, though not as fast as we might wish, is not as slow as we might fear either.

The general ideas of photoionization by a hard spectrum, powered by an accretion disk about a black hole and by non-thermal radiation generated in and near it seem well established. We have many clues and insights into the structure and velocity field, but no complete picture from which detailed predictions can be calculated from first principles. Dust and star formation are clearly important parts of the AGN process, but we do not understand the differences between the situations in which they do and do not lead to it. Much observational evidence suggests that interactions between galaxies can lead to the formation of black holes, fuel them, and refuel existing black holes. But we cannot calculate the evolution of a massive black hole and its associated AGN as we can calculate the evolution of a star.

The Hubble space telescope, if it performs up to its specifications, will surely increase our understanding of AGNs. So will large ground-based telescopes, optimized for seeing. Significantly better angular resolution should reveal much of the structure of the NLR. The new x-ray observatory ROSAT and the planned Gamma Ray Observatory should make possible much better measurements of the high-energy spectra of AGNs. Infrared imaging cameras and millimetre-wave interferometers will enable us to make much more complete studies of the dust and molecular gas in these objects. Supercomputers will provide more detailed simulations of many more cases of interactions between galaxies, and the perturbations of the velocity fields of the gas within them. I expect rapid progress in understanding the physical nature of AGNs in the coming decade.

Acknowledgments

This review was written during my sabbatical leave at the Institute for Advanced Study in 1989–90. I am most grateful to the Institute, its director, Marvin L. Goldberger and its professor in astronomy, John N. Bahcall, for their hospitality and support. I am also grateful to my home institution, the University of California, Santa Cruz, for its continued support of my research, and also to the National Science Foundation for its support, most recently under grant AST 86-11457.

References

Adams T F 1977 *Astrophys. J. Suppl.* **33** 19
Aldrovandi S M V and Contini M 1984 *Astron. Astrophys.* **140** 368
—— 1985 *Astron. Astrophys.* **149** 109
Almog Y and Netzer H 1989 *Mon. Not. R. Astron. Soc.* **238** 57
Antonucci R R J 1983 *Nature* **303** 158
Antonucci R R J, Kinney A L and Ford H C 1989 *Astrophys. J.* **342** 64
Antonucci R R J and Miller J S 1985 *Astrophys. J.* **297** 621
Baade W and Minkowski R 1954 *Astrophys. J.* **119** 206
Baldwin J A, Phillips M M and Terlevich R 1981 *Publ. Astron. Soc. Pac.* **93** 5
Baldwin J A, Wilson A S and Whittle M 1987 *Astrophys. J.* **319** 84
Band D L 1987 *Astrophys. J.* **321** 80
Band D L and Malkan M A 1989 *Astrophys. J.* **345** 122
Barr P and Mushotzky R F 1986 *Nature* **320** 421

Barvainis R 1987 *Astrophys. J.* **320** 537

Begelman M C 1985 *Astrophysics of Active Galaxies and Quasistellar Objects* ed J S Miller (Mill Valley: University Science Books) p 411

Begelman M C, Blandford R D and Rees M J 1984 *Rev. Mod. Phys.* **56** 255

Bieging J H, Blitz L, Lada C J and Stark A A 1981 *Astrophys. J.* **247** 443

Binette L 1985 *Astron. Astrophys.* **143** 334

Binette L, Robinson A and Courvoisier T J L 1988 *Astron. Astrophys.* **194** 65

Bothun G D, Halpern J P, Lonsdale C J, Impey C and Schmitz M 1989 *Astrophys. J. Suppl.* **70** 271

Bregman J N, Glassgold A E, Huggins P J and Kinney A L 1986 *Astrophys. J.* **301** 698

Bridle A H and Perley R A 1984 *Ann. Rev. Astron. Astrophys.* **22** 184

Brocklehurst M 1972 *Mon. Not. R. Astron. Soc.* **157** 211

Byrd G G, Sundelius B and Valtonen M 1987 *Astron. Astrophys.* **171** 16

Byrd G G, Valtonen M J, Sundelius B and Valtaoja L 1986 *Astron. Astrophys.* **166** 75

Caganoff S 1989 Optical emission lines in radio sources of intermediate power *PhD thesis* Australian National University

Carleton N P, Elvis M, Fabbiano G, Willner S P, Lawrence A and Ward M 1987 *Astrophys. J.* **318** 595

Cecil G, Bland J and Tully R B 1990 *Astrophys. J.* **355** 70

Cesar M L, Aldrovandi S M V and Gruenwald R B 1985 *Publ. Astron. Soc. Pac.* **97** 850

Cheng F-Z, Danese L, De Zotti G and Franceschini A 1985 *Mon. Not. R. Astron. Soc.* **212** 857

Clavel J *et al* 1991 *Astrophys. J.* **366** 64

Clavel J and Joly M 1984 *Astron. Astrophys.* **131** 87

Collin-Souffrin S and Dumont S 1986 *Astron. Astrophys.* **166** 13

Collin-Souffrin S, Dumont S, Joly M and Pequignot D 1986 *Astron. Astrophys.* **166** 27

Collin-Souffrin S, Dumont D and Tully J 1982 *Astron. Astrophys.* **106** 362

Collin-Souffrin S, Hameury J M and Joly M 1988 *Preprint*

Contini M and Aldrovandi S M V 1983 *Astron. Astrophys.* **127** 15

Czerny B and Elvis M 1987 *Astrophys. J.* **321** 305

Dahari O 1984 *Astron. J.* **89** 966

—— 1985 *Astrophys. J. Suppl.* **57** 643

Dahari O and De Robertis M M 1988a *Astrophys. J. Suppl.* **67** 249

—— 1988b *Astrophys. J.* **331** 727

De Robertis M M 1985 *Astron. J.* **90** 998

De Robertis M M and Shaw R A 1990 *Astrophys. J.* **348** 421

Dressler A and Gunn J E 1983 *Astrophys. J.* **270** 7

Dressler A and Richstone D O 1988 *Astrophys. J.* **324** 701

Dressler A, Thompson I A and Schectman G A 1985 *Astrophys. J.* **288** 481

Edelson R A 1987 *Astrophys. J.* **313** 651

Edelson R and Malkan M A 1986 *Astrophys. J.* **308** 59

Edelson R A, Malkan M A and Rieke G H 1987 *Astrophys. J.* **321** 233

Elvis M, Green R F, Bechtold J, Schmidt M, Neugebauer G, Soifer B T, Mathews K and Fabbiano G 1986 *Astrophys. J.* **310** 291

Elvis M and Van Speybroeck L 1982 *Astrophys. J. Lett.* **257** L51

Fabbiano G 1988 *Astrophys. J.* **325** 544

Fabian A C, Guilbert P, Arnaud K, Shafer R, Tenant A and Ward M 1986 *Mon. Not. R. Astron. Soc.* **280** 574

Fath E A 1909 *Lick Obs. Bull.* **5** 71

Ferland G J and Mushotzky R F 1984 *Astrophys. J.* **286** 42

Ferland G J and Netzer H 1983 *Astrophys. J.* **264** 105

Ferland G J and Osterbrock D E 1986 *Astrophys. J.* **300** 658

—— 1987 *Astrophys. J.* **318** 145

Ferland G J and Persson S E 1989 *Astrophys. J.* **347** 656

Filippenko A V 1985 *Astrophys. J.* **289** 475

—— 1989 *Active Galactic Nuclei* ed D E Osterbrock and J S Miller (Dordrecht: Kluwer) p 495

Filippenko A V and Sargent W L W 1985 *Astrophys. J. Suppl.* **57** 503

—— 1989 *Astrophys. J. Lett.* **342** L11

Fricke K J and Kollatschny W 1989 *Active Galactic Nuclei* ed D E Osterbrock and J S Miller (Dordrecht: Kluwer) p 425

Fuentes-Williams T and Stocke J T 1988 *Astron. J.* **96** 1235

Gaskell C M 1988 *Astrophys. J.* **325** 114

Gaskell C M and Ferland G J 1984 *Publ. Astron. Soc. Pac.* **96** 393

Gaskell C M and Sparke L S 1986 *Astrophys. J.* **305** 175

Genzel R and Townes C H 1987 *Ann. Rev. Astron. Astrophys.* **25** 377

Gisler G R 1978 *Mon. Not. R. Astron. Soc.* **183** 633

Giuricin G, Mardirossian F, Mezzetti M and Bertotti G 1990 *Astrophys. J. Suppl.* **72** 551

Gondhalekar P M 1990a, private communication

—— 1990b *Mon. Not. R. Astron. Soc.* **243** 443

Goodrich R W and Keel W C 1986 *Astrophys. J.* **305** 148

Green R F, Pier J R, Schmidt M, Estabrook F B, Lane A L and Wahlquist H D 1980 *Astrohys. J.* **239** 483

Greenstein J L and Matthews T A 1963 *Nature* **197** 1041

Greyber H D 1989 *The Center of the Galaxy* ed M Morris (Dordrecht: Kluwer) p 335

Gunn J E 1979 *Active Galactic Nuclei* ed C Hazard and S Mitton (Cambridge: Cambridge University Press) p 213

Halpern J P and Steiner J E 1983 *Astrophys. J. Lett.* **269** L37

Haniff C A, Wilson A S and Ward M J 1988 *Astrophys. J.* **334** 104

Heckman T M 1980 *Astron. Astrophys.* **87** 152

—— 1987 *Observational Evidence of Activity in Galaxies* ed E Ye Khachikian, K J Fricke and J Melnick (Dordrecht: Reidel) p 421

Heckman T M, Blitz L, Wilson A S, Armus L and Miley G K 1989 *Astrophys. J.* **342** 735

Hernquist L 1989a *Ann. NY Acad. Sci.* **571** 190

—— 1989b *Nature* **340** 687

Hubble E 1926 *Astrophys. J.* **54** 369

Huchra J P, Wyatt N F and Davis M 1982 *Astron. J.* **87** 1628

Hummer D G and Storey P J 1987 *Mon. Not. R. Astron. Soc.* **224** 801

Hutchings J B 1983 *Publ. Astron. Soc. Pac.* **95** 799

Hutchings J B, Crampton D and Campbell B 1984 *Astrophys. J.* **280** 41

Kazanas D 1989 *Astrophys. J.* **347** 74

Keel W C 1980 *Astron. J.* **85** 198

Keel W C 1983a *Astrophys. J.* **269** 466

—— 1983b *Astrophys. J. Suppl.* **52** 229

—— 1989 *Astron. J.* **98** 195

Keel W C, de Grijp M H K and Miley G K 1988 *Astron. Astrophys.* **203** 250

Keel W C, Kennicutt R C, Hummel E and van der Hulst J M 1985 *Astron. J.* **90** 708

Kennicutt R C and Keel W C 1984 *Astrophys. J. Lett.* **279** L5

Kinney A L, Blades J C, Bohlin R C and York D G 1991 *Astrophys. J. Suppl.* **75** in press

Koratkar A P and Gaskell C M 1989 *Astrophys. J.* **345** 637

—— 1991 *Astrophys. J. Suppl.* **75** in press

Kormendy J 1988 *Astrophys. J.* **325** 128

Koyama K 1989 *Active Galactic Nuclei* ed D E Osterbrock and J S Miller (Dordrecht: Kluwer) p 167

Krolik J H and Begelman M C 1988 *Astrophys. J.* **329** 702

Krolik J H and Kallman T R 1987 *Astrophys. J. Lett* **320** L5

Krolik J, McKee, C and Tarter C B 1981 *Astrophys. J.* **249** 422

Kruper J S, Urry C M and Canizares C R 1990 *Astrophys. J. Suppl.* **74** 347

Kwan J 1984 *Astrophys. J.* **283** 70

Kwan J and Krolik J H 1981 *Astrophys. J.* **250** 478

Laor A and Netzer H 1989 *Mon. Not. R. Astron. Soc.* **238** 897

Lawrence A and Elvis M 1982 *Astrophys. J.* **256** 410

Lin D N C, Pringle J E and Rees M J 1988 *Astrophys. J.* **328** 103

Lipovetsky V A, Markarian B E and Stepanian J 1987 *Observational Evidence of Activity in Galaxies* ed E Ye Khachikian, K J Fricke and J Melnick (Dordrecht: D. Reidel) p 17

MacAlpine G M 1985 *Astrophysics of Active Galaxies and Quasi-Stellar Objects* ed J S Miller (Mill Valley: University Science Books) p 259

MacAlpine G M and Williams G A 1981 *Astrophys. J.* **45** 113

MacKenty J W 1989 *Astrophys. J.* **343** 125

—— 1990 *Astrophys. J. Suppl.* **72** 231

Malkan M A 1983 *Astrophys. J.* **268** 582
Malkan M A and Filippenko A V 1983 *Astrophys. J.* **275** 477
Maoz D, Netzer H, Leibowitz E, Brosch N, Laor A, Mendelson H, Beck S, Mazeh T and Almoznino
E 1990 *Astrophys. J.* **351** 75
Maoz D, Netzer H, Mazeh T, Beck S, Almoznino E, Leibowitz E, Brosch N, Mendelson H and Laor
A 1991 *Astrophys. J.* **367** 493
Markarian B E, Stepanian J A and Erastova L K 1987 *Observational Evidence of Activity in Galaxies*
ed E Ye Khachikian, K J Fricke and J Melnick (Dordrecht: Reidel) p 25
Mathews W G and Doane J S 1989 *Astrophys. J.* in press
Mathews W G and Ferland G A 1987 *Astrophys. J.* **323** 456
McAlary C W and Rieke G H 1988 *Astrophys. J.* **333** 1
Meixner M, Puchalsky R, Blitz L, Wright M and Heckman T 1990 *Astrophys. J.* **354** 158
Meurs E J A and Wilson A S 1984 *Astron. Astrophys.* **136** 206
Miley G 1980 *Ann. Rev. Astron. Astrophys.* **18** 165
Miley G K, Neugebauer G and Soifer B T 1985 *Astrophys. J. Lett.* **293** L11
Miller J S 1985 *Astrophysics of Active Galaxies and Quasi-Stellar Objects* ed J S Miller (Mill Valley:
University Science Books) p 387
Miller J S 1988 *Active Galactic Nuclei* ed H R Miller and P J Wiita (Berlin: Springer) p 112
Miller J S and Goodrich R W 1990 *Astrophys. J.* **355** in press
Morgan W W and Dreiser R D 1983 *Astrophys. J.* **269** 438
Mushotzky R F 1984 *Advances Space Research* **3** 157
—— 1988 *Active Galactic Nuclei* ed H R Miller and P J Wiita (Berlin: Springer) p 239
Mushotzky R F and Ferland G J 1984 *Astrophys. J.* **278** 558
Netzer H 1989 *Mon. Not. R. Astron. Soc.* **225** 55
Netzer H, Maoz O, Laor A, Mendelson H, Brosch N, Leibowitz E, Almoznino E, Beck S and Mazeh
T 1990 *Astrophys. J.* in press
Netzer H and Wills B J 1983 *Astrophys. J.* **275** 445
Norman C A and Scoville N 1988 *Astrophys. J.* **332** 124
Norman C and Silk J 1983 *Astrophys. J.* **266** 502
Nussbaumer H and Schild H 1979 *Astron. Astrophys.* **75** L17
O'Dell S L, Scott H A and Stein W A 1987 *Astrophys. J.* **313** 164
Oliva E and Moorwood A F M 1990 *Astrophys. J. Lett.* **348** L5
Osterbrock D E 1978a *Proc. Natl Acad. Sci. USA* **75** 540
—— 1978b *Astronomical Papers Dedicated to Bengt Strömgren* ed A Reiz and T Anderson (Copen-
hagen: University Observatory) p 299
—— 1982 *Extragalactic Radio Sources* ed D S Heeschen and C M Wade (Dordrecht: Reidel) p 369
—— 1984 *Quart. J. R. Astron. Soc.* **25** 1
—— 1985 *Astrophysics of Active Galaxies and Quasi-Stellar Objects* ed J S Miller (Mill Valley:
University Science Books) p 111
—— 1989 *Astrophysics of Gaseous Nebulae and Active Galactic Nuclei* (Mill Valley: University
Science Books)
Osterbrock D E and Mathews W G 1986 *Ann. Rev. Astron. Astrophys.* **24** 171
Osterbrock D E and Shaw R A 1988 *Astrophys. J.* **327** 89
Osterbrock D E and Shuder J M 1982 *Astrophys. J. Suppl.* **49** 149
Padovani P, Burg R and Edelson R A 1990 *Astrophys. J.* **353** 438
Peimbert M and Torres-Peimbert S 1981 *Astrophys. J.* **245** 845
Penston M V 1988 *Mon. Not. R. Astron. Soc.* **233** 601
Penston M V and Perez E 1984 *Mon. Not. R. Astron. Soc.* **211** 33P
Perez E, Penston M V and Moles M 1989a *Mon. Not. R. Astron. Soc.* **239** 55
—— 1989b *Mon. Not. R. Astron. Soc.* **239** 75
Persson S E 1988 *Astrophys. J.* **330** 751
Persson S E and McGregor P J 1985 *Astrophys. J.* **290** 125
Peterson B M 1988 *Publ. Astron. Soc. Pac.* **100** 18
Peterson B M *et al* 1991 *Astrophys. J.* **368** 119
Petrosian A R 1982 *Astrofizika* **18** 548 (English translation *Astrophysics* **18** 312)
Phillips M M, Charles P A and Baldwin J A 1983 *Astrophys. J.* **266** 485
Pogge R W 1988a The circumnuclear environment of nearby non-interacting Seyfert galaxies *Thesis*
University of California, Santa Cruz

—— 1988b *Astrophys. J.* **328** 519

—— 1989a *Astron. J.* **98** 124

—— 1989b *Astrophys. J.* **345** 730

—— 1989c *Extranuclear Activity in Galaxies* ed E J A Meurs and R A E Fosbury (ESO Conference and Workshop Proceedings no 32) p 411

Ramaty R and Lingenfelter R E 1982 *Ann. Rev. Nucl. Part. Phys.* **32** 235

Rawlings S, Saunders R, Eales S A and Mackay C D 1989 *Mon. Not. R. Astron. Soc.* **240** 701

Rees M J 1977 *Quart. J. R. Astron. Soc.* **18** 429

—— 1978 *Observatory* **98** 210

—— 1984 *Ann. Rev. Astron. Astrophys.* **22** 471

—— 1987 *Mon. Not. R. Astron. Soc.* **228** 47P

Rees M J, Netzer H and Ferland G J 1989 *Astrophys. J.* **347** 640

Reichert G A, Mushotzky R F, Petre R and Holt S S 1985 *Astrophys. J.* **296** 69

Rieke G H and Lebovsky M J 1979 *Ann. Rev. Astron. Astrophys.* **17** 477

Saikia D J and Salter C J 1988 *Ann. Rev. Astron. Astrophys.* **26** 93

Salzer J J 1989 *Astrophys. J.* **347** 152

Salzer J J and MacAlpine G M 1988 *Astron. J.* **96** 1192

Sanders D B, Phinney E S, Neugebauer G, Soifer B T and Matthews K 1989 *Astrophys. J.* **347** 29

Sanders D B, Scoville N Z, Sargent A I and Soifer B T 1988a *Astrophys. J.* **324** L55

Sanders D B, Soifer B T, Elias J H, Madore B F, Matthews K, Neugebauer G and Scoville N C 1988b *Astrophys. J.* **325** 74

Schmidt M 1963 *Nature* **197** 1040

Schmidt M and Green R F 1983 *Astrophys. J.* **269** 352

Schneider D P, Schmidt M and Gunn J E 1989 *Astron. J.* **98** 1951

Schulz H 1987 *Astron. Astrophys.* **178** 7

—— 1988 *Astron. Astrophys.* **203** 33

Seyfert C K 1943 *Astrophys. J.* **97** 28

Shields G A 1978 *Nature* **272** 706

Shlosman I, Begelman M C and Frank J 1990 *Nature* **345** 679

Shuder J M 1981 *Astrophys. J.* **244** 12

Shuder J M and Osterbrock D E 1981 *Astrophys. J.* **250** 55

Simkin S M, Su H J and Schwarz M P 1980 *Astrophys. J.* **237** 404

Slipher V M 1917 *Lowell Obs. Bull.* **3** 59

Spinoglio L and Malkan M A 1989 *Astrophys. J.* **342** 83

Stasinska G 1984a *Astron. Astrophys.* **135** 341

—— 1984b *Astron. Astrophys. Suppl.* **55** 15

Stauffer J R 1982 *Astrophys. J.* **262** 66

Stocke J, Liebert J, Maccacaro T, Griffiths R E, Steiner J E 1982 *Astrophys. J.* **252** 69

Sun W-H and Malkan M A 1989 *Astrophys. J.* **346** 68

Terlevich R and Melnick J 1985 *Mon. Not. R. Astron. Soc.* **213** 841

Terlevich R, Melnick J and Moles M 1987 *Observational Evidence of Activity in Galaxies* ed E Ye Khachikian, K J Fricke and J Melnick (Dordrecht: Reidel) p 499

Thorne K S 1974 *Astrophys. J.* **191** 507

Tohline J E and Osterbrock D E 1982 *Astrophys. J. Lett.* **252** L49

Toomre A and Toomre J 1972 *Astrophys. J.* **178** 623

Turner T J and Pounds K A 1988 *Mon. Not. R. Astron. Soc.* **232** 463

—— 1989 *Mon. Not. R. Astron. Soc.* **240** 833

Ulvestad J S and Wilson A S 1984a *Astrophys. J.* **278** 544

—— 1984b *Astrophys. J.* **285** 439

—— 1989 *Astrophys. J.* **343** 659

Unger S W, Pedlar A, Axon D J, Whittle M, Meurs E J A and Ward M J 1987 *Mon. Not. R. Astron. Soc* **228** 671

Urry C M 1990 *Multiwavelength Astrophysics* ed F Cordova (Cambridge: Cambridge University Press) p 279

Urry C M, Arnaud K, Edelson R A, Kruper J S and Mushotzky R F 1990 *X-Ray Astronomy* in press

Veilleux S 1988 *Astron. J.* **95** 1695

—— 1989 A study of the structure and kinematics of the narrow-line region in Seyfert galaxies *Thesis* University of California, Santa Cruz

—— 1991a *Astrophys. J. Suppl.* **75** 383
—— 1991b *Astrophys. J.* **368** 158
—— 1991c *Astrophys. J.* **369** in press
Veilleux S and Osterbrock D E 1987 *Astrophys. J. Suppl.* **63** 295
Veron-Cetty M P, Veron P and Tarenghi M 1983 *Astron. Astrophys.* **119** 69
Voit G M and Shull J M 1988 *Astrophys. J.* **331** 197
Vorontsov-Velyaminov B 1977 *Astron. Astrophys. Suppl.* **28** 1
Vrtilek J M 1985 *Astrophys. J.* **294** 121
Vrtilek J M, and Carleton N P 1985 *Astrophys. J.* **294** 106
Walker M F 1968 *Astrophys. J.* **151** 71
Wandel A 1989 *Active Galactic Nuclei* ed D E Osterbrock and J S Miller (Dordrecht: Kluwer) p 62
Wandel A and Petrosian V 1988 *Astrophys. J. Lett.* **329** L11
Wasilewski A J 1983 *Astrophys. J.* **272** 68
Weedman D W 1983 *Astrophys. J.* **266** 479
—— 1986 *Quasar Astronomy* (Cambridge: Cambridge University Press)
Whittle M 1985a *Mon. Not. R. Astron. Soc.* **213** 1
—— 1985b *Mon. Not. R. Astron. Soc.* **213** 33
—— 1985c *Mon. Not. R. Astron. Soc.* **216** 817
—— 1989 *Active Galactic Nuclei* ed D E Osterbrock and J S Miller (Dordrecht: Kluwer) p 349
Wilkes B J 1984 *Mon. Not. R. Astron. Soc.* **207** 73
—— 1986 *Mon. Not. R. Astron. Soc.* **218** 331
Wilkes B J and Elvis M 1987 *Astrophys. J.* **323** 243
Wills B J, Netzer H and Wills D 1985 *Astrophys. J.* **288** 94
Wilson A S 1979 *Proc. R. Soc.* A **366** 461
Wilson A S, Ward M J and Haniff C M 1988 *Astrophys. J.* **334** 121
Wilson A S and Ulvestad J S 1982a *Astrophys. J.* **260** 56
—— 1982b *Astrophys. J.* **263** 576
Wilson A S and Ulvestad J S 1983 *Astrophys. J.* **275** 8
Wu C-C, Boggess A and Gull T R 1983 *Astrophys. J.* **266** 28
Yee H K C 1980 *Astrophys. J.* **241** 894
Zdziarski A A 1986 *Astrophys. J.* **305** 45

Rep. Prog. Phys. **54** (1991) 243–288. Printed in the UK

The detection of high red-shift quasars

S J Warren† and P C Hewett‡

† European Southern Observatory, Karl-Schwarzschild-Straße 2, D-8046 Garching bei München, Germany

‡ Institute of Astronomy, Madingley Road, Cambridge CB3 OHA, UK

Abstract

The detection of high-redshift (i.e. distant) quasars provides direct information on the physical conditions existing when the universe was only 10% of its present age. Quasars are the bright nuclei of certain galaxies and are the only astronomical sources that can currently be found in substantial numbers at redshifts $z > 2$. It is probable that quasars are active for only a small proportion of the life of the host galaxy, in which case a substantial fraction of all galaxies must go through a quasar phase at some time in their history. Consequently quasars may be tracers of the galaxy population at high redshift, less active members of which are too faint for detailed study with present-day instrumentation. One of the main goals of surveys for high-redshift quasars, then, is to map the evolution of the population, in luminosity and number density, back through the history of the universe, thereby furnishing clues both to the nature of the quasar phenomenon itself and to the poorly understood process of how galaxies formed. Of special interest is the possibility of detecting the epoch of the birth of the quasar population.

This review summarises the optical and radio techniques employed in the detection of high-redshift ($z > 3$) quasars and attempts to reconcile the contradictory claims in the literature concerning the evolution of their space density at the highest redshifts. Significant progress in the compilation of samples of high-redshift quasars has come about through the application of new detector technologies at long optical wavelengths, together with the advent of fast, automated microdensitometers for digitising the information on photographic plates. More than thirty quasars with redshifts $z > 4$ have been discovered in the last few years, and substantial samples of quasars with redshifts $z > 3$ have recently been compiled. Simultaneously, advances in computer-based selection and modelling techniques enable the completeness of these quasar samples to be determined. The combination of large area, faint flux-level surveys at long optical wavelengths with automated selection procedures is producing greatly improved limits on the shape and normalisation of the quasar luminosity function at high redshift.

Analysis of the results of recent CCD-grism and wide-field multicolour surveys indicate a decline in the space density of quasars of intermediate luminosity at high redshift. For $M_B = -26.5$ at $z = 4$ the space density is lower by a factor of 3 to 4 relative to $z = 2$. The decline is less steep than previously believed, and is still

0034-4885/91/000243+46$14.00 © 1991 IOP Publishing Ltd

subject to some uncertainty in the corrections for incompleteness. The decline for the rare brightest quasars $M_B < -27.5$, is less marked and the space density may be constant over the range $2 < z < 4.5$. For the faintest quasars $M_B > -25$, larger samples are required to determine if a decline similar to that found at intermediate luminosities occurs.

This review was received in its present form in October 1991.

Contents

1. Introduction

The discovery of quasars, in 1963, presented astrophysicists with one of their greatest challenges: the development of a quantitative model of the quasar central energy source and its interaction with the immediate environment. Quasars are the enormously bright nuclei of certain galaxies. The energy output of some quasars is as great as the integrated light of hundreds of galaxies, yet the size of the central source is comparable to the diameter of the solar system. Because of their extremely high luminosities, they have enabled us to probe the universe to much greater redshifts ($z \equiv$ distance, see § 3) than was previously possible, and thereby have provided unique opportunities to study aspects of the evolution of the universe over a period some 90% of its age (larger distances correspond to earlier epochs because of the light travel time). Quasars are the only astronomical sources that can be detected in substantial numbers at large redshifts, hence their study provides valuable constraints to the very different theoretical models for the growth of structure and formation of galaxies, at early times. Distant quasars are rare and these investigations, therefore, depend critically on the development and exploitation of efficient techniques capable of identifying high-redshift quasars in substantial numbers.

The use of quasars as probes of the evolution of galaxies can be divided into two distinct research areas:

(1) Studies of systems intermediate in distance between ourselves and the quasars. These systems are revealed through the effects of absorption on the spectra of the quasars, which are simply acting as background sources. 'Quasar absorption line' studies have revealed several populations of absorbers, some of which could be the high-redshift counterparts of normal galaxies as seen today. An increase in the maximum redshift of detected quasars offers the opportunity to study the change in space density and the properties of absorption systems that may be the signatures of stages in the formation of galaxies. The subject area is comprehensively covered by a recent collection of review articles (Blades *et al* 1988) and we shall not discuss this rapidly-developing aspect of quasar studies further.

(2) Mapping the space density of quasars themselves as a function of redshift. This entails identifying quasars in significant numbers over a large redshift range, and being able to establish the completeness of the sample in each redshift interval, in order to remove any contribution to the measured evolution in space density caused by the redshift-dependent sensitivity of the survey. Obtaining observational data of the necessary quality and homogeneity is a difficult technical undertaking and much progress in the field is linked to recent improvements in detector technology at long optical wavelengths ($\lambda \sim 600$–900 nm). The combination of technological advances with the instigation of a number of well-defined high-redshift quasar survey programmes has resulted in rapid developments over the last five years, including the discovery of many quasars with redshifts $z > 4$. However, there is apparently little agreement between the conclusions drawn by workers from different research groups regarding the evolution of the space density of quasars beyond $z = 2$; in recent years papers have appeared claiming that the space density of high-redshift quasars increases, while others cited evidence for a sudden and dramatic 'cutoff'.

In this review we attempt to reconcile these contradictory claims, presenting a summary of the results of the different surveys for quasars of redshift $z > 3$. An outline of the properties of quasars and of the basic observational quantities and technical terms is included for those unfamiliar with observational astrophysics and research related to quasars and active galactic nuclei (AGN). Before discussing recent investigations in detail we consider the factors that until the last few years have conspired to make the detection of large samples of quasars at high-redshift such an intractable problem.

2. Quasars

2.1. The standard model

Taking the luminosities inferred from the quasar redshifts, minimum lifetimes from the extent of large-scale radio structure and source size estimates from the observed variability timescales, it is possible to rule out nuclear fusion as a viable energy source. This has led to the development of models with energy generation from the gravitational accretion of matter. The efficiency of mass to energy conversion from accretion onto a massive central object may approach 10%, which contrasts with the extremely low efficiency of nuclear fusion ($\sim 0.1\%$). A model that satisfies qualitatively the observational constraints on the energetics (luminosity $\sim 10^{39}$ J s^{-1}) and source size ($\sim 10^{13}$ m), without requiring unreasonable mass fuelling rates, invokes mass infall via an accretion disk onto a supermassive black hole of mass $\sim 10^8$ M$_\odot$ (1 M$_\odot$ = one solar mass = 2×10^{30} kg). From a theoretical standpoint the formation of a black hole is the inevitable evolutionary end point of mass accretion in such a dense environment, however, there is currently no unambiguous observational evidence for the existence of supermassive black holes in AGN (see McKee 1988 for a summary). Furthermore a quantitative understanding of the relativistic regime close to the central object is lacking, as well as a detailed model describing the geometry of the system and elucidating the importance of the various physical processes responsible for reprocessing, and perhaps generation, of radiation more than several Schwarzschild radii from the central source. A comprehensive review of the progress towards understanding the physics of the central regions of quasars is given by Rees (1984).

2.2. Quasar spectra

Figure 1 shows the rest-frame ultra-violet spectral energy distribution (SED) of a 'typical' quasar. These features appear at longer wavelengths in the observed spectra of distant quasars, stretched by a factor $1 + z$ (as explained in § 3), and for the most distant quasars appear in the visible wavelength range (~ 340 nm to ~ 1000 nm). A wavelength scale in the observed frame corresponding to a quasar redshift of $z = 3$ is also indicated. The spectrum longward of the Lyman-α hydrogen resonance emission line at 121.6 nm has conventionally been characterised as a power-law, with the flux per unit frequency, f_ν, varying as $f_\nu \propto \nu^\alpha$, where α is ~ -1. This power-law extends into the infrared, and in the case of radio-loud quasars out to radio wavelengths, until, in most cases, a flattening of the spectrum is observed. The power-law form of the continuum is usually interpreted as the result of synchrotron processes, with the flattening in the radio or infrared regime due to synchrotron self-absorption (although quasars were first discovered through their radio emission, it is now known that radio-loud quasars form only a small subset of quasars detectable at optical wavelengths). There is also evidence for a substantial additional continuum component present in the rest-frame ultraviolet—the so called 'big blue bump'. This feature can be represented by a thermal black-body spectrum with temperature $\sim$ 30 000 K and accords well with the postulated contribution from an accretion disc surrounding the black hole (Lynden-Bell 1969, Malkan and Sargent 1982). More recently, observations in the far-infrared from the IRAS satellite have shown evidence for another significant continuum component that can be approximated by a black-body spectrum with temperature $\sim$ 1000 K (Sanders *et al* 1989). The origin of this component which peaks in the infrared is believed to be radiation from the quasar that has been absorbed and subsequently re-emitted by dust surrounding the quasar. The origin of quasar continuum radiation is thus not simple but whatever the physical origin of the radiation the key factor is the extended range in frequency over which radiation is observed. A simple power-law is not a good approximation over large wavelength ranges but for restricted ranges in wavelength a power-law often provides an adequate approximation to the form of the continuum. In the rest-frame ultraviolet the quasar 'continuum' comprises a mixture of the underlying power-law and the big blue bump. The presence of the latter can result in a significant flattening of the measured continuum in this region with local values of $\alpha \sim -0.5$ common.

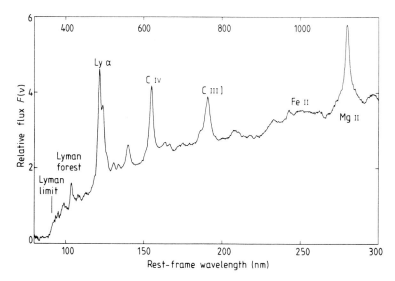

Figure 1. Composite rest-frame ultraviolet quasar spectrum. The principal emission features are indicated together with the onset of the Lyman limit and the region affected by Lyman-α absorption. The wavelength scale at the top of the figure corresponds to the observed-frame wavelengths for a quasar with redshift $z = 3$.

Superposed on the ultraviolet continuum are strong emission lines due to resonance transitions of common elements, notably hydrogen, carbon, oxygen and nitrogen. These lines exhibit extreme doppler broadening (10^3–10^4 km s^{-1} full width at half maximum) and provide the primary means for quasar identification and redshift determination. Ionisation states for elements are denoted using roman numerals with neutral species indicated by 'I', singly ionised species by 'II' etc . Thus, neutral and singly ionised hydrogen are denoted by H I and H II respectively, cf. the perhaps more familiar notation of H and H$^+$. Shortward of the Lyman-α emission line numerous very-narrow absorption lines are observed, due to large numbers of low column density ($\lesssim 10^{20}$ atoms m^{-2}) 'clouds' at redshifts lower than that of the quasar. Neutral hydrogen at a redshift z_{abs} is detected as a Lyman set of absorption lines at observed wavelengths $\lambda_{obs} = \lambda_{Lyman}(1 + z_{abs})$, where λ_{Lyman} is the laboratory value for a particular Lyman-series line. This region of the spectrum is commonly termed the 'Lyman-α forest' and for all but the highest resolution observations the cumulative effect of the absorption by the clouds results in a substantial drop in the apparent quasar continuum shortward of the Lyman-α line. A second major feature is evident at a wavelength $\sim 91.2(1 + z_{abs})$ nm (the Lyman limit) when a high column density ($\gtrsim 10^{21}$ atom m^{-2}) system results in the absorption of practically all ultraviolet photons of shorter wavelength, producing a sudden spectral discontinuity. The observed wavelength of the feature depends on the redshift of the nearest high column density cloud to each quasar, that lies along the line of sight.

2.3. Quasars as probes of evolution

Mapping the evolution of the quasar population with redshift, as a function of luminosity and number density, provides important observational evidence against which theoretical models of the central energy source, and of the development of individual quasars, may be compared. The relevance of the evolution of quasars to that of the majority population of 'normal' galaxies depends on (a) how much may be inferred about the formation and space density of galaxies in general from observations of quasars that have a space density only $\lesssim 1\%$ of that of normal galaxies? and (b) how is the birth, evolution in luminosity, and death of an individual quasar connected with the life history of the host galaxy?

With regard to the first question, that quasars lie in galaxies at all was a matter of controversy in the period following their discovery. This may partly be attributed to historical accident, for the first quasars found were so bright that their relationship to the lower luminosity nearby Seyfert 1 galaxies was not immediately apparent. Seyfert 1 galaxies are spirals, whose nuclei display a strong non-thermal continuum and broad emission lines. It is now generally accepted that quasars, Seyfert galaxies and broad-lined radio-loud giant ellipticals are all manifestations of the same underlying physical phenomenon. Support for this view comes from high-resolution imaging (e.g. Hutchings, Johnson and Pyke 1988) and off-nuclear spectra (Boroson, Oke and Green 1982, Hickson and Hutchings 1987) of 'fuzz' associated with quasars at low redshifts ($z \lesssim 0.7$), which display absorption features characteristic of stars. Thus, the distinction between classes is to some extent an arbitrary one, based on morphology (resolved/unresolved) and/or luminosity, dependent on the brightness of the active nucleus relative to the stars in the host galaxy. For example, Schmidt and Green (1983) considered unresolved (i.e. star-like) objects of nuclear absolute magnitude (§ 3) brighter than $M_B = -23$ to be quasars. The differences in the environment in which the active nucleus resides, in particular the nature of the host galaxy, is believed to explain the very different radio-properties of the different classes.

In § 5.3 evidence is presented that, although only a small fraction of galaxies display quasar activity at any one time, the lifetimes of quasars are short ($\sim 10^8$ yr) compared to the age of the universe ($\sim 10^{10}$ yr), in which case a large fraction of all bright galaxies must go through a quasar phase, in order to explain the numbers of quasars seen at any redshift. This strengthens the argument that the evolution of quasars is relevant to the evolution of galaxies in general.

The disadvantage of using quasars to map evolution in the universe depends to a large extent on the lack of a quantitative understanding of the central sources and also on the connection between the occurrence of the quasar phenomenon and certain critical events (Cavaliere and Szalay 1986). What are the necessary conditions and timescales for formation of a massive central object? Is the formation of a massive central source related to the initial collapse of a protogalaxy or to some other phenomenon? However, it is clear that a massive gravitationally bound system must exist in order for a quasar to be seen, although estimates of this mass are uncertain (e.g. Efstathiou and Rees 1988, Turner 1991). The strength of constraints on particular galaxy formation models derivable from observations of the quasar population is, then, dependent primarily on the masses of quasar host systems at high redshift.

The advantages of using quasars as probes of evolution are considerable. It is difficult to obtain spectra of galaxies of intermediate luminosity at redshifts $z \gtrsim 0.3$. Although a few observations of high-luminosity galaxies in rich clusters at redshifts $z \sim 0.8$ exist, it is not yet possible to observe the bulk of the galaxy population at redshifts $z > 0.5$. In contrast, quasars are typically about 100 times as bright at optical wavelengths as galaxies and there is no reason in principle why they can not be detected out to redshifts $z \geqslant 6$ (if they exist), thereby offering the possibility of investigating the evolution of systems over more than 90% of the time since the Big Bang.

2.4. Alternative models

An alternative to the black-hole picture for the power source of AGN attempts to model the nuclear activity as a consequence of high rates of star formation (Terlevich and Melnick 1985, 1988). Filippenko (1989) has shown that the spectra of certain supernovae closely resemble those of typical Seyfert 1 galaxies and has argued that the ideas of Terlevich and Melnick might be applicable to at least some low-luminosity AGN. Terlevich (1991) has argued further that the starburst model leads to a predicted luminosity function for the cores of elliptical galaxies at redshifts $z = 2$ that resembles closely the measured quasar luminosity function. A review of the strengths and shortcomings of this model is provided by Heckman (1991).

3. The observational quantities

3.1. Magnitudes

The apparent magnitude, m, of a source is a logarithmic measure of the incident energy flux F,

$$m = \text{constant} - 2.5 \log F \tag{1}$$

Apparent magnitudes corresponding to particular passbands are denoted by a subscript, e.g. m_B for magnitudes in the blue B band. The very brightest quasars on the sky have apparent magnitudes $m_B \sim 14$ (equivalent to 1.1×10^{-28} W m^{-2} Hz^{-1}) while the faintest systematic surveys for quasars have now reached magnitudes of $m_B \sim 23$. The five broad passbands in the visual range of the electromagnetic spectrum that are commonly employed extend from the ultraviolet (U, central wavelength $\lambda \sim 360$ nm) through the blue (B, 440 nm), green or 'visual' (V, 550 nm), and red (R, 640 nm) to far-red (I, 790 nm).

The absolute magnitude, M, of an object is the apparent magnitude the object would have if observed from a distance of 10 pc

$$M = m - 5\log(d/10) \tag{2}$$

where d is distance in parsecs (a parsec is the conventional measure of distance, 1 pc = 3.26 lightyear or 3.09×10^{16} m). Absolute magnitudes, therefore, are a logarithmic measure of an objects luminosity L. Subscripts are again employed to denote magnitudes in a particular passband. For a Hubble constant $H_0 = 50$ km s^{-1} Mpc^{-1} (see below), the value we use throughout this review, absolute magnitudes of quasars are typically in the range $-30 < M_B < -23$, while normal galaxies populate the range $-23 < M_B < -16$ and fainter.

3.2. Redshifts and cosmology

The universe is observed to be expanding, and there is strong observational evidence, that on very large scales the universe is homogeneous and isotropic. Combined with the theoretical framework of Einstein's General Theory of Relativity this observational evidence provides the framework for current cosmological models. For an introduction to Big-Bang cosmology see the appendix to the article by Partridge (1988) in this series, which also includes references to more detailed treatments.

In the Big-Bang models, galaxies and quasars are approximately stationary in space which is itself expanding, while the wavelengths of photons emitted from a source are stretched in the same way as the space through which they travel. Photons detected from a distant galaxy have been shifted to longer (i.e. redder) wavelengths by a factor equal to the amount by which the universe has expanded in the time since their emission. This is the origin of the redshift, z, defined by $z = \lambda_{\text{observed frame}} / \lambda_{\text{rest frame}} - 1$. Since radiation from more distant objects was emitted earlier in the history of the universe, redshifts increase monotonically with distance and can be employed as a relative measure of distance.

The present rate of expansion of the universe is measured by the Hubble constant, H_0, where the rate of expansion between two objects, separated by distance d, is $\dot{d} = H_0 \times d$. The rate of expansion of the universe is slowing down due to the gravitational effect of matter, and the rate of deceleration, measured by the parameter q_0, is related therefore to the matter density in the universe. In the simplest models a value of $q_0 = 0$ corresponds to an empty universe with no deceleration, while a value of $q_0 = 0.5$ corresponds to the limiting case where the gravitational effects of matter eventually bring the expansion to a halt. Values of $q_0 > 0.5$ mean that the gravitational effects of the matter eventually overcome the expansion and the universe will re-collapse. There are attractive theoretical reasons for believing that the value of q_0 is exactly 0.5, however, the majority of observational evidence indicates a value of $q_0 = 0.1 - 0.2$. The two cases, $q_0 = 0.0$ and $q_0 = 0.5$, bracket the values favoured by both observational and theoretical studies, and conventionally models using both values are examined. Derived values of physical parameters such as distance or volume are quite different depending on the value of q_0. However, for limited intervals in redshift, $\Delta z \sim 1$, parameters characterising the evolution of the quasar population are relatively insensitive to this parameter. In this review, for simplicity, we present results for a value of $q_0 = 0.5$, but the conclusions regarding the evolution of the space density of quasars at high redshifts are qualitatively similar for a value of $q_0 = 0.0$.

In studying evolution, look-back time is a more useful concept than distance. Extrapolation back in time to the point where spatial intervals between objects reduce to zero (the Big Bang) leads to an age, t_0, for the universe. For $q_0 = 0.0$ this is simply $t_0 = H_0^{-1}$. For models with mass there is some deceleration and t_0 is smaller, the solution for $q_0 = 0.5$ gives $t_0 = \frac{2}{3} H_0^{-1}$ (i.e. 13 Gyr for $H_0 = 50$ km s^{-1} Mpc^{-1}). Look-back time, τ, is defined as the light-travel time of a photon from a distant source, expressed as a fraction of the age of the universe. For $q_0 = 0.0$, $\tau = 1 - (1+z)^{-1}$, and for $q_0 = 0.5$, $\tau = 1 - (1+z)^{-1.5}$. These relationships are graphed in figure 2(*a*). The interest in surveys for quasars of redshift $z > 2$ is immediately apparent: for example, at redshifts $z \sim 4$ we are obtaining information on conditions in the universe when it was only $\sim 10\%$ of its present age.

To allow for the effect of the expansion of the universe, space, or volume densities of objects are calculated as a function of the comoving volume element. Comoving volume elements scale with the expansion of the universe and the comoving space density of a non-evolving population of objects is constant with time—i.e. independent of redshift.

In order to calculate the absolute magnitude of distant sources, such as quasars, equation (2) must be modified to include the effects of space curvature on the distance, and a correction applied to allow for the change in the rest-frame wavelengths of the quasar's SED that appear in the observed frame. For an object with a power-law SED, index α, the absolute magnitude is related to the apparent magnitude and redshift through the expression

$$M_{\mathrm{B}} = m_{\mathrm{B}} - 5\log A(z) + 2.5(1 + \alpha)\log(1 + z) - 5\log(c/H_0) - 25 \qquad (3)$$

where c is the velocity of light, and the bolometric luminosity distance $A(z)$ is given by:

$$A(z) = z \left(1 + \frac{z(1 - q_0)}{[(1 + 2q_0 z)^{0.5} + 1 + q_0 z]} \right) \qquad (4)$$

for $q_0 \geqslant 0$. Figure 2(*b*) indicates how faint in absolute magnitude a survey reaches at a given redshift for a particular limiting apparent magnitude.

3.3. Luminosity functions

The luminosity function, at a particular redshift, is the variation of the comoving space density per unit luminosity as a function of luminosity $\rho = \rho(L, z)$. By determining the luminosity function of quasars at several different epochs, we will be able to map the evolution of the population, and so obtain constraints on the evolution of individual quasars. The observation of a decline in the space density of quasars

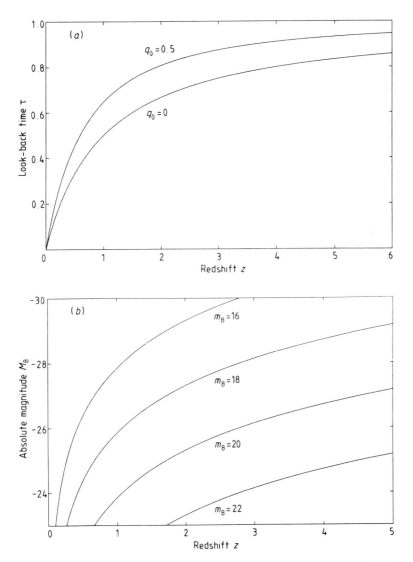

Figure 2. (*a*) Look-back time, τ, as a function of redshift, z, for two values of the deceleration parameter, q_0. (*b*) The limiting absolute magnitude, M_B, as a function of redshift, z, for surveys with limiting apparent magnitudes, $m_B = 16, 18, 20$ and 22. A correction has been applied to the apparent magnitudes to take account of the effects of absorption caused by the Lyman-α forest. A value of $q_0 = 0.5$ has been used and the quasar spectral energy distributions were modelled using a power-law continuum slope of index $\alpha = -0.5$.

beyond a certain redshift would be of great interest as this could imply that we are probing the epoch at which quasars were born.

Figure 3 shows model fits to the luminosity function of nearby galaxies, $z \sim 0$, and for quasars at redshifts $z \sim 0.5$ and $z \sim 2$. It is usual to calculate luminosity functions per unit magnitude, for which the symbol ϕ is employed. Ideally the bolometric magnitude, from the integrated flux over all wavelengths, would be used as a luminosity indicator. Instead, for reasons of practicality, the absolute magnitude of the continuum at blue rest-frame wavelengths M_B is usually used. For the simple case of a power-law luminosity function $\rho \propto L^{\alpha}$ one finds that $\log(\phi) = 0.4 M (1 + \alpha) +$ constant, and by convention it is the parameter α that is usually quoted (unrelated to the α used to parameterise the quasar SED). Figure 3 illustrates a number of critical points:

(1) luminosity functions extend over a significant range in luminosity;

(2) the space density of galaxies today is several orders of magnitude greater than the space density of quasars at any redshift $z < 2$;

(3) quasars possess far greater luminosities than the galaxy population as a whole; and

(4) bright quasars were more common in the past $(z = 2)$ than they are today.

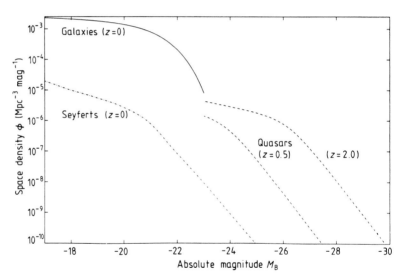

Figure 3. Model fits to the space density, ϕ, per cubic megaparsec, per magnitude, as a function of absolute magnitude, M_B, of normal galaxies $(z = 0.0)$, Seyfert galaxies $(z = 0.0)$ and quasars $(z = 0.5$ and $z = 2.0)$. The normal galaxy data is taken from Efstathiou *et al* (1988), the Seyfert data from Weedman (1986b) and the quasar data from Boyle *et al* (1988).

The ideal method of determining a luminosity function is to compile a sufficiently large sample of objects that the luminosity function may be read off the M–z plane by taking a thin slice at constant redshift and binning the data into intervals of absolute magnitude of arbitrarily small size. For quasars, such a programme is not feasible, because of the size of the sample required, and since at the brightest absolute magnitudes even a survey of the whole sky would contain too few objects. Furthermore, the observed steep decline in the numbers of quasars as a function of increasing luminosity means that any survey of a given area of sky is dominated by quasars at the faint limit of the survey, so that in practice the dynamic range covered by a given survey is rarely more than two magnitudes—i.e. only a small portion of the luminosity function at each redshift is accessible. The complications are increased as the range in apparent luminosity of any survey translates into a different range in absolute luminosity as a function of redshift; see figure 2(b).

A number of statistical techniques have been developed in order to avoid the loss of information consequent on binning small samples, associated with the variation of the luminosity function across the bin. One would like to use a non-parametric method such as that of Efstathiou *et al* (1988), which has been employed in determining the local galaxy luminosity function. They make the assumption that the evolution function is separable from the luminosity function $N(M, z) = \phi(M)\rho(z)$ i.e. the luminosity function retains its shape, but the normalisation varies with redshift. It is now well established that this is not the case for the evolution of the quasar luminosity function (e.g. Boyle *et al* 1988), so the non-parametric methods are of limited utility here. General parametric maximum-likelihood techniques have been developed by Marshall and coworkers (Marshall *et al* 1983, Marshall 1985) and Peacock (1985). For a particular parameterisation of the quasar luminosity function (QLF), and its evolution, the best-fit model is computed, and the goodness of fit between model and data is assessed by use of an appropriate statistic. In such a way it becomes possible to compare the observational data against theoretical models, and to obtain a quantitative idea of the range of theoretical models that are consistent with the data.

An unsatisfactory feature in the calculation of the luminosity function at very high redshifts, $z \sim 4$, is the potential redshift-dependent error introduced in computing absolute magnitudes M_B, in extrapolating from observed wavelengths in the rest-frame ultraviolet, to $\lambda_{rest frame} = 440$ nm; the term $2.5(1 + \alpha)\log(1 + z)$ in equation (3). Different authors have adopted values of the quasar spectral index in the range $-1.0 < \alpha < -0.5$, corresponding to a difference in the computed absolute magnitude $\Delta M_B = 0.9$ magnitudes at $z = 4$. A complete calculation should include the (currently unknown) distribution of α for the quasar population, since objects of the same apparent magnitude but different spectral index sample different parts of the luminosity function. These uncertainties would be reduced by adopting a reference wavelength closer to the rest-frame wavelength that is observed, e.g. $\lambda = 200$ nm rather than 440 nm conventionally employed.

4. Detecting quasars: the observational problem

In this section we consider the difficulties involved in detecting samples of quasars for studying the evolution of the luminosity function. While the prodigious energy output and extreme nature of the quasar rest-frame SEDs should in principle allow the identification of large, well-defined samples, there is a range of practical considerations that make quasar surveys problematic.

At optical wavelengths quasars are rare, they constitute only a small fraction of the objects seen on the sky, the vast majority of objects being normal galaxies, and stars in our own galaxy. Although intrinsically the most luminous astronomical sources known, they appear faint because of their great distances. Except at the lowest redshifts and faintest nuclear luminosities quasars appear as point sources (like stars), because the light from the active nucleus greatly outshines the rest of the galaxy. Table 1 tabulates approximate surface densities for stars, galaxies and quasars as a function of magnitude in the B passband for the most favourable situation i.e. at the Galactic poles; towards the Galactic plane the surface density of stars increases rapidly. Confirmation that a stellar source is a quasar requires detailed spectroscopic information and compilation of quasar samples through complete spectroscopic identification of flux-limited object surveys at optical wavelengths (thereby eliminating the need for any 'selection strategy') is not yet a viable proposition. A highly effective selection strategy for identifying the small proportion of sources that are quasars is essential, and because of the high surface density of potentially contaminating objects any selection technique must have an exceptionally low probability of generating 'false-positives'.

Table 1. Approximate integral surface densities per square degree brighter than a blue apparent magnitude, m_B, for various classes of object. Column 1, stars in our own galaxy, Column 2, galaxies, Column 3, quasars at all redshifts, and Column 4, quasars with redshifts $z > 3$. The surface densities are given for a field at high Galactic latitude where the number of stars is a minimum. The numbers, particularly for the high-redshift quasars, are approximate but give an indication of the problem involved in identifying high-redshift quasars at optical wavelengths.

m_B	Stars	Galaxies	Quasars	Quasars ($z > 3$)
16	200	2	0.02	< 0.001
18	500	40	1	0.02
20	1000	400	25	0.5
22	2000	3000	100	2

Many of the spectral characteristics of quasars in the rest-frame ultraviolet exhibit substantial variations from object to object. Principal among these are:

(1) the range of apparent continuum slopes, values of $\alpha = -0.3$ to -1.0 are common within a range of $\alpha = 0.0$ to -2.0;

(2) variations of an order of magnitude in the strength of the primary emission features relative to the underlying continuum; and

(3) the existence of Broad Absorption Line quasars (BALs), a subset of quasars showing exceptionally strong absorption. In extreme cases BALs possess spectra with virtually no emission features evident and extensive absorption troughs as prominent as emission features in many other quasars.

This extended range in the SEDs of quasars has important ramifications for observational detection programmes, particularly in the optical, and argues against a strategy which targets 'typical' quasars. The variation in the broad-band magnitudes for quasars with different rest-frame SEDs can be large. This is particularly true at high redshifts, since any spectral feature makes a larger fractional contribution to the broad-band flux, because the spectrum has been stretched. Even broad passband widths, $\sim 60 - 160$ nm, correspond to only $\sim 15 - 40$ nm in the quasar rest-frame at redshift $z = 3$. The range of emission and absorption line strengths together with localised variations in the quasar continuum slope combine to give a spread in each colour of typically ±0.5 magnitudes about the mean at a fixed redshift. (Note that 'colour' is the difference between two magnitudes and so measures the flux ratio. Analogous to the description of hot or cool stars, objects that are brighter in the short-wavelength band are referred to as 'blue', and those brighter in the long-wavelength band are 'red'). In addition, as spectroscopic features move through a particular passband rapid changes in the colours of quasars can be expected for only small variations in redshift.

While the total surface density of astronomical sources at faint magnitudes is high (table 1) the range of SEDs of stellar objects observed away from star forming regions in our own galaxy is limited. The spatially unresolved sources consist almost exclusively of stars within our own galaxy. The SED of a star depends primarily on the surface temperature modified by opacity effects in the outer atmosphere. The opacity is sensitive to the temperature, chemical composition (particularly to the abundance of elements other than hydrogen and helium) and surface gravity. The strength of individual absorption features in stars is however generally weak and variations have little effect on observations made at very low spectral resolution. For such observations, significant variations occur only at the extremes of the observed abundance range and are predominantly confined to ultraviolet wavelengths, except for the coolest stars where molecular absorption at longer wavelengths is very strong and highly sensitive to abundance changes. Surface gravity has only a small effect on low resolution observations. Thus the SEDs of the bulk of stellar objects are determined primarily by stellar surface temperature, with a second parameter—chemical abundance—introducing additional variation at ultraviolet wavelengths and also at long wavelengths for the coolest stars.

The combination of the thermal nature of the SEDs of stars, the narrow spread at any particular temperature, and the power-law nature of quasar spectra, means an effective detection strategy for quasars is to obtain low-resolution spectral information of stellar objects, and to identify sources with SEDs *different* to those of all common stars. This is the principle behind the multicolour and ultraviolet excess techniques described below. These methods identify unusual objects, possibly including peculiar stars and galaxies as well as quasars, so that it will be necessary to obtain higher-resolution data to confirm the nature of these candidate quasars. The same approach may also be applied to the analysis of slitless spectroscopic data (e.g. Hewett and Irwin 1987), although in practice much work has concentrated on the identification of a more limited range of quasar SEDs—those exhibiting strong emission lines (e.g. Schmidt *et al* 1986a).

The majority of galaxies can be eliminated from consideration in quasar surveys because they are spatially well resolved. The exceptions include surveys for very nearby systems or for low-luminosity quasars where the luminosity from any associated galaxy may be comparable to, or exceed that of the quasar component. Neither of these situations is relevant for high-redshift searches. The SEDs of most galaxies in the rest-frame optical region are dominated by radiation from stars. However, bright galaxies with redshifts $z \gtrsim 0.2$ can appear only marginally resolved and present a potential source of confusion in quasar searches, since the range of observed SEDs is greater than for stars because of the variation introduced by the spread in redshift.

Observations in wavelength regimes away from the optical offer potential advantages for quasar detection. The SEDs of most stars peak at optical or near-infrared ($< 2 \mu$m) wavelengths and the spectral differences between stars and quasars become progressively greater in moving both to shorter and to longer wavelengths. Surveys in regions of the electromagnetic spectrum that have been opened to observational astronomy in recent decades, firstly radio (from the ground), later X-ray and far infrared (from satellites), offer greater discrimination between stars and quasars than optical searches. In practice the sensitivity and spatial resolution currently achieved in the X-ray and infrared are such that surveys at these wavelengths are not yet competitive at high redshifts. At radio wavelengths quasars cannot be distinguished from radio galaxies on the basis of the radio morphological data alone as many radio-loud quasars have spatially extended radio emission (although as a rule they are more compact), however, a simple spectral-slope criterion does provide an effective discrim-inant. Spectroscopic identification of 'flat-spectrum' ($\alpha_{radio} > -0.5$) radio sources, at least to the bright flux limits of the earlier samples, requires little if any selection strategy since the fraction of objects that are not quasars is small. This advantage is offset by the paucity of quasars detectable by radio techniques.

5. The quasar luminosity function at redshifts $z < 2.2$

5.1. Ultraviolet excess surveys

The ultraviolet excess (UVX) survey technique, which consists of observing spectro-scopically all stellar objects with $m_U - m_B$ colour less than some threshold value (objects bright in U), has proven to be the most efficient means of obtaining large samples of quasars at redshifts $z \lesssim 2.2$. For a power-law SED of spectral index $\alpha = -0.5$ the $m_U - m_B$ colour is ~ -0.8. This can be modified somewhat by the presence of emission or absorption in either passband. However, stars with colour $m_U - m_B < -0.4$ are rare so the simple UVX criterion is highly effective for quasars of low redshift. Based on comparison with quasar samples selected by optical vari-ability, radio or X-ray emission, UVX samples are found to be very nearly complete to a redshift $z = 2.2$ (Véron 1983), and so are suitable for the determination of the QLF. In the redshift range $0.5 < z < 0.9$ some quasars are missed, as the presence of the hydrogen Balmer discontinuity, strong Fe II emission, centred on ~ 300 nm, combined with Mg II (279.8 nm) emission in the B band cause the $m_U - m_B$ colour to redden (e.g. Wampler and Ponz 1985). The UVX technique becomes ineffective be-yond $z \sim 2.2$ as the Lyman-α emission line passes into the B band, and the Lyman-α forest depresses the continuum flux in the U band (figure 1).

Two major surveys, based on U and B photometry from photographic plates, provide the majority of the data on which our knowledge of the QLF at low redshifts is based. The Palomar Bright Quasar survey, or BQS, (Schmidt and Green 1983) covers an area of 10 714 square degrees, one quarter of the whole sky, to an average limiting magnitude $m_B = 16.16$, and comprises 92 quasars. The survey of Boyle *et al* (1988, 1990) (hereafter BSP) covers 11.2 square degrees and reaches $m_B = 20.9$, and includes 351 quasars of redshift $z < 2.2$. BSP have combined these two data sets with others from the literature to produce a total sample of 624 quasars. Their analysis of these data represents the current state of our knowledge of the low-redshift QLF and is summarised below. Another detailed analysis using much the same data has been presented by Hartwick and Schade (1990).

5.2. The luminosity function

BSP show the data are consistent with a model where the shape and normalisation of the luminosity function are invariant with redshift, but for which the luminosity function shifts progressively to brighter magnitudes at higher redshifts—i.e. pure luminosity evolution. They choose to parameterise the QLF by a two power-law form:

$$\phi(M_B, z)\, dM_B\, dz = \frac{\phi^*}{\left[10^{0.4(M_B - M_B(z))(\alpha+1)} + 10^{0.4(M_B - M_B(z))(\beta+1)}\right]}\, dM_B\, dz \qquad (5)$$

and investigate models for luminosity evolution as a power-law function of redshift, $L = L_0(1+z)^{k_L}$, and an exponential function of look-back time, $L = L_0 \exp(k_L \tau)$. BSP apply the maximum-likelihood technique to obtain the best parametric model fit (Marshall *et al* 1983), and test the acceptability of their fits to the data in the M–z plane by means of the two-dimensional Kolmogorov-Smirnov statistic (Peacock 1983, Fasano and Franceschini 1987). The exponential evolution model is rejected at the 1% confidence level, but the power-law evolution with index $k_L = 3.2 \pm 0.1$ ($q_0 = 0.5$) provides a good fit. Figure 4 shows the best-fit model, for several different redshifts. As first noted by Koo and Kron (1982) the steep power-law evident at the brightest magnitudes turns over to a shallower slope at faint magnitudes—values of $\alpha = -3.8 \pm 0.15$ and $\beta = -1.4 \pm 0.2$, the power-law slopes of the luminosity function, are obtained. The strength of the evolution as a function of redshift is remarkable; the characteristic luminosity in the BSP model is a factor ~ 35 brighter at redshift $z = 2$ compared to that at redshift $z = 0$.

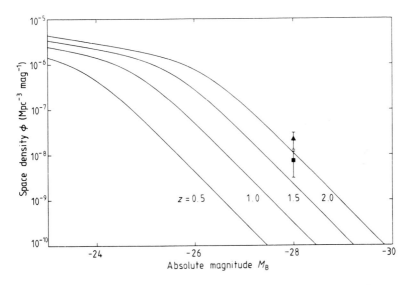

Figure 4. The space density, ϕ, per cubic megaparsec, per magnitude, as a function of absolute magnitude, M_B, for quasars at different redshifts. The model fits derived from BSP are shown. The quasar space densities at absolute magnitude $M_B = -28$, calculated as described in the text (§7) for the surveys of Hazard *et al* (1986) (redshift range $z = 3.3 - 3.9$) (▲) and Mitchell *et al* (1990) (redshift range $z = 3.4 - 4.1$) (■) are also shown.

There is some evidence that the data fall below the model at low redshifts and faint magnitudes. This may reflect incompleteness due to the redder intrinsic $m_U -$

m_B colours of quasars at redshifts $z \sim 0.6$, or because at faint nuclear absolute magnitudes the surrounding galaxy becomes visible causing the quasars to appear both non-stellar and redder. At higher redshifts extrapolation of the model luminosity function to fainter magnitudes shows good agreement with the number of quasars found in the very deep ($m_B = 22.6$) survey of Koo et al (1986) (§ 7.3.1), thereby supporting the pure luminosity evolution model and suggesting that the discrepancy between the model and the BSP data at low redshifts is due to incompleteness. Further evidence comes from the agreement between the extrapolation to $z = 0$ of the BSP model, and the Seyfert luminosity function at the present epoch determined by Cheng et al (1985).

Boyle et al (1991b) have recently extended this analysis of the low-redshift QLF to $z = 2.9$ incorporating the faint $m_{B_J} < 22$ multicolor surveys of Boyle et al (1991a) and Zitelli et al (1991) (B_J is a blue passband close to B). This analysis is discussed in § 7.3.1.3.

5.3. Interpretation of the luminosity function

Although the pure luminosity evolution model is a strikingly simple description of the variation with cosmological epoch of the energy output of the quasar population, the physical significance of this behaviour is not clear. It is possible to conceive of more than one physical model that can reproduce the observed luminosity function and which satisfies, for each epoch and over the range of luminosities, the continuity equation (Cavaliere et al 1971):

$$\frac{dN}{dt} = S(L, t) - \frac{d(\dot{L}N)}{dL}$$

where S is the birth function. The different schemes do make testable predictions, and should be compatible with the constraints applying to galaxy formation, which affects S, as well as models for the quasar fuelling mechanism, which also affects S and controls L and $\dot{L}$. The simplest interpretation of the observed evolution is that it represents the slow dimming, with a timescale of ~ 3 Gyr, of long-lived quasars born before $z = 2$, the fading with cosmological epoch being the result of a dwindling fuel supply. The fuel could be provided through mass loss from young massive stars, or by stellar collisions in a dense star cluster and through tidal disruption of stars by the central object. Theoretical studies (e.g. Duncan and Shapiro 1983, David et al 1987) indicate fuelling rates of the right order and that decay with time.

The long-lived hypothesis implies that supermassive black holes reside in the present-day counterparts of the quasars seen at high redshift, i.e. in the nuclei of the small percentage of galaxies that today exhibit quasar-like activity, principally the Seyferts. By implication any massive central objects present in a larger fraction of galaxies today would have considerably smaller masses. The minimum mass in Seyfert nuclei, then, can be estimated directly from the present-day luminosity, as the sum of the accreted mass, over the redshift interval $0 < z < 2.2$, and the minimum mass associated with the inferred luminosity at $z = 2.2$ i.e. assuming quasars then radiated at the Eddington limit (this is the theoretical maximum luminosity for a given central mass, calculated by equating the gravitational force with the force on matter due to radiation pressure; it corresponds to the minimum central mass for a given luminosity). Consider a present day Seyfert galaxy, of nuclear absolute magnitude M_B, and hence bolometric luminosity $L_{bol} = 10^{(29.78-0.4M_B)}$ W (Weedman 1986a). At a particular epoch the luminosity is related to the mass accretion rate by $L = \dot{m}c^2\eta$, where η is the efficiency, taken to be 10% (Begelman *et al* 1984). For the BSP model the minimum mass accreted is found to be $M_{accr} = 1.9 \times 10^{10}\ 10^{(0.4(-21-M_B))}$ M$_\odot$, to which must be added the minimum mass at $z = 2.2$, $M_{edd} = 4.8 \times 10^8\ 10^{(0.4(-21-M_B))}$ M$_\odot$ (noting that M_B is the present-day absolute magnitude).

Observational constraints on the masses of central objects in nearby galaxies are difficult to obtain, however there is a growing body of evidence that masses for Seyfert nuclei do not greatly exceed $\sim 10^8$ M$_\odot$, too small for the hypothesis that quasars are long lived. For example, estimates of the mass of the nucleus of the nearby bright Seyfert galaxy NGC4151 lie in the range $3 - 10 \times 10^7$ M$_\odot$ (Clavel *et al* 1987, Gaskell 1988), while the nuclear absolute magnitude $M_B = -19$ (for $m_B = 12.5$, Penston *et al* 1974) leads to an estimate of the mass, based on the equations above, of 3×10^9 M$_\odot$. One is led, therefore, to consider an alternative picture in which quasars have shorter lifetimes, but a larger fraction of all galaxies go through a quasar phase, with the evolution of the QLF reflecting the ensemble behaviour of many short-lived events. This picture fits well with the timescale, $\sim 10^8$ yr, associated with galaxy-galaxy interactions which have been proposed as 'triggers' for the fuelling of quasars (Hernquist 1989 and references therein). In this scenario the quasar activity would be recurrent; a larger fraction of galaxies would host lower mass central objects, the relevant factor being the reciprocal of the duty cycle $\delta = \frac{t_{on}}{(t_{off}+t_{on})}$. Evidence that many currently inactive galaxies may have been through a quasar phase comes from the observation of very low-level quasar-like activity in the nuclei of many spiral galaxies (Filippenko and Sargent 1985), and from evidence for central massive objects with masses $\sim 10^7$ M$_\odot$ in currently inactive galaxies (e.g. Dressler and Richstone 1988, Kormendy 1988).

The study by Yee and Green (1987) of the environments of radio-loud quasars at different redshifts may be interpreted as providing additional evidence that the lifetimes of quasars are short. They find that the objects in their sample at redshifts $z > 0.5$ preferentially lie in galaxy clusters, while the quasars at lower redshifts are found in less rich environments. Since the analogues of the higher-redshift quasars are not found in their sample at lower redshifts, they argue that they have died out in the intervening period.

Cavaliere and Padovani (1989) have provided a summary of the current status of estimates of the masses of the central objects in both active and inactive galaxies, discussed in terms of the Eddington ratio, the ratio of the nuclear bolometric luminosity to the Eddington luminosity. They conclude that the observational data can rule out long-lived activity patterns for AGN. Whereas the data favour some form of recurrent activity, the possibility that the quasar phenomenon is a single short-lived event cannot be excluded. They argue that at least 20% of all bright galaxies, and possibly a considerably larger fraction have gone through a quasar phase at some point in their lifetime. One could invoke models that attempt to unify the different AGN on the basis of orientation effects (e.g. Barthel 1989) to argue that the fraction of galaxies that were once quasars is even higher; for example Barthel believes that narrow-lined radio galaxies are active radio-loud quasars that have been excluded from the census of quasars because the bright optical nucleus is obscured.

For a model in which quasars are short-lived, one would expect to see both density and luminosity evolution from the effects of, respectively, the change in the frequency of galaxy-galaxy interactions with cosmological epoch, and the change with time of the fuel supply rate. The short-lived hypothesis is thus more contrived than the long-lived picture, implying a conspiracy of the variations in the fuelling rate with time and luminosity in order to match the simple pure luminosity evolution of the QLF. Cavaliere *et al* (1988) have devised such a model of recurrent interactions that qualitatively reproduces the behaviour of the QLF at low redshifts, although in their model the space density of low-luminosity quasars increases with time, while the observations suggest that the space density remains constant or even declines (as noted by Boyle in the Discussion section following the Cavaliere *et al* article).

Increasing the redshift range over which the behaviour of the QLF is known will reduce the range of theories compatible with the data. Two other fields of research which will contribute significantly to our understanding of the quasar phenomenon are the accumulation of evidence relating to the hypothesis that galaxy interactions trigger quasar activity, and the determination of the masses of the nuclei of a substantial sample of both active and inactive galaxies.

6. Survey techniques for finding quasars of redshift $z > 2.2$

Although the first quasar with a redshift greater than three (OH471, $z = 3.40$, Carswell and Strittmatter 1973) was discovered over 15 years ago, our knowledge of the nature of the QLF at high redshifts has, until recently, remained sketchy. The principal reason for the limited progress is the lack of a simple and effective survey method such as the UVX technique. A number of methods have been applied successfully to finding high-redshift quasars, but as experience with these techniques has grown it has been realised that various selection effects can cause the resulting quasar samples to be very incomplete (Smith 1983), in which case the apparent evolution of the QLF seen in these samples may just reflect the variation in sensitivity of the survey to quasars of different redshifts. In order to be able to draw meaningful conclusions we need to be able to quantify precisely what range of quasar type (luminosity, redshift and spectral energy distribution) is detectable with a particular survey method, and subsequently calibrate out the computed redshift-dependent incompleteness. Two recent technological advances that produce digitised source material have made such surveys feasible. Firstly, the fast-scanning microdensitometers, such as APM at Cambridge (Kibblewhite *et al* 1984, Irwin and Trimble 1984) and COSMOS at Edinburgh (MacGillivray and Stobie 1985), can digitise the information on wide-field astronomical photographic plates in a few hours. Secondly, the development of the charge-coupled device (CCD) as a detector for use on large telescopes has enabled digital surveys to reach faint flux levels. CCDs possess high quantum efficiency, typically $> 50\%$, with a linear response over a wide dynamic range.

The key to establishing the completeness of quasar samples is the application of well-defined and objective criteria for selecting candidate quasars for follow-up spectroscopy. It is then possible to establish whether any particular type of quasar would be detected in a given survey. After describing the various search techniques that have been developed for detecting quasars of high redshift, we consider in § 6.3 the requirements for applying these methods to derive the evolution of the QLF in the presence of selection effects.

6.1. Searches at radio wavelengths

Following the discovery of quasars from the 3C survey of radio sources (flux limited at a frequency of 178 MHz) subsequent optical observations have shown that surveys to bright flux limits (~ 1 Jy $= 10^{-26}$ W m^{-2} Hz^{-1}) contain substantial numbers of quasars. The majority of the remainder of the samples comprise galaxies whose optical spectra show narrow emission lines. Most radio–loud quasars have flatter spectra ($\alpha_{radio} > -0.5$) than radio galaxies so that surveys for quasars have concentrated on samples flux limited at high frequencies, particularly at 2.7 and 5.0 GHz. Radio surveys that cover a sufficiently large area of sky to produce a substantial sample of quasars are a major undertaking. The limited discriminatory power provided by radio morphology and the lack of any redshift indicator means that subsequent optical identification and spectroscopy is required. However, surveys at radio wavelengths have two important advantages. At bright flux limits optical identifications of complete flux-limited radio samples is possible, i.e. the surveys are free of the complex optical selection effects. (Indeed, the optical spectra of complete samples of radio-selected quasars provide one of the best estimates of the distribution of quasar emission-line equivalent-widths for example —information crucial to the interpretation of the slitless spectroscopic surveys. Although there are still potential problems with using such information because it has not been established that the optical properties of radio-loud and radio-quiet quasars are identical.) Secondly, since radiation at radio wavelengths is not scattered by dust, radio-selected high-redshift quasars could be used to test the hypothesis that optical samples are deficient in quasars at the highest redshifts due to obscuration by intervening dust (Heisler and Ostriker 1988). Furthermore it is possible that optically-selected samples are formed of two distinct populations as defined by their radio properties. Peacock *et al* (1986) have suggested that quasars that are 'radio-loud', which they define by radio power $P_{2.7\,GHz} > 10^{24}$ W Hz^{-1} sr^{-1}, reside in elliptical galaxies, while 'radio-quiet' quasars are hosted by spirals. Thus, the evolution of the radio-loud population may differ from that of the radio-quiet population.

At the faint flux limits of current large scale surveys, $S_{2.7\,GHz} \sim 0.1$ Jy, the proportion of low-redshift galaxies in radio samples rises rapidly and some selection strategy will be necessary, based, for instance, on optical morphology and colour, in choosing objects for spectroscopic observation.

6.2. Searches at optical wavelengths

A search for high-redshift quasars at optical wavelengths must be able to discriminate between quasars and stars in an efficient manner, by exploiting their different SEDs, see § 4. A simple approach analogous to the highly successful UVX technique, based on the $m_B - m_V$ colour, is ineffective, as now stars with SEDs that peak in the B band are not rare. Two techniques approach the ideal of obtaining high signal-to-noise ratio SEDs of all the stellar images in a particular field. In slitless spectroscopy, low-resolution (10 nm) spectra are obtained, while multicolour techniques employ broad-band photometry in several passbands (e.g. U, B, V, R and I). The latter method effectively produces spectra of very low resolution, but with higher signal-to-noise ratio than a slitless technique using the same telescope-detector combination, thereby enabling fainter flux limits to be achieved (by typically 1.5 magnitudes relative to the comparable slitless technique).

Surveys at optical wavelengths may be further subdivided according to the class of telescope used. It is possible to exploit the wide-field of view obtained for telescopes of relatively small diameter that employ Schmidt optics. Photographic plates taken with the United Kingdom Schmidt Telescope (UKST), in Australia, cover an area of ~ 36 square degrees, and can detect objects as faint as $m_B = 22.5$. The greater light gathering power of the 4-metre class telescopes allow magnitudes of $m_B \sim 24.5$ to be reached, but at the expense of a much reduced field of view, typically ~ 0.5 square degrees. Use of CCDs as detectors extends the magnitude limit as faint as $m_B = 26$ on 4-metre class telescopes, but the small physical size of CCDs restricts the field of view to only $\lesssim 0.05$ square degrees. The limiting magnitude achieved and the area covered with different telescope and instrument combinations encompass a wide range, and the suitability of a particular configuration will depend on the actual form of the luminosity function.

6.2.1. Objective-prism surveys. By placing a thin prism over the objective of a Schmidt telescope the images at the focal plane are dispersed into low-resolution spectra. The technique was first applied to the detection of quasars by Smith (1975). Many quasars are most easily identified by the presence of one or more strong emission lines, but may also be recognised through the power-law nature of their continua, and, in some cases the presence of broad-absorption features. In contrast to slit spectroscopy the spectra are superposed on a sky background of undiminished intensity. The choice of prism angle is dictated by a balance between resolution and signal-to-noise ratio as well as the necessity of minimising the number of overlapping spectra. A dispersion of ~ 200 nm mm^{-1} at 400 nm is typically employed, which allows the detection of quasars of continuum magnitude $m_B \lesssim 20$ on the fine-grained Kodak IIIa-J emulsion, although this limit depends critically on the type of quasar spectrum, especially the emission line strengths. The detectability of a quasar in a survey based on the identification of emission lines is essentially dependent on the two parameters line flux and redshift. The limiting continuum magnitude will, therefore, be a function of line equivalent width (defined as the line flux divided by the flux per unit wavelength of the underlying continuum). The derivation of space densities of quasars from such a survey is not straightforward. This topic is considered in more detail in discussing the work of Schmidt, Schneider and Gunn (§ 7.2.4).

On IIIa-J emulsion the Lyman-α line (usually the strongest emission line in quasar spectra) is detectable for redshifts $1.8 < z < 3.3$, the limits set in the blue by atmospheric transmission, and at the red end by the emulsion response. At longer wavelengths the use of red-sensitive IIIa-F emulsion could allow the detection of Lyman-α to redshifts as high as $z = 4.7$. At these longer wavelengths a larger-angle prism is required to produce a comparable dispersion to that in the blue. The brighter sky background coupled with the lower sensitivity of the IIIa-F emulsion reduces the limiting magnitude to $m_R \sim 18.5$, and furthermore, the IIIa-F emulsion has a rather uneven wavelength-dependent sensitivity. The limiting magnitude may be improved at the expense of a reduced wavelength range by the use of a suitable filter.

The development of automated detection procedures (Clowes *et al* 1984, Hewett *et al* 1985) now permit large samples of quasars with well-defined objective selection criteria to be produced, and therefore allow the completeness of the samples to be established. The first major survey to result from employing such techniques is the Large Bright Quasar Survey (LBQS Chaffee *et al* 1991), consisting of 1050 quasars brighter than $m_{B_J} \sim 18.7$, $z < 3.4$. Analysis of this sample will establish the nature of the bright end of the QLF in the redshift range $2.2 < z < 3.4$ to significantly greater precision than present estimates.

6.2.2. Slitless surveys on 4-metre class telescopes. By placing a dispersion element directly in front of the focal plane of a 4-metre class telescope the slitless spectroscopic technique may be extended to faint magnitudes, $m_B \sim 21.5$. The technique was introduced by Hoag and Schroeder (1970) and the most notable surveys have been undertaken with the Cerro Tololo 4-metre telescope (Hoag and Smith 1977, Osmer 1982) using a grating-prism (grism) combination, and at the Canada-France-Hawaii telescope (Weedman 1985, Crampton *et al* 1987) using a grating-lens (grens) combination. The technique suffers from many of the same selection effects as the objective-prism technique, with the exception that the dispersion is linear.

The full power of combining a slitless technique on a large telescope with an automated detection procedure has recently been applied in the surveys of Schmidt, Schneider and Gunn (hereafter SSG) using a grism and CCD at the prime focus of the Palomar 5-metre telescope. Most published deep slitless surveys, however, have been based on the visual identification of quasars on photographic plates, and are therefore again liable to charges of subjectivity and incompleteness. The most important of the earlier photographic visual surveys is that of Osmer (1982), § 7.2.1, which produced the first evidence for a decline in the space density of quasars beyond $z = 3$.

6.2.3. Multicolour surveys. The extension of the UVX technique by the addition of broad-band photometry at longer wavelengths permits the detection of quasars of redshift $z > 2.2$. Koo and Kron (1982) have used photographic plates taken with the 4-metre Mayall telescope in the U, B_J (close to B), F (between V and R) and N (close to I) passbands, scanned with a PDS microdensitometer. The information is conventionally displayed in two-colour diagrams, for example by plotting $m_U - m_{B_J}$ against $m_{B_J} - m_V$, as in figure 5: Galactic stars occupy a well-defined sequence, close to the black-body line (§4), and objects lying away from the sequence are selected as candidate quasars. Koo *et al* (1986) argue, on the basis of searches for quasars in the same area of sky employing different techniques, that there will be few quasars that mimic the colours of Galactic stars, and that their samples are $\sim 80\%$ complete. By employing an objective algorithm for the selection of candidates (Warren *et al* 1991a), it is possible to quantify the sample incompleteness, without having to resort to qualitative arguments. The effectiveness of the multicolour technique may be extended by the use of additional passbands, and by including upper limits in the analysis for objects too faint to be detected in particular passbands. The multicolour technique offers the distinct advantage of being able to detect objects whose spectra do not contain strong emission lines.

Following the success of the multicolour technique in identifying quasars at redshifts $z > 4$, Irwin *et al* (1991) have developed a high–redshift analogue of the UVX selection technique. Noting that the discontinuity in flux across Lyman–α caused by the increasingly dense Lyman–α forest lies at wavelengths $\gtrsim 610$ nm at these redshifts, their technique involves searching for objects which are extremely red in $m_{B_J} - m_R$. There are very few Galactic objects at high Galactic latitudes with $m_{B_J} - m_R \gtrsim 3$ and galaxies of intermediate redshifts can be excluded from the candidate list on the basis of their non–stellar appearance. Irwin and McMahon's application of the technique to a large survey area at bright magnitudes $m_R < 19$ has been highly successful, identifying 18 of the 36 quasars with redshifts $z > 4$ known at this time (September 1991).

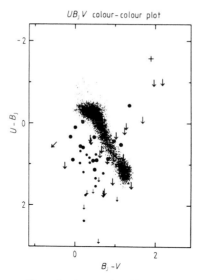

Figure 5. A two-colour U, B$_J$, V projection for approximately 10 000 stellar objects, magnitudes $17 \leqslant m_R \leqslant 19.5$, from the first field of the Warren *et al* (1991a) survey. Superposed on each plot are the locations of 53 redshift $z \geqslant 3.0$ quasars identified in the first field (large symbols) and the second field (small symbols). Quasars detected in all three passbands are represented by filled circles. Arrows denote quasars with only upper limits in one or two of the photometric passbands. The cross indicates the position of the redshift $z = 4.43$ quasar which is based on three upper limits—no detections in the U, B$_J$ or V passbands. Outliers that are UVX were excluded from this survey as these will be almost exclusively low-redshift quasars.

6.2.4. Further optical techniques. Two further optical techniques are available: (1) after many years, extragalactic sources may be identified by an absence of proper motion (Kron and Chiu 1981), and (2) quasars may be selected on the basis of variability, using photometric data from several epochs (Usher *et al* 1983, Koo *et al* 1986, Hawkins 1986). Both techniques offer the potential to obtain quasar samples essentially free of many of the complications involved in surveys which depend on the detailed form of quasar SEDs. However, very long periods of time are required before either method becomes competitive with the more common radio and optical selection techniques. Koo *et al* (1986) concluded that in practice neither technique is as effective as the multicolour method, and that no new class of quasar has been discovered by these methods. While this may have been true of the more limited database on which this conclusion was based, in an encouraging development Majewski *et al* (1991) (§ 7.3.1.2) have recently demonstrated the effectiveness of combining the criteria of variability and absence of proper motion. Their study uses a total of 31 photographic plates, taken over a period of 16 years, so the technique is demanding in terms of resources. However spectroscopic observations of their candidates will provide a valuable estimate of the completeness of multicolour-selected samples.

6.3. Rationale of surveys, and sample incompleteness

The results of a survey may be thought of as the convolution of the quasar luminosity function with the survey 'selection function'. The selection function quantifies the whole of the detection procedure, and is just the array of probabilities for detecting quasars covering the complete range of luminosity, redshift and SED. A full description of the luminosity function must include not only the variation of the space density but also the range of quasar SEDs, as a function of absolute magnitude and redshift. For the modern digital surveys which employ well-defined algorithms for the selection of candidates it is possible to compute the probability of detection of any quasar type, and so to determine the survey selection function. Hence, in principle, by a process of deconvolution the QLF may be derived. More correctly, because of the existence of observational error and the finite sample sizes it is possible to constrain the range in which the QLF must lie. The deconvolution of the selection function may be achieved by convolving hypothetical luminosity functions with the selection function, and finding which are compatible with the observational results, by testing, ideally simultaneously, against the observed M–z distribution as well as the distribution functions for the observed spectral properties (e.g. equivalent widths of emission lines, and continuum spectral indices).

To carry through this methodology is a formidable undertaking and requires a large sample from a survey sensitive to a wide range of quasar SEDs. If the number of quasars detected is small, and if the survey is sensitive only to a narrow range of quasar SEDs, the class of compatible luminosity functions will be large, i.e. the survey will provide only weak constraints.

Several surveys for high-redshift quasars have been based on the detection of emission lines. It may be argued that this is a simple form of selection, making the interpretation of survey results relatively easy. However, the use of more complex selection criteria, particularly those involving a number of different techniques— including selection by extreme colour, and the detection of absorption lines in slitless spectroscopic surveys for example —can greatly extend the range of SEDs to which the survey is sensitive, increasing the fraction of quasars detected and producing tighter constraints on the QLF. Furthermore, for modern digital surveys there is no difficulty in principle in incorporating these selection procedures into the analysis of the luminosity function: the complexity of the selection process is immaterial provided the procedures are precisely specified and that high-precision simulations of the detectability of quasars of particular M, z and SED can be accomplished. This is the basis of the approach developed by Hewett *et al* (1985) and underlies the rationale of the wide-field multicolour surveys of Mitchell *et al* (1990) and Warren, Hewett and Osmer (hereafter WHO)—§ 7.3.2.

7. The quasar luminosity function at redshifts $z > 2.2$

Progress towards understanding the QLF at high redshift has been slow because, in addition to there being no simple technique for finding quasars, high-redshift quasars are far less common than first thought. The early UVX surveys revealed the steep bright-end slope of the QLF (figure 4) and the strong evolution to increased space density at higher redshifts for a particular absolute magnitude. Extrapolating to faint magnitudes and to redshifts $z = 3 - 4$ suggested that faint high-redshift quasars would be common. This led to a series of very deep surveys covering very small areas of sky. These established that the steep power-law of the QLF at bright magnitudes turns over to a flatter slope at faint magnitudes (Koo and Kron 1982), and that the rapid evolution seen at low redshifts does not continue to the highest redshifts, $z > 3.5$ (Osmer 1982). Nevertheless the limited areal coverage of these and later surveys (Crampton *et al* 1987, SSG 1986a,b) has meant that the sample numbers are too small to constrain strongly the form and evolution of the QLF at high redshifts. The surveys were designed to test models in which the quasar space density was expected to be 10–100 times as large as it has actually turned out to be, and while they achieved this goal they were not able to investigate the high-redshift luminosity function in detail. In addition little effort has been devoted to determining the range of luminosity functions and evolutionary models that are consistent with the data. In general it has not been demonstrated whether the surveys are able to distinguish between the alternatives that the evolution of the luminosity function is merely slowing down, or whether the space density declines at high redshift. Furthermore, with few exceptions the incompleteness of the optical surveys has not been considered in a strictly quantitative manner.

In this section we consider the observational data that relates to the luminosity function at high redshifts and attempt to identify the strengths and weaknesses of the different surveys. We describe in some detail the recent surveys of SSG (1986a,b, 1991), WHO (1991a,b,c) and Irwin *et al* (1991) that satisfy the requirements of covering a large area, are sensitive to quasars of bright and intermediate magnitudes, and for which quantifiable selection functions may be calculated. The space density of the faintest quasars at high redshifts is addressed by the work of Koo and Kron (1988) who have undertaken the most detailed study in this magnitude range. In the final section we attempt to draw together the results from the different observational groups.

7.1. Radio surveys

The most complete study of the evolution of extragalactic radio sources at high redshift is the analysis, based mostly on the Parkes Selected Regions ($S_{2.7\,\text{GHz}} > 0.1$ Jy, 216 square degrees) and various samples from brighter Parkes surveys, at 2.7 GHz, by Peacock and colleagues (Peacock and Gull 1981, Peacock 1985, Dunlop and Peacock 1990). Peacock divides the data into flat-spectrum sources $\alpha > -0.5$—predominantly quasars—and steep-spectrum sources $\alpha < -0.5$—predominantly galaxies. The radio luminosity function is modelled as a series expansion with the goodness of fit assessed using a two-dimensional Kolmogorov-Smirnov statistic. The method of analysis allows for samples comprising complete redshift data, as well as source counts. The extensive spectroscopic information now available for the Parkes Selected Regions sample permits a fruitful analysis of these data despite its small size. There are 49 quasars in the Selected Regions but only 3 have redshifts $z > 2.5$.

The flat-spectrum sources display a peak in the comoving space density at a redshift near $z = 2$, with a decline by a factor $\geqslant 3$ by redshift $z = 4$ (Dunlop and Peacock 1990), as illustrated in figure 6. It is of interest that the steep spectrum sources show a similar decline at high redshift (Peacock and Miller 1988, Dunlop and Peacock 1990) suggesting a relationship between the two classes of source—radio galaxy and radio quasar. The extension of this work to fainter flux limits will provide improved statistics, and permit an investigation of the luminosity dependence of the decline.

The largest number of high-redshift quasars detected in a radio investigation comes from the survey of Savage *et al* (1988) based on the large-area brighter Parkes survey, flux limit $S_{2.7\,\text{GHz}} > 0.5$ Jy, which covers some 15 000 square degrees. Optical

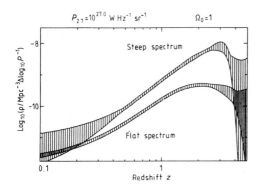

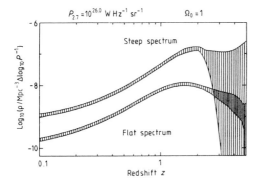

Figure 6. Limits to the space density for steep- and flat-spectrum radio sources at constant radio power, $P_{2.7 \text{ GHz}}$, as a function of redshift, z. Acceptable models are constrained to lie within the shaded regions. Data for two radio powers, $P_{2.7 \text{ GHz}} = 10^{27}$ W Hz^{-1} sr^{-1} and $P_{2.7 \text{ GHz}} = 10^{26}$ W Hz^{-1} sr^{-1}, are shown. The value of the deceleration parameter is $q_0 = 0.5$. The figure is reproduced from Dunlop and Peacock (1990).

spectroscopy of a sample of 403 flat-spectrum sources is 67% complete. Some 3% of the identified quasars have redshifts $z > 3$. Savage *et al* (1988) state that there is no strong evidence for a redshift cutoff in their data. However, a detailed study remains to be undertaken. Completion of the survey will provide a powerful test of the Dunlop and Peacock model.

As explained earlier, the proportion of high-redshift quasars in radio samples decreases for fainter flux limits. McMahon (1991) has demonstrated how the few high-redshift quasars in faint radio samples may be successfully extracted in an efficient manner. Four new radio-loud quasars $z > 3.5$ have so far be found in the $S_{5 \text{ GHz}} > 0.1$ Jy MIT-Green Bank survey, by limiting spectroscopy to objects with red optical colours.

7.2. Slitless surveys

7.2.1. The Osmer IIIa-F grism survey. Osmer (1982) used a red-blazed grism and IIIa-F emulsion combination together with a suitable filter that produced spectra in the wavelength range $570 - 690$ nm, i.e. sensitive to the detection of Lyman-α in the interval $3.7 < z < 4.7$ and C IV in the interval $2.7 < z < 3.5$. Seventeen plates covering a total area of 5 square degrees were taken, and candidates were selected by a visual search for emission lines. No quasars were found in the higher redshift range, while five quasars in the range $2.77 < z < 3.36$ were detected through their C IV emission.

As stated in § 6.2.1 the limiting sensitivity of such a survey is a function of several variables and because of the subjectivity of the search technique it is very difficult to deal with this question in a quantitative manner. Nevertheless Osmer was able to use consistency arguments to show that the complete lack of quasars detected through their Lyman-α emission implies that the space density of quasars declines beyond a redshift $z = 3.5$. He first compared the number of quasars detected by their C IV emission with the results of an earlier survey conducted by Hoag and Smith (1977), sensitive to Lyman-α over the same redshift interval; the relative strength of the Lyman-α and C IV emission lines determines the relative sensitivity of the two surveys. Assuming that the relative strength of the Lyman-α and C IV lines does not evolve with redshift, and using the number of quasars found in the interval $2.5 < z < 3.5$ in the earlier survey, the expected number of quasars in Osmer's survey in the interval $3.7 < z < 4.7$ for no evolution in the comoving space density was calculated to be 9–12. Osmer's results implied, at a significance level of 95%, that the space density in this higher redshift interval is down by a factor $\gtrsim 3$. This important result was the starting point for the new surveys. The method of analysis however naturally prompts the question of whether the apparent decline reflects a true decrease in space density or the evolution of the strength of the Lyman-α emission line.

For reference, and with the caveat that the value is approximate, we can estimate from the details provided in Osmer's paper that this survey has an equivalent limiting magnitude for $3.7 < z < 4.7$ of $m_R \sim 20.5$, i.e. a survey complete to this limiting magnitude would produce a similar number of quasars, assuming that the line properties are similar to those at lower redshifts. This magnitude limit corresponds to an absolute magnitude of $M_B \sim -26$.

7.2.2. The Canada-France-Hawaii Telescope grens survey. From the visual examination of the inner 0.52 square degrees of eight IIIa-J grens plates taken at the Canada-France-Hawaii Telescope, Crampton *et al* (1987, hereafter CCH) have produced a sample of 117 quasars, $m_B \leqslant 20.5$, of which 12 have redshifts $z > 2.2$. Continuum magnitudes are based on PDS scans of Palomar Observatory Sky Survey (POSS) plates, corrected for emission lines in the passband—these magnitudes are liable to substantial errors because of the variable depth of the POSS plates, as the calibration is based on measurements of standard stars on other plates. Apparently no correction has been made for the effect of absorption in the Lyman-α forest, which will affect the estimates of the continuum magnitudes at high redshift. The completeness limit of $m_B = 20.5$ does not appear entirely secure, as the sample contains very few quasars fainter than this limit, yet the very broad distribution of emission line equivalent-widths found in quasar samples implies that strong-lined quasars could be found to continuum fluxes substantially fainter than the limit of completeness. On the other hand it can be argued that only a few weak-lined objects have been missed, since the observed surface density of quasars with $z < 2.2$ agrees closely with that found in UVX samples.

CCH follow the formalism of Schmidt and Green (1983) and adopt a luminosity-dependent density-evolution (LDDE) model for the QLF, where the evolution relative to the local luminosity function is parameterised by $D(M, z) = exp(k(M_0 - M)\tau)$. The function $D(M, z)$ is the evolutionary correction such that $D(M, 0) = 1$. Although the mathematical formalism of the LDDE model is very different to the pure luminosity evolution model of BSP (§ 5) the shape of the QLF at low redshifts is qualitatively similar. To provide an acceptable fit to the high-redshift data CCH have modified the parameter k to introduce a 'soft cutoff' $k' = \frac{k}{exp(2(z-z_c))+1}$. A model is deemed acceptable if the value of the density-weighted V/V_{max} statistic is close to 0.5. (The V/V_{max} statistic quantifies the distribution with distance of objects in a survey volume. A uniform distribution gives a value 0.5. A larger value implies more objects at large distance, a smaller value fewer. Where the evolution is characterised by an evolutionary function, as here, if the function fits the data well, weighting each point by the reciprocal of the evolutionary correction will give $V/V_{max} = 0.5$, see Avni and Bahcall 1980 for details).

CCH obtain a satisfactory fit to their data using the evolutionary parameters k and M_0 found by Schmidt and Green, and a value of $z_c \sim 2.9$. The result implies that the comoving space density of quasars peaks at a redshift $z \sim 1.7$. Although there is no indication of the range of models that provide an acceptable fit to the data, and even though the V/V_{max} test does not employ the full $M - z$ information, a more sophisticated analysis is probably not justified because of the uncertainty in the quasar continuum magnitudes and the sample completeness. The model of CCH predicts that the space density declines more rapidly at bright magnitudes in contrast to the analyses of Osmer and Smith (1980) and Osmer (1980) who concluded that at bright apparent magnitudes, $m_V < 19$, there was no evidence for a decline in the comoving space density over the interval $1.8 < z < 3.0$. An alternative model in which the lack of high-redshift quasars in the CCH sample is explained by a decline in space density of only the faint quasars would reconcile the two data sets. Some support for this idea comes from the fact that the V/V_{max} value for the deeper grism survey discussed by Osmer (1980), for $m_V < 19$, is smaller than the value for the objective-prism survey reported by Osmer and Smith (1980), $m_V < 17.5$.

7.2.3. UK Schmidt Telescope IIIa-F objective-prism survey. Much of the recent interest in searches for quasars of the highest redshift, $z > 3.3$, was inspired by the success of Hazard and McMahon in finding intrinsically bright quasars, $M_B \sim -28$, through the detection of Lyman-α emission on unfiltered UKST IIIa-F objective-prism plates (Hazard and McMahon 1985, Hazard *et al* 1986, hereafter HMS). HMS list a total of 7 quasars with $3.3 \leqslant z \leqslant 3.8$, $m_R \leqslant 18.5$ in two UKST fields, total area $\sim$ 60 square degrees. The comoving space density implied from the HMS work has been estimated here as follows. The broad-band magnitudes of HMS appear to be too bright; each of the three quasars common to the region surveyed by WHO is found to be 0.5 magnitudes fainter. Furthermore, for $z < 3.5$, the continuum magnitudes will be affected by the presence of C IV in the R band. Accordingly we have added 0.5 magnitudes to their estimates, considered the interval $-28.5 < M_B \leqslant -27.5$, and computed the space density for the redshift interval $3.3 \leqslant z < 3.9$ using the $1/V_a$ estimator (Felten 1976)—the redshift upper limit being that quoted by HMS. The space density so derived is plotted in figure 4, and compared with the luminosity function of BSP.

The recently completed multicolour survey of Mitchell *et al* (1990) (§ 7.3.3) produces a lower surface density of quasars. Their estimate of the space density is also shown in figure 4. Both data points are compatible with a space density of 10^{-8} Mpc^{-3} mag^{-1} for $M_B = -28$, $z = 3.5$, similar to the value found for $z = 2$.

7.2.4. The Palomar faint grism survey

7.2.4.1. Methods, results and analysis. SSG have undertaken a series of searches for quasars of high-redshift using a grism-CCD combination with the Hale 5-metre tele-scope. The resolution of the grism spectra is $\sim 10 - 17$ nm, depending on the seeing conditions. The wavelength range, which is slightly different for each survey, covers $450 \lesssim \lambda \lesssim 720$ nm, equivalent to $2.7 < z < 4.9$ for Lyman-α. Two initial searches produced only a small number of quasars. The first probed faint absolute magnitudes (equivalent limiting magnitude $m_R \sim 21.5$) and covered 0.91 square degrees (SSG 1986a) while the second survey (SSG 1986b) reached less deep $m_R \sim 20.0$ and covered 7.84 square degrees. The main survey (SSG 1991) has just been completed, and is of intermediate depth and is much larger, covering 62 square degrees in six separate strips.

The observational techniques and principles of the selection function calculation are described in SSG (1986a). Quasars are selected on the basis of the detection of emission lines at a signal-to-noise ratio > 7 with the additional requirement of a minimum observed equivalent-width of 5 nm. The spatial (x, y) variation of the CCD response is obtained from the sky-background counts, and the wavelength response from observations of standard stars. After sky subtraction and flux calibration, an automated routine is used to search for the presence of emission lines at each pixel in every spectrum. Since the objects are faint, the noise is dominated by the sky background. The signal-to-noise ratio of a line of given flux is, then, a function of the sky-background brightness, the seeing conditions, the CCD response and a wavelength-dependent defocusing. All these quantities can be measured and it is possible to compute a limiting observed line flux, i.e. the weakest line that will satisfy the selection criteria, as a function of wavelength and position on the CCD chip. The selection procedure results in a large number of candidates, of order 1 000 for a single night's observations in transit mode. Of these $\sim 90\%$ are eliminated on visual inspection of the raw data, as due to overlapping spectra, cosmic ray events and the like. The remainder are observed spectroscopically.

Before considering the results of the main survey we summarise briefly the outcome of the two initial searches. The first survey produced only 10 quasars, and the second 8, and the highest redshifts found were respectively $z = 2.66$ and $z = 2.76$. The lack of high-redshift quasars in the first two surveys was initially interpreted as being the result of a sharp cutoff in the space density of quasars at high redshift. SSG (1986a,b) compared the null results at redshifts $z > 2.9$ with the predictions of the model of Schmidt and Green (1983) that fits data at low redshifts, extrapolated with continuing rapid evolution to higher redshifts. On this basis over 30 quasars $z > 2.9$ should have been found in each of the surveys. They concluded that quasars of absolute magnitude $M_B \gtrsim -26$ ($m_B = 20.0$) suffer a sharp cutoff at or below a redshift $z = 3$. However this interpretation is not unique, and the results have not been shown to be inconsistent with a moderate decline in space density beyond $z = 2$ and SSG's conclusion based on their more recent work is indeed that the decline is gradual. Therefore the earlier suggestion of the existence of a sharp cutoff may be disregarded.

The third survey has produced a sufficiently large sample of high-redshift quasars to allow a detailed investigation of the form of the evolution of the QLF beyond $z = 2$. Spectra have been obtained for all 1660 candidate lines selected from the grism spectra, and emission lines confirmed in 1058 of these. Many of the objects are emission-line galaxies or low-redshift quasars. There are 141 quasars of redshift $z > 2$, including nine with redshifts $z > 4$, with the highest redshift recorded being $z = 4.73$ (Schneider *et al* 1989b). In some cases more than one emission line was detected in a quasar: of the 141 quasars $z > 2$, there are 90 detections of Lyman-α and 72 detections of C IV. The sample of high-redshift quasars is the largest so far compiled.

SSG (1991) have presented a preliminary analysis of their high-redshift results. The calculation differs from the usual procedure for computing luminosity functions. The detectability of a quasar in their surveys depends on the object line flux. Hence their measure of the quasar luminosity is the line luminosity, and they compute a line luminosity function (number of objects per unit volume per unit line luminosity) rather than a continuum luminosity function. When comparing the results of SSG to those of other surveys it should be born in mind that the same evolution as found for the continuum luminosity function is expected only if the distribution of line equivalent widths does not vary with redshift, which is at present an untested assumption. SSG use their sample to compute the evolution of the line luminosity function over the redshift interval $2.0 < z < 4.7$. At the end, as an external check of their survey, they compare their results at $z = 2$ with the model of BSP by attempting to compute the continuum luminosity function from their line luminosity function.

To simplify the analysis SSG assume that over the limited luminosity range of the sample the line luminosity function is adequately described by a power law, of invariant slope, and a redshift-dependent normalisation $\rho(z)$. As the limiting line luminosity as a function of redshift is known for each of the six survey strips, the slope is easily derived, from the relative numbers of objects as a function of luminosity above the limiting luminosity. The conversion of the results for the C IV line to the reference line Lyman-α was established using quasars in which both lines were detected in the grism spectra. The variation of the normalisation is computed in bins, from the number of objects detected, the survey volume in the redshift interval considered, and the limiting line flux. The results are presented in terms of the integral space density of quasars brighter than a reference line luminosity, $L_{Lyman-\alpha} > 44.75$, and are reproduced in figure 7. The computed space densities from this preliminary analysis are approximately constant over the interval $2.0 < z < 3.0$ and are declining by a factor of about 3.2 per unit redshift interval beyond $z = 3$.

7.2.4.2. Discussion. The striking result displayed in figure 7 provides us with the clearest picture so far of the variation in quasar space density at the highest redshifts. This result however is still preliminary and there remain a number of questions to be addressed. For example the analysis takes no account of the absorption of the blue wing of the Lyman-α line by Lyman-α forest lines (§ 2.2). The number of absorption lines in quasar spectra evolves rapidly with redshift (see e.g. Murdoch *et al* 1986) so that more and more of the Lyman-α line is eaten away at high redshift. This would cause an apparent deficit of high-redshift quasars, relative to the true number, in which case it may be that the decline is not as steep as illustrated. In addition the influence of line profile on the detectability of a line has yet to be addressed. Many of the spectra from the SSG survey show narrow Lyman-α lines, while objects from other surveys apparently display a broader range of line profiles.

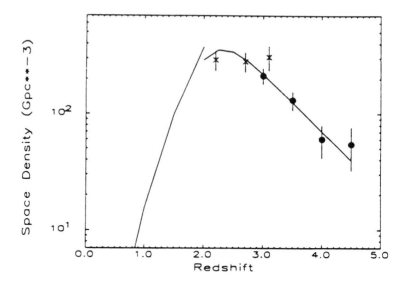

Figure 7. The results of a preliminary analysis of the sample from SSG's third survey, taken from SSG (1991), showing the integrated space density per cubic Gigaparsec for quasars $\log L_{Ly\alpha} > 44.75$. Crosses mark the results from 72 detection of C IV, and filled circles the 90 Lyman-α detections. The line for $z < 2$ is the integrated space density $M_B < -26$ from the model of BSP. The line for $z > 2$ is a tentative model in which the density declines by a factor of 3.2 per unit redshift for $z > 3$.

An external check of the SSG survey is possible through a comparison at $z = 2$ with the results of BSP. The problem of comparing the results of a line-flux limited survey with the predictions of a continuum luminosity function model was first considered by SSG (1986a). The problem can be solved if the equivalent-width distribution at high redshift is known. At a fixed redshift the limiting continuum flux for a line of given equivalent width is just the survey limiting line flux at that redshift divided by the equivalent width. The prediction then involves a double integral over the equivalent-width distribution and the continuum luminosity function (with an appropriate k-correction term to convert from $\lambda = 121.6$ nm to $\lambda = 440$ nm for M_B). In their analysis SSG (1991) adopt an equivalent-width distribution taken from other surveys in the literature. They find a discrepancy of over 0.5 mag, in that the line flux limit of their survey implies an equivalent limiting continuum absolute magnitude $M_B \sim -26.5$ at $z = 2$, while the number of quasars found corresponds to a limit $M_B \sim -26$ (figure 7). The cause of the discrepancy is not immediately apparent, but may lie in the adopted equivalent-width distribution. Meanwhile it would be possible to derive the continuum luminosity function directly from their data, since the continuum magnitudes are known for all their quasars.

Schneider *et al* (1991) report on the first success of a new multicolour survey aimed at extending this work to higher redshifts still. Because of the dramatic drop in flux across the Lyman-α emission line due to the increasingly dense Lyman-α forest, the multicolour technique is very effective for redshifts $z > 4$. The new survey at Palomar uses four passbands covering 600 nm to 800 nm, designed to detect quasars over the redshift range $4.0 < z < 5.5$. In the first three square degrees of the survey they discovered a quasar of redshift $z = 4.90$, currently the most distant object known.

7.3. Multicolour surveys

7.3.1. Narrow-angle photographic surveys.

7.3.1.1. Mayall 4-metre survey: methods, results and analysis. Estimates of the space density of quasars of the faintest luminosity come from deep photographic multicolour surveys, in particular the studies by Koo and Kron of selected areas covering ~ 0.6 square degrees and reaching $m_{B_J} = 22.5$ (Koo and Kron 1982, Koo *et al* 1986) using photographic plates taken at the prime focus of the Mayall 4-metre telescope at Kitt Peak. Other surveys, to $m_{B_J} = 22$, employing similar plate material from the Anglo-Australian Telescope and the 3.6-metre telescope at the European Southern Observatory are discussed in §7.3.1.3. Based on their own surveys and published results of surveys at brighter magnitudes, Koo and Kron (1988) have proposed a model for the evolution of the QLF that shows no decline in space density at any luminosity out to redshifts $z \sim 4$. We summarise here the survey and analysis of Koo and Kron and then re-examine the observational basis for their model.

Candidate quasars in selected area SA57 (area 0.29 square degrees) were chosen by identifying objects, with colours different to those of normal stars, from an inspection of UB_JF two-colour diagrams. This colour system is close to UBR (see § 6.2.3 for more details), and we assume $m_B = m_{B_J} + 0.1$. Follow-up spectroscopy has resulted in the identification of 30 quasars, redshifts $0.9 < z < 3.1$; spectroscopy of the 77 candidates is complete to $m_B = 21.1$, and half-complete and unbiased by redshift to $m_B = 22.6$. On the basis of colour, photometric variability, and proper-motion information, combined with the available spectroscopy, they divide the remaining candidates into categories—quasar, star, narrow-line galaxy—providing an estimate of the number-apparent magnitude relation, $N(m)$, for the survey. They estimate the sample incompleteness to be $\lesssim 20\%$, due to quasars with colours that lie within the stellar locus.

Koo and Kron choose to investigate the QLF by deriving a function directly from their data, rather than adopting the approach of Peacock and others in which the predictions of a range of models are compared to the observational data. Koo and Kron's data suggest that the shape of the number-redshift relation, $N(z)$, does not change with apparent magnitude. With this assumption and adopting smooth functional forms for the $N(z)$ and $N(m)$ relations (to avoid difficulties associated with binning small samples), the number density of quasars is then defined anywhere in the M–z plane. A synthetic quasar SED is employed to compute the corrections for the presence of emission and absorption features in the blue B_J passband (the k-corrections). This completes the information required to compute the comoving space densities, which are provided in seven redshift bins $-0.1 < \log(z) < 0.6$, centred on apparent magnitudes $m_B = 20.1, 21.1, 22.1$ (their table 4). No quasars were found in the highest-redshift bin, $3.16 < z < 3.98$, and the space densities quoted are upper limits computed on the basis of one quasar having been found in this interval. The absolute magnitude limit of the survey brightens rapidly over the redshift interval $3 < z < 4$ because the sample is limited at blue wavelengths where the effects of absorption in the Lyman-α forest become increasingly important. At $z = 3$ the survey reaches significantly further into the luminosity function than the first survey of SSG, whereas by $z = 4$ the SSG survey is more sensitive.

At faint magnitudes the model QLF of Koo and Kron is independent of redshift (by assumption) and is described by a power-law of slope $\beta = -1.9$. The Koo and Kron results for $z = 1.8$, show a similar normalisation, but an appreciably steeper faint-end slope, compared to the BSP model ($\beta = -1.4$). Although the gradient of the faint end of Koo and Kron's QLF is not well determined because of the small sample size, this steeper slope is apparently confirmed by the results of an extension of this work (Kron *et al* 1991).

Using data from surveys with brighter limiting magnitudes Koo and Kron (their figure 7) show that the shallow faint-end slope becomes much steeper at bright magnitudes, with the location of the transition point shifting to higher luminosities at earlier times; in their model the space density of the faintest quasars is then invariant with redshift, while the numbers of bright quasars increase continuously to $z = 4$. The model approximates the pure luminosity evolution model of BSP at low redshifts, $z \leqslant 2.2$, and at higher redshifts the predictions lie close both to the upper limits computed from their own survey at redshifts $3.16 < z < 3.98$ and to their calculation of the space density at bright magnitudes, $M_B < -28$, based on the HMS data (§ 7.2.3). This leads Koo and Kron to consider a model in which the lifetime of a quasar decreases with luminosity, $t_L = t_0 (L/L_0)^{-1/n}$, so that the brighter quasars fade more quickly: the turnover magnitude of the QLF represents the luminosity of the brightest surviving quasars at that epoch. An acceptable fit using the data in the range $1 < z < 3$ is found for a range of birth epochs $3.9 < z < 13$, $q_0 = 0.5$, and for a value $n \sim 2.6$.

7.3.1.2. Mayall 4-metre survey: discussion. The model of Koo and Kron provides an alternative view of the evolution of the quasar population. In fact, it is not in particularly good agreement with the low-redshift data, $z \leqslant 2.2$ (Marshall 1988), but a restatement of the same appealing physical picture could no doubt be made that would give an acceptable fit. The extrapolation to higher redshifts rests on a comparison with only very limited observational data and at bright magnitudes the model is probably inconsistent with observations. For the redshift interval $2.51 < z < 3.16$ the model lies well above the data point of Osmer and Smith (1980) (Koo and Kron's figure 7), while in the highest redshift interval, as has been suggested in §7.2.3, the HMS point should be shifted to a fainter magnitude, as well as a lower space density if the results of Mitchell *et al* (1990) are included. It has already been argued that the space density of bright quasars, $M_B \sim -28$, is quite similar at $z = 3.5$ to that at $z = 2$ (figure 4), rather than increasing significantly at high redshift.

At faint magnitudes $M_B \lesssim -25$ the deep narrow-angle photographic surveys provide the only available information on the form of the QLF at redshifts $z \sim 3$. Does the survey of Koo and Kron support their contention that there is no decline in space density for faint high-redshift quasars? Scaling the number of detected quasars to allow for the candidates for which spectra have yet to be obtained, and applying their 20% correction for quasars hidden in the stellar locus, the numbers of quasars in the SA57 sample are estimated to be 6 and 0 for the redshift ranges $2.51 < z < 3.16$ and $3.16 < z < 3.98$ respectively. The predictions based on the evolution model of Koo and Kron are 6 for the first redshift bin, and 2–3 for the second. Consider now an alternative picture where there is a strong decline in space density at high redshift.

We have taken the QLF model prediction of Koo and Kron at $z = 2.5$, preserving this shape but reducing the space density in a linear fashion from 100% at $z = 2.5$ to 10% at $z = 4$, with k-corrections taken from their paper. The predicted numbers of quasars in the two redshift bins are 5 and 1 respectively. It is clear that the current survey cannot distinguish between two very different pictures, and that a survey over a much larger area of sky, preferably based on a red limiting magnitude, is required in order to establish whether or not the space density of faint quasars declines beyond $z = 2$.

In conclusion, the continuing evolution to higher space densities beyond $z = 2$ displayed at the bright end of Koo and Kron's luminosity function is not supported by the data, whereas the areal coverage of their original survey is insufficient to provide strong constraints on the behaviour of the QLF at high redshifts for the faintest absolute magnitudes.

Kron *et al* (1991) report on the current status of an extension of this work to cover an area four times as large, but now limited on m_F, and employing a well-defined selection algorithm, as opposed to their previous subjective selection of candidates. Kron *et al* believe that their sample is unbiased with respect to redshift, but calculations of the expected colours of high-redshift quasars (Giallongo and Trevese 1990, WHO 1991c) suggest that the multicolour technique is significantly incomplete in certain redshift intervals, particularly $2.4 < z < 2.9$. Majewski *et al* (1991) (§ 6.2.4) have identified quasars using the same plate material, by selecting as candidates objects that show no proper motion, but display luminosity variations at different epochs. Their candidate list in SA57 contains 20 objects not found in the list of Koo and Kron (1988). Five have been observed spectroscopically and four are quasars, one with a redshift $z = 3.54$. This sample, will provide an indication of the redshift distribution of objects missed by the multicolour selection, and an estimate of completeness.

7.3.1.3. Other faint narrow-angle surveys. Two other surveys, in southern Galactic fields, that are similar in scope to that of Koo and Kron, but reach 0.5 mag. less deep, have recently appeared. Boyle *et al* (1991a) used U, B_J, F and N plates to select candidates in fields covering a total area of 0.85 square degrees, to $m_{B_J} = 22.0$, and Zitelli *et al* (1991) used U, B_J and F plates and covered 0.69 square degrees to $m_{B_J} = 20.85$ and 0.35 square degrees to $m_{B_J} = 22.0$. The latter survey selected candidates also from grism plates, and used photometry at three epochs to identify additional candidates through their variability. Zitelli *et al* found a higher surface density of quasars $2.2 < z < 2.9$, 28.8 ± 8.4 per square degree, $m_{B_J} < 22.0$, than Boyle *et al* , 9.0 ± 3.7 per square degree. Only about one half of the quasars found by Zitelli *et al* in this redshift range were obvious outliers in their two-colour diagrams. Two of the others were only identified through their grism spectra, and the remainder lay in the border of the stellar locus. Thus, it is possible that there remain a number of quasars hidden in the stellar locus, and that the true surface density is substantially higher than the lower limit provided by this survey.

Boyle *et al* (1991b) have incorporated these two new surveys into the extensive database used by BSP, to extend the analysis of the QLF to cover the entire redshift range $0 < z < 2.9$. The k-corrections are those computed by Cristiani and Vio (1990) for their composite spectrum. This spectrum has a spectral index $\alpha \sim -0.7$, somewhat steeper than the slope of $\alpha = -0.5$ traditionally used in computing absolute magnitudes. A model, very similar to that of figure 4, evolving out to $z \sim 2$, but with constant space density over the interval $2.0 < z < 2.9$, fits the data satisfactorily. They then consider the possibility that the space density remains constant out to $z = 4.5$, and compare the predictions of this model against the results of a number of high-redshift samples. They conclude that, with the exception of the third SSG survey, the observational data at redshifts $2 < z < 4.5$, although indicating a modest decrease in space density, are consistent with a constant comoving space density. With regard to the k-corrections, if a shallower slope $\alpha = -0.5$ is used one finds a steeper decline, and the opposite for a steeper slope $\alpha = -1.0$. This analysis was based on preliminary data from the various high-redshift surveys, and has been superseded by the more complete results reported recently by these groups (SSG 1991, WHO 1991c, Irwin *et al* 1991, § 7.2.4, 7.3.2, 7.3.3 respectively).

7.3.2. Wide-angle photographic surveys to intermediate depth. WHO (1991a, b, c) have applied the multicolour technique to APM scans of UKST plate material. Using pairs of plates in each of the six passbands U, B_J, V, OR, R, I, they have undertaken a survey of two fields (OR and R are two slightly different red passbands). Using six passbands rather than three means the survey is sensitive to a broader redshift band and to a wider range of quasar SED. The effective survey area is 45.7 square degrees covering the magnitude range $16 \leqslant m_{OR} \leqslant 20$, making the survey of order 100 times as large in area as that of Koo and Kron, but reaching to a magnitude limit about 2 magnitudes brighter. The number of stellar objects in the survey is 140 000 and substantial effort has been devoted to the reduction of the raw APM scans in order that the number of images with spurious colours, and which thereby become false quasar candidates, is reduced to an absolute minimum.

The methods developed by WHO differ from those of Koo and Kron in a number of respects. Firstly, in the WHO survey all objects whose OR band magnitudes lie within the range $16 \leqslant m_{OR} \leqslant 20.0$ are included, irrespective of whether the objects appear in the other four passbands—high-redshift quasars, $z > 3$, may very likely be missing from the U passband due to Lyman-α absorption. Secondly WHO employ an objective algorithm to identify candidates so that the selection criteria are precisely specified. The six broad-band magnitudes define a six-dimensional parameter space and outliers are identified by quantifying the degree of isolation of an object from the locus of common stars by measuring the six-dimensional distance to its 10th nearest neighbour. Objects for which this distance exceeds the chosen threshold value are observed spectroscopically. The method utilises all the magnitude information simultaneously and is therefore more effective than a technique that employs two-colour diagrams, which are simply projections of the six-dimensional data set. Figure 5 shows the locations of the quasars of redshift $z > 3$ identified by WHO, in a UB_JV two-colour diagram. A number of the high-redshift quasars identified in the six-dimensional search lie in the stellar locus in this plot. The importance of including objects that are missing from one or more passbands (quasars indicated by arrows) is also evident from this diagram. Some 30% of the objects of the total OR band limited sample in each field were not detected in the U passband.

Initially candidates were observed by selecting in a subjective manner from the 1000 or so most distant outliers in each field. Guided by these results a set of precise selection criteria based on nearest-neighbour distance was specified, and spectra were obtained for all but a very small fraction of the candidates satisfying these criteria. With due allowance for the few objects for which spectra were not obtained, this subsample may then be used in computing the luminosity function, by correcting for incompleteness as appropriate for the known selection criteria. In all, 100 new quasars $z > 2.2$ were found, of which 49 have $z \geqslant 3.0$, including 3 with redshifts $z > 4$ (Warren *et al* 1987a,b). With the addition of 14 previously known quasars, which also meet the final selection criteria, the sample for the luminosity function calculation contains 85 quasars $z > 2.2$. The M_B–z distribution for this sample is shown in figure 8. Also shown, as open symbols, are bright quasars from the surveys of Mitchell *et al* (1990) (145 square degrees) and Osmer and Smith (1980) (340 square degrees) which were included in the luminosity function calculation, to improve the estimate of space densities at bright absolute magnitudes $M_B \sim -28$.

The use of precise selection criteria means that the selection function for the survey can be determined. For any particular quasar SED, synthetic broad-band colours may be computed, and the quasar can be introduced as an artificial point in the six-dimensional magnitude distribution of all the objects in a particular survey field. The candidate selection procedure is then run to determine whether the artificial quasar would have been picked out. A large number of realisations are undertaken for each quasar SED, varying the colours in a statistical manner as appropriate for the effect of absorption by intervening hydrogen clouds, and random photometric error. The probability of detection is simply the number of realisations in which the quasar is successfully selected divided by the total number of realisations. This procedure is carried out for a range of redshifts, absolute magnitudes and quasar SEDs. The array of detection probabilities for all quasar types defines the selection function. This approach takes account of all aspects of the selection, including the actual number and distribution of objects in the six-dimensional magnitude space for each survey field. There are two additional advantages to this procedure, that the Eddington-type bias (1913) and the k-corrections due to the presence of emission or absorption in the R passband are incorporated automatically.

WHO (1991a) present contour plots of detection probability as a function of apparent magnitude and redshift, for both fields, for nine SEDs, comprising three bins of spectral index, and three of line strength, that encompass the properties of the sample. The detection probabilities at faint magnitudes are very low, in many cases less than 30%. A revised calculation (WHO 1991c) incorporates the effects of variability, since the plates were taken over a period of seven years, and an improved absorption model (Møller and Warren 1991), and yields somewhat higher probabilities. The overall selection function for the survey, weighting as appropriate over the two fields and the different spectral bins, is illustrated in figure 8.

WHO (1991c) have used this sample to investigate the evolution of the luminosity function over the range $2.2 < z < 4.5$. The calculation proceeds by finding a luminosity function, coupled with the true relative numbers of objects in the nine SED bins, which, convolved with the selection function reproduces the observed distribution of objects in the M_B–z plane (figure 8) and the observed relative numbers of objects of different SED. In the figure it is notable that the brightest quasars are found at high redshifts. At the same time there are noticeably fewer quasars $z > 3.5$ than $z < 3.5$, at the same absolute magnitude, even though the detection probabilities are similar. Both these observations are confirmed by a quantitative analysis. A two power-law pure luminosity evolution model, evolving to brighter luminosities at higher redshifts, fits the data satisfactorily over the range $2.2 < z < 3.3$. The evolution is quite modest, and is rather similar to an extrapolation of that found at lower redshifts for models in which the evolution function is an exponential in look-back time τ (much more modest evolution than the power-law model of figure 4).

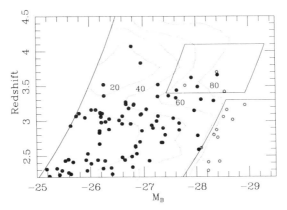

Figure 8. The absolute magnitude M_B against redshift z diagram for redshift $z \geqslant 2.2$ quasars in the WHO (1991b) multicolour sample used in the calculation of the QLF. The curved line to the left is the absolute magnitude limit, M_B, corresponding to WHO's apparent magnitude limit. Survey detection probabilities are shown as contours at intervals of 20%. Also shown, as open symbols, are objects from the Mitchell *et al* (1990) survey, which covered the region outlined at upper right, and from the Osmer and Smith (1980) survey which covered the region outlined at lower right.

The model overpredicts the number of quasars found $z > 3.3$, and it appears that the space density peaks near a redshift $z = 3$. To quantify the decline at high redshift WHO predict the number of quasars that would have been seen in their survey in the interval $3.5 < z < 4.5$ if the luminosity function stops evolving at $z = 3$ and remains constant thereafter. Compared to the eight quasars found the expected number is 41.5, $q_0 = 0.5$. On this basis they conclude that at the 95% confidence level the integrated space density of quasars brighter than $M_B = -26.3$ at $z = 4$ is lower than the corresponding space density at $z = 3$ by a factor > 2.9. The corresponding value for $q_0 = 0.0$ is a factor > 2.1. These limits are very similar to the values originally calculated by Osmer (1982) (§ 7.2.1). Qualitatively the results are also similar to the decline found by SSG (§7.2.4). However the WHO survey has low completeness, and the estimate requires extensive modelling of properties of the quasar population which are not yet well established at these high redshifts. This result must be considered subject to some uncertainty. (The relative insensitivity of the result to different values of q_0, as discussed in §3, is typical for calculations at these redshifts. The explanation is that the result depends on the change with redshift of the comoving volume and of the quasar luminosity, and the two effects largely cancel each other out.)

7.3.3. Wide-angle photographic surveys at bright magnitudes. Mitchell *et al* (1990) have undertaken a multicolour survey for bright, $17 < m_R < 18.5$, high-redshift quasars, $3.4 < z < 4.1$, that employs similar plate material and procedures to those of WHO. Pairs of plates in the five UKST standard passbands in 8 fields have been scanned by the COSMOS measuring machine. The survey covers ~ 240 square degrees, but the effective area is 145 square degrees as the stringent criteria used in separating stellar from non-stellar objects results in the exclusion of many stellar images from their multicolour catalogue. The aim of this is to remove any objects for which the photometry may be in error because, for example, merged with a close companion in one passband but not in another, and therefore possessing spuriously interesting colours. An initial candidate list was produced by identifying outliers to the stellar locus, from throughout the multicolour parameter space. Follow-up spectroscopy, however, was limited to candidates with colours which it was believed were compatible with their being high-redshift quasars, based on the computed synthetic colours of a wide range of model quasar SEDs (Miller and Mitchell 1988) and the known colours of several of the quasars discovered by WHO.

The survey has produced three new bright quasars within their magnitude and redshift limits, with redshifts $z = 3.42, 3.51$ and 3.70. Mitchell *et al* have estimated the completeness of this sample as close to 50%, using these quasars and 11 from the WHO sample to investigate the proportion of quasars that lie within the stellar locus in each of their fields as a function of magnitude. The computed space density is plotted in figure 4. Consistency with the result of HMS (modified as discussed in § 7.2.3) is just achieved, and, as suggested in § 7.2.3, it is likely that the true space density lies between the two estimates.

Irwin *et al* (1991) have undertaken a survey for bright quasars $z > 4$, over a very large area of sky, 2000 square degrees, using UKST plates scanned by the APM. The survey has been highly successful, resulting in the discovery of 18 quasars $z > 4$. For the first 1000 square degrees the redshift range $4 < z < 5$ was targeted, using single plates in the B_J, OR and I passbands, selecting candidates brighter than $m_I = 19$. The survey exploits the fact that the $m_{B_J} - m_{OR}$ colour of quasars becomes rapidly redder over the redshift range $3 < z < 4$ due to the increasing absorption in the B_J band from the Lyman-α forest and Lyman-limit systems. At redshifts beyond $z = 3.8$ the colours of quasars become increasingly well separated from the colour locus of stars. The main contaminants in this work are elliptical galaxies at moderate redshift $z \sim 0.3$ which, because they are distant, can be misclassified as stellar. The ability to discriminate against these galaxies sets the survey magnitude limit.

It was subsequently realised that the I plate information becomes progressively redundant towards higher redshifts. Consequently a second 1000 square degrees was surveyed, using only B_J and OR plates, limited to $m_{OR} < 19$. Objects with $m_{B_J} - m_{OR} > 3$ were selected, targeting quasars with redshifts $z > 4.2$. At high galactic latitudes the only other objects with such red colours are exceedingly rare, a few carbon stars and low–luminosity halo M dwarfs.

Irwin *et al* (1991) have estimated the incompleteness of their sample by computing the locus in their two-colour and one-colour selection spaces corresponding to quasars with the expected mean Lyman-α forest absorption at each redshift. On this basis they infer an integrated space density $M_B < -27.8$ of $6 \times 10^{-9} \text{Mpc}^{-3}$ at $z = 4.3$, which is virtually identical to that found at $z = 2$ at these bright magnitudes. Combined with the results of HMS and Mitchell *et al*, above, this would suggest that the comoving space density of the rare brightest quasars is approximately constant over the redshift interval $2 < z < 4.5$. The Irwin *et al* result at $z > 4$ should be considered as preliminary, however, as the photometric calibration of their fields is not yet complete. In addition, consideration of the distribution of quasar colours, rather than the mean colour at any redshift, may lead to a revision of the completeness estimate.

In the same paper these authors call into question the existence of a decline in space density at fainter absolute magnitudes, presenting again Boyle's analysis of the WHO and SSG survey data, discussed earlier (§7.3.1.3). The median equivalent width of the quasars found by Irwin *et al* is smaller than that for the sample of SSG, which leads then to suggest that the shortfall in numbers in the SSG sample relative to $z = 2$ comes from weak-lined quasars that have been missed. However, it can be shown that a line-flux limited quasar sample that is complete will have a larger median equivalent width than a continuum-magnitude limited sample. Therefore there is no particular reason to believe that SSG have missed quasars above their line-flux limit.

7.4. The effects of dust

The initial lack of success in finding quasars of redshifts $z > 3$ led Ostriker and Heisler (1984) to investigate the possibility that extinction by dust in intervening galaxies could be responsible for the decline in the observed numbers of quasars at high redshifts. The average extinction suffered by a quasar, at a particular observed wavelength, would increase rapidly with redshift, because of the crowding together of galaxies at earlier times and because the opacity of dust increases with decreasing wavelength. Heisler and Ostriker (1988) report on a substantially refined version of their model which includes a more sophisticated treatment of the extinction due to lines of sight passing through hypothesized dusty galaxies. They conclude that dust is a viable mechanism for producing the observed behaviour of the QLF at high redshifts and that the true space density of quasars may not decline until much higher redshifts.

The evolution of the luminosity function itself cannot be used to distinguish between the two possibilities that the decline is due to obscuration by dust, or reflects a true decrease in space density. However the effect of dust on the SEDs of quasars at high redshift does provide a test. Heisler and Ostriker stress that in a quasar survey limited on magnitude there is a strong bias towards observing those quasars which are least affected by obscuration, i.e. the quasars that have not been dimmed substantially. Even allowing for this bias the effects should be detectable as a significant steepening in the average continua slopes of quasars at high redshifts. A strong argument against significant obscuration by dust comes from the lack of evidence for reddening effects in quasars with redshifts $z > 4$ and absolute magnitudes of -28.5 to -25.0 (Schneider *et al* 1989a)— there should be strong correlation between absolute magnitude and the observed continuum slope under the dust hypothesis.

A further test of the model of Heisler and Ostriker is possible at radio wavelengths. Quasar radio-fluxes are unaffected by dust, so the observation by Peacock (1985) of a decline at high redshifts is evidence against the dust explanation. Advocates of dust, however, could argue that the postulated association of radio-quiet and radio-loud quasars with different host galaxies (Peacock *et al* 1986) means there is no a priori reason to believe the evolution of both classes of quasar should be similar.

Disregarding, for the present, the evidence from the radio data, the plausibility of dust accounting for the observed decrease depends on the amount of dust in galaxies at high redshift. In some high-redshift quasars Lyman-α absorption lines broadened by radiation damping are observed (Wolfe *et al* 1986). It is estimated that the total mass of neutral hydrogen in these damped Lyman-α systems is comparable to that in all forms of luminous matter at the present epoch. They are believed to be the signatures of the disks of spiral galaxies at an early stage of evolution and recent attention has focussed on potential obscuration effects associated with these systems. Fall and Pei (1989), building on earlier work by Wright (1981, 1986), have compared the spectroscopic properties of quasars, redshifts $2.75 \leqslant z < 3.4$, with and without damped Lyman-α systems present in their spectra. The derived dust-to-gas ratio depends on the shape of the extinction curve, i.e. on the properties of the dust, which are unknown, but Fall and Pei find the mean optical depth in dust to quasars at high redshift is a factor > 2 lower than that adopted by Heisler and Ostriker, and they argue that dust associated with damped Lyman-α systems is not responsible for the observed behaviour of the QLF at high redshifts. The analysis of Fall and Pei (1989) was based on an upper limit to the observed reddening in quasars with damped Lyman-α systems, from measuring spectra not well suited to the task of determining spectral indices. Fall *et al* (1989) have since found that there is a small but significant difference in the spectra of quasars with damped Lyman-α systems compared to those without such systems, based on spectra obtained by Sargent *et al* (1989). The observed reddening is less than the upper limit assumed in the earlier analysis, strengthening the conclusion that dust in the damped Lyman-α systems is not responsible for the decline.

7.5. The high-redshift luminosity function

In the preceding sections we have attempted to draw attention to the relative strengths and weaknesses of the major surveys for high-redshift quasars. In this section we summarise the main conclusions that can be drawn from this work, and list surveys in progress which can be expected to improve our understanding of the high-redshift QLF in the near future.

There is good agreement between most of the different groups that the space density of quasars is approximately constant or rising slightly between redshifts $z = 2$ and $z = 3$. There appears to be little change in the space density of the brightest quasars $M_B \sim -28$ over the entire redshift range $2.0 < z < 4.5$. At $z = 4$ the luminosity function rises towards fainter luminosities, but less steeply than at $z = 2$, and the space density of quasars of intermediate luminosity $M_B \sim -26.5$ is lower at $z = 4$ than at $z = 2$ by a factor of three to four. At fainter absolute magnitudes $M_B \sim -25$ the current surveys have insufficient areal coverage to distinguish between constant space density $2 < z < 4$ or a modest decline. There remain uncertainties in the estimates of the completeness of the high-redshift samples, and the size of the samples is substantially smaller than that on which our understanding of the QLF for $z < 2$ is based.

A number of surveys are underway that are aimed at improving our knowledge of the QLF at high redshifts, particularly at the faint end. These include the extension of the work of Koo and Kron (Kron *et al* 1991), and a number of new surveys including those described by McMahon and Irwin (1991, $m_B - m_R$ colour selection using deep 4-metre photographic plates), Osmer *et al* (1991, six passband multicolour using a 2048×2048 CCD), and the multicolour survey of Schneider *et al* (1991). At somewhat lower redshifts the analysis of the LBQS (Chaffee *et al* 1991) will improve significantly our knowledge of the form of the QLF at bright magnitudes for all redshifts $z < 3.3$. In fact the QLF $z < 2$ is not well determined at bright magnitudes and the LBQS together with the Edinburgh Multicolour Survey (Goldschmidt and Miller 1991) will produce an improved determination at low redshifts, against which to compare the high redshift data. At radio wavelengths the wide-field survey of Savage *et al* (§ 7.1) at bright flux levels, and new surveys at faint flux levels (< 0.1 Jy) will provide improved statistics and the possibility of investigating any luminosity-dependence of the decline noted by Dunlop and Peacock. Although large samples based on the optical identification of radio catalogues are difficult to compile because of the low surface density of sources, the systematic investigation of radio sources coincident with faint red optical counterparts offers one possible strategy for detecting quasars with redshifts as high as $z \sim 5$ (e.g. McMahon 1991).

As at low redshifts (§ 5.3) the interpretation of the evolution of the QLF at high redshifts should be approached with caution, recalling that, in the absence of additional constraints, many different physical models can produce a particular form for the evolution. Thus it would be premature to conclude, in contrast to the predictions of Efstathiou and Rees (1988), that there are fewer faint quasars at high redshift relative to bright quasars. For example it is conceivable that the QLF retains the same shape at high redshifts but shifts to higher luminosities and lower space densities. One could argue that the continued luminosity evolution reflects the same epoch-dependent change in the fuelling mechanism seen at lower redshift, and that the function quantifying the shift to lower space densities of the whole luminosity function describes a luminosity-independent decline in the space density of all quasars. A better understanding of the physical significance of the evolution of the QLF will come about not only through a more precise quantification of the QLF itself but also improvements in our knowledge of the lifetimes and the fuelling mechanism of quasars, and the relationship between the formation of an active galactic nucleus and the collapse of a protogalaxy. Progress in the understanding of how galaxies formed and how quasars evolve is, then, a symbiotic process.

The investigation of the QLF at high redshifts has developed from a point where the problem was simply one of detection to a stage where relatively large samples of quasars can be compiled in a homogeneous and quantifiable fashion. As confidence in the results of the new generation of surveys grows and discussion moves away from technical details of selection and incompleteness, more attention can be paid to the theoretical interpretation of the evolution of the QLF. With the key differences between the predictions of the different theoretical models identified, a new series of observational tests involving larger samples covering greater ranges in magnitude, redshift and quasar SEDs will doubtless begin.

Acknowledgments

We are grateful to David Koo, Lance Miller, Maarten Schmidt and Don Schneider all of whom provided preprints or information on work prior to publication. John Peacock and Maarten Schmidt were both kind enough to allow us to present data from their recent work and these appear as figures 6 and 7 respectively. Richard Green, Margaret Harding, Lance Miller, Pat Osmer, Martin Rees and Peter Shaver provided valuable comments on earlier drafts of this review that resulted in improved clarity and presentation.

References

Avni Y and Bahcall J N 1980 *Astrophys. J.* **235** 694
Barthel P D 1989 *Astrophys. J.* **336** 606
Begelman M C, Blandford R D and Rees M J 1984 *Rev. Mod. Phys.* **56** 255
Blades J C, Turnshek D A and Norman C A eds 1988 *QSO Absorption Lines: Probing the Universe, Space Telescope Science Institute Symp. Ser.* **2** (Cambridge: Cambridge University Press)
Boroson T A, Oke J B and Green R F 1982 *Astrophys. J.* **263** 32
Boyle B J, Jones L R and Shanks T 1991a *Mon. Not. R. Astron. Soc.* **251** 482
Boyle B J, Jones L R, Shanks T, Marano B, Zitelli V and Zamorani G 1991b *The Space Distribution of Quasars* (*Astron. Soc. Pacific Conf. Series* **21**) ed D Crampton p 191
Boyle B J, Shanks T and Peterson B A 1988 *Mon. Not. R. Astron. Soc.* **235** 935
Boyle B J, Shanks T and Peterson B A 1990 *Mon. Not. R. Astron. Soc.* **243** 1
Carswell R F and Strittmatter P A 1973 *Nature* **242** 394

Cavaliere A, Giallongo E, Padovani P and Vagnetti F 1988 *Proc. Workshop on Optical Surveys for Quasars (Astron. Soc. Pacific Conf. Series* **2**) ed P S Osmer, A C Porter, R F Green and C B Foltz p 311

Cavaliere A, Morrison P and Wood K 1971 *Astrophys. J.* **170** 223

Cavaliere A and Padovani P 1989 *Astrophys. J. Lett.* **340** L5

Cavaliere A and Szalay A S 1986 *Astrophys. J.* **311** 589

Chaffee F H, Foltz C B, Hewett P C, Francis P J, Weymann R J, Morris S L, Anderson S F and MacAlpine G M 1991 *Astron. J.* **102** 461

Cheng F Z, Danese L, De Zotti G and Franceschini A 1985 *Mon. Not. R. Astron. Soc.* **212** 857

Clavel J, Altamore A, Boksenberg A, Bromage G E, Elvius A, Pelat D, Penston M V, Perola G C, Snijders M A J and Ulrich M H 1987 *Astrophys. J.* **321** 251

Clowes R G, Cooke J A and Beard S M 1984 *Mon. Not. R. Astron. Soc.* **207** 99

Crampton D, Cowley A P and Hartwick F D A 1987 *Astrophys. J.* **314** 129

Cristiani S and Vio R 1990 *Astron. Astrophys.* **227** 385

David L P, Durisen R H and Cohn H N 1987 *Astrophys. J.* **316** 505

Dressler A and Richstone D O 1988 *Astrophys. J.* **324** 701

Duncan M J and Shapiro S L 1983 *Astrophys. J.* **268** 565

Dunlop J S and Peacock J A 1990 *Mon. Not. R. Astron. Soc.* **247** 19

Eddington A S 1913 *Mon. Not. R. Astron. Soc.* **73** 359

Efstathiou G, Ellis R S and Peterson B A 1988 *Mon. Not. R. Astron. Soc.* **232** 431

Efstathiou G and Rees M J 1988 *Mon. Not. R. Astron. Soc.* **230** 5p

Fall S M and Pei Y C 1989 *Astrophys. J.* **337** 7

Fall S M, Pei Y C and McMahon R G 1989 *Astrophys. J. Lett.* **341** L5

Fasano G and Franceschini A 1987 *Mon. Not. R. Astron. Soc.* **225** 155

Felten J E 1976 *Astrophys. J.* **207** 700

Filippenko A V 1989 *Astron. J.* **97** 726

Filippenko A V and Sargent W L W 1985 *Astrophys. J. Suppl.* **57** 503

Gaskell C M 1988 *Astrophys. J.* **325** 114

Giallongo E and Trevese D 1990 *Astrophys. J.* **353** 24

Goldschmidt P and Miller L 1991 *The Space Distribution of Quasars (Astron. Soc. Pacific Conf. Series* **21**) ed D Crampton p 212

Hartwick F D A and Schade D 1990 *Ann. Rev. Astron. Astrophys.* **28** 437

Hawkins M R S 1986 *Mon. Not. R. Astron. Soc.* **219** 417

Hazard C and McMahon R G 1985 *Nature* **314** 238

Hazard C, McMahon R G and Sargent W L W 1986 *Nature* **322** 38

Heckman T M 1991 *Massive Stars in Starbursts* ed N Walborn and C Leitherer in press

Heisler J and Ostriker J P 1988 *Astrophys. J.* **332** 543

Hewett P C and Irwin M J 1987 *Pattern Recognition Lett.* **5** 113

Hewett P C, Irwin M J, Bunclark P S, Bridgeland M T, Kibblewhite E J, He X T and Smith M G 1985 *Mon. Not. R. Astron. Soc.* **213** 971

Hernquist L 1989 *Fourteenth Texas Symposium of Relativistic Astrophysics* Ann. New York Acad. Sci. 571 (New York Acad. Sci.) p 190

Hickson P and Hutchings J B 1987 *Quasars, IAU Symp. No. 119* ed G Swarup and V K Kapahi (Dordrecht: Reidel) p 117

Hoag A A and Schroeder D J 1970 *Publ. Astron. Soc. Pacific* **82** 1141

Hoag A A and Smith M G 1977 *Astrophys. J.* **217** 362

Hutchings J B, Johnson I and Pyke R 1988 *Astrophys. J. Suppl.* **66** 361

Irwin M J, McMahon R G and Hazard C 1991 *The Space Distribution of Quasars (Astron. Soc. Pacific Conf. Series* **21**) ed D Crampton p 117

Irwin M J and Trimble V 1984 *Astron. J.* **89** 93

Kibblewhite E J, Bridgeland M T, Bunclark P S and Irwin M J 1984 *Proc. Astronomical Microdensitometry Conf.* NASA Conf. Publ. 2317, ed D A Klinglesmith (NASA)

Koo D C and Kron R G 1982 *Astron. Astrophys.* **105** 107

—— 1988 *Astrophys. J.* **325** 92

Koo D C, Kron R G and Cudworth K M 1986 *Publ. Astron. Soc. Pacific* **98** 285

Kormendy J 1988 *Astrophys. J.* **325** 128

Kron R G, Bershady M A, Munn J A, Smetanka J J, Majewski S and Koo D C 1991 *The Space Distribution of Quasars (Astron. Soc. Pacific Conf. Series* **21**) ed D Crampton p 32

Kron R G and Chiu L-T G 1981 *Publ. Astron. Soc. Pacific* **93** 397

Lynden-Bell D 1969 *Nature* **223** 690

MacGillivray H T and Stobie R S 1985 *Vistas in Astronomy* **27** 433

Majewski S R, Munn J A, Kron R G, Bershady M A, Smetanka J J and Koo D C 1991 *The Space Distribution of Quasars (Astron. Soc. Pacific Conf. Series* **21**) ed D Crampton p 55

Malkan M A and Sargent W L W 1982 *Astrophys. J.* **254** 22

Marshall H L 1985 *Astrophys. J.* **289** 457

—— 1988 *Proc. Workshop on Optical Surveys for Quasars* (*Astron. Soc. Pacific Conf. Series* **2**) ed P S Osmer, A C Porter, R F Green and C B Foltz p 11

Marshall H L, Avni Y, Tananbaum H and Zamorani G 1983 *Astrophys. J.* **269** 35

McKee C F 1988 *Proc. 3rd George Mason Astrophysics Workshop: Supermassive Black Holes* ed M Kafatos (Cambridge: Cambridge University Press)

McMahon R G 1991 *The Space Distribution of Quasars (Astron. Soc. Pacific Conf. Series* **21**) ed D Crampton p 129

McMahon R G and Irwin M J 1991 *The Space Distribution of Quasars (Astron. Soc. Pacific Conf. Series* **21**) ed D Crampton p 391

Miller L and Mitchell P S 1988 *Proc. Workshop on Optical Surveys for Quasars (Astron. Soc. Pacific Conf. Series* **2**) ed P S Osmer, A C Porter, R F Green and C B Foltz p 114

Mitchell L, Miller P S and Boyle B J 1990 *Mon. Not. R. Astron. Soc.* **244** 1

Moller P and Warren S J 1991 *The Space Distribution of Quasars (Astron. Soc. Pacific Conf. Series* **21**) ed D Crampton p 96

Murdoch H S, Hunstead R W, Pettini M and Blades J C 1986 *Astrophys. J.* **309** 19

Osmer P S 1980 *Astrophys. J. Suppl.* **42** 523

—— 1982 *Astrophys. J.* **253** 28

Osmer P S, Green R F, Porter A C, Hall P and Warren S J 1991 *The Space Distribution of Quasars (Astron. Soc. Pacific Conf. Series* **21**) ed D Crampton

Osmer P S and Smith M G 1980 *Astrophys. J. Suppl.* **42** 333

Ostriker J P and Heisler 1984 *Astrophys. J.* **278** 1

Partridge R B 1988 *Rep. Prog. Phys.* **51** 647

Peacock J A 1983 *Mon. Not. R. Astron. Soc.* **202** 615

—— 1985 *Mon. Not. R. Astron. Soc.* **217** 601

Peacock J A and Gull S F 1981 *Mon. Not. R. Astron. Soc.* **196** 611

Peacock J A and Miller L 1988 *Proc. Workshop on Optical Surveys for Quasars (Astron. Soc. Pacific Conf. Series* **2**) ed P S Osmer, A C Porter, R F Green and C B Foltz p 194

Peacock J A, Miller L and Longair M S 1986 *Mon. Not. R. Astron. Soc.* **218** 265

Penston M V, Penston M J, Selmes R A, Becklin E E and Neugebauer G 1974 *Mon. Not. R. Astron. Soc.* **169** 357

Rees M J 1984 *Ann. Rev. Astron. Astrophys.* **22** 471

Sargent W L W, Steidel C C and Boksenberg A 1989 *Astrophys. J. Suppl.* **69** 703

Saunders D B, Phinney E S, Neugebauer G, Soifer B T and Matthews K 1989 *Astrophys. J.* **357** 291

Savage A, Jauncey D L, White G L, Peterson B A, Peters W L, Gulkis S and Condon J J 1988 *Proc. Workshop on Optical Surveys for Quasars (Astron. Soc. Pacific Conf. Series* **2**) ed P S Osmer, A C Porter, R F Green and C B Foltz p 204

Schmidt M and Green R F 1983 *Astrophys. J.* **269** 352

Schmidt M, Schneider D P and Gunn J E 1986a *Astrophys. J.* **306** 411

—— 1986b *Astrophys. J.* **310** 518

—— 1991 *The Space Distribution of Quasars (Astron. Soc. Pacific Conf. Series* **21**) ed D Crampton p 109

Schneider D P, Schmidt M and Gunn J E 1989a *Astron. J.* **98** 1507

—— 1989b *Astron. J.* **98** 1951

—— 1991 *Astron. J.* **102** 837

Smith M G 1975 *Astrophys. J.* **202** 591

—— 1983 *Proc. 24th Liege Astrophysics Colloq., Quasars and Gravitational Lenses* ed J P Swings (Liege: Institut d'Astrophysique) p 4

Terlevich R 1991 *Proc. Workshop on Relationship between Active Galactic Nuclei and Starburst Galaxies* ed ???? in press

Terlevich R and Melnick J 1985 *Mon. Not. R. Astron. Soc.* **213** 841

—— 1988 *Nature* **333** 239

Turner E L 1991 *The Space Distribution of Quasars (Astron. Soc. Pacific Conf. Series* **21**) ed D Crampton p 361

Usher P D, Warnock A and Green R F 1983 *Astrophys. J.* **269** 73
Véron P 1983 *Proc. 24th Liege Astrophysics Colloq., Quasars and Gravitational Lenses* ed J P Swings (Liege: Institut d'Astrophysique) p 210
Wampler E J and Ponz D 1985 *Astrophys. J.* **298** 448
Warren S J, Hewett P C, Irwin M J, McMahon R G, Bridgeland M T, Bunclark P S and Kibblewhite E J 1987a *Nature* **325** 131
Warren S J, Hewett P C and Osmer P S 1991b *Astrophys. J. Suppl.* **76** 23
—— 1991c *The Space Distribution of Quasars* (*Astron. Soc. Pacific Conf. Series* **21**) ed D Crampton p 139
Warren S J, Hewett P C, Osmer P S and Irwin M J 1987b *Nature* **330** 453
—— 1991a *Astrophys. J. Suppl.* **76** 1
Weedman D W 1985 *Astrophys. J. Suppl.* **57** 523
—— 1986a *Quasar Astronomy* (Cambridge: Cambridge University Press) p 155
—— 1986b *Structure and Evolution of Active Galactic Nuclei* ed G Giuricin, F Mardirossian, M Mezzetti and M Ramella (Dordrecht: Reidel) p 215
Wolfe A M, Turnshek D A, Smith H E and Cohen R D 1986 *Astrophys. J. Suppl.* **61** 249
Wright E L 1981 *Astrophys. J.* **250** 1
—— 1986 *Astrophys. J.* **311** 156
Yee H K C and Green R F 1987 *Astrophys. J.* **319** 28
Zitelli V, Mignoli M, Zamorani G, Marano B and Boyle B J 1992 *Mon. Not. R. Astron. Soc.* in press

Rep. Prog. Phys. **53** (1990) 421–481. Printed in the UK

Large-scale motions in the universe: a review

David Burstein

Department of Physics, Arizona State University, Tempe, AZ 85287, USA

Abstract

The expansion of the universe can be retarded in localised regions within the universe both by the presence of gravity and by non-gravitational motions generated in the post-recombination universe. The motions of galaxies thus generated are called 'peculiar motions', and the amplitudes, size scales and coherence of these peculiar motions are among the most direct records of the structure of the universe. As such, measurements of these properties of the present-day universe provide some of the severest tests of cosmological theories. This is a review of the current evidence for large-scale motions of galaxies out to a distance of $\sim 5000 \, \text{km s}^{-1}$ (in an expanding universe, distance is proportional to radial velocity). 'Large-scale' in this context refers to motions that are correlated over size scales larger than the typical sizes of groups of galaxies, up to and including the size of the volume surveyed. To orient the reader into this relatively new field of study, a short modern history is given together with an explanation of the terminology. Careful consideration is given to the data used to measure the distances, and hence the peculiar motions, of galaxies. The evidence for large-scale motions is presented in a graphical fashion, using only the most reliable data for galaxies spanning a wide range in optical properties and over the complete range of galactic environments. The kinds of systematic errors that can affect this analysis are discussed, and the reliability of these motions is assessed. The predictions of two models of large-scale motion are compared to the observations, and special emphasis is placed on those motions in which our own Galaxy directly partakes.

This review was received in March 1989.

0034-4885/90/040421+61$14.00 © 1990 IOP Publishing Ltd

289

Contents

1. Introduction

The motion of the Earth in space is the sum of many separate motions. The Earth rotates on its axis at an equatorial rate of 0.46 km s^{-1}; it revolves around the Sun at a speed of 30 km s^{-1}; the Sun orbits the centre of the Milky Way Galaxy at a speed of $\sim$220 km s^{-1}. Our Galaxy, together with our nearest neighbouring galaxies (the 'Local Group' of galaxies) has a net motion relative to other more distant galaxies. On the largest scales, all groups of galaxies are moving away from one another due to the expansion of the universe. It is known that some of these motions are gravitationally driven (e.g. the revolution of the Earth about the Sun) and that some are due to initial conditions (e.g. the rotation of the Earth or the initial expansion of the universe). The physical origins of some motions are still unknown. Our current understanding of 'large-scale motions' in the universe falls into this last category.

If the expansion of the universe were absolutely smooth and homogeneous, we, along with galaxies and stars, would not exist. The presence of mass and the associated force of gravity locally retards the expansion, so that one can directly observe this expansion only on size scales large compared to the sizes of galaxies, typically $\sim$0.1 megaparsec (1 megaparsec [Mpc] $\simeq 3 \times 10^{24}$ cm).

The term 'large-scale motions' is used to describe the motions of galaxies, relative to the overall expansion of the universe, that are coherent over size scales much larger than the typical sizes of galaxy groups ($\sim$2 Mpc). These motions could be primarily driven by gravity or they could be the result of non-random motions imprinted on regions of galaxies during the first epochs of galaxy formation. The amplitudes and sizes of such coherent, large-scale motions are therefore among the most sensitive probes of both the large-scale structure of the universe and of the initial conditions of galaxy formation (Peebles 1980).

Davis and Peebles (1983), reviewing evidence for large-scale motions in the universe, concentrated mainly on the large-scale velocity field believed to be generated by the nearest large mass concentration to our Local Group—the Virgo cluster of galaxies (see also Yahil 1985). Several recent reviews have considered the circumstantial evidence for large-scale motions that is implied by the large-scale clustering properties of galaxies (Oort 1983, Bahcall 1987, 1988, Rood 1988). Much has happened in this field of study since the discovery that large-scale coherence on >20 Mpc scales does exist in the motions of galaxies (Aaronson *et al* 1986, Burstein *et al* 1986, Collins *et al* 1986a, b). A popular discussion of these recent developments has been given by Dressler (1987b), while more specialised reviews have been made by Gunn (1989), Rubin (1988, 1989) and Dressler (1989).

Section 2 of the present review introduces the 'jargon' of the field, as the terminology used to describe large-scale motions is non-intuitive. The history of this subject over the past 40 years is of relevance not only to document what has happened, but also as a means of learning how to approach this subject in the future (§ 3). The reader is referred to the reviews of Davis and Peebles (1983) and Rubin (1988, 1989) for historical accounts from different perspectives.

The principal intent of this review is to present the current evidence for large-scale motions emphasising the data available as of mid-1989. The acquisition of these data has required the work of many different astronomers, and has occupied a significant fraction of the observing time on the major telescopes of the world for the past decade. Not all astronomical data are of equal quality, however; a critical assessment of the accuracy of these data, and of the associated systematic errors, is needed to use these data in an optimal manner (§ 4). The observational evidence for large-scale motions is presented primarily in graphical form in § 5. The motions that are observed are compared in § 6 to a model based only on the observed motions, and to a model based on a recent determination of the density distribution of galaxies. The main conclusions of this review are summarised in § 7 together with desiderata for the future.

2. A brief orientation in space, distance, velocity and mass

2.1. Distance and velocity

The only component of the three-dimensional space velocity that can be observed for (almost) all galaxies is the radial component; transverse motion for an individual galaxy cannot be measured (the distances to galaxies are simply too large). In a smoothly expanding universe, distance is directly proportional to radial velocity for distances small compared to the size of the universe, i.e. $V = H_0 r$ (Hubble 1929). The constant H_0 is termed the Hubble constant, in honour of the discoverer, and one is free to specify distances either in terms of a length (typically measured in Mpc) in which case the value of H_0 must be specified, or directly in terms of velocity (measured in km s^{-1}). As the actual value of the Hubble constant is a constant source of debate (cf Aaronson *et al* 1986, Kraan-Korteweg *et al* 1988), units of velocity (km s^{-1}) will be the preferred distance measure for this review. Simply divide by your favourite value of H_0 to obtain a distance in Mpc (e.g. if $H_0 = 75$ km s^{-1} Mpc^{-1}, a distance of 3000 km s^{-1} corresponds to a distance of 40 Mpc. Values of H_0 currently quoted by astronomers range from 40–100 km s^{-1} Mpc^{-1}). The use of a velocity to measure distance is not a straightforward concept, but it does emphasise that large-scale motions are measured *relative* to the overall expansion.

In order to determine the degree of smoothness in the expansion, one must be able to determine an independent distance, r, to a galaxy, predict its expansion velocity and compare that to its observed velocity, V. In a smoothly expanding universe V and r will be related strictly by H_0. The extent to which r does not predict V is the measure of the 'peculiar motion' of a galaxy or a group of galaxies, relative to the idealised smooth expansion. If both V and r are measured in units of km s^{-1}, the difference $(V - r)$ is defined as the peculiar velocity, also in units of km s^{-1}. Although only the radial component of peculiar velocity can be measured for any single galaxy, the combination of peculiar velocities of many galaxies spread out over the sky is the most direct method we have to measure the coherent motions of galaxies on large scales.

The radial velocity that is measured for any galaxy is measured by an astronomer on a moving Earth. If the rotation of the Earth and the revolution of the Earth around the Sun are removed, the resulting radial velocity is termed 'heliocentric'. If the motion of the Sun (due to a combination of the Sun's rotation around the centre of our home Galaxy, and the motion of our Galaxy relative to the other galaxies in the Local Group) is removed, the resulting radial velocity is termed 'Local Group' (V_{LG}). These two velocity reference frames have been found to differ by 300 ± 20 km s^{-1} in amplitude (figure 1; the value used here comes from de Vaucouleurs *et al* 1976). A third velocity frame of reference (also illustrated in figure 1) will be defined in § 3.3.2.

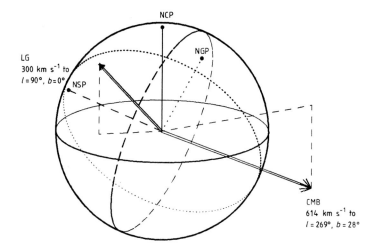

Figure 1. A schematic representation of three spherical coordinate systems important for extragalactic astronomy. The pole and equator for each coordinate system are drawn and the pole is labelled. The Celestial coordinate system (north pole labelled NCP), based on the projection of the Earth's equator and pole on the sky, is drawn with a full curve. In order to give the illusion of three dimensions, the darker half of the equator is in the foreground, the fainter half in the background. The Galactic coordinate system, based on the equator and north pole of our galaxy (north pole labelled NGP, with galactic latitude = b and galactic longitude = l), is drawn with a dotted curve. The Supergalactic coordinate system, based on the planar distribution of galaxies near us (north pole labelled NSP), is drawn with a broken curve. Two motions are represented by vectors originating at the centre of the sphere: the motion of the Sun with respect to velocity reference frame formed by the nearest galaxies (the Local Group vector, with an amplitude of 300 km s^{-1} towards $l = 90°$, $b = 0°$) and the motion of the Local Group with respect to the velocity reference frame defined by the cosmic background radiation (614 km s^{-1} towards $l = 268°$, $b = 29°$). The component of each vector in the celestial equator is also given. (Figure kindly supplied by Nigel Sharp.)

2.2. Space

Two coordinate systems used by astronomers to measure positions on the sphere of the sky for extragalactic objects are illustrated schematically in figure 1 (relative to the Earth-centred coordinate system):

(i) Galactic coordinates (b = latitude; l = longitude), with the plane of the equator defined by the plane of the disc of our own Galaxy, and the poles defined by the local perpendicular to that plane. Figure 2(a) (from Sharp 1989) illustrates the distribution of galaxies on the sky in this projection. The use of galactic coordinates emphasises the diminishing number of galaxies that are seen towards the Galactic equator, due to the obscuration caused by dust in the plane of the disc of our Galaxy. This 'zone of avoidance' severely impedes the systematic study of those galaxies situated within a band of width 30° on the sky centred on the Galactic equator (analogous to trying to look at the mountains around Los Angeles on a smoggy day, or the setting Sun on a slightly foggy day in London).

(ii) Supergalactic coordinates (B, L), with the planar distribution of nearby galaxies defining the equator, and the poles defined by the local perpendicular (de Vaucouleurs

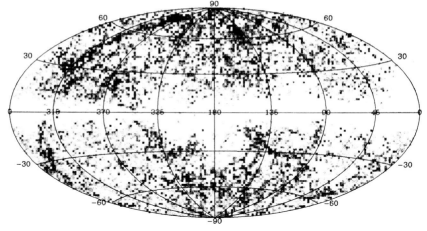

Galactic coordinates

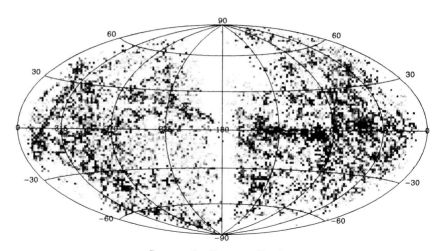

Supergalactic coordinates

Figure 2. The two-dimensional distribution of galaxies in space (figures from Sharp (1989)). Rather than plot individual galaxies, these Aitoff projections are plots of galactic density weighted by the amount of light produced per unit area. A total of 34 712 galaxies are included in these diagrams (which exclude the 17 nearest galaxies for clarity). (*a*) The distribution of galaxies in Galactic coordinates. Galactic latitude runs from −90° to +90° and longitude from 0° to 360°, and is centred on 180° in this projection. The marked absence of galaxies near the Galactic equator is the well known 'zone of avoidance', and is caused by the obscuration of dust in the plane of the disc of our Galaxy. (*b*) The same distribution of galaxies, but this time in Supergalactic coordinates (defined analogously to Galactic coordinates) centred at a longitude of 180°. The distinctive thin, narrow distribution of bright galaxies at Supergalactic longitudes between 0° and 180° defines the Supergalactic plane (de Vaucouleurs 1953). Note that the zone of avoidance runs nearly perpendicular to the Supergalactic plane.

and de Vaucouleurs 1964). Figure 2(b) illustrates the distribution of galaxies on the sky in this projection. Many nearby galaxies have positions that lie near the Supergalactic plane (which is how it was originally defined; e.g. de Vaucouleurs 1953, Tully and Fisher 1987). Fortunately for extragalactic astronomy, the Galactic plane and the Supergalactic plane are oriented nearly at right angles to one another (figure 1, and compare figure 2(b) with figure 2(a)).

2.3. Mass

Not all matter that emits electromagnetic energy can be detected at every wavelength here on the Earth, and not all matter that exists emits electromagnetic radiation that can be detected by us. Baryonic matter, that is atoms and ions (Primack *et al* 1988) can emit electromagnetic energy of any wavelength. This kind of matter constitutes not only stars but also dust and gas at temperatures ranging from 10 K to $>10^7$ K. The existence of matter that does not emit detectable radiation ('dark matter') is inferred from the motions of visible, 'luminous' matter (Kormendy and Knapp 1985), but the kind(s) of particle that comprise(s) this dark matter is (are) currently unknown.

The sum total of baryonic matter in the universe is less than 14% of the total mass required to keep the universe from expanding forever (Primack *et al* 1988). Only a fraction of this baryonic matter is luminous. Hence, if the universe has enough mass to barely keep it from expanding forever (as favoured by current cosmological theories of Big Bang Inflation) most of the matter in the universe must be of a form that is both dark and non-baryonic.

If non-baryonic matter is the dominant form of matter in the universe, it is especially important to evaluate the peculiar velocities of galaxies in a manner that does not depend on direct knowledge of the distribution of baryonic matter (assuming these motions are gravitational in origin). Such studies are emphasised in the present review. Furthermore, comparison of the peculiar motions predicted from the density distribution of visible matter, to those that are actually observed, is one of the best methods to deduce the distribution of dark matter on the largest scales. An initial attempt at this more complex comparison is made in this review.

3. A historical perspective

3.1. 1948–1970

The discovery of the expansion of the universe (Hubble 1929) was such a large leap in our understanding of the structure of the universe that it is perhaps natural that this expansion was first viewed as being smooth and homogeneous. Moreover, until accurate distances to galaxies could be obtained independent of these expansion velocities, no examination of peculiar motions could be attempted.

Gamow (1948) questioned whether significant deviations from a smooth expansion could be observed. An answer to this question was sought by Vera Rubin in a Master's thesis (an abstract of which was published (Rubin 1951)), which investigated the possible rotation of the universe using the magnitudes and radial velocities of 108 galaxies. Similar studies by Ogorodnikov (1952), de Vaucouleurs (1958) and de Vaucouleurs and Peters (1968) were the only such investigations over a 20 year period.

Each of these early studies searched for evidence of rotation of the universe under the same assumption that galaxies of similar appearance had the same absolute luminosities. Although some evidence of rotation was found, it was also realised by the mid-1960s that galaxies of similar optical appearance do not all have the same absolute luminosities. In addition, galaxies near our own are not distributed randomly in space, but are part of a large, planar distribution of galaxies, commonly termed the 'Local Supercluster' (de Vaucouleurs 1958). As such, reliable detection of large-scale motions had to await better means of measuring distances of galaxies.

In 1965 Penzias and Wilson discovered the existence of isotropic radiation in the universe having a black-body temperature now estimated to be 2.75 K (Lubin and Villela 1986, Partridge 1988). It is generally believed that this cosmic background radiation (CBR) is the temperature of the universe when it became transparent to radiation (the 'time of recombination'), redshifted by a factor of 1000 and cooled to the currently observed value. So defined, the CBR is an absolute velocity frame of reference against which absolute space motions in the universe can be measured.

3.2. 1970-1978

Tests of the degree of isotropy in the expansion of the universe on size scales of tens of Mpc require estimates of distances to galaxies that are independent of their observed radial velocities (i.e. r determined independent of V). This in turn necessitates the measurement of one or more physical parameters of galaxies that are distance-dependent (such as luminosity or size). The apparent faintness of the galaxies ensures that the acquisition of data will be slow and require the largest available telescopes. The development of more sensitive detectors has greatly aided the data-gathering effort over the past 20 years, but to this day such a project remains time- and telescope-intensive.

At the beginning of the 1970s, the physical parameters of a galaxy that could be easily measured were the same as in the 1950s—its total luminosity and/or its charac-teristic size. Practitioners either selected a certain type of galaxy with a well defined optical appearance that correlates with its absolute brightness, or used the characteristic luminosity of many galaxies in the same cluster (i.e. the luminosity function of these galaxies; see below).

The luminosities of giant elliptical galaxies were used by Sandage in his investiga-tions (Sandage 1972, 1973a, b); luminosity functions of cluster galaxies were used by Abell and collaborators (Abell and Eastmond 1968, Bautz and Abell 1973). The principal aim of almost all of the investigations at this time was to measure combinations of the local expansion rate of the universe (H_0), the rate at which the expansion is decreasing (q_0) and the average density of matter in the universe (Ω_0). None of these studies had the wide spatial distribution of galaxies on the sky with which to search for large-scale motions, although Sandage (1972) speculated on possible reasons for the then-apparent absence of motion of the Local Group towards the nearby Virgo cluster.

The one notable exception during this period of time was the survey of Rubin *et al* (1976a, b), who used the brightnesses of giant Sc spiral galaxies to purposefully search for evidence of large-scale motions in the universe. Rubin *et al* did indeed find evidence for a large-scale, coherent motion, which they interpreted in terms of a net motion of 600 km s^{-1} amplitude, for galaxies in the distance range of 3500-6500 km s^{-1} relative to the Local Group velocity frame of reference. This result was initially discounted, as few astronomers were willing to believe that the Local Group could have a motion this large (Rubin, private communication).

In one of the classic coincidences of modern science, at nearly the same time as the work of Rubin *et al* was published, researchers at Princeton and Berkeley (Corey and Wilkinson 1976, Smoot *et al* 1977) confirmed the existence of a dipole anisotropy in the CBR temperature at a level of $\Delta T/T = 10^{-3}$ (marginally detected earlier, e.g. Conklin (1969)). As we view it, the CBR is a little hotter in one direction and a little cooler in the other. This dipole anisotropy in temperature was interpreted as the Doppler shift of a black-body spectrum due to the motion of the observer relative to the CBR. If the CBR is a remnant from the origin of the universe, then the amplitude of this anisotropy implies a peculiar velocity of the Local Group, with respect to the CBR, of 600 km s^{-1} towards $l \sim 270°$, $b \sim 30°$.

The interpretation of the dipole anisotropy as a motion of 600 km s^{-1} of the Local Group was generally accepted, and further refined by more observations (Smoot and Lubin 1979, Cheng *et al* 1979). The Rubin *et al* result, although it predicted a motion of similar amplitude, proved more controversial, as this motion (towards $l = 160°$, $b = -10°$) is directed nearly orthogonal to the direction of the CBR dipole anisotropy. Although several concerns were raised about the proper handling of the data (e.g. Schechter 1977, Fall and Jones 1976), this result was never shown to be due either to error or to misinterpretation of the data. Since the kinds of concerns raised then can be rightfully expressed for all of the data used in this review they are discussed separately in § 4.2. However, the net effect of this debate in the late 1970s was to place the result of Rubin *et al* in the astronomical 'limbo' reserved for controversial results of uncertain reliability.

At about the same time, two techniques for measuring the distances to galaxies were proposed that compare a distance-independent kinematic property of galaxies to a distance-dependent property. Tully and Fisher (1977) showed that the maximum rate of rotation in a spiral galaxy was well correlated with the absolute luminosity of the galaxy (TF method). Intrinsically brighter spirals rotated faster than intrinsically fainter spirals. Faber and Jackson (1976) following up a suggestion made by Fish (1964), showed that the amplitude of the random motions of stars inside elliptical galaxies (termed the velocity dispersion) is correlated with the absolute luminosities of the galaxies (FJ method).

3.3. 1979-1983

3.3.1. Virgocentric infall. The TF relation was applied almost immediately to samples of galaxies to obtain distances independent of expansion velocities, in attempts to evaluate H_0 (Tully and Fisher 1977, Sandage and Tammann 1976). With the discovery that the TF relationship had smaller intrinsic scatter if one used near-infrared H-band luminosities (centred at a wavelength of 1.6 microns (Aaronson *et al* 1979a)), its use in determining the distances of galaxies greatly increased. Over the next three years Marc Aaronson, John Huchra, Jeremy Mould and collaborators applied this technique to ~300 spiral galaxies with distances less than 3000 km s^{-1} (Aaronson *et al* 1979b, 1980, 1981, 1982a, b, Mould *et al* 1980, Aaronson and Mould 1983). At the same time, de Vaucouleurs and others investigated the motions of galaxies using both the TF relation and the FJ relationship (Schechter 1980, de Vaucouleurs and Peters 1981, de Vaucouleurs *et al* 1981, 1982, Tonry and Davis 1981a, b, Hart and Davies 1982, Bottinelli *et al* 1983, Dressler 1984).

These many investigations established the existence of a net peculiar motion of the Local Group towards the Virgo cluster, generally termed Virgocentric infall. (Of course the Local Group is not literally falling into the centre of the Virgo cluster, its expansion velocity is merely retarded.) Davis and Peebles (1983) summarised these many estimates and made the then reasonable assumption that the Local Group's motion towards the Virgo cluster is produced solely by the Virgo cluster, which in turn defines the centre of the Local Supercluster. As explained in that review, this assumption, together with the implicit assumption that these motions are gravitationally driven, can be used to extract useful dynamical information about the Local Supercluster as well as estimates of the mean density of the universe, Ω_0.

3.3.2. The necessary implications of the CBR motion of the Local Group. If the CBR anisotropy dipole reflects the absolute space motion of the Local Group with respect to the average motion at least of the nearby universe, then, at minimum, the study of nearby large-scale motions requires the following question to be answered: 'On what size scales is the Local Group's CBR motion either correlated or generated (if the source of motion is gravity)?' Combining the CBR dipole anisotropy as currently measured with the Local Group motion of de Vaucouleurs *et al* (1976), the CBR motion of the Local Group is 614 ± 50 km s^{-1} towards Galactic coordinates $l = 269° \pm 5°$, $b = 28° \pm 5°$ (Lubin and Villela (1986); the new results of Boughn *et al* (1989) are in excellent agreement with these values).

Davis and Peebles (1983) were among the first to implicitly pose this question (their equation (9)). They divided the CBR anisotropy dipole into three components, with one component within the Supergalactic plane directed towards the Virgo cluster (V_v), one directed perpendicular to the Virgo direction but within the Supergalactic plane (V_s) and one perpendicular to the Supergalactic plane (V_z). The value for the CBR velocity dipole cited above, expressed in this coordinate system, is $V_v = 418 \pm 25$ km s^{-1}, $V_s = 277 \pm 25$ km s^{-1} and $V_z = -355 \pm 25$ km s^{-1}.

The amplitudes of these three components clearly show that Virgocentric motion cannot be the sole gravitational source of the Local Group's CBR motion. The motion of the Local Group perpendicular to both the Virgo cluster and to the Supergalactic plane (V_z) is nearly as large as that of Virgocentric motion (V_v) itself. The component of motion directed towards the Virgo cluster accounts for only 46% of the total amplitude of the CBR vector. Although similar conclusions can be drawn from equation (9) of Davis and Peebles (1983), their review placed its greatest emphasis on measurements of Virgocentric infall. The number and origin of the motions that combine to give rise to the full CBR dipole anisotropy were not then clear nor, as this review will detail, are they yet clear.

3.4. 1984-1988

Whether by coincidence, or by direct influence of the review by Davis and Peebles, from late 1983 to early 1984 a number of workers noted that much of the Local Group's motion was directed away from the Virgo cluster (de Vaucouleurs and Peters 1984, Sandage and Tammann 1984, Yahil 1985, Davies and Stavely-Smith 1985, Tammann and Sandage 1985, Aaronson and Mould 1985). Tammann and Sandage (1985) emphasised that, if one subtracted a motion of 250 km s^{-1} towards the Virgo cluster from the CBR dipole, the remaining vector of $\sim$450 km s^{-1} pointed in the general direction of the Hydra cluster of galaxies ($l \sim 270°$, $b \sim 10°$; cf also Davies and Stavely-Smith (1985)). Shaya (1984) pointed out that the much more extended supercluster formed by Hydra and Centaurus (a cluster $\sim$40° in angle away from Hydra) was the nearest large mass concentration to the Local Supercluster and, as such, would reasonably be expected to exert a strong gravitational pull on the Local Supercluster.

Aaronson *et al*, as first reported by Aaronson and Mould (1985) and published by Aaronson *et al* (1986), used the TF relation to measure the distances of 148 galaxies in 10 distant clusters (mean distance ~6500 km s^{-1}) in their continuing effort to determine H_0. They found that the motions of these clusters defined a velocity reference frame at rest with respect to the CBR. Aaronson *et al* concluded that their result would plausibly fit into the already proposed motion towards the Hydra–Centaurus Supercluster if all of the Local Group's motion were generated by matter lying within a distance of 5000 km s^{-1} of the Local Group.

A radically new view was suggested by the first results of a study of the distances to 385 elliptical galaxies, which showed that the galaxies in the Centaurus clusters were moving *away* from the Local Group (not towards), at a velocity even *greater* than that of the CBR velocity of the Local Group (Burstein *et al* 1986). This study was also the first to propose that motions of galaxies be viewed in the velocity rest frame of the CBR, which can be plausibly defined as the *fundamental* velocity rest frame of the universe. The seven astronomers involved in this study, including the author of this review, were quickly dubbed the 'Seven Samurai (7S)' (by Amos Yahil in the summer of 1986,) a name that has stuck. The 7S study used a revised version of the Faber–Jackson distance estimator for elliptical galaxies that measures a characteristic diameter (termed D_n) of the galaxies instead of a total luminosity. (This method, termed the D_n-σ relation, was discovered at the same time by Djorgovski and Davis (1986, 1987) and originally suggested by de Vaucouleurs and Olson (1982).)

The initial interpretation of the 7S results was that the motions of all galaxies within a radius 6000 km s^{-1} of the Local Group could be reasonably modelled in terms of a dipole moment, termed a 'bulk' motion, with an amplitude of ~600 km s^{-1} (Dressler *et al* 1987a). Supporting evidence for such a bulk motion was presented by Collins *et al* (1986a, b) and was marginally consistent with the original result of Rubin *et al* (1976a, b). However, the direction of this bulk motion (Dressler *et al* 1987a) was not towards the direction of the CBR dipole (in Galactic coordinates $l = 269°$, $b = 28°$) but 40° away, towards the direction of $l = 312°$, $b = 6°$.

It was soon shown, however, that motions of galaxies within a distance of 5000 km s^{-1} are too complicated to be adequately described by a simple dipole moment vector (§ 5.2). Lilje *et al* (1986) used the Aaronson *et al* (1982a, b) distances for spiral galaxies having distances <3000 km s^{-1} to show that a significant quadrupole moment existed in the motions of galaxies in this volume of space. Remarkably, the direction of the expansion component of the quadrupole moment vector they found points in the same direction ($l = 308°$, $b = 13°$) as the dipole moment vector of the 7S bulk motion. The 7S collaboration found essentially the same quadrupole moment in the motion of their own data set, with very similar amplitudes and directions as those derived by Lilje *et al* (Lynden-Bell *et al* 1988; cf § 5.4.1. of this review).

If gravity is the origin of these motions, the existence of a quadrupole moment in the nearby velocity field implies the presence of a large, perturbing mass lying outside this volume of space. Lilje *et al*, following this line of argument, assumed (i.e. did not derive) a Virgocentric infall of 250 km s^{-1} and located the centre of a larger, perturbing mass in the direction $l = 288°$, $b = -9°$, in a direction between the Hydra and Centaurus clusters.

The 7S collaboration, using only their own data, solved simultaneously for both Virgocentric infall and motion towards a more distant perturbing mass. This analysis located the central position of that mass in the direction $l = 310°$, $b = 9°$, centred at a distance of 4300 ± 500 km s^{-1} (Lynden-Bell *et al* 1988); i.e. behind the Centaurus cluster. As this perturbing mass was estimated to have a total mass of $5 \times 10^{16} M_0$ (comparable to that of the largest-known associations of clusters of galaxies) and it appears to be distributed over nearly a steradian of the sky (Dressler 1988) it was given the dramatic name 'Great Attractor' by Alan Dressler.

Because motions induced by both the Great Attractor region and the Virgo cluster are directed primarily within the plane of the Local Supercluster, these mass concentrations cannot give rise to the CBR V_z component of 350 km s^{-1}. Moreover, it was found by both Peebles (1988) and Faber and Burstein (1989) that the full CBR motion of the Local Group was shared by essentially all galaxies within a distance of 750 km s^{-1} (§ 5.3.). Thus, the minimum empirical model of motions involving the Local Group must include at least three components, independent of the assumption that all such motions are gravitationally driven.

Following the initial investigation of Aaronson *et al* (1986), several groups also tried to measure the net motion of the Local Group with respect to very distant (distances >6000 km s^{-1}) galaxies, with mixed results (due to the fact that peculiar velocities of a given amplitude are smaller fractions of these larger expansion velocities). The results found by the 7S collaboration (Lynden-Bell *et al* 1988), Lucey and Carter (1988) and Aaronson *et al* (1989) are all in agreement with the original Aaronson *et al* (1986) result that there is no evidence of large-scale motions over these very large distances. This is at variance to the conclusions of Collins *et al* (1986a, b) and James *et al* (1987), who find a net motion of galaxies on these larger scales.

3.5. *The current situation and the intent of this review*

In the above description the situation becomes confusing towards the end; it was, and is still, not clear to the participants. The central questions are the following.

(i) On what size scales is the CBR motion of the Local Group generated (if the motions indeed originate with gravity)?

(ii) The CBR motion predicts that the Local Group is moving towards the Virgo Cluster with a velocity of over 400 km s^{-1}. How much of this motion is due to the gravitational pull of the Virgo cluster itself?

(iii) The component of motion of the Local Group that appears to be induced by the Great Attractor (535 km s^{-1} in the model of Faber and Burstein (1989)) is only 50% larger than the component of motion perpendicular to the Supergalactic plane (350 km s^{-1}). What is causing this other motion?

(iv) Is it possible that these motions are artefacts of the varying physical properties of galaxies, which might lead to different calibrations of the TF or D_n-σ relations in different environments? To what extent do random and/or systematic observational errors colour the current view of these motions?

(v) To what extent are motions gravitationally driven by mass that can be seen? Is there evidence that the region of the Great Attractor is as massive as is required to produce the observed motions?

(vi) Is gravity responsible for these motions or must one invoke other kinds of non-gravitational motions (e.g. explosions) to explain them?

The emphasis in this review will be to present the best available data in a manner that can address these questions.

4. Choice of data and effects of systematic errors

4.1. The available data on large-scale motions

Current methods used to derive peculiar motions (i.e. distances) of galaxies have a wide range in intrinsic accuracy, and often there is little or no overlap between many of these data samples to permit verification. In addition, some workers cite only large-scale motion solutions, and do not publish peculiar velocities for individual galaxies and groups. This latter means of communication does not permit an external evaluation of the accuracy of the data that are used.

As discussed in § 4.2, the size of random errors directly affects the size of systematic errors. It is therefore important for any comprehensive discussion of large-scale motions to critically review the available data and use only those data of the highest quality. This review has therefore applied three selection criteria to the available data.

(i) We will not consider large-scale motions of galaxies that are cited simply as dipole or multipole solutions, without data provided for the individual galaxies that went into these solutions (Hart and Davies 1982, Collins *et al* 1986a, b, Visvanathan 1986, James *et al* 1987).

(ii) Distances are to be measured via an objective relationship between a distance-independent parameter (e.g. rotation velocity of a spiral galaxy) and a distance-dependent parameter (e.g. luminosity of the spiral galaxy). This excludes methods of measuring distance that assume galaxies of a given morphological class are of the same absolute luminosity (Sandage 1972, Rubin *et al* 1976a, b, Weedman 1976, Abell and Eastmond 1968, Bautz and Abell 1973, Jackson 1982; see the review of Davis and Peebles (1983) for a complete reference list on these kinds of investigations). These latter methods can be too easily biased by subjective classification errors, the effects of which are difficult to assess.

(iii) Multiple cross-checks between different data sets are important tests of systematic errors, the dominant source of error in this field of study. Only those data samples will be considered that provide enough data on individual galaxies in many different areas of the sky to permit mutual verification. This excludes distances derived from supernovae (Branch 1982, Sandage and Tammann 1982) as well as direct comparisons between only one or two clusters (other references used by Davis and Peebles (1983); Pierce and Tully 1988)).

Finally, if data from more than one method are available for an individual galaxy, only data that are demonstrably the most accurate will be used or, if of comparable quality, an average distance will be used.

Given these constraints, as of mid-1989 the published investigations from which one can study the motions of individual galaxies include: (1) The Aaronson *et al* series of papers (Aaronson *et al* 1979a, b, 1980, 1981, 1982a, b, 1986, 1989, Bothun *et al* 1984, 1985, Mould *et al* 1980, Bothun and Mould 1987, Aaronson and Mould 1983, 1986) which give near-IR Tully–Fisher parameters for spiral galaxies. (2) The D_n–σ parameters for elliptical galaxies obtained by the 7S collaboration (Dressler *et al* 1987b, Davies *et al* 1987, Burstein *et al* 1987a, b, Faber *et al* 1989) as well as more recently by Lucey and Carter (1988) and Dressler and Faber (1990). (3) The investigations by de Vaucouleurs, Bottinelli and collaborators using the blue magnitude TF relation for spiral galaxies (de Vaucouleurs and Peters 1984, Bottinelli *et al* 1983, 1984, 1986, 1987, 1988a, b). (4) The investigations by Huchtmeier, Kraan-Korteweg and collaborators using the blue Tully–Fisher relation (Kraan-Korteweg 1983, 1985, 1986, Kraan-Korteweg et al 1984, 1986, 1988, Huchtmeier and Richter 1986a, b, Richter and Huchtmeier 1984, Richter et al 1987). (5) The investigations by Davies and Stavely-Smith (1985) and Stavely-Smith and Davies (1987, 1988, 1989) using several kinds of methods related to the Tully–Fisher method.

The accuracies with which these data samples predict the distances of galaxies have been ranked, based both on their quoted errors and a separate analysis for this review. The most accurate method to predict the distances of galaxies appears to be that of the near-IR TF relation for well observed spiral galaxies (Aaronson and colleagues), with an estimated accuracy of 16% or better in distance (<0.35 mag scatter in absolute magnitude; see § 6.5). The next most accurate distance estimator is the D_n–σ relation for elliptical and S0 galaxies (21% error in distance). The least accurate distances considered here are those derived from the blue Tully–Fisher relation for intrinsically bright galaxies, as well as the more poorly observed data of the Aaronson et al sample (errors of 25% in distance).

All of the remaining methods used to measure distances appear either to have observational errors in excess of 35% in distance, have likely sources of systematic errors (Burstein et al 1988, 1989a, Burstein and Raychaudhury 1989) or duplicate more accurate observations, and consequently are not used in this review. These latter data include: the blue Tully–Fisher relation for intrinsically fainter spiral and irregular galaxies; substitution of neutral hydrogen flux or diameter for luminosity in Tully–Fisher relations; colour–magnitude relations for elliptical or spiral galaxies; the Hubble type of a galaxy and the sizes of internal rings in spiral galaxies (cf de Vaucouleurs and Peters 1984).

The above restrictions, much more stringent than have been applied before, nonetheless leave a homogeneous data sample that includes: 281 galaxies from the best-observed subset of Aaronson et al data; 544 galaxies from the elliptical and S0 galaxies observed by the 7S, Lucey and Carter and Dressler and Faber; 287 galaxies of the more poorly observed Aaronson et al sample; and 76 galaxies observed by de Vaucouleurs et al with the blue Tully–Fisher relation that have $V_{LG} < 1400$ km s^{-1} and are not also observed by the Aaronson et al group (there is an overlap of over 133 galaxies between these two samples). This yields a total of 1188 galaxies, with 898 (or 76%) having observed Local Group radial velocities less than 5000 km s^{-1}.

As stated in § 1, this review is based on the efforts of many astronomers over many years to study the distances of galaxies. Distances are known for about 70% of the galaxies with $V_{LG} < 3000$ km s^{-1} whose distances can be measured by present techniques (e.g. about 50% of all spiral galaxies can be used for the TF relation). Of these, perhaps 30% (20% of all possible) have the most accurate distances that are likely to be obtained, with more accurate data being continuously added.

4.2. Sources of systematic bias

4.2.1. Selection (Malmquist) bias—'I see you'. Consider the following thought problem: populate the universe with galaxies drawn randomly from a single, Gaussian-peaked luminosity function that has a mean absolute magnitude M_0 and a measured dispersion in absolute magnitude of e_0. (In practice, since apparent luminosity l decreases as r^2 it is more convenient to express astronomical luminosities in terms of a logarithmic quantity, magnitude, defined by historical convention to be $m = -2.5 \log l + \text{constant}$.) Furthermore, assume that these galaxies are distributed smoothly in space, with density being independent of distance r. The number of galaxies will increase as $r^2 \, dr$, so for any given radius there will always be more galaxies at larger radii than at smaller radii.

Galaxies are then chosen strictly on the basis of *apparent* magnitude (i.e. how bright the galaxies appear to us here on Earth). Galaxies of apparent magnitude m_1 (corresponding to apparent luminosities l_1) could be at distances $r_1 \pm r_{\mathrm{disp}}$ due to the fact that these galaxies have a *measured* dispersion in magnitude e_0 (r_{disp} = function of (r_1, e_0)). This measured dispersion is a combination of intrinsic scatter and measurement error. Since $r_1 + r_{\mathrm{disp}}$ is larger than $r_1 - r_{\mathrm{disp}}$ there are always more galaxies that are brighter than average included in the sample compared with galaxies that are fainter than average. This is true for any apparent magnitude m_1 and, hence, at *any* distance r_1. It is also true for any sample of galaxies at any wavelength observed if an observed dispersion in absolute magnitude exists. (Absolute magnitude is defined as the apparent magnitude galaxies would have if they were all at the same distance from us.)

In a *smooth, homogeneous distribution* of galaxies as a function of distance the observed mean *absolute* magnitude M_1, corresponding to a sample of galaxies *chosen* to have a mean apparent magnitude m_1, will always be *brighter* than the 'true' mean absolute magnitude M_0 of these same galaxies. The difference in these two mean values is proportional to the square of the measured dispersion (cosmic + observed), $(e_0)^2$, placing a premium on having accurate observations. The predicted absolute magnitudes are brighter by the same amount both for individual galaxies and for the whole ensemble of galaxies (but only in this special case).

This selection bias was first pointed out by Eddington (1913) in the simple case outlined above (but using stars, not galaxies, as test objects). However, it was put into a more general mathematical description by Malmquist (1920, 1922) for an arbitrary density distribution. As a result this kind of systematic effect is called Malmquist bias by most modern workers (cf Mihalas and Binney 1981).

Unfortunately, the term 'Malmquist bias' has been interpreted in almost as many different ways as there are astronomers who have attempted to use it. It is, to this reviewer, one of the most misunderstood concepts in astronomy. As defined here, Malmquist bias is the result of a geometrical requirement (volume increases as $r^2 \, dr$) combined with the facts that galaxies are not perfectly observed nor perfectly made in nature, and galaxies must be found by their apparent luminosities.

Even in the total absence of observational error, Malmquist bias is dependent both on the cosmic dispersion in luminosity for a given kind of object and the real space distribution of the objects belonging to that class. Observational errors simply serve to increase the RMS scatter. Since the size of this systematic error (in km s^{-1}) is proportional to distance (errors in magnitudes are logarithmic errors) it is the least serious for measurement of peculiar velocities of nearby galaxies and the most serious for the furthest galaxies in any sample. One can therefore study the motions of nearby galaxies with data of relatively poor observational accuracy, but one is constrained to study more distant galaxies only with data of the highest accuracy. Fortunately, this kind of bias is reduced for galaxies located together in space (as in a cluster) as the combined error in distance is reduced by $n^{1/2}$ for n galaxies and the Malmquist bias error reduced by n (cf Lynden-Bell *et al* 1988).

The net effect of Malmquist bias in a smooth, homogeneous density distribution is to rescale the galaxies in distance, but to preserve the homogeneity and smoothness (i.e. it will produce an apparent change of the Hubble constant H_0). However, the true distribution of galaxies in the universe is not homogeneous and smooth (as defined by radial velocities; e.g. de Lapparent *et al* (1986), Tully and Fisher (1987)). Galaxies are often found in dense regions bordered by underdense regions, leading to density discontinuities.

Galaxies on the inside of such a boundary (relative to us) will have distances estimated to be too small, as density increases rapidly with distance; galaxies on the outer edge of this boundary will have distances estimated to be too large, as density decreases rapidly with distance. The net result is that any sharply bounded distribution of galaxies in real space will be *observed* to be a more fuzzy bounded feature on the basis of distance measurements. If all galaxies in a region of space have similar 'true' distances and similar radial velocities, but errors in their measured distances, their predicted radial peculiar velocities ($V - r$) will scatter with r^{-1} dependence from the centre of this region. Unfortunately, many realistic mass distributions for clusters of galaxies yield true velocity fields of the form r^{-k}, where k is close to unity. Thus, it is inherently difficult to separate motions of galaxies that are gravitationally attracted to distant clusters from the observational errors associated with the measurements.

Finally, if a smooth and homogeneous distribution of galaxies is sampled in an inhomogeneous manner (e.g. such as that dictated by the obscuration of dust in our own Galaxy) spurious net motions could result (Fall and Jones 1976). The same could also result if the density distribution of galaxies differed systematically over the sky (Szalay 1989). It was the possibility of this kind of systematic error that caused Fall and Jones to doubt the Rubin *et al* (1976a, b) result.

4.2.2. Use of models to predict distances—'I tell you where to be'. If one predicts the distances of galaxies on the basis of a velocity field model the resulting prediction of absolute magnitude is not directly affected by Malmquist bias (Schechter 1977). This fact has been used by a number of workers (e.g. Aaronson *et al* 1982a, 1989, Tully 1988) to investigate various mathematical models of the velocity field. In this method one uses the radial velocity of a galaxy and its distance-dependent property (i.e. luminosity or diameter), combined with the model velocity field, to predict the distance-independent property of the galaxy (e.g. rotation velocity for spiral galaxies). The technique works only in those regions of the model velocity field for which distances are one-to-one correlated with radial velocity (e.g. infall regions into clusters cannot be used because galaxies falling into the centre from opposite sides can have the same apparent radial velocity). As pointed out by Lynden-Bell *et al* (1988, Appendix C of that paper), this method is not entirely free of the effects of Malmquist bias, but it is less affected than direct estimates of distance as determined in § 4.2.1.

Unfortunately, the model itself introduces a new bias into this process due to the fact that these velocity field models are not perfect. This 'model-bias' is systematic as a function of distance and position on the sky, corresponding to the (unknown) difference between the model that is used and the true velocity field. The standard analysis associated with this method cannot uncover this 'model-bias' nor the manner in which the velocity field model is in error.

4.2.3. Selection bias for cluster galaxies. Galaxies in individual clusters are often used to calibrate the Tully–Fisher relationship (for spiral galaxies) and D_n–σ relationship (for elliptical galaxies) in zero point and/or slope (Aaronson *et al* 1986, Dressler *et al* 1987a). Observational errors and intrinsic dispersion will always combine to produce a range of values of the distance-dependent parameter (e.g. luminosity) at a given value of the distance-independent parameter (e.g. rotation velocity). Furthermore, for practical reasons galaxies can be studied only down to a given faintest level, known as the 'magnitude cut-off.' These two observational facts can produce a bias in the derived slope and zero point of either the Tully-Fisher or $D_n-\sigma$ relationship, as discovered separately by Teerikorpi (1987) and by Faber in Lynden-Bell *et al* (1988).

If, for example, one assumes that the observed scatter in the relationship is Gaussian about the distance-independent parameter, the imposition of a magnitude cut-off ensures that only the bright end of this distribution is observed for the apparently fainter galaxies. The seriousness of the effect is proportional to the size of the dispersion (disappearing for a perfectly dispersionless relationship), leading one to derive the wrong slope and zero point for the relationship. Applying this wrong calibration to galaxies not in these clusters will then lead to incorrect distances and incorrect peculiar velocities. The data used in the present review have been modelled for this kind of bias when required.

5. The observational evidence for large-scale motions

5.1. The distribution of galaxies in space

The best current picture of the space distribution of galaxies within a distance of 3000 km s^{-1} is given by Tully and Fisher (1987). In their *Nearby Galaxies Atlas* the distribution of galaxies is colourfully broken down into a hierarchical structure of groups and clusters on relatively small scales, with 'clouds', 'superclusters', 'filaments' and 'spurs' on larger scales. Distances for all galaxies with known radial velocities less than 3000 km s^{-1} (over 2700) are estimated in that book. Faber and Burstein (1989) display the spatial distribution of only those galaxies having independently predicted distances in an analogous manner. A comparison of the maps in the Tully and Fisher book with the maps in the Faber and Burstein paper shows that the galaxies with independent distances are a fair sampling of the galaxies that can be *seen*.

Based on the observed data, distances and the resulting peculiar motions are corrected for a variety of statistical effects. All data are placed on the same relative distance scale, which is calibrated by a derived distance to the Coma cluster of galaxies, 7380 km s^{-1} (Faber and Burstein 1989). Malmquist bias corrections are applied to the observed distances under the assumption of a smooth, homogeneous distribution of galaxies. Although this assumption is wrong in detail in specific regions it is better than making no Malmquist bias correction at all (Lynden-Bell *et al* 1988) in most directions (see, however, the discussion in § 5.5.3).

Figure 3 shows the distance distribution of galaxies projected on the plane of the Local Supercluster (de Vaucouleurs and de Vaucouleurs 1964), usually termed the '*X*-*Y*' projection. The different symbols represent the different data sets that are used (see figure caption). Galaxies placed within a given cluster have similar measured distances. The main density features seen in these diagrams deserve some discussion.

The apparent 'butterfly' shape in the distribution of galaxies is due to obscuration from dust in the Galactic plane, which is nearly parallel to the $Y = 0$ km s^{-1} line. Within an X distance of 2000 km s^{-1} more galaxies are seen at positive Y than at negative Y, an asymmetry which is probably real. The most prominent clusters—Virgo, Ursa Major, Hydra, Centaurus, Fornax, Eridanus, Perseus-Pisces and Pavo-Indus-Telescopium (all named for the constellations in which they appear)—are identified.

Most of the galaxies within a distance of 2000 km s^{-1} at positive X values, including our own, lie in a very flattened system which we view face-on in this projection. It is

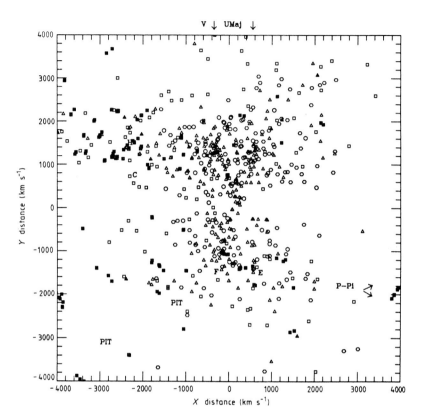

Figure 3. The projection onto the Supergalactic plane, in distance (measured in units of km s^{-1}; see text), of the 898 galaxies used in this review having $V_{LG} \leq 5000$ km s^{-1}. In this figure and those following all distances are corrected for Malmquist bias as described in the text. The Supergalactic plane coordinates are common termed X and Y. Galaxies measured by different observers are denoted by separate symbols: open triangles denote Aaronson *et al* 'good' spiral galaxies; open hexagons represent Aaronson *et al* 'fair' spiral galaxies; full triangles represent Aaronson *et al* clusters (several spiral galaxies in a single cluster); open squares represent elliptical and S0 galaxies with distances determined by the D_n-σ method and not placed in a cluster; full squares represent elliptical and S0 galaxies in clusters and open circles represent de Vaucouleurs and Peters spiral galaxies. Many of the principal regions of galaxy clustering are noted. Two regions are so densely populated in these graphs that they are referred to at the top of the figure: $V =$ Virgo cluster; UMaj = Ursa Major cluster (at $Y \simeq +1300$ km s^{-1}). Other regions are noted at their approximate positions: P-Pi = Perseus-Pisces (mostly off the map); PIT = Pavo-Indus-Telescopium; C = Centaurus; H = Hydra; E = Eridanus; F = Fornax. See Faber and Burstein (1989) for views in the other two Supergalactic planes.

because we are situated at near one end of this system that a thin plane is prominent in maps of the nearby galaxy distribution. This thin plane was used by de Vaucouleurs and de Vaucouleurs (1964) to define the 'plane of the Local Supercluster' (cf figure 2(*b*)).

Also not readily discernible in this projection is the prominent *absence* of galaxies at positive Supergalactic latitudes (the positive 'Z' direction) near the Local Group. The extent of this 'Local Void' has been mapped by Tully and Fisher (1987), and its presence near the Local Group must affect the dynamical motion of the Local Group in a significant manner: if regions of overdensity (e.g. clusters) exert increased gravitational pull, regions of large underdensity (voids) create a 'negative' effect in the overall gravity field and tend to 'push' mass away from the void.

In the discussion that follows, regions in this volume of space will be assigned the names of 'clouds' or clusters, following the nomenclature devised by Tully and Fisher (1987). This serves both to orient one in space (by comparison to the maps in that book) and to define the kinds of peculiar motions that exist for well defined regions of galaxies.

5.2. Choice of velocity reference frame

For the past 50 years the primary choice of velocity reference frame has been the Local Group (LG) velocity reference frame. The discovery of the cosmic background radiation (CBR) dipole anisotropy gives astronomers another choice, assuming that this dipole anisotropy is a measure of the total space motion of the Earth relative to the rest frame of the universe.

One can perform a straightforward test as to whether the LG or CBR velocity reference frame is to be preferred in the analysis of the motions of other galaxies: assume that the distances of galaxies are predicted perfectly from the Hubble equation, $r = V/H_0$, with V being either V_{LG} or V_{CBR} (i.e. as if there were no peculiar motions). The apparent CBR velocity of the Local Group implied by the dipole anisotropy is large, 614 ± 50 km s^{-1} (cf § 3.3.2). If all galaxies within a certain volume of space shared the CBR motion of the Local Group then V_{LG} would yield the best 'Hubble' distances. If, on the other hand, other galaxies did not share the CBR motion of the Local Group V_{CBR} would yield the best Hubble distances. The large amplitude of the Local Group's CBR velocity, combined with accurate distances to relatively nearby galaxies, should be able to clarify which velocity reference frame, if either, is an inappropriate choice.

Figures 4(*a*) and (*b*) plot the Tully–Fisher relations that are observed for the Aaronson *et al* 'good' data, the data set having the smallest observational errors. The absolute magnitudes of galaxies are calculated according to distances predicted from $r = V/H_0$ using either V_{LG} or V_{CBR} (with H_0 arbitrarily set to 50 km s^{-1} Mpc^{-1}). To avoid confusion due to *a priori* known, localised motions of galaxies the 136 galaxies used in this test have $900 \leq V_{LG} \leq 3000$ km s^{-1} (a different sample from that used by Faber and Burstein (1989) for a similar graph) and are not part of the Virgo cluster proper. The incorrect addition of a 600 km s^{-1} velocity vector would lead to errors of ~20–60% in distance for these galaxies, resulting in errors of size 0.5–1.0 magnitude in absolute magnitude.

The formal RMS scatter in both diagrams is the same, 0.59 mag (compared to an expected observational error of <0.35 mag; see later discussion), although the scatter in figure 4(*b*) might appear to some to be smaller than in figure 4(*a*). This similarity implies that a dipole transformation does not model all of the large-scale motions in

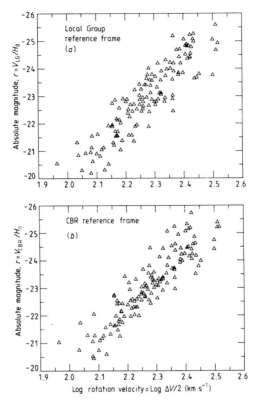

Figure 4. Predictions of the near-infrared Tully–Fisher relation for 136 spiral galaxies from the Aaronson *et al* 'good' sample, including only those galaxies with $900 < V_{LG} < 3000$ km s^{-1} and excluding members of the Virgo cluster. Distances are predicted based on the radial velocity of a galaxy and a smooth Hubble flow model ($r = V/H_0$; H_0 arbitrarily set to 50 km s^{-1} Mpc^{-1}). Absolute magnitudes, based on the observed magnitude of a galaxy and its predicted distance, are plotted on the vertical axis. The speed at which the disc of the galaxy rotates (rotation velocity $\Delta V/2$) is plotted on the horizontal axis. (*a*) Radial velocities V_{LG} are used to predict distances. (*b*) Radial velocities V_{CBR} are used to predict distances. The measured RMS scatter in both figures is the same (0.59 mag), despite the fact that V_{CBR} can differ from V_{LG} by ± 600 km s^{-1}, leading to predicted differences in absolute magnitude of 0.5 magnitude and more.

this volume of space. A significant fraction of the peculiar motions of galaxies, out to at least 3000 km s^{-1}, should be due to the same common physical source (otherwise the LG reference frame would not have predicted reasonable distances). Sources of motion must also exist within this volume of space that are not simply due to a net translation of the whole volume (otherwise the CBR reference frame would have worked significantly better). This review chooses the CBR as the preferred reference frame for measuring the radial velocities of galaxies because this permits viewing their motions from a velocity reference frame that is plausibly assumed to be an absolute rest frame.

It is worthwhile to emphasise again, when viewing the graphs that follow, that we are restricted by geometry and distance to measuring only the radial component of motion for any particular galaxy. It is only by measuring that radial component for many galaxies located in all directions that one can develop a more realistic picture of the three-dimensional peculiar motions that exist.

5.3. Motions of galaxies in the vicinity of our own: the 'Local Anomaly'

Since the motions of the galaxies nearest to the Local Group are those that are the best-determined, a detailed investigation of the motions of the nearest galaxies can offer unique insights into the kinds of motions that can exist elsewhere in this volume of space. Figure 5(*a*) plots the peculiar velocities, measured in the Local Group velocity reference frame, versus observed distance for 55 galaxies from all of the data sets with observed distances ≤850 km s^{-1}. If a galaxy completely shares the CBR motion of the Local Group its peculiar velocity in the LG frame of reference should lie near zero (cf discussion of previous section). Galaxies are identified as to whether or not they are part of the Coma-Sculptor Cloud, the cloud of galaxies in which, according to Tully and Fisher (1987), the Local Group resides.

All galaxies within a distance of 500 km s^{-1} share the CBR motion of the Local Group. The observed dispersion of ~150 km s^{-1} among the peculiar velocities is due in part to the 'random' (i.e. uncorrelated) motions of other galaxies, combined with the peculiar velocity of the Local Group relative to these other galaxies. A reasonable estimate of the random velocity dispersion for these galaxies is ~100 km s^{-1} (cf Faber and Burstein 1989). The motions of galaxies begin to deviate from that of the Local Group motion beyond a distance of 600–750 km s^{-1} (in a systematic manner; see below) but all of the Coma-Sculptor galaxies can be described as having similar observed peculiar motions relative to the CBR.

The existence of this 'quiet' flow of nearby galaxies has been known for over two decades (e.g. Sandage 1972) but its significance relative to the other large-scale motions around it has only recently been appreciated (Richter *et al* 1987, Dressler 1987b, Peebles 1988, Faber and Burstein 1989, who term the motion of the Coma-Sculptor Cloud the 'Local Anomaly'). The three-dimensional form of the Coma-Sculptor Cloud is that of a very narrow, thin sausage-shaped distribution fully contained within the plane of the Local Supercluster (Tully and Fisher 1987). The long axis of this sausage extends nearly 750 km s^{-1} in the direction $l = 202°$, $b = 65°$ (Raychaudhury 1989) and ~400 km s^{-1} in the opposite direction. The −350 km s^{-1} 'Z' component of motion of the Local Group (cf § 3.3.2) is shared, at least in projection, by the whole Cloud. However, the planar distribution of the Coma-Sculptor Cloud means that the full space motions of the Cloud in the Z direction, perpendicular to this plane, are unknown.

Tully (1988) has proposed that the Coma-Sculptor Cloud is undergoing internal collapse due to its own self-gravity. Figure 5(*b*) plots the peculiar velocities of the Coma-Sculptor galaxies, as a function of distance projected along the long axis of the Coma-Sculptor Cloud which points towards $l = 202°$, $b = 65°$ (Raychaudhury 1989). If motions are viewed along a vector to a specific direction, we choose to take the vantage point of the CBR (i.e. remove ourselves from the motion) and flip the signs of distance and peculiar velocities in the backward-looking direction. As with the data plotted in figure 5(*a*), the peculiar velocities used in figure 5(*b*) are relative to the Local Group velocity reference frame.

The present data do indeed indicate that galaxies at the edge of the Cloud are moving towards the centre of the Cloud. This compression-like motion reaches a midpoint at a distance of ~+150 km s^{-1} from the Local Group, with a net peculiar

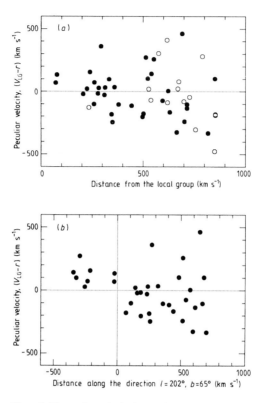

Figure 5. The motions of galaxies, as viewed in the Local Group velocity reference frame, for those galaxies nearest to the Local Group. If all galaxies shared the CBR motion of the Local Group there would be little significant peculiar motions observed in these diagrams. (*a*) 55 galaxies with measured distances and $V_{LG} < 850$ km s^{-1}, with peculiar velocity ($V_{LG} - r$) plotted versus observed distance, r, from the Local Group. Those galaxies that are members of the Coma–Sculptor Cloud (the Cloud of galaxies in which the Local Group resides) are denoted by full circles, the other galaxies by open circles. Coma–Sculptor galaxies tend to share the CBR motion of the Local Group. (*b*) The peculiar velocities ($V_{LG} - r$) for only the Coma–Sculptor galaxies, plotted as a function of the projected distance towards the direction $l = 202°$, $b = 65°$ (defined by the thin, narrow spatial distribution of galaxies in this Cloud). Galaxies with distances opposite to this direction, relative to the Local Group, have the sign of both distance and peculiar velocity reversed. This is in order to have motions presented as if viewed from outside this region of space, with the Local Group located at the origin. When viewed in this manner, a net compressional motion (i.e. galaxies moving towards each other) is observed within the Cloud.

velocity of ~ -100 km s^{-1} with respect to the Local Group (i.e. reflecting the 'random' motion of the Local Group relative to the rest of the Cloud). There are two possible physical sources for this compression-like motion. As suggested by Tully (1988), it could be due to a gravity-induced condensation of the mass distribution within the Coma-Sculptor Cloud itself. Alternatively, this compression could be a part of a more general motion involving all galaxies within the plane of the Local Supercluster, and the apparent midpoint within the Coma-Sculptor Cloud would be illusory. This particular issue will be further explored in §§ 5.5.2 and 6.4.1.

To summarise, examination of the motions of the most nearby galaxies has shown the following: (1) A thin, planar structure of galaxies ~ 1400 km s^{-1} in length can move as a coherent body in space. (2) The full space motion of this coherent body can be in a direction not dictated by the apparent distribution of mass within that body. (3) The random motions of galaxies relative to this 'bulk' motion are 100 km s^{-1} or less.

Galaxies not in the Coma-Sculptor Cloud (i.e. most galaxies with observed distances) do not share the full CBR motion of the Local Group. The information on motions that are shared by these galaxies in the volume of space surveyed can be conveyed by graphs having a number of different formats and quantified by dipole and quadrupole solutions of average motion (cf Lilje *et al* 1986, Lynden-Bell *et al* 1988, Faber and Burstein 1989).

5.4. The evidence for a shared, global flow

5.4.1. Motion in the Supergalactic plane: observations and reliability. Graphs of the motions of galaxies found within a certain solid angle, relative to a well defined plane, can be made in a straightforward manner with the Local Group at the centre of the graph (cf Lynden-Bell *et al* 1988, Faber and Burstein 1989). Four angular cuts of 45° angle each cover the full sky. Figure 6 plots the observed motions in the CBR velocity reference frame for those galaxies in one such angular cut, namely galaxies found within $\pm 22.5°$ of the Supergalactic plane (whose pole is at $l = 227°$, $b = -6°$). For the purposes of this review, this particular angular cut is the most informative as it includes the Supergalactic plane in which most of the nearby galaxies lie (Tully and Fisher 1987). The motions of galaxies within other angular cuts are given in Faber and Burstein (1989).

Peculiar velocities $(V_{CBR} - r)$ are represented as a line in figure 6, joining the observed distance (r) with the observed radial velocity (V_{CBR}). These lines are, of necessity, radial from the point of origin (us). Motions that are directed towards us are represented as open circles plotted at the observed distances, joined by dotted lines to the position given by the radial velocities. Motions that are directed away from the Local Group are represented by full circles and full lines in the same format. The motions of clusters are represented either as a single point or as a close grouping of points as noted. An upper limit of ~ 5000 km s^{-1} is placed on the distances of galaxies shown in these diagrams because Malmquist bias corrections and random errors become too large beyond these distances and the available volume coverage much less (§ 5.6).

The overall impression of motion in this plane is one of a net flow towards the upper left of the diagram (towards the Centaurus region) combined with a net inward motion on either side of the Supergalactic plane from the Local Group. The motions of all galaxies in this figure appear to be involved in this global flow to a greater or lesser degree. (For identification of individual regions in figure 6 please refer to figure 3.)

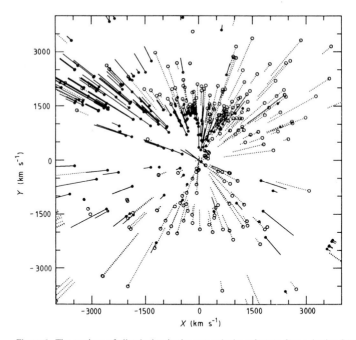

Figure 6. The motions of all galaxies, in the CBR velocity reference frame, having Super-galactic latitudes within $\pm 22.5°$ of the Supergalactic equator and $V_{LG} < 5000\,\mathrm{km\,s^{-1}}$. Peculiar velocities that are directed away from the Local Group (positive motion) are plotted as full dots at the predicted distances, together with full lines of length proportional to the total (not projected) peculiar velocities. Peculiar velocities that are directed towards the Local Group (negative motion) are plotted as open circles at the predicted distances, together with dotted lines of length proportional to the peculiar velocities. Galaxies in clusters are given the net peculiar motion of the cluster. Positive peculiar velocities are predominantly in the upper left quadrant (the Hydra–Centaurus region; cf figure 3), while negative peculiar motions dominate the other three quadrants. The overall impression is that of a coherent flow in this volume of space, in both direction and distance.

Separate dipole and quadrupole solutions for the two largest and most accurate galaxy samples support this interpretation for the motions of galaxies with $V_{LG} < 3200\,\mathrm{km\,s^{-1}}$: 217 spiral galaxies of the Aaronson *et al* 'good' data set and 179 elliptical galaxies of the original 7S collaboration. As given in table 1, the two data sets agree that there is a net dipole motion of amplitude $500 \pm 50\,\mathrm{km\,s^{-1}}$ in the direction $l = 314° \pm 15°$, $b = 14° \pm 5°$. They also agree on the relative directions and amplitudes of the quadrupole moments, and in both samples the long axis of the quadrupole moment points in a similar direction to the dipole moment vector.

The very good agreement in observed motions between the two data sets is all the more remarkable due to the very different nature of these data (Gunn 1989). (1) The Aaronson *et al* spiral galaxies are disc-dominated, rotationally supported systems of stars and gas. The elliptical galaxies used by the 7S collaboration, on the other hand, owe their intrinsic shapes more to the degree of anisotropy of the orbits of their stars than to any rotational motions by these stars (cf Davies *et al* 1985). (2) Most of the Aaronson *et al* galaxies are located in directions primarily perpendicular to the dipole

Table 1. Comparison of dipole and quadrupole solutions from two different data samples for galaxies with distances less than 3200 km s^{-1}.

Data sample		Aaronson *et al*	7 Samurai
Type of galaxies used		Spirals	Ellipticals
Distance method used		Near-IR TF	D_n-σ
Number of galaxies		217	179
Dipole direction	l	$300° \pm 6°$	$329° \pm 14°$
	b	$13° \pm 3°$	$14° \pm 7°$
Dipole amplitude		482 ± 41 km s^{-1}	514 ± 95 km s^{-1}
Quadrupole moment			
Q1: Relative amplitude		0.912 ± 0.024	0.840 ± 0.045
Direction	l	$204°$	$188°$
	b	$54°$	$63°$
Q2: Relative amplitude		0.980 ± 0.036	1.003 ± 0.067
Direction	l	$239°$	$239°$
	b	$-31°$	$-18°$
Q3: Relative amplitude		1.108 ± 0.035	1.157 ± 0.050
Direction	l	$319°$	$322°$
	b	$17°$	$20°$

moment vector and are closer than 2000 km s^{-1} in distance (figure 3). Most of the ellipticals are observed in directions along the dipole moment and are at distances greater than 1500 km s^{-1}. (3) Most of the Aaronson *et al* galaxies are found in regions of sparse density such as loose associations and clouds that are not necessarily gravitationally bound. Many ellipticals are found in gravitationally bound dense clusters, for which a reliable average peculiar motion can be obtained from an average of the motions of individual members assigned to the same distance.

The possibility that the observed large-scale motions could be systematically in error, and therefore spurious, has been raised (Djorgovski *et al* 1989, Silk 1989). However, detailed tests of the physical properties of galaxies have *not* discovered significant sources of systematic error in either the Tully–Fisher relation or the D_n-σ relation (Aaronson and Mould 1983, Lynden-Bell *et al* 1988, Burstein *et al* 1988, 1989, 1990, Burstein 1989). In a separate problem, the absence of galaxies in the 'zone of avoidance' might sufficiently modify the observed distribution of galaxies to produce similar, but incorrect dipole moments from different data samples (Szalay 1989). Yet, obscuration by dust in our Galaxy is not likely to produce the similar quadrupole moments in the two separate samples, nor to have the long axis of the quadrupole moment line up with the axis of the dipole moment. Finally, the fact that galaxies of all kinds, in all environments, over such a large region of space, share a motion that is *not* simply dipolar in nature argues very strongly that such large-scale motions are real (cf discussion in Burstein *et al* (1990)).

5.4.2. The overall flow as seen in a cosine diagram. The previous graph presents evidence of a global flow as a function of both direction and distance. A classical test for the existence of an overall velocity flow is to plot peculiar velocity against the cosine of the angle made by a galaxy relative to the predicted apex of the motion. In this kind of diagram, peculiar velocities that are not directly related to the flow appear as 'noise'.

Figure 7 shows the cosine diagram for motions of 675 galaxies from all data samples that lie within a distance of 3500 km s^{-1}. The choice of distance or radial velocity

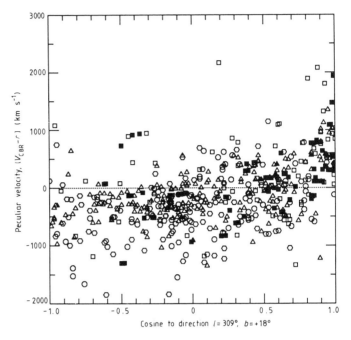

Figure 7. The peculiar velocities, in the CBR velocity reference frame, of those galaxies having distances less than 3500 km s^{-1} from the Local Group, plotted versus the cosine of the angle between each galaxy and the direction towards the dipole moment, $l = 309°$, $b = 18°$. The same symbols as in figure 3 are used to represent galaxies from the different data samples. The coherent flow implied in figure 6 is now more evident, and is seen to vary differentially as a function of angle towards the dipole moment direction. The motions of all galaxies in this volume of space appear to partake of this global flow, but much of the 'noise' relative to this flow is due to other motions of a more local nature (implied by the comparison made in figure 4). It is important to note that the peculiar velocities measured by all data samples used in figures 6 and 7 are in substantial agreement with the direction and size of peculiar motions in this volume of space.

range is somewhat arbitrary (e.g. see Lynden-Bell *et al* (1988), Burstein *et al* (1988) or Burstein (1990) for different velocity limits) but yields similar views for samples restricted to $V_{CBR} < 5000$ km s^{-1}. The choice of apex direction ($l = 309°$, $b = 18°$) is obvious, being near the direction of both the dipole moment vector and the long axis of the quadrupole moment as defined in the more detailed modelling by Faber and Burstein (1989).

A large-scale, coherent motion of galaxies clearly exists in this volume of space, at least out to a distance of 3500 km s^{-1}. The existence of this large-scale motion has captured the intense interest of cosmologists since its discovery by the 7S collaboration (Burstein *et al* 1986, Dressler *et al* 1987a, Lynden-Bell *et al* 1988, Faber and Burstein 1989). Moreover, the pattern of flow is one of acceleration towards the apex; average peculiar motion varies slowest with cosine angle at the antapex and increases quickly towards the apex at positive cosine angles. It is because of this pattern of acceleration that the existence of a very large, massive object centred at the apex was proposed by

Lynden-Bell (1987) and Lynden-Bell *et al* (1988) and dubbed the 'Great Attractor' by Alan Dressler.

Although this large-scale flow appears to be shared by almost all galaxies in this volume of space, it cannot be the only coherent kind of motion that exists if the motion of the Coma–Sculptor Cloud is typical. These other kinds of motions are better pictured in other ways.

5.5. Motions of galaxies in, and within, specific regions

The sky can be divided into about two dozen two-dimensional regions that contain almost all galaxies with distances <5000 km s^{-1} (cf Tully and Fisher 1987). Faber and Burstein (1989) have pictured the motions of galaxies in many of these regions by plotting peculiar velocity, $V_{CBR} - r$, versus distance, r, for the galaxies and groups in each region. Since observed distance can also be defined as the sum of the peculiar velocity of the galaxy and V_{CBR}, the errors in the two quantities are correlated along a line of slope -1. These errors can either be of random origin or of Malmquist bias-type systematic origin. The fact that either kind of error produces a line of slope -1 further compounds the problem of separating motions produced by the gravity field of a cluster from motions that are artefacts of observational error (§ 4.2.1). Most galaxies in clusters are given the distance of the cluster if so identified by the source of distance data. The exceptions to this are the Aaronson *et al* galaxies in the Ursa Major cluster.

5.5.1. Motions in six representative regions not containing a dense cluster.

Figures 8(*a*)–(*f*) present the peculiar motions of galaxies in six different directions plotted as $(V_{CBR} - r)$ versus r. A line of slope -1 is given in figure 8(*e*) to represent the correlated effects of errors in this kind of diagram. These six regions have motions that are representative of two kinds of 'regional' motion that have been found (Faber and Burstein 1989). (1) Relatively quiet motions, showing a small internal velocity dispersion over a region ~ 1000–1500 km s^{-1} in extent, but possibly large gradients in peculiar velocity across the region; i.e. much like the motion seen for the Coma–Sculptor Cloud. Examples of this kind of motion are found in Camelopardalis (figure 8(*d*)), Sculptor (figure 8(*e*)), N1023 + Cetus (figure 8(*f*)), Serpens–Ophiuchus (figure 8(*c*)), as well as in Sextans and in Leo (not pictured). (2) Regions in which the differential peculiar motions are too noisy to be well defined. These regions include the more distant parts of Serpens–Ophiuchus (figure 8(*c*)) and Camelopardalis (figure 8(*d*)), most of Pavo-Indus–Telescopium (figure 8(*b*)) as well as Dorado (not pictured). It is not yet clear whether these 'noisy' regions are the result of observational errors, too sparse sampling of the existing motions or simply intrinsically large motions within these regions. The motions in Pegasus (figure 8(*a*)) are too ill-defined for classification as either kind of motion. The following sections cover in greater detail the kinds of motions that occur in the third kind of region—regions of apparent gravitational infall.

Independent of the kind of differential motion within a region, galaxies in each region tend to share in the same kind of net motion relative to V_{CBR}, just as those galaxies within the Coma–Sculptor Cloud share the same net motion. The net motions of these other regions, which can be observed only in one dimension relative to the CBR velocity reference frame, range from -600 km s^{-1} (e.g. Pegasus or N1023 + Cetus) to over $+1000$ km s^{-1} (see discussion below). Since only the one-dimensional component of these net motions can be determined, it is now seen that the 600 km s^{-1}

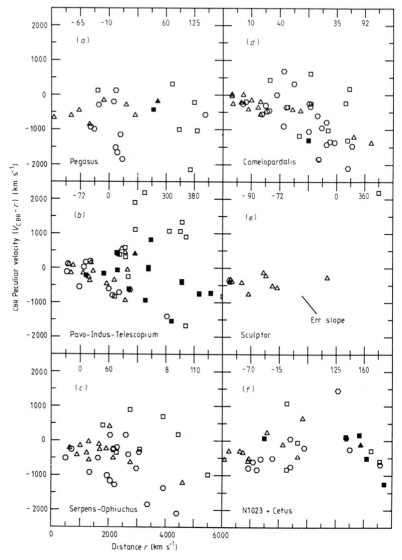

Figure 8. The peculiar motions of galaxies, in the CBR velocity reference frame, as a function of distance in six representative directions, chosen not to contain a dense cluster. Symbols used to represent data for galaxies from different samples are the same as in figure 3. The numbers at the top of each figure give the range in Galactic latitude (left side) and longitude (right side) for each region except for (*d*), which is in Supergalactic latitude and longitude. A line showing the correlated error in both coordinates (plotted here as 500 km s^{-1}) is given in (*e*). Note the coherence of motions of galaxies within some regions, 'noisy' motions in other regions, and the net motion of each region relative to the CBR velocity reference frame.

three-dimensional velocity of the Coma–Sculptor Cloud is not particularly unusual in this volume of space.

5.5.2. The peculiar velocity fields in the directions of Virgo and Ursa Major. The model of large-scale motions that has dominated this field of study for the past decade is that of Virgocentric infall, produced by the centralised gravitational pull of the Virgo cluster. The inferred motion of the Local Group relative to the CBR reference frame, which has the Local Group moving at over 400 km s^{-1} towards the Virgo cluster, served to strengthen the support for this model (cf Davis and Peebles 1983). Indeed, for many astronomers the term 'motion towards the Virgo cluster' is identified with models of 'Virgocentric infall motion' (see Pierce and Tully 1988, Kraan-Korteweg *et al* 1988, Huchra 1989, for some recent examples).

Nonetheless, it has also been recognised that the current Virgocentric infall models do not work very well outside the immediate vicinity of the Virgo cluster (e.g. de Vaucouleurs and Peters 1984, Aaronson 1986, Peebles 1988, Faber and Burstein 1989). The principal purpose of this section is to review the available evidence relating to the peculiar velocity field of galaxies in the vicinity of the Virgo cluster.

To do this, the available data are subjected to a double selection process. Only galaxies within a projected 30° cone of the centre of the Virgo cluster (taken to be in the direction $l = 284°$, $b = 74°$) are considered. Of this set, only those galaxies are used for which at least half of their observed motion is directed towards the Virgo cluster. Peculiar velocities are defined in the CBR velocity reference frame ($V_{CBR} - r$) and the motions are viewed from the vantage point of the CBR (as in figure 5(b)). The forward direction points towards the Virgo cluster, the backwards direction has the signs of distance and peculiar velocity reversed.

Figure 9(a) plots the motions of galaxies towards the Virgo cluster for the galaxies in the Aaronson *et al* 'good' sample and the galaxies in the 7S sample. The projected distance towards the direction of the cluster is used in place of the true distance to more directly view motion towards the cluster. Data from the noisier Aaronson *et al* 'fair' and de Vaucouleurs and Peters data sets show similar, but less well defined trends (not shown), while the data from Lucey and Carter, and Dressler and Faber, do not cover this region of the sky. The motion of the Local Group relative to the CBR in this direction is given by an X.

There is a net motion of galaxies relative to the Virgo cluster (the position of which is noted), roughly similar on either side of the cluster. These motions do not have the overall 'signature' of gravitational infall into a central cluster: if galaxies are gravitationally attracted only to a cluster in the centre, the motion will be Keplerian, with the highest coherent peculiar velocities seen just outside the core of the cluster and the lowest peculiar motions occurring furthest from the cluster. Instead, those galaxies observed to have the highest peculiar motions relative to the Virgo cluster are among the *furthest* from the cluster, and those galaxies with the lowest peculiar velocities relative to the cluster have projected distances among the *closest*!

This kind of motion is not a pattern of Keplerian infall to a point mass, but more like the gravitational collapse of a disc (or sphere) of uniform density: for a sphere of uniform density $M \sim r^3$ and the force of gravity increases linearly with radius. Peculiar motions of test particles (i.e. galaxies) in this density field would also increase linearly with radius outwards. The region of compression-like motion extends from the Local Group to a distance of 2000 km s^{-1}, 650 km s^{-1} in distance past the Virgo

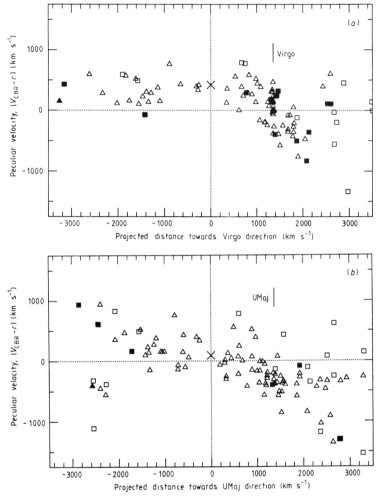

Figure 9. The peculiar velocities, in the CBR velocity reference frame, of those galaxies having the most reliable distance estimates, in the directions of the two nearby clusters that dominate the Local Supercluster: (*a*) the Virgo cluster and (*b*) the Ursa Major cluster. Symbols for the data from different galaxy samples are the same as in figure 3. In each diagram, only those galaxies are included that satisfy two selection criteria: to be within a 30° cone of the direction of the cluster and to have their motion towards the cluster comprise at least half of the observed, radially directed peculiar velocity. The component of motion of the Local Group in each direction is noted with an X and the observed distance of each cluster is noted. The horizontal axis plots the projected distance of each galaxy towards the direction of the cluster in order to emphasise those motions directed to cluster. As with figure 5(*b*), galaxies in the opposite direction from the cluster have the signs of distance and peculiar velocity reversed, so as to view the motions from the 'outside'. Note the *lack* of a Keplerian-like pattern of motion directed towards each cluster and the increase of peculiar motion away from each cluster. This latter kind of motion is discussed more fully in the text.

cluster. Galaxies on the opposite side of the Local Group from the Virgo cluster, and beyond the Coma–Sculptor Cloud (where the plane of the Local Supercluster does not appear to extend; cf § 5.1), do not have observable peculiar motions that are part of this internal compression.

The compressional motions of galaxies within the Coma–Sculptor cloud are now seen to be part of this larger compressional motion within the plane of the Local Supercluster. This coordinated motion cannot be due solely to the gravitational pull of the Virgo cluster. The fact that the pattern of peculiar motions towards the Virgo cluster does not follow the classic pattern of 'Virgocentric infall' (cf Aaronson *et al* 1982b) was first shown in detail by Faber and Burstein (1989), although hinted at by previous workers (de Vaucouleurs and Peters 1984, Aaronson 1986, Lilje *et al* 1986).

If the observed motions towards the Virgo cluster are not due solely to the gravitational pull of the Virgo cluster itself, what other gravitational sources of this motion exist? A reasonable place to look within the Supergalactic plane is towards the Ursa Major cluster, the other principal mass concentration near the Local Group in this direction (Tully and Fisher 1987). The motions of galaxies towards the Ursa Major direction of $l = 140°$, $b = 62°$ are plotted in figure 9(b), selected and plotted in the same manner as in figure 9(a). Peculiar motions of galaxies relative to the centre of the Ursa Major cluster (noted on the graph) are seen that are similar to, but somewhat noisier than, motions of galaxies towards the Virgo cluster.

The apparent region of compressional motion also extends about 2000 km s^{-1} and includes galaxies on the opposite side of the Local Group from the Ursa Major cluster (not explicitly shown here). Moreover, the net peculiar motion of the centre of the Ursa Major cluster is ~ -350 km s^{-1} relative to the CBR velocity reference frame and over 400 km s^{-1} different from that of the Virgo cluster (cf Pierce and Tully 1988).

This large net motion between the Virgo cluster and the Ursa Major cluster has been interpreted in terms of the classic Virgocentric infall pattern (e.g. Pierce and Tully 1988). Again, however, the pattern of peculiar motions in figure 9(b) is not that of Keplerian infall to a nearby centralised object. Neither do the motions towards the Ursa Major cluster look much like compression, at least when viewed in this manner. Here is a case where the amplitude of motions within a region and line-of-sight range of motion due to the global flow are of comparable size, requiring a global model of this flow to better understand the regional motions (§ 6.4).

5.5.3. The observed velocity field towards Centaurus. All available evidence points to the direction $l = 309°$, $b = 18°$, near the Centaurus clusters, as the direction towards which the overall flow in our volume of space is directed. This region of the sky is in the southern hemisphere of the Earth and, as such, has not been well studied until recently (owing to the north–south asymmetry in large telescopes before ~ 1970). Over the past decade a number of authors have suggested that the clusters of galaxies in Centaurus, together with clusters in Hydra, form another Supercluster of galaxies similar to our own Local Supercluster (Chincarini and Rood 1979, Chincarini 1982, Winkler 1983, Fairall and Winkler 1984, Fairall 1984, 1988, Hopp and Materne 1985, Dickens *et al* 1986, Lucey *et al* 1986a, b, Da Costa *et al* 1986, 1987, Tully and Fisher 1987, Chincarini and Vettolani 1988).

The net motions of galaxies towards this direction were first pointed out by the 7S collaboration, and initially expressed in terms of an overall 'bulk' flow (Burstein *et al* 1986, Dressler *et al* 1987a). Later, when it was realised that a differential flow existed in this volume of space (cf figure 7) it was proposed that this motion was gravitational in origin, generated by the large, overdense region termed the Great Attractor (Lynden-Bell 1987, Lynden-Bell *et al* 1988, Dressler 1988, Faber and Burstein 1989; § 3.4). Spurred on by this prediction, several groups have since measured the distances and peculiar velocities of more galaxies in this direction, making it now one of the best studied regions in the sky (Lucey and Carter 1988, Aaronson *et al* 1989, Dressler and Faber 1990; see summary by Burstein *et al* (1990)).

The investigations of Lucey and Carter and of Aaronson *et al* concentrated on measuring the motions of seven clusters in the region $260° < l < 340°$, $-30° < b < 50°$. There is substantial overlap in the clusters studied by these two collaborations, and with those studied before them by the 7S collaboration. Lucey *et al* (1986a) found that the velocity field in the direction of the Centaurus clusters ($l = 302°$, $b = 22°$ subtending an area 3° in radius) was rather complicated, with apparently two subgroupings of radial velocity centred at ~ 3000 km s^{-1} (Cen30) and 4500 km s^{-1} (Cen45). Following that paper, the galaxies in the Centaurus clusters have usually been divided into two clusters on the basis of Local Group radial velocities V_{LG}. Those galaxies having $V_{LG} < 4200$ km s^{-1} are put in the Cen30 cluster, those with $V_{LG} > 4200$ km s^{-1} are put in the Cen45 cluster.

Using these definitions, the 7S collaboration measured very large outward (positive) peculiar motions for both the Cen30 ($+1110$ km s^{-1}) and the Cen45 ($+1660$ km s^{-1}) clusters, based on data for 5 and 3 galaxies in each cluster respectively (Lynden-Bell *et al* 1988, Faber *et al* 1989). In contrast, Lucey and Carter (1988) find these two clusters to have peculiar motions of $+240$ and $+2620$ km s^{-1} (based on data for 15 and 6 elliptical and S0 galaxies respectively), while Aaronson *et al* (1989) find them to have peculiar motions of $+490$ and $+590$ km s^{-1} (10 and 6 spiral galaxies respectively). Some of the differences in quoted peculiar motions are due to the different handling of observational corrections within the data (≤ 300 km s^{-1}).

In a separate programme of study Dressler and Faber (1990), guided by the redshift survey of Dressler (1988), have obtained distances for 136 elliptical galaxies (91 newly observed) spread over the region $290° < l < 350°$, $-35° < b < 45°$. Burstein *et al* (1990) have combined all of the available data in the larger region $260° < l < 360°$, $-40° < b < 50°$ in a self-consistent manner in an attempt to clarify the kinds of motions that exist in this direction.

With a total of 47 galaxies with measured distances Burstein *et al* find that the radial velocities of the galaxies are poorly correlated with their distances. Cen30 and Cen45 galaxies occupy the same range of distance (except for the Cen45 spirals, which may be somewhat more distant). This is in substantial agreement with the most recent analysis of Lucey *et al* (1990), based on a similar sample of Centaurus galaxies. Burstein *et al* find that, independent of whether the Centaurus galaxies are defined as one or two clusters, the Centaurus galaxies have a net peculiar motion of $\sim +800$ km s^{-1}. The previous disagreements about the motions of the two Centaurus clusters now appear to result from selecting different subsets of galaxies from a larger sample in which galaxies have a poor correlation of distance with radial velocity.

The peculiar motions of all galaxies with positions lying within a circle of 40° diameter, centred on the direction ($l = 309°$, $b = 18°$), are shown in figure 10. The median peculiar motion for the 47 galaxies in the Centaurus clusters defines one point, using the net peculiar motion obtained by Burstein *et al* (1990). The Malmquist bias correction for the Centaurus cluster is negligible ($\ll 100$ km s^{-1}) owing to the large number of galaxies in that cluster. Peculiar motions for 75 other galaxies are used in

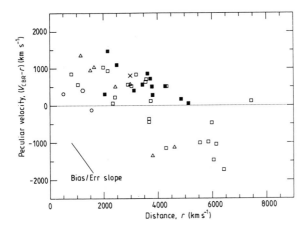

Figure 10. The peculiar motions of galaxies, in the CBR velocity reference frame, for galaxies lying within an angle of 20° of the dipole moment direction $l = 309°$, $b = 18°$. The symbols used for each data set are as in figure 3. In this diagram the two Centaurus clusters, Cen30 and Cen45, are represented as a single point with their total net motion denoted by an X. Note the difference in vertical scale between this figure and figure 9; the positive peculiar velocities seen in this direction are about four times greater in amplitude than seen towards the Virgo and Ursa Major clusters. The form of motion, however, is like that seen towards these other two clusters, being one of increasing motion from the centre of infall. The peculiar motions in this direction appear to cross over to negative motions at a distance of 4000–5000 km s^{-1}, although the interpretation of the apparent negative peculiar velocities beyond this distance is open to question (due to uncertain Malmquist bias corrections).

this diagram: 59 elliptical and S0 galaxies (including 12 smaller clusters) and 16 spiral galaxies.

If the Malmquist bias correction that is used is taken at face value (see below), the form of infall motion in this region ($\sim$40 Mpc in size if $H_0 = 50$ km s^{-1} Mpc^{-1}) is like that seen towards the Virgo and Ursa Major clusters, one of compression rather than of gravitational collapse to a central point. If the Malmquist bias correction is valid, infall from the 'back' side of the Great Attractor appears to now be detected, with the centre of infall at a distance of $\sim$4500 km s^{-1}. The amplitude of motion in this direction is up to about four times the size of compression-like motions observed towards the Virgo and Ursa Major clusters. Moreover, the 800 ± 50 km s^{-1} peculiar velocity of the Centaurus cluster galaxies is inconsistent with this cluster(s) being the principal gravitational source of motion in this direction.

Five pieces of evidence point to a very large, distributed mass in the Great Attractor region as being a plausible gravitational source of the global flow in our volume of space. The directions of both the dipole moment and the long axis of the quadrupole moment point in this direction. The cosine diagram shows that a differential flow accelerates towards this direction. Motions of very large amplitude exist within this extended region of space, implying the existence of a large mass. A large overdensity of galaxies is known to exist in this direction (e.g. Lynden-Bell *et al* 1988) despite the fact that the region is bisected by the 'zone of avoidance'. Finally, most of the galaxies in this extended region have the same range of radial velocities between 2000 and

6000 km s^{-1} (Dressler 1988, Fairall 1988). All of these facts lend support to the Great Attractor model of these motions proposed by Lynden-Bell *et al* (1988), which is one of the two models of these motions that is evaluated in § 6.

Burstein *et al* find that a further understanding of peculiar motions on the 'other side' of the Great Attractor region is hampered by a lack of detailed understanding of the space distribution of galaxies at distances 3500–6000 km s^{-1}. If part of this region is a dense region of galaxies, surrounded by a relative underdensity of galaxies on the side away from us (cf Melnick and Moles 1987), it is a good example of an 'edge' of the kind discussed in § 4.2.1. If true, the application of a uniform Malmquist bias correction to the galaxies in this direction at these distances would be wrong, creating an artificially larger infall at distances ~4500 km s^{-1} and greater. The distances for many more galaxies, out to distances of ~10 000 km s^{-1}, will be needed to make further progress in understanding the motions within the centre of the Great Attractor region.

5.6. Other possible large-scale motions in other volumes of space

There is little evidence of coherent, large-scale motions of galaxies on scale sizes greater than 15 000 km s^{-1} (i.e. for distances >7500 km s^{-1}) from the peculiar velocity studies used for this review (Aaronson *et al* 1986, 1989, Lynden-Bell *et al* 1988, Lucey and Carter 1988). In particular, the Coma Cluster of galaxies, if placed at the distance of 7380 km s^{-1} used in this review, has a minimal peculiar motion relative to the cosmic background radiation (≤200 km s^{-1}; Faber and Burstein 1989). In stating these results, however, it should also be stressed that relatively little of the sky has been mapped for peculiar velocities at these much greater distances.

Even if there is no large-scale coherence of motions on scales of ~15 000 km s^{-1}, one might expect to find large-scale coherence of motions within other large regions of space similar in structure to that of our own. The nearest other such region is that of the Perseus–Pisces Supercluster (Haynes and Giovanelli 1986, 1989, Lynden-Bell *et al* 1988), which also happens to lie at nearly the opposite side of the sky from the dipole direction at $l = 309°$, $b = 18°$, and at about the same distance from us as the centre of infall in the Great Attractor region. Unfortunately, the Perseus–Pisces region is even more heavily affected by obscuration due to Galactic dust than is the Centaurus region. Galaxies appear fainter and the measured distances are more uncertain. Nonetheless, there are indications in the current data that a large-scale flow may be occurring within the Perseus–Pisces Supercluster (cf the peculiar velocities given in Faber *et al* (1989)).

It is also somewhat puzzling that the gravitational field of the Perseus–Pisces Supercluster is not more evident in the peculiar velocities within our local volume of space (e.g. Yahil 1989). However, the data in the literature are too few to give anything more than hints of the structure of the peculiar velocity field in this direction at these distances (cf figure 6, near $X \sim +3000$ km s^{-1}, $Y \sim 0$ km s^{-1}). It has thus been recognised that this is a region deserving of much new work with several ongoing research programmes being conducted in this part of the sky.

Current observations also suggest that the Pavo–Indus–Telescopium region, which has galaxies with distances ranging from 3000 to 10 000 km s^{-1} in distance, is a region containing large, internal motions of uncertain coherence (§ 5.5.1). All other known regions of possible interest are at least twice as far away as the edge of the volume surveyed, including the region in Hercules at a distance of ~10 000 km s^{-1} (Aaronson *et al* 1986, Geller and Huchra 1989, Freudling *et al* 1989), a more distant region in Pisces (Batuski and Burns 1985a, b, Tully and Fisher 1987) and clusters in the direction of the Great Attractor but at distances of $\sim 14\,000\,\text{km s}^{-1}$ (Scaramella *et al* 1989).

Increasing the distances to be investigated also increases the size of errors in peculiar motions in direct proportion. More galaxies must be observed with higher accuracy even though the galaxies appear smaller and fainter. The large-scale motions that exist in these more distant regions of space will be difficult to measure in the kind of detail in which we can understand the motions of closer objects.

6. Models of large-scale motions

6.1. Kinds of models and their uses

The large-scale peculiar motions that are observed could have evolved from a primordial, random noise fluctuation spectrum, imprinted on the distribution of matter at the beginning of the universe, upon which gravitational forces have amplified the motions with time (cf Peebles 1980, Davis and Peebles 1983). This has to be at least partly true (galaxies and clusters of galaxies exist) but the extent to which it *must* be true for *all* observed motions needs to be understood. Alternatively, some of the peculiar motions that are seen today might be due to large, non-random, non-gravitational perturbations of the early universe (such as 'contagious' explosions of the kind suggested by Ostriker and Cowie (1981)).

The only reasonable way to conclude that primordial, *non-gravitational* motions could be the source of all present-day motions is to rule out the viability of gravitational interactions in explaining these motions. This, in turn, requires understanding the manner in which both luminous and dark matter (both being sources of gravity) are distributed in space. We see the luminous matter (by definition; § 2.3); it is the distribution of the dark matter (both baryonic and non-baryonic) that is *a priori* unknown. A thorough theoretical understanding of the observed motions requires a series of laborious, interactive steps between theory and observation, of which measurements of peculiar motions are simply the first act (an act, it is worthwhile remembering, which has taken 10 years and over 1000 nights of large telescope time to stage).

It is only since December 1987 that, given a self-consistent set of peculiar velocities for 785 galaxies, groups of theorists have been analysing the observed motions in ways that can be reliably compared with one another (Bertschinger and Juszkiewicz 1988, Kaiser 1988, Kaiser and Lahav 1989a, b, Szalay 1989, Groth *et al* 1989, Bertschinger and Dekel 1989, Yahil 1989, Strauss and Davis 1988, 1989, Gorski *et al* 1989, Vittorio and Juszkiewicz 1989, Regos and Szalay 1989). A full review of the results from these comparisons between theory and observation is beyond the scope of the present review. It is also somewhat premature, as many further studies will be done in the near future by these and other workers in the field.

The results of two kinds of models are worth presenting here. (1) A model of motions based on the observed peculiar velocities of that subset of galaxies that have measured distances. This kind of model makes the fewest *a priori* assumptions about the physical origins of motions. If these motions are assumed to be gravitational in origin, the distribution of matter in this volume can be predicted. (2) Under the assumption that these galaxies are test particles of the whole gravitational field (which may be primarily dark matter), a semi-empirical velocity field model can be derived that is based on one's best guess of the density distribution of *all* galaxies. This model is then compared with the observed peculiar motions. It is by performing these two separate analyses that a better understanding can be obtained of the roles (or non-roles) played by dark matter and/or non-random fluctuations in producing the observed peculiar motions.

6.2. The Great Attractor model

A model of the observed large-scale flow based on the hypothesis that it is caused by the gravity of a 'very massive object' was first developed by Donald Lynden-Bell (1987) and is now known as the Great Attractor model (§ 3.4). This model has been refined over the past two years (Lynden-Bell *et al* 1988, Faber and Burstein 1989) as a greater understanding of these large-scale motions has been acquired.

Historically, the analytic forms of empirical velocity field models have been kept simple. The most commonly used model is based on that of the gravitational field due to a centrally condensed, spherically symmetric mass (Aaronson *et al* 1982a, 1989, Pierce and Tully 1988, Lynden-Bell *et al* 1988). It is the intent of the present section to illustrate the degree to which this kind of model can fit the observed peculiar velocities, based on the combined velocity field model proposed by Faber and Burstein (1989).

This velocity field model has three independent components.

(i) The Great Attractor (GA), which is pictured as a spatially extended, spherically symmetric, centrally concentrated mass located in the general direction of the dipole moment. From maximum-likelihood fits to the observations, Faber and Burstein find that the centre of infall for the Great Attractor is at a distance of 4200 ± 300 km s^{-1}, and that the mass associated with the central 'core' is distributed to a distance of ~ 1500 km s^{-1} from the centre. This core has an angular size of $\sim 40°$ as projected on the sky. The mathematical form of peculiar velocities generated by the mass of the Great Attractor, as a function of distance from the infall centre is given by

$$u_A = v_A(r_A/d_A)[(d_A^2 + c_A^2)/(r_A^2 + c_A^2)]^{(n_A+1)/2} \tag{1}$$

where u_A is the predicted peculiar velocity (in three dimensions) radially directed toward the infall centre, r_A is the distance of a galaxy from the infall centre, d_A is the distance of the infall centre from the Local Group, v_A is the peculiar velocity generated by the Great Attractor at the position of the Local Group, expressed in the CBR velocity reference frame and c_A is the core radius of the centralised mass of the Great Attractor. The exponent n_A controls the radial dependence of the peculiar velocity field: at small values of r_A, $u_A \sim 0$, while at large r_A, $u_A \sim (r_A)^{-n_A}$ (formula taken from Faber and Burstein 1989). The direction of the infall centre, in terms of Galactic coordinates l_A and b_A, is also a free parameter.

The values of these parameters are found from a maximum-likelihood method originally devised by Lynden-Bell, which uses the 'I see you' method of predicting distances (§ 4.2.1) and hence must predict Malmquist bias corrections to the distances of each galaxy or group. Current values for the GA parameters are: $d_A = 4200$ km s^{-1}; $v_A = 535$ km s^{-1}; $n_A = 1.7$; $l_A = 309°$; $b_A = 18°$ and $c_A = 1430$ km s^{-1}.

(ii) Gravitational infall into the Virgo cluster of galaxies (V_{inf}) formulated along lines similar to that of Aaronson *et al* (1982a):

$$u_V = v_V[(d_V^2 + c_V^2)/(r_V^2 + c_V^2)]^{n_V/2}. \tag{2}$$

The parameter d_v and the direction to the Virgo cluster centre of infall, in Galactic coordinates l_V, b_V, are fixed from the data ($d_v = 1350$ km s^{-1}, $l_V = 284°$, $b_V = 74°$). The other parameters are variable and, although relatively ill-determined from the data (Faber and Burstein 1989; see also discussion below), are given as: $v_V = 130$ km s^{-1}, $n_V = 1.0$, $c_V = 0.10$ (again, radial velocities expressed in the CBR velocity reference frame).

(iii) The Local Anomaly (LA). Both the Virgo cluster and the Great Attractor lie within or near the Supergalactic plane. The full CBR motion of the Coma–Sculptor cloud, of which the Local Group is a part, has a significant component of motion perpendicular to the Supergalactic plane (V_Z in the terminology of § 2.1). Motion in this direction must be generated in a manner unrelated to the above two mass concentrations. Faber and Burstein (1989), in the absence of any form of physical model for this motion, have a 'switch' which, when turned on, requires all galaxies lying within distances of 750 km s^{-1} of the Local Group to share the full CBR motion of the Local Group. (This is equivalent to viewing the peculiar motions for these galaxies *alone* in the Local Group velocity frame as in § 5.3).

Of the three parts to this model (GA, V_{inf} and LA), only the velocity field generated by the Great Attractor is relevant to all galaxies in this volume of space to distances of ~4000 km s^{-1} from the Local Group. The current model of Virgo cluster infall attempts to map the motions of that subset of galaxies that lie within a distance of 1000 km s^{-1} of the Virgo Cluster (including the Coma–Sculptor Cloud). The Local Anomaly 'switch' attempts to match only the peculiar motion of the galaxies in the Coma–Sculptor Cloud. Thus the three parts of this model attempt to map only those motions important to the Local Group.

Since there are motions within this volume of space in which the Local Group does not partake (e.g. motions within regions other than the Coma–Sculptor Cloud) this model is necessarily incomplete. It is also possible that these three motions of the Local Group, currently modelled as separate components, are coupled in a more complicated manner than is currently allowed (e.g. the velocity gradient within the Coma–Sculptor Cloud (§ 5.3 above) is part of a more general compressional motion). It is therefore instructive to see how well the present, relatively 'simple' velocity field model can map the observed peculiar velocity field within a distance of 5000 km s^{-1}.

6.3. The velocity field predictions of the full $GA + V_{inf} + LA$ model

The predictions of this model are made by calculating the peculiar velocity of a galaxy, or group of galaxies, relative to the velocity field predicted by the model. In general, two kinds of methods have been used for this comparison. In the first method, the observed distance of a galaxy is used and the peculiar motion of the model is predicted at that distance. The resulting 'model' peculiar velocities are analogous to the observed peculiar velocities (e.g. $V_{CBR} - r$), which are defined relative to the velocity field model of smooth, homogeneous expansion (§5.2).

The other procedure predicts the distance the galaxy *would* have if the model peculiar velocity *were* correct. If the observed distance of the galaxy is in error, and the galaxy really was at the distance given by the model, the model predicts a different peculiar velocity for this galaxy. This, in turn, results in a new and different predicted distance. The way to resolve this inconsistency is to perform successive iterations until the radial velocity predicted by the model equals the combination of model-predicted distance plus the peculiar motion at the model distance. Methods along this line have been commonly employed in actual use of these velocity field models to predict the distances of galaxies from radial velocities alone.

Unfortunately, both methods have drawbacks to their use. The first method does not attempt to resolve the inconsistency between observation and model and can lead to correlated errors in regions of infall in a model (Strauss 1989). The second method generates its own problems, much along the lines of those that exist with the 'I tell you where to be' set of models (§ 4.2.2). If the velocity field model is wrong, both the model distance *and* the resulting peculiar velocity are wrong. Since it is likely that the velocity field models are wrong in systematic ways, the resulting errors will be systematic in nature with the systematic error in distance compounding the problem.

This review will use 'the distance the galaxy would have if the model peculiar velocity were correct at the measured distance', here termed d_M, as the test of degree of success of a velocity field model. This is done primarily because it is in keeping with the spirit of how distances are estimated in a smoothly expanding universe.

Figure 11 plots the near-infrared Tully–Fisher relation for the 136 Aaronson *et al* 'good' spiral galaxies that were used in figure 4, using the GA + V_{inf} model distances d_M (the LA part of the model does not apply here, as $900 < V_{LG} < 3000$ km s^{-1} for these data). The scatter in this diagram is much reduced by the use of this velocity field model, the RMS 'noise' of the near-IR TF relationship now measuring 0.43 mag. This corresponds to an effective reduction of 0.40 mag in scatter from the near-IR TF relationship predicted by either LG or CBR smooth expansion model (figure 4). As more formally defined by Faber and Burstein (1989), the GA + V_{inf} + LA model fits the observed peculiar motions of all Aaronson *et al* 'good' galaxies with an improvement of 75 in likelihood (i.e. of extremely high significance) over the predictions of a simple smooth expansion (even taking into account the 13 free parameters). The fact that additional coherent motions remain relative to the GA + V_{inf} model (see below) means that the real observational error for these data is ≪ 0.43 mag. (A current estimate is < 0.35 mag; cf § 4.5.)

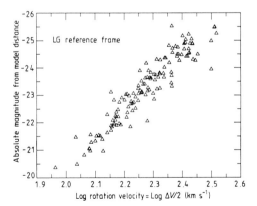

Figure 11. The near-infrared Tully–Fisher relationship for the same 136 galaxies in the Aaronson *et al* 'good' sample as in figure 4, this time determined from the distances predicted by the GA + V_{inf} velocity field model (the LA part of the model is inoperative for $V_{LG} > 750$ km s^{-1}). The marked reduction in scatter is evident both by eye and by direct calculation (see discussion in text).

Figures 12 and 13 show the degree to which the three-component model fits the motions of galaxies as a function of direction and of distance. Figure 12 plots the 'model' peculiar velocities of galaxies in an angular planar cut directly analogous to figure 6. Figure 13 plots the model peculiar velocities of galaxies as a function of the cosine to the GA direction, directly analogous to figure 7. It is evident that the Great Attractor part of the model removes most of the large-scale coherent motion out to distances of ~3500 km s^{-1} from the Local Group, while the Virgo infall part of the model is somewhat less successful (cf the Local Supercluster region in figure 12). In addition, significant residual motions remain in most directions.

Not surprisingly, galaxies in each region tend to share the same kind of residual motion because the principal component of motion removed for most regions is that of the global Great Attractor velocity field. In addition, a net offset in zero point of ~100 km s^{-1} in the model fit is observed in figure 13, reflecting the fact that motion of the Local Group relative to the Coma–Sculptor Cloud, of a similar amplitude, is not currently modelled.

6.4. Motions relative to the Great Attractor model alone

Since it is clear that other regional motions exist that do not involve the Local Group, it is worthwhile to view all of these motions with just the global component of the

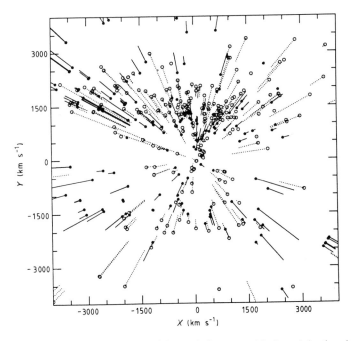

Figure 12. The predicted motions of those galaxies presented in figure 6, i.e. those lying within ±22.5° of the Supergalactic plane, but this time relative to the distance predictions of the GA + V_{inf} + LA velocity field model. The form of presentation is as in figure 6. The velocity field model removes most of the global flow from the motions of these galaxies, but it is apparent that other, more localised motions remain.

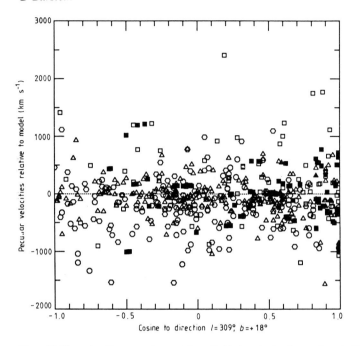

Figure 13. The cosine diagram analogous to figure 7, this time plotting the peculiar velocities of galaxies relative to the GA + V_{inf} + LA velocity field model as a function of cosine to the direction $l = 309°$, $b = 18°$. Symbols represent the same data samples as in figure 3. This model velocity field removes most of the differential flow along this direction. The fact that the net motion of these galaxies is not quite zero with respect to the model reflects the fact that the ~100 km s^{-1} 'random' motion of the Local Group, with respect to other galaxies in the Coma–Sculptor cloud, has not been modelled.

flow removed. In the present model this can be done by viewing motions relative to the velocity field generated by the Great Attractor model alone. Much of the discussion in this section parallels the analysis of §§ 5.4 and 5.5 using the same subsets of galaxies (e.g. only the most reliable data in most circumstances).

6.4.1. Motion towards the Virgo and the Ursa Major clusters. Figures 14(*a*) and (*b*) plot the model peculiar velocities of galaxies observed in the directions of the Virgo and the Ursa Major clusters (this time relative to the CBR velocity reference frame) in a manner directly analogous to that used in figures 9(*a*) and (*b*). The differential motion of the Great Attractor across the line of sight has removed much of the apparent 'noise' in the direction towards the Ursa Major cluster. In both figures the velocity fields relative to the GA model give little indication of an extensive region of gravitational infall into either cluster (cf discussion in § 5.5.2). Instead, the dominant motion present in both figures is once more one of compression. The motions in the direction of Ursa Major are now seen to be remarkably like the motions seen towards the Virgo cluster.

6.4.2. Compressional motion within the plane of the Local Supercluster. The similarity of peculiar motions of galaxies in the Virgo and Ursa Major directions, with the global

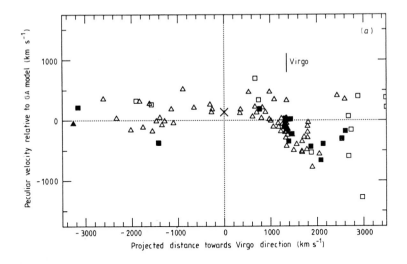

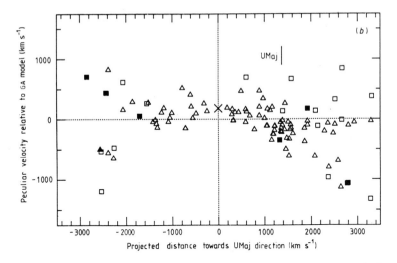

Figure 14. The peculiar motions of galaxies with respect to only the Great Attractor (GA) part of the full GA + V_{inf} + LA velocity field model, in the directions of (a) the Virgo cluster and (b) the Ursa Major cluster. This figure is analogous to figure 9, constructed in the same manner and including galaxies and galaxy samples selected in the same manner. The compression-like motions in both directions are now much more apparent once the global differential flow (modelled by the Great Attractor) is removed. The apparent lack of Keplerian infall directed towards the centre of either cluster is accentuated, especially towards the Virgo cluster. The form of compressional infall patterns in both directions is now seen to be very similar despite the different directions of these two clusters.

flow of the GA model removed, leads one to ask the following question: to what extent are the compressional motions in these two directions shared by all galaxies within the plane of the Local Supercluster? Figure 15(*a*) plots the peculiar motions of those galaxies located within a 120° wide cone centred on $l = 220°$, $b = 80°$ and lying within 25° of the Supergalactic plane (i.e. including essentially the whole plane of the Local Supercluster). Due to the wide viewing angle, true distance is used for this diagram. Otherwise figure 15 is constructed analogously to figures 14 and 9, in that the motion is viewed from the vantage point of the CBR (with signs of distance and peculiar velocity reversed in the backwards direction).

The GA model effectively removes the large-scale flow that exists across the plane of the Local Supercluster, leaving a net motion of galaxies that is one of compression within this plane. The Virgo and Ursa Major clusters lie near the middle of this compression, but there is no well defined 'centre'. Rather, these two clusters define a large central region within the plane, towards which all galaxies in the plane of the Local Supercluster are moving.

Similar kinds of motions are *not* seen in the opposite direction from the axis defined by the Virgo and Ursa Major clusters (negative distances). This is consistent with the fact that the plane of the Local Supercluster is only well defined in the Virgo-Ursa Major direction (cf Tully and Fisher 1987; figure 2(*b*)). In addition, it appears that galaxies that are more distant than the Local Supercluster (at distances $\geqslant 3000$ km s^{-1} in figure 15(*a*)) also do not participate in this compressional motion. The compressional motion within the Coma-Sculptor Cloud (§ 5.3.) is part of this larger compressional motion within the Local Supercluster and is probably not internally generated within this Cloud (as previously suggested by Tully 1988).

6.4.3. Motion perpendicular to the plane of the Local Supercluster. Figure 15(*b*) plots the model peculiar motions of galaxies in the same 120° cone as in figure 15(*a*), but this time including only those galaxies having an angle greater than 25° relative to the Supergalactic plane. The sense of construction of figure 15(*b*) is the same as figure 15(*a*). If infall towards the plane of the Supercluster is occurring, we would be restricted by geometry to observe only a small fraction of it in the radially directed motion of any galaxy that is not directly above (or below) the Local Group. As the region of space just 'above' and 'below' the Coma-Sculptor Cloud is relatively void of galaxies (e.g. the Local Void of Tully and Fisher 1987), evidence of infall towards the plane of the Local Supercluster will be apparent to us only for relatively distant galaxies that lie outside this plane.

It may therefore be significant that the net motion of galaxies grows more negative with distance for galaxies lying in the direction of the Local Supercluster in figure 15(*b*), as if one were viewing a greater percentage of infall motion. The same effect is not seen in the opposite direction despite a similar distribution of galaxies with distance.

6.4.4. Historical misinterpretation of motion towards the Virgo Cluster. The motions of galaxies within the Local Supercluster are not directed to the Virgo cluster *per se*, but rather towards a central, wide region of the Supercluster that is apparently coincident with the positions of both the Virgo cluster and the Ursa Major cluster. The residual motion of these two clusters relative to the Great Attractor velocity field is small (~ -150 km s^{-1}) but significant, and plausibly indicates an additional net motion of the Supercluster as a whole. Galaxies within the plane of the Supercluster are falling

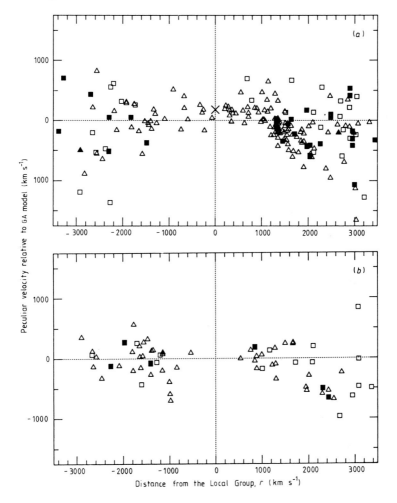

Figure 15. The peculiar motions of galaxies with respect to only the Great Attractor part of the velocity field model. Symbols are the same as in figure 3. (*a*) the motions of those galaxies that lie in a 120° wide 'slice' centred on the direction $l = 225°$, $b = 80°$, that also lie within 25° of the Supergalactic plane. Included in this slice are effectively all galaxies within the Supergalactic plane, including the Virgo and Ursa Major galaxies of figures 14 and 9. The construction of the diagram is the same as in those two previous figures (with signs of distance and peculiar velocity reversed in the opposite direction) except that observed distance is used for the horizontal axis (due to the wide viewing angle). The pattern of compressional infall is present within the whole plane of the Supercluster and is of comparable amplitude over a wide range in viewing angle. (*b*) the motions of galaxies that lie within the same 120° cone centred on $l = 225°$, $b = 80°$, but that lie outside an angle of 25° relative to the Supergalactic plane. Motions of galaxies in the direction of the Local Supercluster tend to become more negative with distance, as one would expect if these galaxies are falling towards the plane of the Local Supercluster. Such motions are not seen in the opposite direction, where the distribution of galaxies is not very flattened within the Supergalactic plane.

towards this central region with velocities that increase with distance outward. The Local Group, lying near the outside of this Supercluster, has one of the largest relative velocities to its centre, ~ 300 km s^{-1}, but only 150 km s^{-1} of net peculiar motion relative to the CBR.

Thus, of the 421 km s^{-1} CBR motion of the Local Group towards the Virgo cluster, ~ 150 km s^{-1} is directed towards the central region of the Local Supercluster (including the Virgo cluster) with probably $\ll 100$ km s^{-1} of that motion generated by the mass of the Virgo cluster itself. The remaining ~ 250 km s^{-1} motion of the Local Group towards the Virgo cluster is its component of motion towards the Great Attractor in this direction.

Models of spherically symmetric infall into the Virgo cluster (generally termed Virgocentric infall) have been the most commonly used models over the past ten years (see discussions in § 3.3.1, § 5.5.2 and Huchra 1989). Now that it is apparent that the peculiar velocity field within the Local Supercluster is not due to infall to any one cluster, more realistic models of this motion are needed (cf § 7).

6.4.5. Motion in and around the Great Attractor direction. Figures 16(a)-(f) plot the *observed* peculiar motions of galaxies ($V_{\mathrm{CBR}} - r$) as a function of distance, in a manner after Burstein *et al* (1990). Figure 16(a) plots the motions of galaxies within a 10° radius of the centre of the Great Attractor (at $l = 309°$, $b = 18°$); figures 16(b)-(f) plot the motions of galaxies within annuli of 10° width centred on this direction, ranging in average distance from 15° radius to 55° radius. All data samples are used for these diagrams.

Superimposed on the observed peculiar motions is the predicted velocity field due to the gravitational field of the spherically symmetric Great Attractor model, assuming an angular distance from the Great Attractor centre that runs through the midway point of each annulus (or the centre itself). As in figure 10 the motions of the Centaurus cluster(s) are plotted as a single point.

These graphs illustrate several important aspects about the model compared to the observed velocity field. (1) The Great Attractor model works surprisingly well in modelling motions of galaxies 30°-60° from the GA centre. (2) More evidence of infall is seen in the outer two annuli (10°-30°) than in the central region of 10° radius containing the Centaurus clusters (cf figure 16(b), (c) and figure 16(a)). Infall of galaxies to these clusters, coupled with the infall of the clusters themselves to a more distant point, makes interpretation of the motions in this centre region especially difficult. (3) If nominal Malmquist bias corrections are used for these galaxies the motions within the centre of the Great Attractor are consistent with compression toward a rather large, diffuse region, rather than a gravitational attraction to a unique centre. This picture might change somewhat if the form of Malmquist bias currently used is in error, but the general picture would probably be preserved. (4) Few galaxies are observed further than distances of ~ 5000 km s^{-1} in the vicinity of the Great Attractor out to 40° in radius. (5) The motions within the core of the Great Attractor plausibly derive from the same physical source as the motions out to at least a radius of 60° from the centre of this core. The motions within the Local Supercluster are of much smaller amplitude, as can be seen both in figure 16(f) (in which part of the plane of the Local Supercluster is plotted) and by comparing the vertical scales of figure 16 with figure 15.

Residual 'regional' motions of amplitude 200 km s^{-1} or greater relative to the GA model are seen in figures 16(b)-(f), much of which may be attributed to the separate

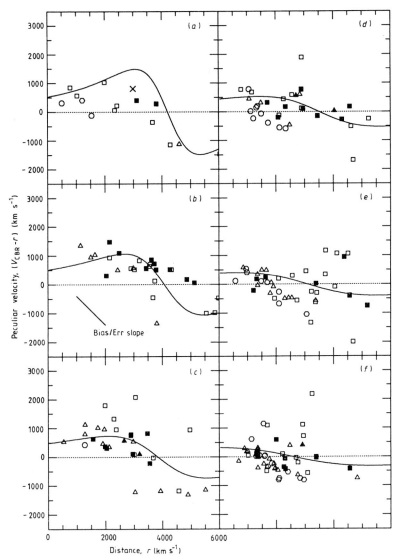

Figure 16. The observed peculiar motions of galaxies from all data sets, in the CBR velocity reference frame, selected according to angular distance from the dipole direction $l = 309°$, $b = 18°$. The centre cone has a radius of $10°$ in (a), the outer annuli have widths of $10°$, and are centred at (b) $15°$, (c) $25°$, (d) $35°$, (e) $45°$ and (f) $55°$ in angle from the dipole direction. Each graph plots observed peculiar motion versus distance for all galaxies found in the particular annulus. Symbols are the same as in figure 3, except that the net motion of the two Centaurus clusters in (a) is represented by an X. The prediction of peculiar motion from the Great Attractor model, for the midpoint of each annulus, is given as a full curve in each graph. A correlated error slope corresponding to an error of 1000 km s^{-1} is shown in (b). Note that the compressional infall pattern is similar in the inner three annuli and, if anything, is more pronounced in (b) and (c) than in the centre zone (a). Significant motions remain relative to the GA model on a region-by-region basis, indicative again of the existence of localised motions in addition to the global flow pattern.

motions of regions like that of the Coma–Sculptor Cloud. As shown in more detail in Faber and Burstein (1989), the amplitude of these regional motions (correlated over a size scale of $\sim 1500 \text{ km s}^{-1}$) is $100–200 \text{ km s}^{-1}$ relative to the GA model. For comparison, the -350 km s^{-1} V_Z component of the Coma–Sculptor Cloud, whose origin cannot be accounted for by either the GA or Virgo infall models, translates to a one-dimensional motion of $350 \times 3^{-1/3} \simeq 200 \text{ km s}^{-1}$. Moreover, the Local Supercluster itself appears to have a net motion (one-dimensional) of 150 km s^{-1} relative to the GA model.

6.4.6. Assessment of the Great Attractor model. Viewing the motions of galaxies relative to the Great Attractor model has clarified the kinds of motions that are seen within the volume of space to distances of 4000 km s^{-1} while at the same time highlighting the limitations of this kind of simple model. Models of spherically symmetric infall towards a central point do not adequately describe the compression-like motions within large volumes of space that are definitely observed within the Local Supercluster and may be observed in the Great Attractor region.

The motions of galaxies in this volume of space are correlated within regions of scale size $\sim 1500 \text{ km s}^{-1}$. Individual regions move relative to each other, with a part of these motions due to a global differential flow present in the whole volume of space and a part due to motions probably generated by the gravitational mass of closer regions. The Great Attractor model does work well in matching the dominant acceleration field of the global flow far from the GA itself. In this context, the best determined peculiar motion of any region, that of the Coma–Sculptor Cloud, is fairly typical both in terms of amplitude and scale of coherence. These points will be discussed further in § 7.

6.5. The peculiar velocity field predicted by the density distribution of galaxies

The direction and radial velocity of every galaxy in a given volume of space gives an *apparent* density distribution that is a convolution of the expansion of the universe with the peculiar motions that exist within this volume. If luminous matter is an accurate tracer of total mass (within a scale factor; as explained in § 2.3, this is an assumption, not a proven fact) then one should be able to reconstruct a self-consistent density field that could predict the peculiar motions of galaxies in this volume of space.

Alternatively, if radial velocities are not available for every galaxy (as is common), a further assumption can be made that the amount of mass is proportional to the amount of light for all galaxies. Since both the force of gravity and apparent luminosity decrease with distance as inverse-square laws, the net gravitational field on the Local Group produced by galaxies can be calculated from the positions and apparent luminosities of galaxies alone, in terms of multipolar moments.

These techniques have been tried over the past several years using data from various galaxy catalogues. More positions of galaxies being known than their radial velocities, multipole moments have been calculated by several groups (Yahil *et al* 1986, Meiksin and Davis 1986a, b, Lahav 1987, Harmon *et al* 1987, Villumsen and Strauss 1987, Lahav *et al* 1988). Ofer Lahav and Donald Lynden-Bell have been able to extend this kind of method to extract spatial information about the origin of the monopole, dipole and quadrupole moments of the gravity field (Lahav *et al* 1988, Lynden-Bell and Lahav 1989, Lahav 1989, Lynden-Bell *et al* 1989b). Detailed discussion of the results of these papers is beyond the scope of the present review, although they are factored into the summary discussion in § 7.

One of the most comprehensive efforts mounted to measure the density distribution of galaxies is the research programme of Marc Davis, Michael Strauss, Amos Yahil and John Huchra (Strauss and Davis 1988, 1989, Yahil 1989, Strauss 1989). These workers have obtained radial velocities for all galaxies observed by the Infrared Astronomical Satellite (IRAS) to have fluxes brighter than a well defined limit in the far-infrared (at a wavelength centred on 60 μm); a total of over 2500 galaxies.

The advantage in using the IRAS survey is that it is one of the most homogeneous surveys of galaxies that covers the complete sky. The homogeneity of the sample over the sky is critical if one is to avoid most problems with selection effects (§ 4.2) and Galactic obscuration (the galaxy is transparent in the far-infrared). The principal disadvantage of using the IRAS survey is that not all kinds of galaxies emit the same level of far-infrared emission (which comes principally from dust radiation at temperatures of 10–50 K). In particular, those galaxies that comprise much of the luminous matter in the densest regions of space—elliptical and S0 galaxies—contain relatively little dust and produce proportionally much less far-infrared radiation. Nonetheless, a complete, homogeneous survey of the radial velocities of galaxies is so important for a calculation of the density field that the advantages of the 'IRAS' galaxy sample substantially outweigh its disadvantages.

Graphs of the motions of galaxies, as predicted by the density distribution of galaxies in the IRAS survey, have been kindly supplied to this review by Michael Strauss. Peculiar motions are predicted from the observed density assuming linear theory, that the IRAS galaxies are tracers of mass and that there are no large contributions to our local gravity from distances greater than $10\,000$ km s^{-1} (Strauss 1989). These graphs have been made in a manner that permits easy comparison with the graphs of observed peculiar motions in § 5.

The predicted peculiar velocities of galaxies in the Supergalactic X-Y plane are plotted in figure 17, analogous to figure 6. The predicted velocities of galaxies as a function of the direction $l = 309°$, $b = 18°$ are plotted in figure 18, in a cosine diagram analogous to figure 7. The larger number of IRAS galaxies in the Strauss *et al* sample compared with the number of galaxies with measured distances is evident. The more homogeneous distribution of IRAS galaxies within the Supergalactic plane is also apparent (although artificial galaxies have been placed within the centre of the 'zone of avoidance' to avoid discontinuities in the data). The reader should keep in mind that the motions pictured in figure 17 and following are the predictions of a model, with the apparent scatter due to motions that are predicted, rather than to observational errors. As such, the point-to-point scatter will appear much less than for the observational data. Unless otherwise stated, the IRAS predicted peculiar velocities are given in the CBR velocity frame of reference.

Comparison of figure 17 with figure 6 shows that, while the peculiar motions of galaxies in most regions are in the same sense of direction in both diagrams, the net amplitudes of these motions are not the same in these two diagrams. In particular, the overall impression of a general flow in this volume of space is not present in the predicted motions of galaxies based on the density distribution of IRAS-bright galaxies. Substantial disagreements between the predicted peculiar velocities of Strauss *et al* and the observed peculiar velocities are seen in the Ursa Major region (IRAS motions are predicted to be nearly at rest in figure 17), Pavo–Indus–Telescopium region (motions predicted to be outwards in figure 17) and towards the Virgo Cluster.

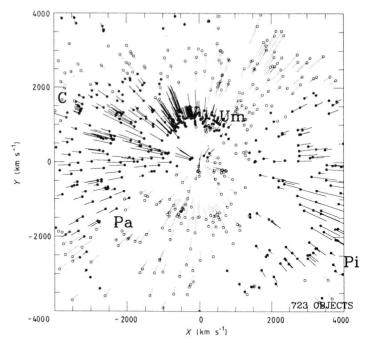

Figure 17. The peculiar motions, in the CBR velocity reference frame, of galaxies in the IRAS sample of Strauss *et al*, lying within ±22.5° of the Supergalactic plane. The peculiar motions are predicted by the model of the density distribution of galaxies developed by Strauss *et al* (1989), based on their sample of IRAS-observed galaxies. This figure, and the ones to follow, have been kindly provided by Michael Strauss for this review. This figure is constructed in the same format as figure 6, but the galaxies are those used by Strauss *et al* in their survey, as opposed to the galaxies having measured distances. The Strauss *et al* sample has a greater number of galaxies and more uniform coverage of space in this diagram (although many of the points plotted within the centre of the zone of avoidance are artificial) compared with galaxies used for figure 6. While there are coherent flows predicted by the Strauss *et al* density field model, it does not predict the kind of global flow over the whole volume that is seen in the direct observations.

The lack of a global flow in the predicted motions of galaxies in figure 17 is confirmed in the cosine diagram of figure 18. Whereas the cosine diagram of figure 7 gives a strong impression of an accelerated flow towards the apex, the IRAS survey predicts a bifurcated flow towards both the apex *and* the antapex. This results in the wrong sense of predicted motions for galaxies with cosines of −0.5 and less (i.e. in the direction of the Perseus–Pisces supercluster). Moreover, the amplitudes of motions towards the apex are typically a factor of two too small relative to the motions that are observed.

Figure 19 plots the predicted motions of IRAS survey galaxies in six regions not containing dense clusters, for direct comparison with the analogous regions presented in figure 8. In all six of these regions, the *trends* of peculiar velocities with distance predicted by Strauss *et al* (i.e. decreasing, level or rising) match the trends of observed motions of galaxies along the line of sight. The IRAS predictions, however, do not give correctly the amplitude of motion in most cases. For example, if one compares the

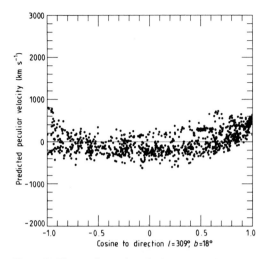

Figure 18. The peculiar motions, in the CBR velocity reference frame, of IRAS-observed galaxies, as predicted by the Strauss *et al* density field model, plotted in a cosine diagram analogous to figure 7. The motions of galaxies predicted by Strauss *et al* predict a flow towards the apex, but with an amplitude about two times too small. More importantly, they predict that the overall flow bifurcates near the position of the Local Group, becoming more positive towards the antapex. This predicted form of motion is quite different from the more coherent differential acceleration over the whole field seen in figure 7.

two diagrams for the N1023 + Cetus region (figure 19(f) with figure 8(f)) the observed peculiar motions are over 300 km s^{-1} offset relative to the model, yet the trends of motion with distance are remarkably similar.

Further differences between the Strauss *et al* predictions and the observations are seen in the motions of galaxies within the Local Supercluster. The predicted motions from the Strauss *et al* survey towards the Virgo and the Ursa Major clusters are given in figures 20(a) and (b) to be compared with figures 9(a) and (b). The selection of galaxies in figure 20 was performed in the same manner as in figure 9. The IRAS survey does not predict the rather strong compressional flow that is actually observed in the direction of the Virgo cluster. Moreover, the net motions of galaxies at a distance of 1500 km s^{-1} in figure 20(a) are about 300 km s^{-1} higher than the observed motions at this distance. Differences in net motion up to 500 km s^{-1}, in the same sense as in the Virgo direction, are also seen towards the Ursa Major cluster (cf figure 20(b) with figure 9(b)). The Ursa Major cluster is predicted to be at rest in this velocity reference frame while the observed motion of the cluster is ~ -400 km s^{-1}.

The peculiar velocities of galaxies predicted by the IRAS survey in the direction of the Great Attractor, $l = 309°$, $b = 18°$, are presented in figure 21 in a manner analogous to figure 16. This figure plots the peculiar velocities of galaxies predicted from the IRAS density distribution versus their predicted distances, separated in 10° annuli centred on this direction. Although the form of predicted velocity field varies little within 30° of the centre direction (figures 21(a)-(c)), the predicted amplitudes of motions are too small by up to a factor of three relative to the observations (cf figure 21(b) with figure 16(b)).

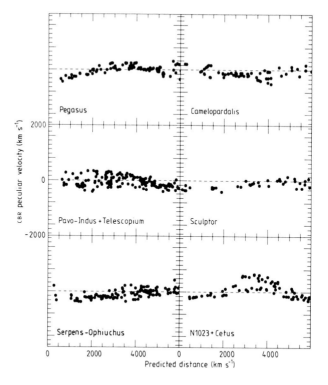

Figure 19. The peculiar motions, in the CBR velocity reference frame, as predicted by Strauss *et al*, plotted as a function of predicted distance in the six directions used in figure 8. As with the previous two figures, this figure uses the galaxies in the Strauss *et al* sample. In contrast to the absence of a coherent, global flow in their predictions the Strauss *et al* model predicts reasonably well the *differential* motions of galaxies within each of these regions. For example, the rise and fall of motions in the N1023 + Cetus direction (f) are similar in both this figure and in figure 8(f), but the net motion of the whole region differs by 300 km s^{-1}. From this and other, more direct, comparisons it is apparent that the Strauss *et al* model matches the motions of galaxies within regions not dominated by dense clusters, but does not match the overall motion of a region relative to the CBR velocity reference frame.

In sum, the motions predicted by the IRAS survey, based on the apparent density distribution of infrared-luminous galaxies, succeed in matching motions *within* regions of ~1500 km s^{-1} in size, but fail to predict the large-scale coherent flow that exists in the observations over a distance of 7000 km s^{-1}. Ironically, this is nearly completely opposite to the sense in which the Great Attractor model, which is constructed to fit this large-scale flow but cannot fit the smaller-scale motions, fails. This key difference between the two models is also apparent in other, separate comparisons in which the IRAS model predicts the peculiar motions of galaxies with observed distances (as has been done in the previous section for the Great Attractor model).

It is instructive to have the IRAS survey predict the near-infrared Tully–Fisher relationship for the same sample Aaronson *et al* 'good' galaxies, in the same manner

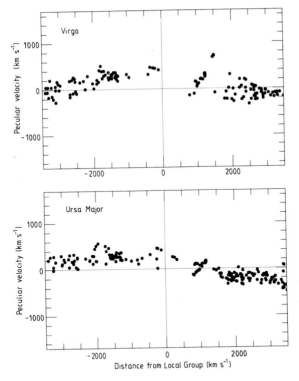

Figure 20. The peculiar motions, in the CBR velocity reference frame, as predicted by Strauss *et al*, for IRAS-selected galaxies in the directions of the Virgo and Ursa Major clusters, selected in angle as in figures 9 and 14. The IRAS sample does not predict the compression-like infall that is actually observed towards the Virgo cluster nor the range of motions observed towards the Ursa Major cluster. In addition, the net motions of both clusters are predicted to be too positive by 300–500 km s^{-1}.

as employed in figures 4 and 11 (see Strauss 1989 for alternative examples of such comparisons). This relationship is shown in figure 22, and has a measured RMS scatter of 0.43 mag for 136 galaxies, the same scatter as found in figure 11 for the predictions of the Great Attractor model. It is known that the residual scatter in both cases is not due to random errors, but due to regional and global motions that each model is not matching. In such a situation it is not unreasonable to assume that the scatter introduced by wrong velocity field models is comparable to that of the observations, indicating that the true observational scatter in the TF relation is $0.43 \times 2^{-1/2}$ or ~ 0.30 mag. In this sense, the discussion of peculiar velocities and velocity field models has come full circle: as deduced in the absence of all models in § 5.2 and confirmed by the detailed models of this section, the peculiar motions of galaxies in this volume of space are generated partly on a large scale and partly on smaller scales. The large-scale source of motion affects all galaxies in a spatially coherent manner, while smaller scale motions of comparable amplitude exist for most localised regions.

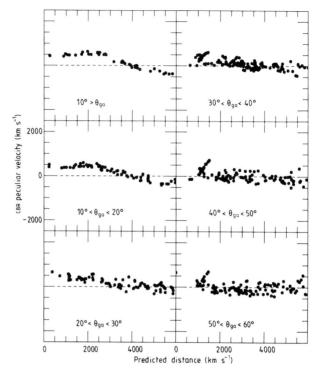

Figure 21. The peculiar motions, in the CBR velocity reference frame, as predicted by Strauss *et al*, for the IRAS-selected galaxies in the directions around the dipole direction, $l = 309°$, $b = 18°$, selected in annuli as in figure 16. Although a large, extended region of infall is predicted for galaxies in the inner three annuli, the amplitude of infall is more than a factor of two lower than that observed. This is particularly evident in the 15° ($10° < \theta_{ga} < 20°$ in this figure) and 25° ($20° < \theta_{ga} < 30°$) annuli.

7. Summary and desiderata

7.1. Reliability of measurement of large-scale motions

All astronomy is an attempt to understand the universe by a series of successive approximations to the truth. As radial velocities of galaxies have become more accurate, and the number of galaxies measured more numerous, efforts to derive their distances have also become more successful. Hence, earlier uncertain conclusions are gradually giving way to more certain results. This review was written with the intent of presenting the evidence on large-scale motions as derived from the best available data. These data show the following.

(i) Remarkable agreement in the observed peculiar motions of galaxies of all kinds and in all environments. Galaxies in regions ~ 1500 km s^{-1} in size share in motions that arise both locally and on a much larger scale (§ 6). Our own galaxy is in such a region, the origin of whose large-scale motion (see below) appears to be both multicomponent and very typical.

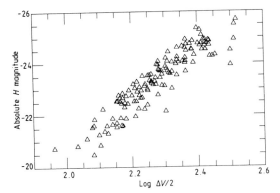

Figure 22. The near-IR Tully-Fisher relation as determined from distances predicted by the Strauss *et al* density field model, for the same 136 spiral galaxies of the Aaronson *et al* 'good' sample used in figures 4 and 11. The RMS scatter in this diagram has the same formal scatter (0.43 mag) as the Tully-Fisher relationship predicted by the GA + V_{inf} velocity field model. This comparison again implies that localised motions (fitted by the Strauss *et al* model but not by the GA model) and the global flow that exists everywhere in this volume (fitted by the GA model but not by the Strauss *et al* model) contribute equally to the total observed scatter in the Tully-Fisher relationship predicted from a smooth Hubble flow model (figure 4).

(ii) The dipole anisotropy of the cosmic background radiation yields a velocity reference frame in which coherent, differential motions of galaxies can be reliably measured. Demonstrable differential motions exist on all size scales. The largest-scale motions are not simply 'bulk' dipole motions of the kind that would be produced if the CBR dipole anisotropy were an artefact caused by an asymmetry in the temperature of the early universe (an explanation suggested by Gunn 1989).

(iii) In order for systematic differences in the properties of galaxies to produce the motions that are observed, errors would have to scale in distance radially from the Local Group in opposite ways in different parts of the sky, over nearly an order of magnitude in distance. The coherence of this large-scale motion, over many smaller regions with different kinds of internal motions that are reasonably predicted by the IRAS galaxy density field, would require a 'geocentric' kind of systematic error of such amplitude as to not be plausible. Finally, tests for these kinds of systematic errors have not yet uncovered any that would produce such global effects (§ 5.4.1).

7.2. What are the components of the CBR motion of the Local Group?

At least four different components can currently be identified that give rise to the observed CBR motion of the Local Group (Burstein 1990).

(i) Global, coherent flow on a scale of at least ~ 7000 km s^{-1} (i.e. within a distance of 3500 km s^{-1} from the Local Group). On the large scale, this flow is reasonably modelled by a gravitational acceleration field directed towards a large extended concentration of groups and cluster of galaxies termed the Great Attractor. The amplitude of this component at the position of the Local Group is currently estimated to be ~ 500 km s^{-1}.

(ii) A coherent, compression-like flow of scale $\sim 2000\,\mathrm{km\,s^{-1}}$, oriented in the direction of the Virgo and the Ursa Major clusters, involving all galaxies in the plane of Local Supercluster and directed within the plane and parallel to it. The amplitude of this component, relative to the midpoint of the plane, is $\sim 300\,\mathrm{km\,s^{-1}}$. However, since the Local Supercluster itself is moving at $\sim -150\,\mathrm{km\,s^{-1}}$ relative to the CBR velocity reference frame the net motion of the Local Group due to this component is $\sim +150\,\mathrm{km\,s^{-1}}$.

(iii) The motion of the Local Group perpendicular to the Supergalactic plane (the 'Z' component of the Local Anomaly) cannot originate with either of the above components of motion (both of which are directed primarily within the Supergalactic plane). This 'Z' motion is shared by the whole Coma–Sculptor Cloud. Although this component might be due to a 'push' by the underdense region of the Local Void (Lynden-Bell and Lahav 1989), its physical origin is currently unknown. The amplitude of this component is $\sim 350\,\mathrm{km\,s^{-1}}$.

(iv) The Local Group has a 'cosmic' dispersion relative to other nearby groups within the Coma–Sculptor Cloud. This motion is likely generated by mutual gravitational interactions among the nearest groups (distances of $\sim$few hundred $\mathrm{km\,s^{-1}}$). The amplitude of this component is $\sim 100\,\mathrm{km\,s^{-1}}$, or of comparable size to the *observed* Local Supercluster motion of the Local Group (but not the total Supercluster-generated motion).

A valid question at this point is, 'Is it realistic to break the origin of the CBR motion of the Local Group into four distinct parts, or do the components of motion have continuous, correlated origins?' The evidence does point to an interdependence of the motions: the perpendicular to the apparent compression of the Local Supercluster, from which the long axis of the quadrupole moment of the velocity field is derived, points in the direction of the dipole vector, which is generated by the global flow. The compression produced by the differential flow towards the Great Attractor, combined with compressional motion generated within the Local Supercluster, results in the measured short-axis components of the quadrupole moment. Apparent condensation within the Coma–Sculptor Cloud is part of the more general compression within the plane of the Supercluster as well as part of the compression produced by global differential flow. Voids, such as the Local Void, will produce other, yet-unmodelled motions in other nearby regions.

The sum of the motion vectors generated by all four components is constrained to reproduce the observed vector of $614\,\mathrm{km\,s^{-1}}$ amplitude of the Local Group. The general amplitudes and ranges of distances over which these motions are generated are in approximate agreement with a recent detailed examination of the motion generated by optically observed galaxies (Lynden-Bell *et al* 1989b) and IRAS galaxies (Strauss 1989). It is significant that no motions are observed that are plausibly generated on a scale larger than $\sim 7000\,\mathrm{km\,s^{-1}}$, a point which is explored further below. The estimates given for the relative sizes of the components of the Local Group's motion are indicative for the present, as we now know that the models used to derive these values are incorrect in form (see below).

7.3. Compressional motions and the apparent lack of large scale gravitational influence from dense clusters

The pattern of motion internal to the two large regions of mass overdensity that have been investigated—the Local Supercluster and the Great Attractor region—is very different from the kinds of motions that had, up until very recently, been assumed to take place. Infall motions are indeed present within each region, with the amplitude of infall within the Great Attractor region being up to about four times that within the Local Supercluster. This infall is *not*, however, directed centrally to the obviously dense cluster that sits within each region—either the Virgo Cluster in the Local Supercluster or the Centaurus cluster(s) in the Great Attractor.

Rather, galaxies over a wide region appear to share the same pattern of infall that *increases* with distance from a midpoint of motion rather than decreasing with increasing distance (as would be the case for Keplerian infall to a central mass). The infall pattern is like that which would be observed for a sphere (or disc) of uniform mass density condensing under its own self-gravity. Moreover, the amplitude of this distinctive infall pattern is relatively the same over a wide viewing angle, requiring that the infall motion be directed towards a diffuse central region. The peculiar motions of the dense clusters participate in this compression-like infall pattern, with little observed infall centred on the dense clusters themselves.

In the case of the Local Supercluster, both the Virgo cluster and the Ursa Major cluster appear to lie within the region to which galaxies are moving. In the case of the Great Attractor, the Centaurus clusters have very large motions both relative to one another and towards a more distant central region. The velocity fields predicted from the density distribution of IRAS-selected galaxies do not predict the amplitude of compressional infall in the Great Attractor region, but come closer for motions within the Local Supercluster. That the Great Attractor region does not appear to contain dense clusters at its centre of motion could be real, or an artefact of obscuration of the real centre by dust within our own Galaxy.

The extent to which the form of compression-like infall motions was unexpected is revealed in the fact that no analytic model yet exists for this kind of motion. As a direct result, the analytic model for this review is a spherically symmetric velocity field model that is demonstrably incorrect in form but the best available approximation. How is this compressional motion *within a plane* (not *to* a plane) generated—is it due to gravity or to initial conditions? Is it possible that the apparently dense clusters are not as massive as they seem, or is it that the surrounding mass is much greater than it seems (i.e. what is the role of dark matter)? If these compression-like motions are not entirely due to gravity, to what extent is the global flow produced by gravity alone?

In a related question, what gives rise to the very large motions that are seen for some regions? Galaxies in the region around the N1600–N1700 clusters of galaxies (the ellipticals in figure 13 at a cosine of -0.4 and $+1200$ km s^{-1} peculiar motion) are moving at a large peculiar velocity that is out of step with the motions of others around them. Galaxies moving with very large velocities are observed in Pavo–Indus–Telescopium (figure 8(b)). Are these very large velocities generated by gravity or initial conditions? The answers to these questions are central both to understanding these large-scale motions and to the origin of structure in the universe.

7.4. Use of velocity field models to predict distances

For many astronomers, the most practical aspect of the study of large-scale motions is the production of an analytic model of these motions that can be used to reliably predict the distances of any galaxy, given a measured radial velocity. Until the 1980s the only available model was that of a smooth Hubble expansion, $r = V/H_0$. During the 1980s, models of centrally concentric, spherically symmetric infall into the Virgo cluster (e.g. Aaronson *et al* 1982a) came into common use.

As shown in this review, the dominant motions of galaxies are neither Virgocentric nor spherically symmetric, but a mix of motions generated on several different size scales. The Local Group is indeed moving towards the Virgo cluster, but this motion is not produced solely by the gravitational field of that cluster. For reasons detailed above, use of these monolithic Virgocentric infall models leads to systematically wrong estimates of distances. Instead, a multi-component model of the velocity field should be developed that incorporates as many known motions as is practical. The Great Attractor model detailed in § 6 is a first step. A non-spherical model, incorporating an analytic approximation to the motions that are mapped by the IRAS survey-determined density distribution, would substantially improve the predictive accuracy. It is expected that new analytic techniques of recovering the gravitational density field from peculiar motions, perhaps along the lines suggested by Bertschinger and Dekel (1989), will lead to better analytic velocity field models.

Finally, practitioners should take note that the distances of galaxies within the Coma-Sculptor Cloud are best predicted by their Local Group velocity, independent of any velocity field model. This is because essentially all galaxies out to a distance of ~ 750 km s^{-1} share in the CBR motion of the Local Group, and the amplitudes of local 'random' motions of galaxies are comparable to the more general compressional motions within the Cloud.

7.5. A view towards the future with an eye towards the past

One can now more fully appreciate the extent to which our knowledge of large-scale motions in the universe has been dictated by our particular place in the universe. Galaxies out to a distance of 750 km s^{-1} from the Local Group are moving apart due to the expansion of the universe in a relatively smooth and 'quiet' fashion. Relative to this expansion velocity, the motion of these nearby galaxies is that of a 'bulk' motion. It is with these nearby galaxies that the fundamental discovery of the expansion of the universe was made (Hubble 1929), and with which studies of large-scale motions were performed up until twenty years ago.

What if our galaxy were situated elsewhere in this volume of space, perhaps in the vicinity of the Great Attractor? Infall motions of galaxies in that region are of amplitude ~ 1400 km s^{-1} over distances of 1500 km s^{-1}. Such an infall pattern effectively counter-acts the radial velocities produced by the expansion of the universe on that scale. As a result, we would *not* observe the radial velocities of galaxies to be correlated with distance and, hence, would not observe the expansion of the universe out to distances of ~ 1000 km s^{-1}! How different would our understanding of the universe be if . . . ?

If the cosmic background radiation defines the fundamental rest frame we can measure the full space motion of the Local Group with respect to the universe as a whole. This review has shown that this motion can be broken down into at least four coupled components, of which the largest in amplitude is effective over a scale of ~ 7000 km s^{-1} in size. On the basis of the peculiar velocity field alone, there is no evidence that the full space motion of the Local Group is produced by motions on scale sizes larger than this.

Interestingly, the separate studies of the dipole and quadrupole moments generated by the distribution of galaxies on the sky come to similar conclusions, albeit with some significant differences in detail (Lynden-Bell and Lahav 1989, Lahav *et al* 1988;

Lynden-Bell *et al* 1989b, Strauss and Davis 1988, 1989, Strauss 1989). The temptation is strong to combine these two results and to state now that all of the peculiar motions that are observed are plausibly generated by gravity associated with mass that can be observed. However, the most ambitious predictions yet made of peculiar motions based on the observed density distribution of galaxies (Strauss *et al*, § 6.5) predict neither the kinds of large-scale coherence in the motions of galaxies that are actually observed, nor the very large amplitudes of these motions.

As discussed by Strauss (1989), the velocity field predictions of the density model derived from the IRAS survey are complicated by a number of factors, both observational and theoretical, that could lead to the apparent disagreement with observations. Assuming that these difficulties can be overcome, either with an expanded IRAS sample or an alternatively selected, large sample of galaxies, certain basic questions remain to be answered. What is the role of dark matter in determining the velocity field? Is the dark matter distribution in space different from that of the luminous matter? Are the observed motions more dependent on initial conditions than on the subsequent gravitational interactions?

These questions still do not have ready answers. It remains for the theorist to devise suitable ways to interpret the data relating to peculiar motions, and for the observer to provide more accurate data worthy of being modelled, before these basic questions can be answered. This is, after all, a review of the beginning of a new field of scientific endeavour rather than a summary of an accepted body of knowledge.

Acknowledgments

Discussions with many colleagues, too numerous to mention, have helped shape this review over the past year. The author would like to especially thank Vera Rubin for providing several careful readings of the manuscript and for being willing to discuss any and all questions. Michael Strauss and the anonymous referee provided insightful comments on the first version of this review. As always, conversations with Sandra Faber, Donald Lynden-Bell and Alan Dressler have been of great help. Ofer Lahav has provided many insightful comments. The author profited greatly during the formative stages of this review by being a Senior Visiting Scientist with the Institute of Astronomy, Cambridge, UK for the summer of 1988, and later as a visitor to the University of California, Santa Cruz. He gratefully acknowledges the hospitality of both the institutions. Two astronomers went out of their way to provide figures for this review. The author would especially like to thank Nigel Sharp for providing the excellent pictures of the sky, and Michael Strauss for his fine effort of preparing graphs of the IRAS survey results that could be compared to the peculiar velocity observations.

Addendum

A.1. Update as of Spring 1991

In the nearly two years since the original review (Paper 1) was written, the typical reader might be surprised to learn that relatively little new direct data on large-scale motions has been added. Yet, this is probably as it should be. It took well over a decade for astronomers to assemble the data base of 1000 galaxy distances that was used for Paper 1. Although many groups are actively working on expanding our knowledge of galaxy distances, and the peculiar motions derived from these measurements, the vast majority of these studies are still in progress. As such, the overall view of large-scale motions in the universe is sensibly the same today as it was two years ago, with perhaps more surety of the existence of the 'Great Attractor' (GA) as a distinct mass entity, mostly through the efforts of Alan Dressler and Sandra Faber (Dressler and Faber 1990).

That relatively little data has been published during this time period is not to imply that this field has stagnated. Quite the opposite. Our understanding and interpretation of large-scale motions has progressed significantly on several fronts during the intervening time. There are many groups using established techniques to map large-scale motions, both to check previous results and over even larger regions of space. Studies of larger volumes of space require much more time; expanding our knowledge over twice the distance requires measuring the distances to eight times as many galaxies. New, and potentially more accurate techniques of measuring galaxy distances are being developed that hold the promise of better defining the large-scale velocity field. Theorists have developed a number of analytical tools both to quantify the observations better in terms of physical quantities (such as inferred mass and mass density), as well as to compare observations better with theory. The existence of one 'Great Attractor' has naturally led to the search for other such structures in the universe, with possible candidates suggested.

This addendum will endeavour to summarize these ongoing activities in a brief, but evaluative, manner consistent with the attitude taken in Paper 1. Distances will be quoted in units of radial velocity, as explained in Section 2.1 of Paper 1.

A.2. Ongoing surveys of large-scale motions

Many groups are working on measuring the distances to galaxies, both in attempts to measure the local rate of expansion of the universe (the 'Hubble constant'), as well as to measure peculiar motions. It would neither be reasonable nor informative to just mention all of these studies, even if this writer could be sure he was aware of every study. For example, to this writer's best knowledge, there are ongoing at least 13(!) surveys of galaxy distances that include ~200 or more galaxies. Even more studies involve galaxy samples of fewer galaxies. Perhaps the primary reason that so many surveys can be done simultaneously is the ubiquitousness of high-sensitivity charge-coupled devices (CCDs) for both imaging and spectroscopy. The high efficiency of such devices means that massive surveys can be completed today in a fraction of the time required for the surveys described in Paper 1. Within three years' time, it is possible that that number of galaxies with measured distances will increase from 1000 to 3000–4000.

In order to give the reader a flavour for the work being done today, this review will summarize the progress of ten of the larger surveys. Five of these programmes involve work related to the PhD theses of young astronomers, most of them having been completed within the last six months. None of these workers has yet published their data, principally owing to the typically long time lag between thesis completion and publication in a refereed journal (up to 3 years in many cases!).

Jeff Willick (1990) has published a short paper summarizing the peculiar motions for 320 spiral galaxies in a large region centred on Galactic coordinates $l \sim 120°$,

$b \sim -30°$ (*cf* figure 2 of Paper 1), and mostly at distances of 4000–8000 km s^{-1}. This direction is commonly referred to as Perseus–Pisces, although the sub-region surveyed by Willick is primarily in the Pisces section. Willick finds that the galaxies in the Pisces region have a net motion of -450 km s^{-1} (*i.e.* towards us). Since the Pisces region lies opposite in the sky from the GA (*cf* figure 3 of Paper 1), such motion is also in the general direction of the GA. The meaning of the Pisces motions is ambiguous, however, owing to the fact that at the distances surveyed by Willick, other known mass concentrations could also produce similar kinds of motions (in particular, the region centred around the Perseus cluster). In absence of detailed studies of large-scale motions in these other regions, interpretation of Willick's result has been uncertain.

Stephane Courteau (1991), working with Sandra Faber, has surveyed the distances of spiral galaxies in a larger part of the Perseus–Pisces region than was covered by Willick. Ming-Sheng Han (1991), working with Jeremy Mould, has concentrated on accurately remeasuring the peculiar motions of the clusters originally studied by Aaronson *et al*, as summarized in Paper 1 (Section 4.1). Both studies are literally being completed as this is being written.

Other studies involving PhD theses that were completed within the last two years include those of Michael Pierce (1989) working with Brent Tully, and Wolfram Freudling (1990) working with Martha Haynes. Freudling has measured relative peculiar motions for about 150 spiral galaxies in an even more distant region centred on the Hercules cluster (at $\sim 10\,000$ km s^{-1}). These data are even more problematic to interpret than Willick's data, both due to lack of tie-in to the local velocity field, and even more incomplete knowledge of other mass concentrations at these distances. Pierce has accurately redetermined the distances to about 400 nearby galaxies, in order to define the local velocity field better; the data from this survey have not yet been made available.

Five surveys are being done by teams of astronomers that include many of the 'game players' from earlier studies of large-scale motions. All of the original 'seven samurai' are involved in one or more surveys. Some of the astronomers involved in the Aaronson *et al* studies are doing a variety of surveys (*e.g.* Jeremy Mould, mentioned above), most of which include relatively few galaxies. Other pioneers of the study of large-scale structure in the universe are now involved in measuring the peculiar velocities of the structures they first defined in terms of radial velocities.

Martha Haynes, Riccardo Giovanelli, Luiz da Costa, John Salzer, Wolfram Freudling and Gary Wegner are involved in an all-sky survey of the peculiar motions of Sc galaxies. This survey covers galaxies at distances from nearby to ~ 6000 km s^{-1}. Loosely associated with this survey is a separate survey of peculiar motions in the Southern Celestial Hemisphere by Donald Mathewson and his collaborators. Each of these surveys includes over 1000 galaxies, and, when both are completed, they will have a significant impact on our knowledge of peculiar motions out to distances of ~ 6000 km s^{-1}.

Two groups are involved in mapping the peculiar motions of clusters of galaxies through D_n–σ measurements for elliptical and S0 galaxies. The author is involved in one group, together with Roger Davies, Gary Wegner, Robert McMahan, Matthew Colless, Edmund Bertschinger, Glenn Baggley and Roberto Saglia. We are measuring the peculiar motions of elliptical-dominated clusters at distances of 7000–13 000 km s^{-1}. The clusters were chosen primarily on the basis that they define two large structures at those distances (Wegner *et al* 1991). One of these structures is part of the

'Great Wall' as defined by John Huchra and Margaret Geller (Geller and Huchra 1989a). Our survey has been in progress for over four years, and is still two years from completion. The other group includes Donald Lynden-Bell, Roberto Terlevich, John Lucey, David Carter, Richard Ellis and collaborators. These workers have defined a sample of elliptical-rich clusters (with some overlap with the sample of the previous group to provide eventual normalization) that does not go as deep as the aforementioned survey, but covers nearly the whole sky. Several papers resulting from this survey have already been published (Lucey and Carter 1988; Lucey *et al* 1990; Lucey *et al* 1991a; Lucey *et al* 1991b) detailing observations for several clusters.

Alan Dressler and Sandra Faber have determined distances for both elliptical and spiral galaxies in the general direction of the GA. The data from two of the papers resulting from this work were included in Paper 1 (Dressler and Faber 1990; Burstein *et al* 1990). Additional data not included in Paper 1 were published in Dressler and Faber (1990) and Dressler *et al* (1991). Distances for both kinds of galaxies lend support to the concept of the GA as a very large, somewhat overdense region subtending a solid angle of nearly a steradian from our perspective, centred in the direction $l \approx 315°$, $b \approx 15°$ at a distance of ≈ 4500 km s^{-1}.

In particular, the Dressler–Faber survey has apparently detected infall motions of galaxies on the opposite side of the GA. As detailed in Paper 1, detection of this 'backside infall' into the GA region is required if the GA is to be considered a distinct entity. With these additional data, large-scale motions in the nearby universe yield evidence for the existence of the GA in many ways (*cf* §5.4 of Paper 1): a) from direct observation of acceleration towards this direction (figure 7, Paper 1); b) from the quadrupolar tidal field that its gravity produces in the motions of more nearby galaxies (table 1, Paper1); c) from optical maps of the distribution of nearby galaxies (Lynden-Bell *et al* 1988); d) from observation of backside infall; and e) from maps of the distribution of infrared-bright galaxies (Scharf 1991).

As various groups try to push measurements of peculiar motions to even further distances, two thoughts should be kept in mind. First, if distance measurements have an accuracy of $\approx 20\%$ per galaxy (3–4 times better for clusters of galaxies), there is a limit to the distances at which peculiar motions of ≈ 500 km^{-1} can be determined. Current surveys are near those limits. Second, if the means by which we measure the distances of galaxies are flawed by systematic problems (§3), those problems will produce artifacts of apparently large peculiar motions at large distances. Hence, these distant surveys are also among the best means of testing the accuracies of the measuring sticks!

A.3. New and possibly improved methods of distance estimation

Two new methods have been developed to measure the distances of elliptical and S0 galaxies, and new ways are being implemented to measure distances via 'old' techniques (old, in this case, is more than 4 years old).

John Tonry and Donald Schneider (1988), building on an idea first put forward by William Baum and Martin Schwarzschild (1955), showed that one could determine the luminosity function of the brightest stars in a galaxy by quantifying the fluctuations seen within the image of the galaxy. This technique makes use of the fact that the image of a galaxy is inherently comprised of the $\sim 10^{11}$ stars in the galaxy, hardly any of which 'shadow' each other. Of these stars, perhaps $\sim 10^6$ stars are bright enough to be seen as individual stars against the background of fainter stars. In galaxies as old as elliptical galaxies, the brightest stars are evolved stars whose outer envelopes have greatly expanded and become cool (hence the term red giants).

If a galaxy is close enough to us, we can actually see these stars individually. This is termed 'resolving' the galaxy into stars, the ability of which is dependent both on the galaxy's distance and the angular resolution capability of the telescope-imaging combination. If the galaxy is too far away to be resolved, these brightest stars still produce sub-arcsecond fluctuations superimposed on the smoother image produced by the fainter stars. Baum and Schwarzschild first presented the idea of measuring the luminosity function from these fluctuations at a time when such measurements were beyond the capability of available instrumentation. Tonry and Schneider used CCDs and multiple, high signal-to-noise images to show that this technique is now viable.

In a series of papers, Tonry and associates (Tonry *et al* 1988; Tonry *et al* 1989, 1990; Tonry and Schechter 1990; Tonry 1991) have employed this 'surface brightness fluctuation (SBF)' method to measure the distances to ≈ 40 galaxies. More measurements are currently being taken. The predicted internal accuracy for this method is $\approx 8\%$ which, if verified by external comparisons, would make this the most accurate distance estimation method known for individual galaxies that are unresolved into individual stars.

George Jacoby, Robin Ciardullo, Holland Ford and associates have discovered that an even later stage of stellar evolution, termed the 'planetary nebula' stage, can be used to measure galaxy distances (Jacoby 1989; Jacoby *et al* 1989; Ciardullo *et al* 1989a; Ciardullo *et al* 1989b; Jacoby *et al* 1989; Jacoby *et al* 1990; Ciardullo *et al* 1991). Their method of measuring distances also relies on determining a luminosity function for their objects. However, as opposed to the SBF method, the 'planetary nebula luminosity function (PNLF)' method uses observations of individual objects. As with the SBF method, the distances measured by the PNLF method appear to be more accurate than the $D_n-\sigma$ method for estimating the distances to ellipticals, with an accuracy of better than 10%. Unfortunately, the PNLF technique is even more time-consuming than the SBF technique, and even fewer galaxies have had their distances measured by this method (≈ 20 galaxies).

Both the SBF and PNLF methods hold much promise, but also require quite intensive observations to obtain distances for relatively few galaxies. Moreover, they apply primarily to dust-free elliptical galaxies, and are difficult to apply to spiral galaxies. Methods to improve distance estimates to spiral galaxies have centred on improving Tully–Fisher measurements, along the line discussed by Pierce and Tully (1988). Essentially all of the current surveys involving spirals that were quoted in §2 above employ these new methods. Improvement of the $D_n-\sigma$ method for estimating distances of elliptical galaxies is also being sought by measuring D_n-type parameters in other wavelength bands, through more accurate determinations of σ, and through possible improvements to the method itself (Lucey *et al* 1991a; Lucey *et al* 1991b).

In such an atmosphere of discovery, it is inevitable that new methods will be proposed that, in fact, do not yield better distance estimates. George Djorgovski and Reinaldo de Carvalho (1990) have proposed a series of distance estimation methods for elliptical galaxies that use integrated measures of the stellar populations of these galaxies. Their tests involve only galaxies in the Virgo and Coma clusters. Unfortunately, these workers made an error in comparing the accuracy of their methods to those of the 7S survey: they assumed that the $D_n-\sigma$ error for galaxies in these two clusters is 21%, the value quoted for the 7S survey as a whole. Instead, the error quoted for galaxies in the Virgo cluster is 17%, and for the Coma cluster, 15%

(Lynden-Bell *et al* 1988). When compared to the correct error estimates, only two of the 15 methods tried by Djorgovski and de Carvalho are even as accurate as the $D_n-\sigma$ method, a result which is not significant. This writer concludes that none of the methods suggested in that paper yield significantly-improved distances to ellipticals, a conclusion supported by other studies (Lynden-Bell *et al* 1989a; Burstein *et al* 1988).

It is the considered opinion of this writer that the minimum acceptable internal accuracy for distances to galaxies is 21% for both elliptical and spiral galaxies. Accuracies significantly better than these are now routinely obtained from a variety of methods.

As the distances to more galaxies are measured in more than one way, or in the same manner by more than one group, it becomes more probable that the real errors of distance estimates (both internal and external) will be understood. Such an understanding is needed, as all of these distance estimation methods depend, at a basic level, on a certain regularity in the physical properties of stellar populations in galaxies. Astronomers cannot *a priori* predict such correlations between galaxies containing billions of stars, yet such correlations appear to exist. In such a situation, it is always possible that systematic errors yet unknown could produce differences in stellar populations that can masquerade as peculiar velocities.

Searches for such problems have so far yielded mixed results for elliptical galaxies. Detailed examination of the 7S data set has yielded negative results (Burstein *et al* 1990, which counter the claims made by Djorgovski *et al* 1989). Tonry (1991) has presented evidence of systematic differences between distances estimated from the SBF method and those from the $D_n-\sigma$ method. However, such differences are at the level of 2–3 standard deviations, marginally significant in a sample of 25 galaxies. SBF distances will have to be measured to many more galaxies before such differences can be quantitatively defined.

Possible systematic problems with the Tully–Fisher relation for spirals were briefly discussed in Paper 1 (§4.1), and have been further studied by this writer and others. No significant problems have been found as the accuracy of the method has increased from 23% errors quoted in the original Aaronson *et al* surveys to $\approx 12–16\%$ typically quoted today (Pierce and Tully 1988; Freudling 1990). A paper that critically examines the pros and cons of each of the major distance-measuring methods (not all of which are discussed here) is currently in preparation, co-authored by nine workers in this field: David Branch, Robin Ciardullo, Roger Davies, William E Harris, George Jacoby, Michael Pierce, Christopher Pritchet, John Tonry and Douglas Welch (who nickname themselves the 'Mudville Nine')

A.4. New methods of interpretation and analysis

In Paper 1 this writer attempted to make a quantitative comparison between the peculiar velocity field as directly observed, and that which is predicted from the observed density distribution of galaxies (Paper 1, §6.5). Only by such studies can we answer such questions as: a) Is there 'dark' matter clustered on size scales different from visible matter? b) Do all of the peculiar motions we observe have to be gravity-driven? c) Can some of the observed motions be a product of initial conditions in the early universe? The comparison made in Paper 1 was with the first predictions by Michael Strauss and Marc Davis (1989), based on their survey (with Amost Yahil) of galaxies selected from the IRAS survey.

In the intervening two years, Strauss, Davis, Yahil and John Huchra have realized, with the help of Edmund Bertschinger and Avishai Dekel, that interpretation of their data in terms of a predicted density field was much trickier than they thought. As such, the paper detailing this comparison is still in the process of being written.

During the past year another of the density distribution of IRAS galaxies was finished (Saunders *et al* 1991). This 'QDOT' survey sampled the redshifts of galaxies more deeply in space than the Strauss *et al* survey, but for only one out of every six possible galaxies that could be surveyed. (The name QDOT comes from the initials of the participants' institutions: Queen Mary and Westfield College, University of Durham, Oxford University and the Canadian Institute for Theoretical Astrophysics, Toronto). These workers present a picture of the density distribution of galaxies out to distances of $\sim 10\,000-15\,000$ km s^{-1} (Saunders *et al*). On the scale of the volume for which the velocity field is known from peculiar motion studies, the agreement between the predicted motions and observed motions is relatively good on the whole (Kaiser *et al* 1991), although more detailed analysis is still needed.

[As a political aside, it should be noted that the announcement by Michael Rowan-Robinson (1991) of the QDOT team, that the 'Great Attractor is "dead"', was much more dramatic than correct. In this writer's opinion, Rowan-Robinson 'buried' the idea of a 'dark' GA, an idea which none of the discoverers of the GA ever had. The GA has been defined in terms of the galaxies that we see since shortly after its predicted existence (see discussion above). Rather, the region defined by the QDOT team as encompassing the Virgo–Ursa-Major–Hydra–Centaurus–Pavo–Indus–Telescopium clusters very nicely outlines the major clusters that are within the defined volume of the GA (Burstein *et al* 1990), as well as those clusters whose motions are most influenced by the mass of the GA. The existence of the GA in the QDOT survey was shown by Caleb Scharf (1991) and Donald Lynden-Bell (1991). A rose by any other name . . . .]

A programme to do a similar kind of study for the density distribution of optically-selected galaxies is currently underway, involving most of the Strauss *et al* group, together with Ofer Lahav. This survey makes good use of the ongoing compendium of radial velocities assembled by John Huchra (which he terms the 'ZCAT'), and is more restricted by the limited optical data for large samples of galaxies.

In a different vein, the POTENT method developed by Bertschinger and Dekel has more than lived up to the promise outlined at the end of Paper 1 (§7.4). In a series of papers, both published and in various stages of preparation (Bertschinger and Dekel 1989; Dekel *et al* 1990; Bertschinger *et al* 1990; Bertschinger and Dekel 1991; Dekel and Bertschinger 1991), these workers have applied POTENT to all existing peculiar velocity data, in various combinations with observers in this field (including this writer). Briefly, the method derives the density field from peculiar velocities under the assumption that no vorticities exist within the volume surveyed (in technical terms, that the velocity field is 'curl-free'). Due to the need to average over the peculiar motions of more than one galaxy per volume, the POTENT method necessarily smooths the density field on size scales ranging from 500–1500 km s^{-1}.

One of the most visually striking pictures to emerge from the study of peculiar motions is given by Bertschinger *et al* (1990), and is reproduced in figure A.1 of this addendum. This picture shows the predicted density distribution of matter within the Supergalactic plane and distances of ≈ 4000 km s^{-1}. It is a suitable companion to figure 6 of Paper 1, in which the peculiar motions of galaxies in the Supergalactic plane are displayed. As pictured in figure A.1, the GA shows itself to be a 'mountain', surrounded on our side by various 'foothills', of which the Virgo and Ursa-Major clusters are but two. This picture gives approximately the correct current view of the GA, in that it is pictured as a large, diffuse, somewhat overdense region of matter. Rather than the GA being comprised of 'extended haloes of rich galaxy clusters' (Saunders *et al* 1991), it is more correct to say that the GA is a very large, somewhat overdense region of galaxies in which some rich clusters are found. More will be said on this point at the conclusion of this addendum.

MASS-DENSITY IN SUPERGALACTIC PLANE Wv, Rw=R5

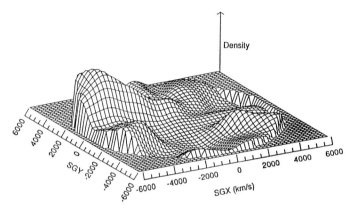

Figure A.1

A.5. In search of other great attractors

The existence of the GA begs two questions: Is this the largest mass structure that influences the gravity field near our Galaxy? Is the GA typical of such large mass structures in the universe? Answers to both of these questions have been actively sought ever since the existence of the GA was reasonably established.

Roberto Scaramella, Guido Chincarini, Paolo Vettolani, Giovanni Zamorani and collaborators (Scaramella *et al* 1989) have shown that one of the largest concentrations of rich galaxy clusters lies at a distance of 14000 km s^{-1} in the same direction as the GA. They have proposed that this 'Shapley Concentration' (named after Harlow Shapley, who did pioneering studies of the southern sky) can produce significant peculiar motions at the position of our Galaxy. A detailed study of the distribution of galaxies in the GA region, done as a PhD thesis by Somak Raychaurdhury, with Donald Lynden-Bell (Raychaudhury 1989), has shown that the Shapley Concentration can produce, at most, 15% of the GA gravity at our position. Both of the above studies assume that light traces mass in making their predictions.

The QDOT survey has identified other large-scale structures that can be viewed as separate from the local, GA-dominated region. One of these is the Hercules region, galaxies in which have measured peculiar motions from Freudling (1990), and which is one of the regions being surveyed by Wegner *et al* (1991).

A.6. Final remarks

In the past two years, this writer has come to realize the degree to which old prejudices die hard in astronomy, as is true in most other human endeavours. Ironically, the prejudice that is the hardest to die is that which is one of the most common misunderstandings about science for non-science college students: the difference between mass and density.

Clusters of galaxies are very dense in mass, but are not large in volume. High density means that galaxies will have high virial motions within the cluster, which produce 'fingers of God' when radial velocity is plotted versus position on the sky (since such a localized velocity distribution will point radially away from us). Yet, even a factor of 100 overdensity in the centre of a cluster cannot compete in total mass with a volume 10 000 times larger that is perhaps 1/50th of this density.

In both the Local Supercluster and in the GA region itself, it is the distributed mass throughout each region that dominates the mass of the superclusters. The well-known, dense clusters within these superclusters are the regions of highest density, but do not contribute much to the total mass. Hence, the large-scale motions that are generated both within these superclusters, and by these superclusters, are *not* motions of localized infall into the dense clusters, but rather motions generated by the much more diffuse mass of the supercluster. Whereas such localized motions can be reasonably approximated by model of spherical infall to a cluster, the motions generated by the more distributed mass are harder to model.

Unfortunately, radial velocity surveys can more easily discover the 'fingers of God' than they can uncover the much larger volume that provides the bulk of the mass. Clusters of galaxies are much easier to identify than galaxies diffusely spread over a large region. In the current redshift surveys, a region such as the GA, which does not have many clusters, would be difficult to detect at large distances.

Finally, mention should be made that even the GA 'mountain' may be, itself, part of a much more massive 'mountain range'. The pencil-beam redshift surveys of Broadhurst *et al* (1990) and Szalay *et al* (1991) yield '*prima facie*' evidence of 'sheets' of galaxies at intervals of 12 000 km s^{-1} that could be periodic. Three similar kinds of structures are evident in the colour pictures of galaxy density presented by the QDOT team in their *Nature* paper (Saunders *et al* 1991). The degree to which such structures are real is yet to be determined, as well as whether our observations have yet to yield a 'fair' picture of the distribution of mass in the universe. Two years after the final thoughts were written for Paper 1, the sentiments written there are as true now as then: we have only begun this adventure.

References

Aaronson M 1986 *Galaxy Distances and Deviations from Universal Expansion* ed. B F Madore and R B Tully (Boston: Reidel) pp 55–62
Aaronson M *et al* 1989 *Astrophys. J.* **338** 654–76
Aaronson M, Bothun G D, Mould J R, Huchra J, Schommer R and Cornell M 1986 *Astrophys. J.* **302** 536–63
Aaronson M, Dawe J, Dickens R J, Mould J and Murray J 1981 *Mon. Not. R. Astron. Soc.* **195** 1P–8P
Aaronson M, Huchra J and Mould J 1979a *Astrophys. J.* **229** 1–13
Aaronson M, Huchra J, Mould J, Schechter P L and Tully R B 1982a *Astrophys. J.* **258** 64–76
Aaronson M *et al* 1982b *Astrophys. J. Suppl.* **50** 241–61

Aaronson M and Mould J R 1983 *Astrophys. J.* **265** 1–17

—— 1985 Pierce Prize Lecture delivered at *166th AAS Meeting, Tucson, AZ, January 1985*

—— 1986 *Astrophys. J.* **303** 1–9

Aaronson M, Mould J and Huchra J 1979b *Astrophys. J.* **237** 655–65

Aaronson M, Mould J, Sullivan W T III, Schommer R A and Bothun G D 1980 *Astrophys. J.* **239** 12–37

Abell G O and Eastmond T S 1968 *Astron. J.* **73** S161

Bahcall N A 1987 *Comment. Astrophys.* **11** 283–308

—— 1988 *Ann. Rev. Astron. Astrophys.* **26** 631–86

Batuski D J and Burns J O 1985a *Astron. J.* **90** 1413–24

—— 1985b *Astrophys. J.* **299** 5–14

Baum W A and Schwarzschild M 1955 *Astron. J.* **60** 247–53

Bautz L P and Abell G O 1973 *Astrophys. J.* **184** 709–18

Bertschinger E and Dekel A 1989 *Astrophys. J. Lett.* **336** L5–8

—— 1991 *Large-Scale Structure and Peculiar Motions in the Universe* ed. D W Latham and L N da Costa (San Francisco: Astron. Soc. of Pacific) in press

Bertschinger E, Dekel A, Faber S M, Dressler A and Burstein D 1990 *Astrophys. J.* **364** 370–305

Bertschinger E and Juszkiewicz R 1988 *Astrophys. J. Lett.* **334** L59–62

Bothun G D, Aaronson M, Schommer R, Huchra J and Mould J 1984 *Astrophys. J.* **278** 475–85

Bothun G D, Aaronson M, Schommer R A, Mould J R, Huchra J and Sullivan W T III 1985 *Astrophys. J. Suppl.* **57** 423–72

Bothun G D and Mould J R 1987 *Astrophys. J.* **313** 629–43

Bottinelli L, Fouque P, Goughenheim L, Paturel G and Teerikorpi P 1987 *Astron. Astrophys.* **181** 1–13

Bottinelli L, Goughenheim L, Paturel G and de Vaucouleurs G 1980 *Astrophys. J. Lett.* **242** L153–6

—— 1983 *Astron. Astrophys.* **118** 4–20

—— 1984 *Astrophys. J.* **280** 34–40

Bottinelli L, Goughenheim L, Paturel G and Teerikorpi P 1986 *Astron. Astrophys.* **156** 157–71

—— 1988a *Astrophys. J.* **328** 4–22

Bottinelli L, Goughenheim L and Teerikorpi P 1988b *Astron. Astrophys.* **196** 17–25

Boughn S, Cottingham D, Fixsen D and Cheng E 1989 Personal Communication

Branch D 1982 *Astrophys. J.* **258** 35–40

Broadhurst T J, Ellis R S, Koo D C and Szalay A S 1990 *Nature* **343** 726–9

Burstein D 1982 *Astrophys. J.* **253** 539–51

—— 1988 *Towards Understanding Galaxies at High Redshift* ed. R G Kron and A Renzini (Dordrecht: Kluwer) pp 93–102

—— 1989 *The World of Galaxies* ed. H G Corwin and L Bottinelli (London: Springer) pp 547–66

—— 1990 *Large Scale Structures and Peculiar Motions in the Universe* ed. D Latham and L N Da Costa in press

Burstein D, Davies R L, Dressler A, Faber S M, Lynden-Bell D, Terlevich R J and Wegner G 1986 *Galaxy Distances and Deviations from Universal Expansion* ed. B F Madore and R B Tully (Boston: Reidel) pp 123–30

—— 1987a *Large-Scale Structures of the Universe (IAU Symp. 130)* ed. J Audouze *et al* pp 177–80

—— 1988 *Towards Understanding Galaxies at High Redshift* ed. R G Kron and A Renzini (Dordrecht: Kluwer) pp 17–22

—— 1989 *Large-Scale Structure and Motions in the Universe* ed. G Giuricin *et al* (Dordrecht: Kluwer) pp 179–96

Burstein D, Davies R L D, Dressler A, Faber S M, Stone R P S, Lynden-Bell D, Terlevich R J and Wegner G 1987b *Astrophys. J. Suppl.* **64** 601–42

Burstein D, Faber S M and Dressler A 1990 *Astrophys. J.* **354** at press

Burstein D and Raychaudhury S 1989 *Astrophys. J.* **343** 18–25

Cheng E S, Saulson P R, Wilkinson D T and Corey B E 1979 *Astrophys. J. Lett.* **232** L139–43

Chincarini G 1982 *The Large-Scale Structure of the Universe* (*Rio de Janiero Summer School, University of Oklahoma preprint*)

Chincarini G and Rood H J 1979 *Astrophys. J.* **230** 648–54

Chincarini G and Vettolani P 1988 *Large-Scale Structures in the Universe: Observation and Analytic Method* ed. W C Seitter *et al* (London: Springer) pp 209–16

Ciardullo R, Jacoby G H and Ford H C 1989 *Astrophys. J.* **344** 715–25

Ciardullo R, Jacoby G H, Ford H C and Neill J D 1989 *Astrophys. J.* **339** 53–69

Ciardullo R, Jacoby G H and Harris W E 1991 *Preprint*

Collins C A, Joseph R D and Robertson N A 1986a *Galaxy Distances and Deviations from Universal Expansion* ed. B F Madore and R B Tully (Boston: Reidel) pp 131–34
—— 1986b *Nature* **320** 506–8
Conklin E K 1969 *Nature* **222** 971–2
Corey B E and Wilkinson D T 1976 *Bull. Am. Astron. Soc.* **8** 351
Courteau S 1991 *Unpublished PhD Thesis* University of California, Santa Cruz
Da Costa L N, Numes M A, Pellegrini P S, Willmer C, Chincarini G and Cowan J J 1986 *Astron. J.* **91** 6–12
Da Costa L N, Willmer C, Pellegrini P S and Chincarini G 1987 *Astron. J.* **93** 1338–49
Davies R L, Burstein D, Dressler A, Faber S M, Lynden-Bell D, Terlevich R J and Wegner G 1987 *Astrophys. J. Suppl.* **64** 581–600
Davies R L, Efstathiou G, Fall S M, Illinsworth G and Schechter P L 1985 *Astrophys. J.* **266** 41–57
Davies R D and Stavely-Smith L 1985 *The Virgo Cluster of Galaxies* ed. O G Richter and B Binggeli (ESO: Garching) pp 391–6
Davis M and Peebles P J E 1983 *Ann. Rev. Astron. Astrophys.* **21** 109–30
de Lapparent V, Geller M J and Huchra J P 1986 *Astrophys. J. Lett.* **302** L1–5
Dekel A and Bertschinger E 1991 *Large-Scale Structure and Peculiar Motions in the Universe* ed. D W Latham and L N da Costa (San Francisco: Astron. Soc. of Pacific) in press
Dekel A, Bertschinger E and Faber S M 1990 *Astrophys. J.* **364** 349–69
de Vaucoulerus G 1953 *Astron. J.* **58** 30–2
—— 1958 *Astron. J.* **63** 253–66
de Vaucouleurs G, Buta R, Bottinelli L, Goughenheim L and Paturel G 1982 *Astrophys. J.* **254** 8–15
de Vaucouleurs G and de Vaucouleurs A 1964 *Reference Catalog of Bright Galaxies* (Austin: University of Texas Press)
de Vaucouleurs G, de Vaucouleurs A and Corwin H G 1976 *Second Reference Catalog of Bright Galaxies* (Austin: University of Texas Press)
de Vaucouleurs G and Olson D 1982 *Astrophys. J.* **256** 346–69
de Vaucouleurs G and Peters W L 1968 *Nature* **220** 868–74
—— 1981 *Astrophys. J.* **248** 395–407
—— 1984 *Astrophys. J.* **287** 1–16
de Vaucouleurs G, Peters W L, Bottinelli L, Goughenheim L and Paturel G 1981 *Astrophys. J.* **248** 408–22
Dickens R J, Currie M J and Lucey J R 1986 *Mon. Not. R. Astron. Soc.* **220** 679–711
Djorgovski S and Davis M 1986 *Galaxy Distances and Deviations from Universal Expansion* ed. B F Madore and R B Tully (Boston: Reidel) pp 135–8
—— 1987 *Astrophys. J.* **313** 59–68
Djorgovski S J and De Carvalho R 1990 *Astrophys. J. Lett.* L37–40
Djorgovski S, De Carvalho R and Han M-S 1989 *The Extragalactic Distance Scale* ed. S van den Bergh and C J Pritchet (Provo: Astron. Soc. of the Pacific) pp 329–41
Dressler A 1980 *Astrophys. J. Suppl.* **42** 565–609
—— 1984 *Astrophys. J.* **281** 525–34
—— 1987a *Astrophys. J.* **317** 1–10
—— 1987b *Sci. Am.* **257** 46–55
—— 1988 *Astrophys. J.* **329** 519–26
—— 1989 *Proc. 14th Texas Relativistic Symp.* in press
Dressler A and Faber S M 1990 *Astrophys. J.* **354** at press
Dressler A, Faber S M and Burstein D 1991 *Astrophys. J.* **368** 54–9
Dressler A, Faber S M, Burstein D, Davies R L, Lynden-Bell D, Terlevich R J and Wegner G 1987a *Astrophys. J. Lett.* **313** L37–42
Dressler A, Lynden-Bell D, Burstein D, Davies R L, Faber S M, Terlevich R J and Wegner G 1987b *Astrophys. J.* **313** 42–58
Eddington A S 1913 *Mon. Not. R. Astron. Soc.* **73** 359–60
Faber S M and Burstein D 1989 *Large-Scale Motions in the Universe* ed. V C Rubin and G Coyne (Princeton: Princeton University Press) pp 115–67
Faber S M and Jackson R E 1976 *Astrophys. J.* **204** 668–83
Faber S M, Wegner G, Burstein D, Davies R L, Dressler A, Lynden-Bell D and Terlevich R J 1989 *Astrophys. J. Suppl.* **69** 763–808
Fairall A P 1984 *Publ. Dept. Astron. Univ. Cape Town* No. 6
—— 1988 *Mon. Not. R. Astron. Soc.* **230** 69–78

Fairall A P and Winkler H 1984 *Clusters and Groups of Galaxies* ed. F Mardirossian, G Giuricin and M Mezzetti (Dordrecht: Reidel) 23–8

Fall S M and Jones B J T 1976 *Nature* **262** 457–60

Fish R A 1964 *Astrophys. J.* **139** 284–305

Freudling W 1990 *Unpublished PhD Thesis* Cornell University

Freudling W, Haynes M F and Giovanelli R 1989 *The Extragalactic Distance Scale* ed. S van den Bergh and C J Pritchet (Provo: Astron. Soc. of Pacific) pp 361–2

Gamow G 1946 *Nature* **158** 549

Geller M and Huchra J 1989 *Large-Scale Structure and Motions in the Universe* ed. G Giuricin *et al* (Dordrecht: Kluwer) pp 3–18

—— 1989a *Science* **246** 897–903

Giovanelli R and Haynes M P 1983 *Astron. J.* **88** 881–908

Gorski K, Davis M, Strauss M A, White S D M and Yahil A 1989 *Astrophys. J.* **344** 1–19

Groth E J, Juszkiewicz R and Ostriker J P 1989 *Astrophys. J.* **346** 558–65

Gunn J E 1989 *The Extragalactic Distance Scale* ed. S van den Bergh and C J Pritchet (Provo: Astron. Soc. of Pacific) pp 344–59

Han M-S 1991 *Unpublished PhD Thesis* California Institute of Technology

Harmon R T, Lahav O and Meurs E J A 1987 *Mon. Not. R. Astron. Soc.* **228** 5p–10p

Hart L and Davies P D 1982 *Nature* **297** 191–6

Haynes M P and Giovanelli R 1984 *Astron. J.* **89** 758–800

—— 1986 *Galaxy Distances and Deviations from Universal Expansion* ed. B F Madore and R B Tully (Boston: Reidel) pp 117–22

—— 1989 *Large Scale Motions in the Universe* ed. V C Rubin and G V Coyne (Princeton: Princeton University Press) pp 31–70

Hopp U and Materne J 1985 *Astron. Astrophys. Suppl.* **61** 93–106

Hubble E 1929 *Proc. Nat. Acad. Sci.* **15** 168–73

Huchtmeier W K and Richter O-G 1986a *Astron. Astrophys. Suppl.* **63** 323–45

—— 1986b *Astron. Astrophys. Suppl.* **64** 111–30

Huchra J 1989 *The Extragalactic Distance Scale* ed. S van den Bergh and C J Pritchet (Provo: Astron. Soc. of Pacific) pp 257–80

Jackson R E 1982 *PhD Thesis* University of California, Santa Cruz

Jacoby G H 1989 *Astrophys. J.* **339** 39–52

Jacoby G H, Ciardullo R, Ford H C and Booth J 1989 *Astrophys. J.* **344** 704–14

Jacoby G H, Walker A R and Ciardullo R 1990 *Astrophys. J.* **365** 471–7

James P A, Joseph R D and Collins C A 1987 *Mon. Not. R. Astron. Soc.* **229** 53–9

Kaiser N 1988 *Mon. Not. R. Astron. Soc.* **231** 149–68

Kaiser N, Efstathiou G, Ellis R, Frenk C, Lawrence A, Rowan-Robinson M and Saunders W 1991 *ICAT preprint*

Kaiser N and Lahav O 1989a *Large-Scale Motions in the Universe* ed. V C Rubin and G Coyne (Princeton: Princeton University Press) 337–84

—— 1989b *Mon. Not. R. Astron. Soc.* **237** 129–62

Kormendy J and Knapp J 1987 ed. *Dark Matter in the Universe (IAU Symp. 117)* (Boston: Reidel)

Kraan-Korteweg R C 1983 *Astron. Astrophys.* **125** 109–11

—— 1985 *The Virgo Cluster of Galaxies* ed. O G Richter and B Binggeli (Garching: ESO) pp 397–406

—— 1986 *Astron. Astrophys. Suppl.* **66** 255–79

Kraan-Korteweg R C, Cameron L M and Tammann G A 1986 *Galaxy Distances and Deviations from Universal Expansion* ed. B F Madore and R B Tully (Boston: Reidel) pp 65–72

—— 1988 *Astrophys. J.* **331** 620–40

Kraan-Korteweg R C, Sandage A and Tammann G A 1984 *Astrophys. J.* **283** 24–32

Lahav O 1987 *Mon. Not. R. Astron. Soc.* **225** 213–20

—— 1989 *Proc. XIth Cracow Summer School on Morphological Cosmology, August 1988, Preprint*

Lahav O, Rowan-Robinson M and Lynden-Bell D 1988 *Mon. Not. R. Astron. Soc.* **234** 677–702

Lauberts A 1982 *The ESO/Uppsala Survey of the ESO (B) Atlas* (Garching: ESO)

Lilje P B, Yahil A and Jones B J T 1986 *Astrophys. J.* **307** 91–6

Lubin P and Villela T 1986 *Galaxy Distances and Deviations from Universal Expansion* ed. B F Madore and R B Tully (Boston: Reidel) pp 169–76

Lucey J R, Bower R G and Ellis R S 1991a *Mon. Not. R. Astron. Soc.* **249** 755–62

Lucey J R and Carter D 1988 *Mon. Not. R. Astron. Soc.* **235** 1177–201

Lucey J R, Currie M J and Dickens R J 1986a *Mon. Not. R. Astron. Soc.* **221** 453–72

—— 1986b *Mon. Not. R. Astron. Soc.* **222** 427–47

Lucey J R, Gray P M, Carter D J and Terelevich R J 1991b *Mon. Not. R. Astron. Soc.* **248** 804–15

Lucey J R, Lahav O, Lynden-Bell D, Terlevich R J and Melnick J 1990 *Large Scale Structures and Peculiar Motions in the Universe* ed. D Latham and L N Da Costa in press

Lynden-Bell D 1987 *Quart. J. R. Astron. Soc.* **28** 186–96

—— 1991 as quoted in *Sky and Telescope 81* 355–6

Lynden-Bell D, Burstein D, Davies R L, Dressler A, Faber S M, Terlevich R J and Wegner G 1989a *The Extragalactic Distance Scale* ed. S van den Bergh and C J Pritchet (Provo: Astron. Soc. of Pacific) pp 307–16

Lynden-Bell D, Faber S M, Burstein D, Davies R L, Dressler A, Terlevich R J and Wegner G 1988 *Astrophys. J.* **326** 19–49

Lynden-Bell D and Lahav O 1989 *Large-Scale Motions in the Universe* ed. V C Rubin and G Coyne (Princeton: Princeton University Press) 199–217

Lynden-Bell D, Lahav O and Burstein D 1989b *Mon. Not. R. Astron. Soc.* **241** 325–45

Malmquist K G 1920 *Medd. Lund.* Ser. II. Band 3, No 22, 1–68

—— 1922 *Medd. Lund.* Band V, No 100, 1–52

Meiksin A and Davis M 1986a *Galaxy Distances and Deviations from Universal Expansion* ed. B F Madore and R B Tully (Boston: Reidel) pp 147–50

—— 1986b *Astron. J.* **91** 191–8

Melnick J and Moles M 1987 *Rev. Mex. Astron. Astrophys.* **14** 72–6

Mihalas D and Binney J 1981 *Galactic Astronomy* 2nd edn (San Francisco: Freeman)

Mould J, Aaronson M and Huchra J 1980 *Astrophys. J.* **238** 458–70

Nilson P 1973 *Uppsala General Catalogue of Galaxies, Uppsala Astron. Ann. Obs.* vol. 6

Ogorodnikov K F 1952 *Prob. Cosmogony* **1** 147–88

Oort J H 1983 *Ann. Rev. Astron. Astrophys.* **21** 373–428

Ostriker J P and Cowie L L 1981 *Astrophys. J. Lett.* **243** L127–32

Partridge R B 1988 *Rep. Prog. Phys.* **51** 647–706

Peebles P J E 1980 *The Large-Scale Structure of the Universe* (Princeton: Princeton University Press)

—— 1988 *Astrophys. J.* **332** 17–25

Penzias A A and Wilson R W 1965 *Astrophys. J.* **142** 419–21

Pierce M J 1989 *Unpublished PhD Thesis* University of Hawaii

Pierce M J and Tully R B 1988 *Astrophys. J.* **330** 579–95

Primack J R, Seckel D and Sadoulet B 1988 *Ann. Rev. Nucl. Part. Sci.* **38** 751–807

Raychaudhury S 1989 *Nature* **342** 251–5

—— 1990 *Preprint*

Regos E and Szalay A S 1989 *Astrophys. J.* **345** 627–36

Richter O-G and Huchtmeier W K 1984 *Astron. Astrophys.* **132** 253–64

Richter O-G, Tammann G A and Huchtmeier W K 1987 *Astron. Astrophys.* **171** 33–40

Rood H J 1988 *Ann. Rev. Astron. Astrophys.* **26** 245–94

Rowan-Robinson M 1991 as quoted in *Sky and Telescope 81* 355–6

Rubin V C 1951 *Astrophys. J.* **56** 47–8

—— 1988 *Gerard and Antoinette de Vaucouleurs: A Life for Astronomy* ed. M Capaccioli and H G Corwin (Singapore: World Scientific) in press

—— 1989 *Le Monde des Galaxies, A Symposium Honoring G and A de Vaucouleurs* ed. H G Corwin and L Bottinelli, to be published

Rubin V C, Ford W K Jr, Thonnard N, Roberts M S and Graham J A 1976a *Astron. J.* **81** 687–718

Rubin V C, Thonnard N, Ford W K Jr and Roberts M S 1976b *Astron. J.* **81** 719–37

Sandage A 1972 *Astrophys. J.* **178** 1–24

—— 1973a *Astrophys. J.* **183** 731–42

—— 1973b *Astrophys. J.* **183** 743–57

Sandage A and Tammann G A 1976 *Astrophys. J.* **210** 7–24

—— 1982 *Astrophys. J.* **256** 339–45

—— 1984 *Large-Scale Structure of the Universe, Cosmology and Fundamental Physics. First ESO-CERN Symp.* ed. G Setti and L Van Hove (Geneva: ESO-CERN) pp 127–49

Saunders W, Frenk C, Rowan-Robinson M, Efstathiou G, Lawrence A, Kaiser N, Ellis R, Crawford J, Xia X-Y and Perry I 1991 *Nature* **349** 32–8

Scaramella R, Baiesi-Pillastrini G, Chincarini G, Vettolani G and Zamorani G 1989 *Nature* **338** 562–4

Scharf C 1991 as quoted in *Sky and Telescope 81* 355–6

Schechter P L 1977 *Astron. J.* **82** 569–76

—— 1980 *Astron. J.* **85** 801–11

Sharp N 1989 unpublished

Shaya E J 1984 *Astrophys. J.* **280** 470–75

Silk J 1989 *Preprint*

Smoot G F, Gorenstein M V and Muller R A 1977 *Phys. Rev. Lett.* **39** 898–901

Smoot G F and Lubin P M 1979 *Astrophys. J. Lett.* **234** L83–6

Stavely-Smith L and Davies R D 1987 *Mon. Not. R. Astron. Soc.* **224** 953–85

—— 1988 *Mon. Not. R. Astron. Soc.* **231** 833–71

—— 1989 *Mon. Not. R. Astron. Soc.* **241** 787–826

Strauss M A 1989 *PhD Thesis* University of California, Berkeley

Strauss M A and Davis M 1988 *Large-Scale Structures of the Universe (IAU Symp. 130)* ed. J Audouze *et al* pp 191–201

—— 1989 *Large-Scale Motions in the Universe* ed. V C Rubin and G Coyne (Princeton: Princeton University Press) 255–74

Szalay A 1989 *Large-Scale Motions in the Universe* ed. V C Rubin and G Coyne (Princeton: Priceton University Press) 323–38

Szalay A S, Ellis R S, Koo D C and Broadhurst T J 1991 *Preprint*

Tammann G A and Sandage A 1985 *Astrophys. J.* **294** 81–95

Teerikorpi P 1987 *Astron. Astrophys.* **173** 39–42

Tonry J L 1991 *Astrophys. J. Lett.* **373** L1–4

Tonry J L, Ajhar E A and Luppino G A 1989 *Astrophys. J. Lett.* **346** L57–60

—— 1990 *Astron. J.* **100** 1794–8

Tonry J and Davis M 1981a *Astrophys. J.* **246** 666–79

—— 1981b *Astrophys. J.* **246** 680–95

Tonry J L, Luppino G A and Schneider D P 1988 *The Extragalactic Distance Scale: Proceedings of the ASP 100th Anniversary Symposium* ed. S van den Bergh and C J Pritchet ASP Conf. Series 4 213–7

Tonry J L and Schechter P L 1990 *Astron. J.* **100** 1416–23

Tonry J L and Schneider D P 1988 *Astron. J.* **96** 807–15

Tully R B 1988 *Nature* **334** 209–12

Tully R B and Fisher J R 1977 *Astron. Astrophys.* **54** 661–73

—— 1987 *Nearby Galaxies Atlas* (New York: Cambridge University Press)

Villumsen J V and Strauss M A 1987 *Astrophys. J.* **322** 37–47

Visvanathan N 1986 *Galaxy Distances and Deviations from Universal Expansion* ed. B F Madore and R B Tully (Boston: Reidel) pp 99–104

Vittorio N and Juszkiewicz R 1989 *Large-Scale Structure and Motions in the Universe* ed. G Giuricin *et al* (Dordrecht: Kluwer) pp 241–58

Weedman D W 1976 *Astrophys. J.* **203** 6–13

Wegner G, Burstein D, Colless M, Davies R L, Bertschinger E and McMahan R 1991 *Large-Scale Structure and Peculiar Motions in the Universe* ed. D W Latham and L N da Costa (San Francisco: Astron. Soc. of Pacific) in press

Willick, J 1990 *Astrophys. J. Lett.* **351** L5–8

Winkler H 1983 *Mon. Not. Astron. Soc. South Africa* **42** 74–88

Yahil A 1985 *The Virgo Cluster of Galaxies* ed. O G Richter and B Binggeli (Garching: ESO) pp 359–74

—— 1989 *Large-Scale Motions in the Universe* ed. V C Rubin and G Coyne (Princeton: Princeton University Press) pp 219–53

Yahil A, Walker D and Rowan-Robinson M 1986 *Astrophys. J.* **301** L1–6